NEW MEXICO'S MAGNIFICENT
SANDIA MOUNTAIN

~

THE COMPLETE GEOLOGICAL STORY

NEW MEXICO'S MAGNIFICENT
SANDIA MOUNTAIN

~

THE COMPLETE GEOLOGICAL STORY

DIRK VAN HART

SUNSTONE
PRESS

SANTA FE

Sunstone books may be purchased for educational, business, or sales promotional use.
For information please write: Special Markets Department, Sunstone Press,
P.O. Box 2321, Santa Fe, New Mexico 87504-2321.
Printed on acid-free paper
∞

———————————————————

Library of Congress Cataloging-in-Publication Data

Names: Van Hart, Dirk, 1941- author.
Title: New Mexico's magnificent Sandia mountain : the complete geological
    story / Dirk Van Hart.
Description: Santa Fe : Sunstone Press, [2023] | Includes bibliographical
    references and index. | Summary: "The magnificent Sandia Mountain forms
    an enormous rampart towering over the city of Albuquerque, New Mexico.
    Regionally, the feature's distinctive "whale back" profile utterly
    dominates the horizon within a huge area of central New Mexico. This
    book provides the complete geologic story of the mountain's origin-a
    story given within the context of the greater American Southwest. The
    text is richly illustrated, producing a reader-friendly narrative
    understandable to the non-geologist. The mountain and its surroundings
    are the end-products of a long sequence of geologic events spanning a
    vast period of 1.7 billion years, but the uplift we call today's Sandia
    Mountain was formed quite recently. In this way it differs in origin
    from the Rocky Mountains, which are located nearby but are much older.
    Paradoxically, then, what we see today is a relatively new mountain made
    from very old rocks"-- Provided by publisher.
Identifiers: LCCN 2023003032 | ISBN 9781632933928 (paperback)
Subjects: LCSH: Physical geology--New Mexico--Sandia Mountains. | Sandia
    Mountains (N.M.)
Classification: LCC QE28.2 .V36 2023 | DDC 557.89--dc23/eng20230415
LC record available at https://lccn.loc.gov/2023003032

———————————————————

WWW.SUNSTONEPRESS.COM
SUNSTONE PRESS / POST OFFICE BOX 2321 / SANTA FE, NM 87504-2321 /USA
(505) 988-4418

CONTENTS

PART I: BUILDING A MOUNTAIN 79

1. Back to the Beginning 79

2. Emplacement of the Sandia Granite 91

3. Unroofing of the Sandia Granite and Its Subsequent Reburial 112

Reburial

4. Laramide Compression and the Rocky Mountains   136

5. Volcanic Flareup   166

6. Neogene Tension and Creation of a Mountain   179

LIST OF FIGURES

PART I: BUILDING A MOUNTAIN

1. Back to the Beginning: Supercontinent of *Rodinia*

2. Emplacement of Sandia Granite

3. Unroofing of Sandia Granite and its Subsequent Reburial

4. Laramide Compression and the Rocky Mountains

5. Volcanic Flareup

6. Neogene Tension and Creation of a Mountain

7. The Rio Grande

PART II: PLACES OF SPECIAL INTEREST

8. Northern Nose/Placitas Area

9. Rincón Ridge and Juan Tabó Canyon

12: Tijeras Canyon and the East Mountain

PART III: CONTEXT

15: The *Really* Big Picture

APPENDIX I: Select Previous Geologic Work

APPENDIX II: Geologic Time – History of a Concept

PREFACE

Philosophy of This Book

This book deals with Sandia Mountain, that massive landform that forms the towering eastern backdrop for the city of Albuquerque in central New Mexico. The mountain provides a focal point for a population approaching a million people, or about 40% of the population of New Mexico.

The book is about the mountain's geology and how the mountain formed. It is neither a travelogue nor a guide book aimed at the outdoorsman. These have already been written by others.

Neither is the book a rehash of the large volume of prior geologic work produced over more than a century, which is more directed toward the technical person. Two examples of the latter type are the classic, *Geology of Sandia Mountain and Vicinity, New Mexico*, by V.C. Kelley and S.A. Northrop, 1975, and the slightly less technical *Albuquerque: Its Mountains, Valley, Water and Volcanoes*, by V.C. Kelley, 3rd edition 1982.

Three Important Notes (*Don't Skip These*)

Note 1. ***This book is for the non-technical reader as well as for the geologist***. The book is not elementary, but it is *almost* entirely free of jargon and technical gobbledygook. In short, my intent is for it to be "user-friendly." At first some of the figures may seem a bit foreign and intimidating, but I believe that in a short time the reader will become comfortable with them.

Consider this little chestnut by the Nobel-Prize-winning American chemist, Linus Pauling (1901–1994), that sets our stage:

"The satisfaction of curiosity is one of the greatest sources of happiness in life."

I say amen to that!

Note 2. This book is about geology, and that moniker indicates it's about rocks. To many (except geologists) the subject of rocks can appear rather "dry" and maybe tedious. But rocks are more than inanimate objects (paintings are also inanimate objects!). They tell stories—if one knows how to read them.This book strives to help in that regard, and it will strongly emphasize the landscape forms that rocks produce—those images we live with and see every day.

Note 3. At first glance the large size of the book may be a bit intimidating. Even though all the sections do follow a logical and chronological order, each can be read out of sequence and some even *skipped over and dealt with at a later time when the interest calls*. The reader should therefore consider the chapters as a series of closely related vignettes. Consider the book a compendium of what we know or think we know about the mountain and its origin.

Organization

I've constructed this book to gather up and synthesize all the geologic literature relevant to the Sandia Mountain and to boil it down into a digestible form, even for the non-geologist. I've placed very heavy emphasis on the visual experience by way of 207 figures. Geology is an extremely visual science, and in this book the written word is intended to support the figures, not the other way around.

Time and again I will stress *context* to address the "big picture." To appreciate the importance of the big picture and context I suggest imagining an ant who has spent its entire life on the back of an elephant (*Figure P-1*). The ant's entire world is a monotonous series of gray hills and valleys (the elephant's skin), but the ant has no idea that it resides on the back of a huge animal. In order for us (the ants) to understand the wonderful story of Sandia Mountain (the elephant) we must stand back, take in the regional setting, and see the mountain as part of a whole, i.e., the "big picture." This is vitally important! In geology context is everything, and the more the better!

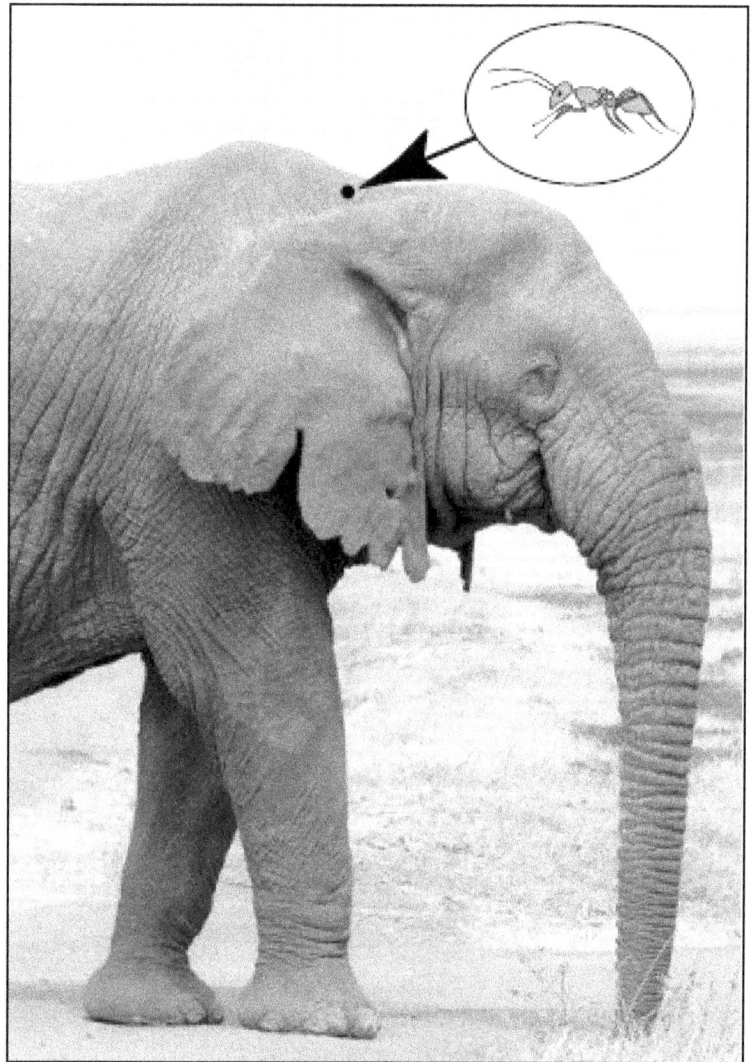

Figure P-1. Perspective

I will construct the mountain step by step, from the very beginning to our present edifice. After an introduction to the mountain, including some obligatory geologic essentials (INTRODUCTION), we start with PART I: BUILDING A MOUNTAIN, where we'll visit the ancient assemblage of western North America (*Chapter 1*), the intrusion of the Sandia Granite forming the core of the mountain (*Chapter 2*), the unroofing, exposure, and subsequent reburial of the Sandia Granite (*Chapter 3*), the regional and pervasive "Laramide" compressive event that created the Rocky Mountains and shoved around what would become the future Sandia Mountain (*Chapter 4*), the regional, so-called volcanic flareup that modified the area (*Chapter 5*), the coupled opening of the Rio Grande rift with the uplift of today's Sandia Mountain (*Chapter 6*), and the evolution of the Rio Grande in its valley (*Chapter 7*).

In PART II: PLACES OF SPECIAL INTEREST we'll visit some intriguing features on the mountain's northern nose (*Chapter 8*), at Rincón Ridge and Juan Tabó Canyon (*Chapter 9*), in the so-called Juan Tabó Cul-de-Sac (*Chapter 10*), along the Sandia foothills (*Chapter 11*), those in Tijeras Canyon and on the east side (*Chapter 12*), and those in the Ortiz Igneous Belt east of the mountain (*Chapter 13*). I've devoted an entire chapter to the economically important Cretaceous-age rocks on both sides of the mountain (*Chapter 14*).

Finally, in PART III: CONTEXT, we'll call on some of the new ideas about the really big picture (*Chapter 15*). This latter chapter may be heavy lifting for the casual reader but, although not essential to the previous narrative, it will flesh out the context.

Tangents

Occasionally I just can't help myself and veer off onto a number of tangents. I warn the reader ahead of time using either of these two icons:

"In the weeds." This is where I felt the need, to add an esoteric or perhaps semi-technical explanation about a subject and therefore risk wandering into the "tall weeds." This book contains many of these in the attempt to allow the reader to (temporarily) bypass, or even skip them. The information in these asides is not essential, but will be interesting and appealing to some readers.

"Query." In a few places I introduce a non-technical story of some sort, usually one with a geological twist. I alert the reader with this "query" icon.

I have included six appendices with additional information. APPENDIX I lists select previous work, underscoring the fact that much of this book is indeed a compilation of the work of others. Also, I've highlighted a few of the authors whose names come up again and again in the literature. There are Dr. Karl E. Karlstrom, Professor of Geology at the University of New Mexico, and his students. Their "big picture" insights provide much inspiration and guidance for this book. Right behind them are Dr. Shari A. Kelley, who has supplied a plethora of critical dates of geological events, Dr. Steve Cather, who always thinks "outside the box," and Dr. Sean D. Connell, who unraveled much of the more recent history of the Sandia Mountain's flanks.

APPENDIX II deals with geologic, "deep" time and the fascinating evolution of the understanding of this vital concept. A glossary follows as APPENDIX III, defining some of the unfamiliar terms used in this book. I've tried to keep the technical jargon to the barest minimum and to talk in simple English, but a few terms, unfortunately, are unavoidable. When I introduce a term for the first time in the text, I put it in *italics* as a warning. APPENDIX IV is a lexicon that gives the origin of geological formation names. Finally, APPENDIX V gives the geographic coordinates of a select number of interesting features.

Throughout the text the reader will encounter many citations enclosed in parentheses. These give the source of information mentioned in the preceding sentence. These can be cumbersome, but it is proper to give them so that my statements can be verified. All these citations are listed in the REFERENCES CITED section at the back of the book.

Despite the fact that almost all the contemporary geological literature uses the metric system (meters, kilometers, degrees Celsius) for measurement, I suspect that most readers of this narrative are most familiar and comfortable with the British system (inches, feet, miles, degrees Fahrenheit). Therefore, I will stick to the British system.

ACKNOWLEDGMENTS

I am indebted to the following people who provided me with guidance, information, and/or material:

Dr. Bruce D. Allen, geologist, New Mexico Bureau of Geology and Mineral Resources, Albuquerque.
Cynthia Benedict, U.S. Forest Service (retired).
Dr. Bruce A. Black, petroleum geologist, Farmington (retired).
Elaine D. Briseño, staff writer, *Albuquerque Journal*.
Sandra Arazi-Coambs, archeologist, Sandia and Mountainair Zone, U.S. Forest Service.
Rick Holben, historian, East Mountain Historical Society.
Dr. Adam S. Read, geologist, New Mexico Bureau of Geology and Mineral Resources, Albuquerque.

Finally, to two who provided the absolutely-vital peer reviews:

James Neal, a retired geologist, who spent his distinguished career with Sandia National Laboratories.
Erich Thomas, another retired geologist, who gave the manuscript a thorough hard edit, vastly improving it and rescuing me from glaring and embarrassing glitches that I'd become blind to.

INTRODUCTION

Sandia Mountain: What is it? Where is it?

Sandia Mountain forms the eastern margin of the City of Albuquerque, New Mexico's largest city with more than a third of the state's two million population (*Figure 0-1*). It forms one of two of the state's super-impressive urban backdrops. The other is provided by the Organ Mountains, about eight miles east of the City of Las Cruces in southern New Mexico. The name *Sandia* is the Spanish word for watermelon. To some unfettered imagination the mountain must have seemed like a lengthwise-quartered watermelon, with its western face reflecting the soft pink glow of the setting sun.

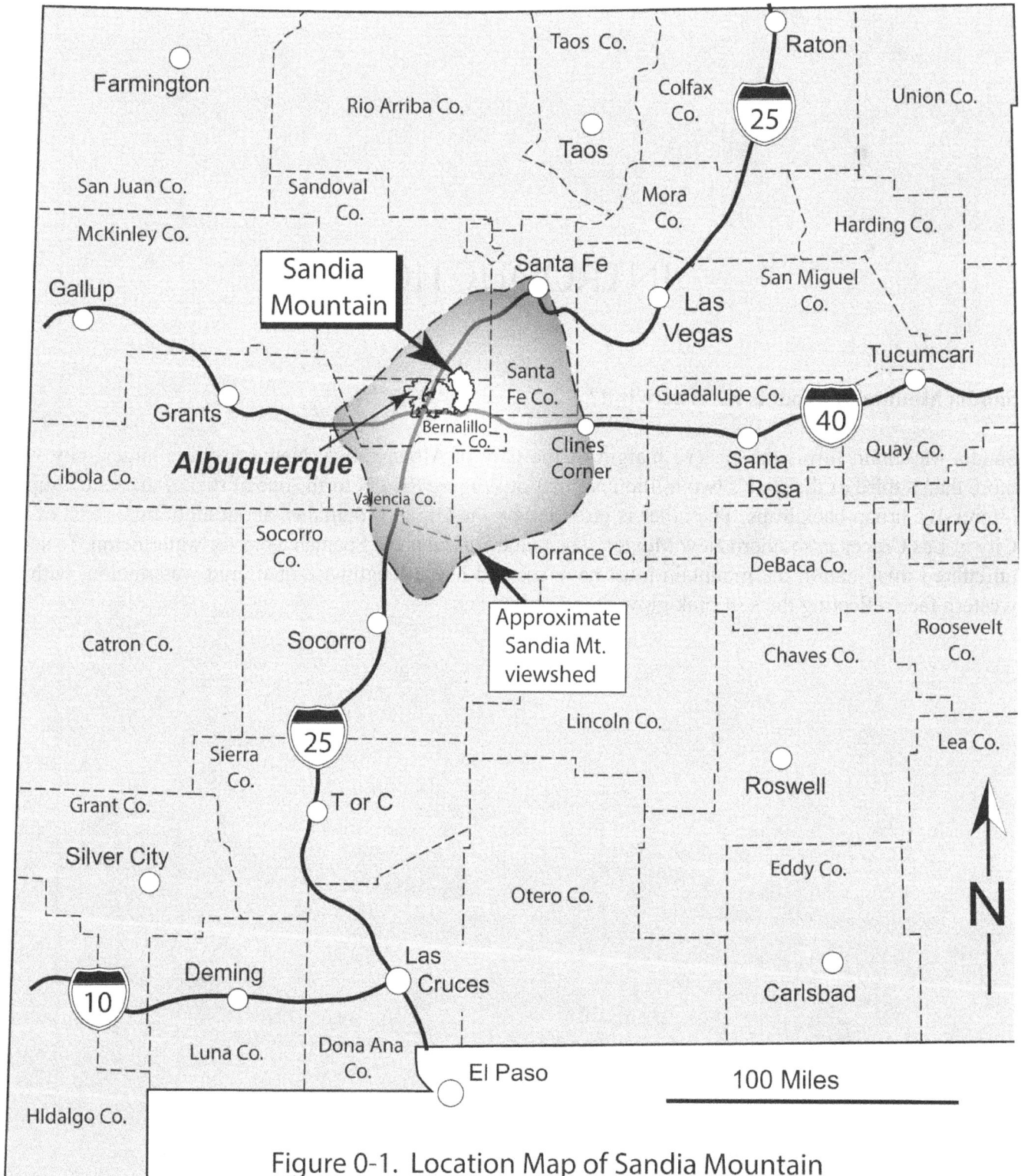

Figure 0-1. Location Map of Sandia Mountain

But just what *is* Sandia Mountain? *Figure 0-2* is from a classic old textbook (Lobeck 1939) used to illustrate a "block-faulted" mountain, which in the very barest terms is what Sandia Mountain is. This basic design is frequently employed to provide the public with a "geology at a glance" image. However, it is difficult to erase such an image from the mind if that is all the observer ever sees. The mountain is so much more complex, and so much more interesting.

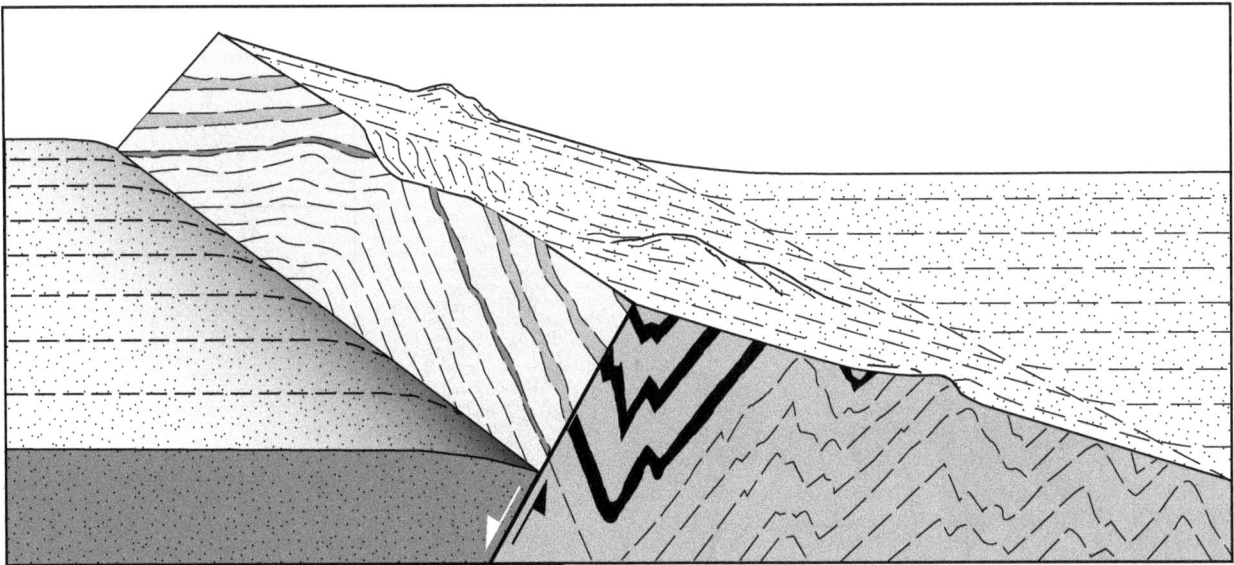

Figure 0-2. "Block-Faulted" Mountain (Modified from Lobeck 1939)

Let's now take our initial close look at Sandia Mountain, first viewing it in its regional, state-wide context, and then zooming in on the mountain itself. Sandia Mountain is by far the highest point in a broad area of high elevation exceeding 5000 ft in the center of New Mexico (*Figure 0-3*). The mountain is not a particularly large feature in a regional frame, but in its field of view, i.e., its *viewshed*, spans a very broad swath indeed (*Figure 0-1*).

Continental
Divide

Highest point
Wheeler Peak
13,160 ft

Sandia
Mt.

Lowest point 2841 ft
Red Bluff Reservoir

N

Elevation zones

③ ④ ⑤ ⑥ ⑦ ⑧ ⑨

3000' 4000' 5000' 6000' 7000' 8000' 9000' 10,000'

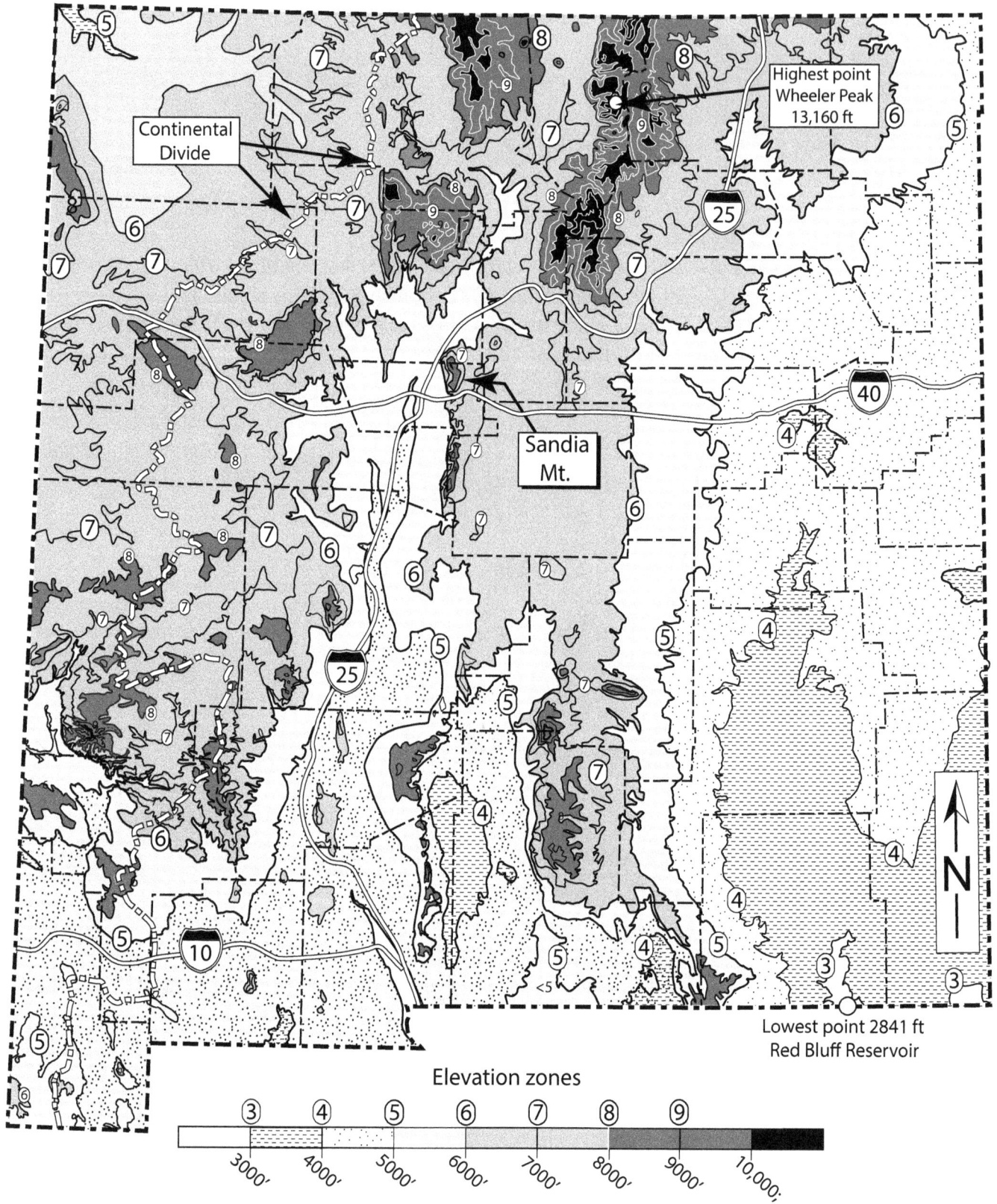

Figure 0-3. Topographic Map of New Mexico (CI = 1000 Ft)

Published state maps give the impression that Sandia Mountain is the northern end of a 75-mile-long continuous range of mountains, extending from Albuquerque south to almost Socorro. This is not really the case. A northeast-trending feature, the Tijeras fault zone (*Figure 0-4*), provides a sharp dividing line separating Sandia Mountain to the north from other mountains to the south that seemingly are connected but have very different ages and histories. Sandia Mountain is younger than them by roughly 20-25 million years. The Manzanita Mountains, located just south of the Tijeras fault zone and lying east of Kirtland Air Force Base, is a semi-flat-capped but deeply eroded platform, topping out at about 7600 ft, with a highest point at about 7900 ft. South of the Manzanitas are the more robust and complex Manzano Mountains, with high points of between about 9500 and 10,100 ft.

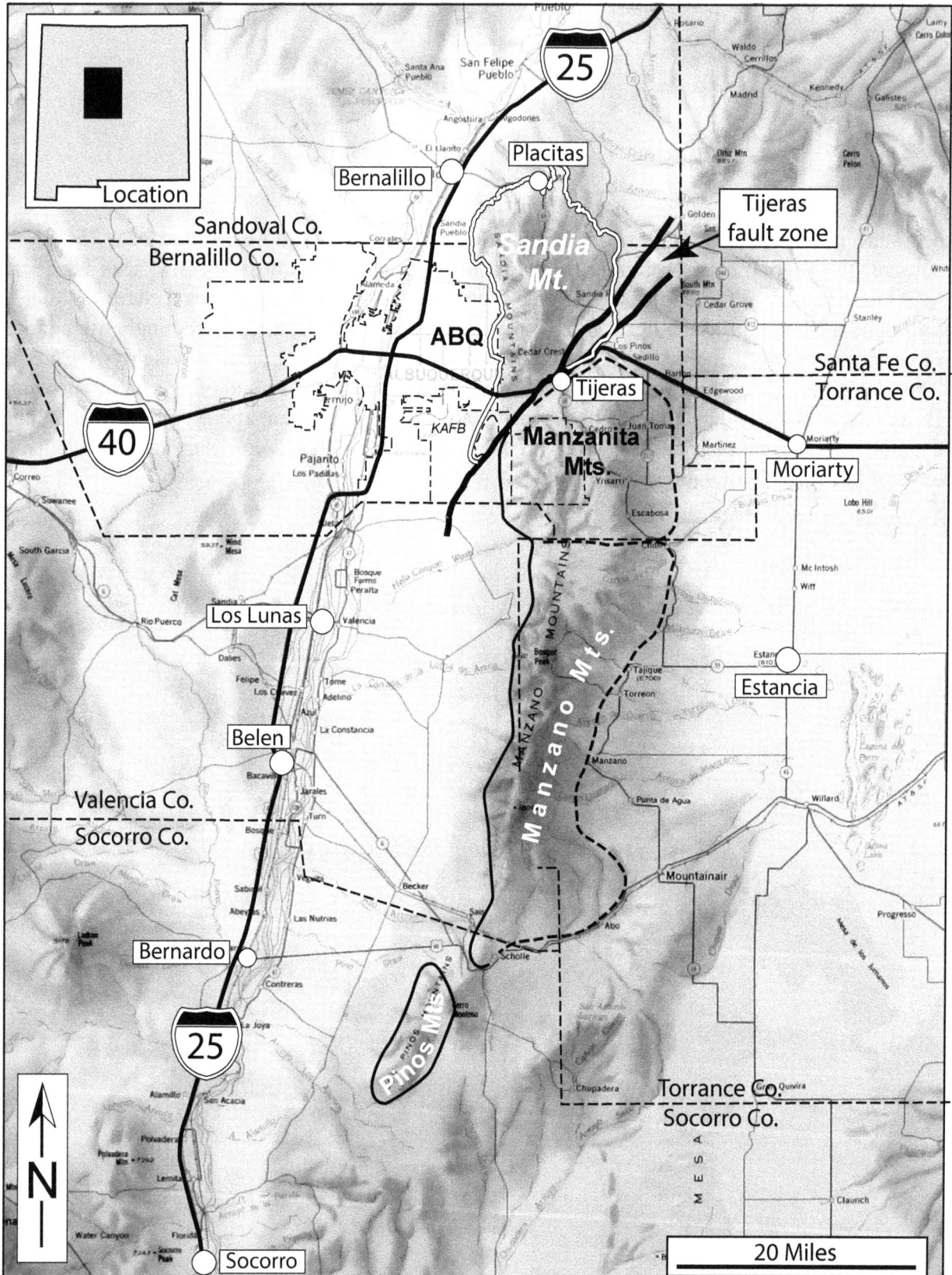

Figure 0-4. Shaded Relief Map of Sandia Mountain and Mountain Chain to South
(Shaded Relief Map Modified from Raven Maps and Images 1991)

Sandia Mountain can be very simply described as an elongated north-south dome, measuring about 22 miles in length by a maximum width of roughly eight miles. It has a broken off northern end and west side, and a gently-inclined "dip slope" or ramp-like flexure down to the east and northeast. This gentle eastern flexure creates a quandary of how to define where the mountain ends. I loosely select this boundary as a combination of the 6500-ft elevation contour on the west, north and northeast, the southern strand of the Tijeras fault zone basin on the southeast, and highway I-40 on the south except where a lobe of the mountain extends into Kirland Air Force Base (*Figure 0-5*). Admittedly this is messy, but a limit needs to be made somewhere. The boundary between the east and west sides of the mountain is the much-more-obvious topographic ridgeline called Sandia Crest, which splits the feature in half. On the north end the crest splits into two prongs before becoming indistinct.

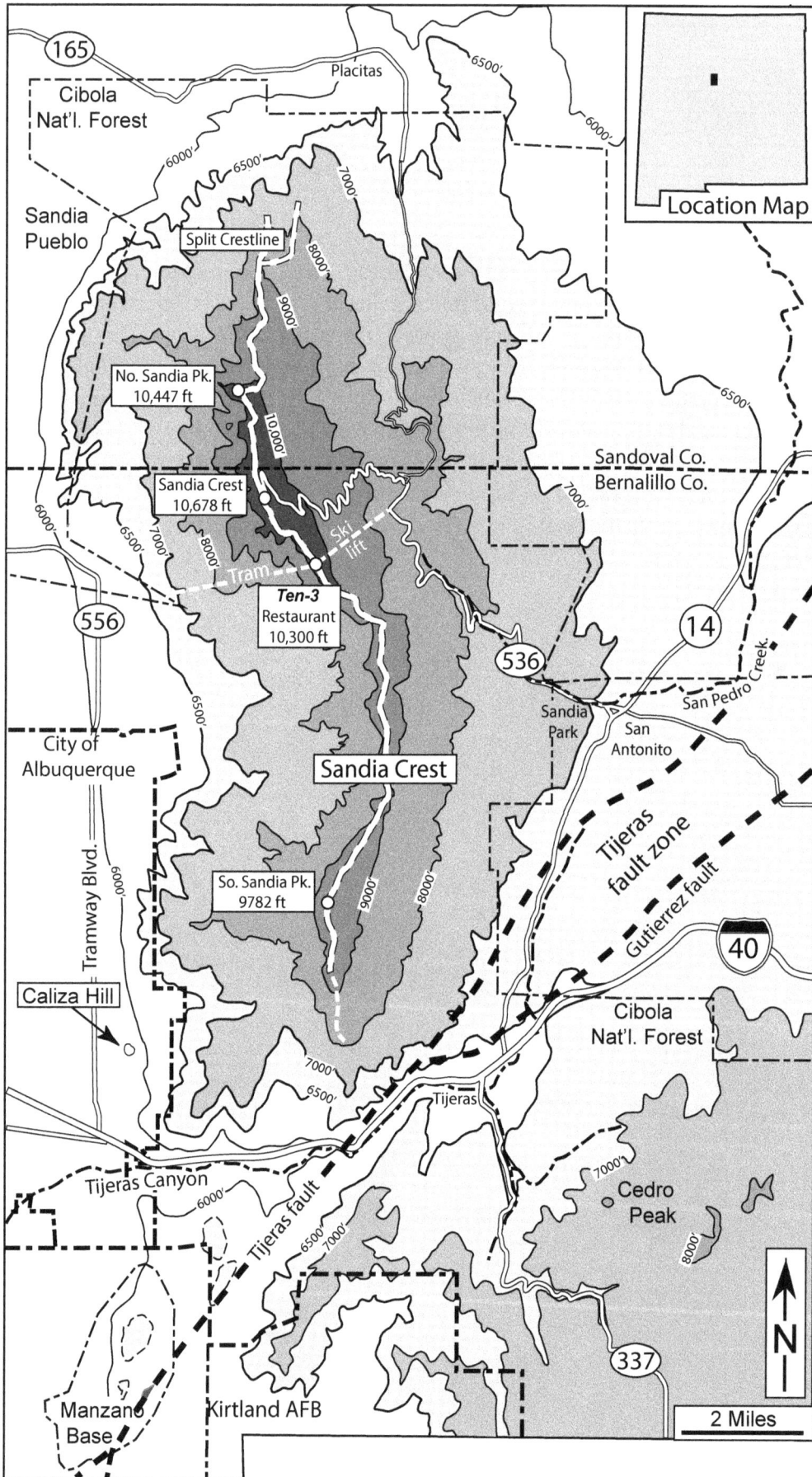

Figure 0-5. Topographic Map of Sandia Mountain

Three high points punctuate the crest. From north to south these are North Sandia Peak at 10,447 ft, the highest point on Sandia Crest at 10,678 ft, and South Sandia Peak at 9782 ft. All the literature I've encountered refers to *Sandia Mountains* in the plural. All three high points on the mountain's crest though are simply protuberances on a single ridge, rather than individual edifices (*Figure 0-5*). To put a fine point on it, according to "rules" set by the U.S. Geological Survey for naming peaks, a summit must be a mile from its neighboring summit and have a separating saddle at least 500 ft lower. To me then the term "Sandia Mountains" does not meet these tough requirements to be considered a *range* of mountains. Sandia Mountain is a single feature. Henceforth in this book I'll refer to "*Sandia Mountain*" in the singular.

In map view and satellite images the mountain displays a wavy "S-shaped" form (*Figures 0-5, 0-6* and *0-7*). From a distance in just about any direction Sandia Mountain appears as is a huge "whaleback" hump (*Figure 0-8*).

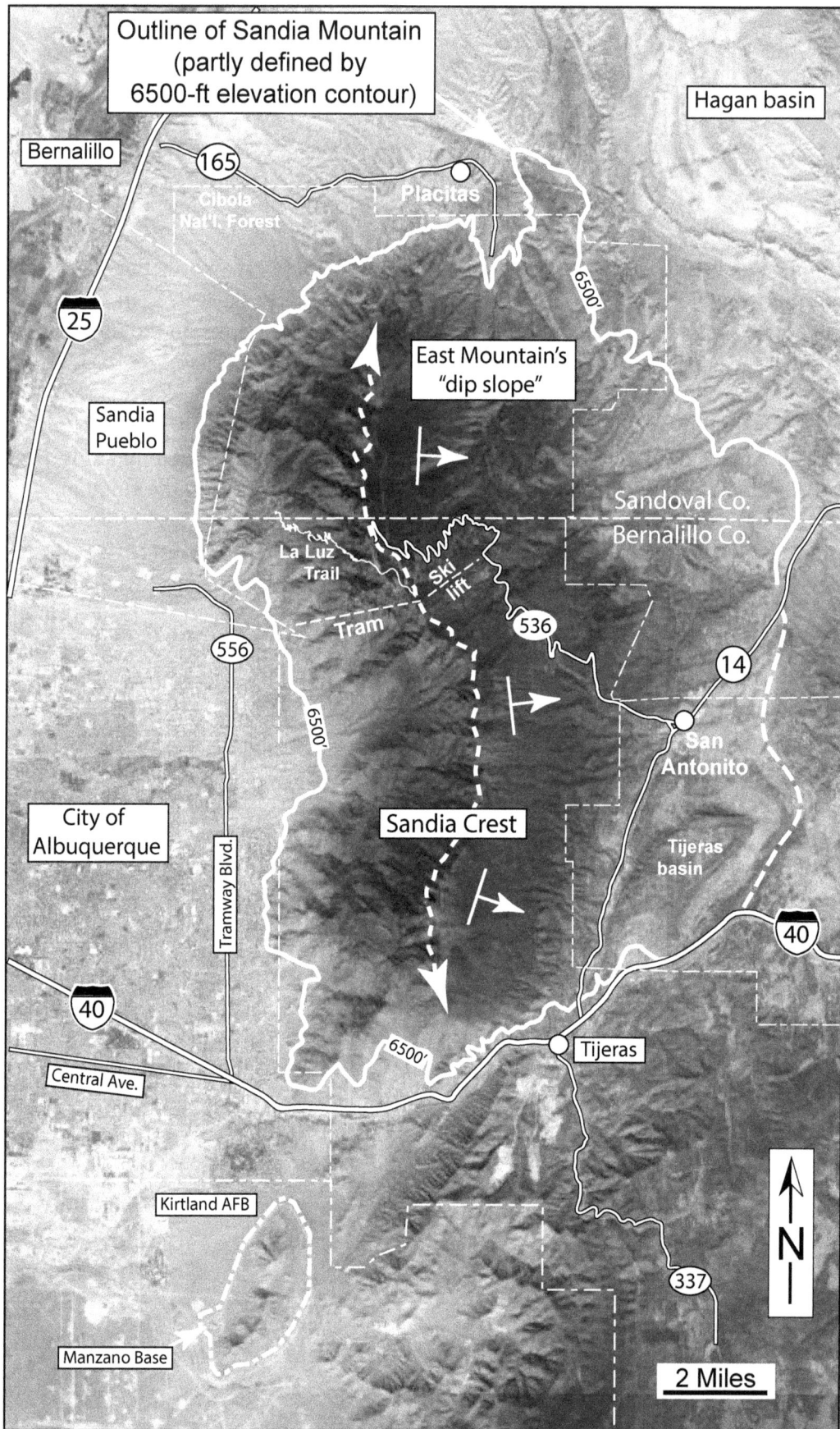

Figure 0-6. *Google Earth* Image of Sandia Mountain

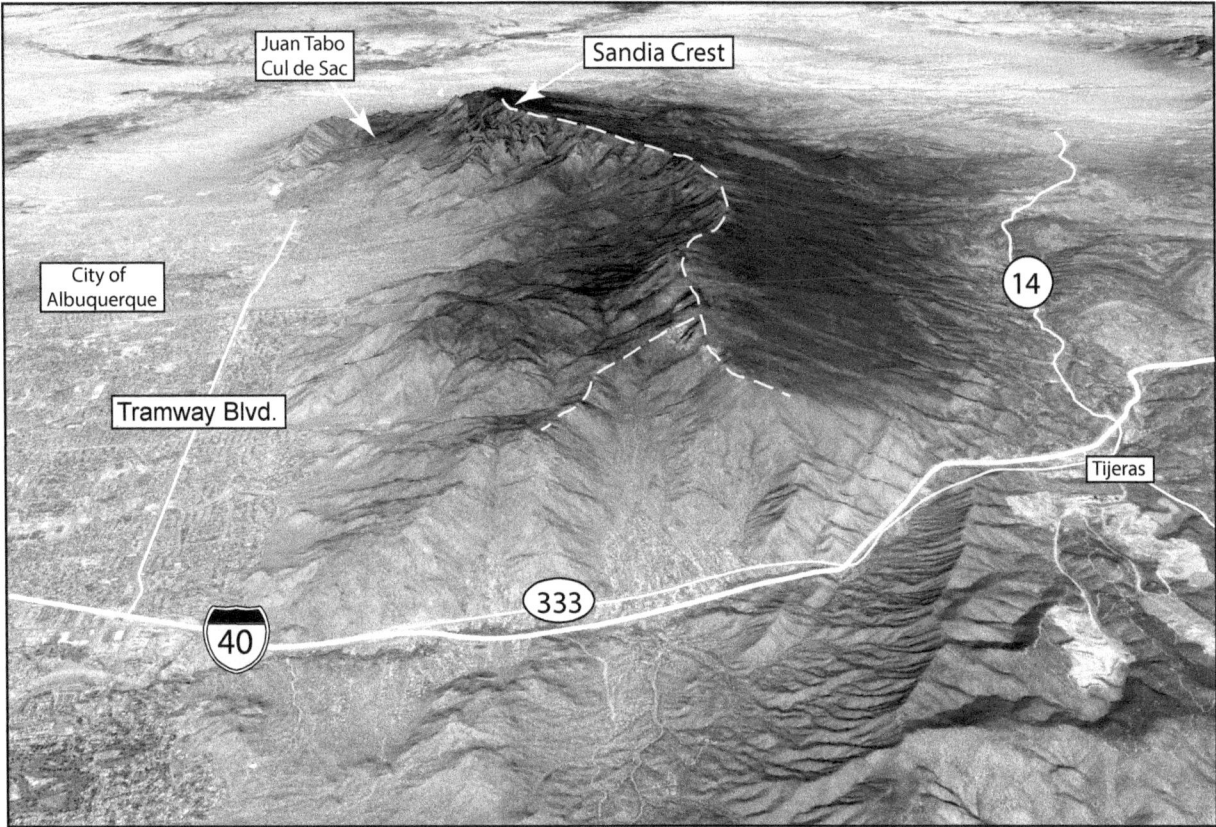

A. North Image

A. South-Southeast Image

Figure 0-7. Oblique *Google Earth* Images of Sandia Mountain

A. Lithograph of Southwest View from Santa Fe to Ortiz Mountains (Left) and Sandia Mountain (Right) (Modified from Gregg 1968; original by Brothers R.H. Kern or E.M. Kern, J.H. Simpson Expedition, 1849)

B. Same View in "A" Above Compressed Vertically 50% to Appear Natural

C. South-Southeast View of Sandia Mountain (Photograph by Author)

Figure 0-8. Distant Views of Sandia Mountain

I've subdivided the mountain into five geographic sectors (*Figure 0-9*): 1) the northern nose in the area of the village of Placitas; 2) Rincón Ridge on the northwest, 3) the Juan Tabó Cul-de-Sac east of Rincón Ridge; 4) the western foothills extending south into Kirtland Air Force Base; and 5) Tijeras Canyon on the south along with the "East Mountain," that broad area east and downslope from Sandia Crest. I've included another area, 6) the so-called Ortiz Mountains or Ortiz Igneous Belt to the east, which although not part of Sandia Mountain, is proximal to it.

Figure 0-9. Southeast Oblique *Google Earth* Image of Sandia Mountain Showing Main Sectors

Infrastructure

The main access to Sandia Crest is the all-weather Sandia Crest Byway, NM-536, which runs for 13.3 switchbacky miles from the community of San Antonito at NM-14 down below on the east, rising 3814 ft to the crest where there are an overlook, gift shop and cafe. The Sandia Peak ski lift on the east side and the extremely popular Sandia Peak Tramway on the west side meet at the crest at 10,300 ft. That convergence point houses the spanking-new *Ten-3* restaurant (*Figures 0-5* and *0-6*). Also, there are numerous foot-trails, the most impressive of which is the La Luz Trail on the northwest side. This hugely popular and world-class hike (a slog!) ascends 3290 feet over 7.5 excruciating miles to the upper terminal of the Sandia Peak Tramway on the Sandia Crest (*Figure 0-6*).

Some Initial Questions

Several questions emerge at the very start. First, what happened to the rest of the mountain that broke off on the west side? Why does the west side seem to consist of several giant topographic stairsteps down below the crest? Why does the northern cul-de-sac seem like an isolated and detached bowl? Why is the north end near the village of Placitas so different from everything else? Why does the mountain ramp so gently down to the east and then flatten out? And, not at all least, why is the Sandia Granite, the rock forming the core of the mountain and originally produced several miles below the surface (more in *Chapter 2*), now sticking more than 10,000 feet up in the air? Just asking such things stokes the juices of curiosity. This is obviously not a simple mountain! Its history is indeed multifaceted, fascinating, but *completely understandable*. This book will plumb that bizarre history with all its twists and turns. And we'll start from the beginning.

But first let's get the obligatory primer on geology out of the way.

Some Geologic A-B-Cs

To get started on our study of Sandia Mountain we must consider some geologic essentials: 1) types of rock, 2) bodies of rock, 3) types of geologic illustrations, and 4) geologic time (plus standard abbreviations used for geologic time and age).

Types of Rock. The fodder for geologists is the *rock*. The term is definitely not a synonym for "stone," which is something landscaped with, built with, or thrown. Rock (except Earth's core) is what Earth of made of. A rock is defined as a naturally-occurring, solid aggregate of one or more *minerals*. A mineral, in turn, is defined as a naturally occurring, inorganic, chemical element or compound, with an orderly internal atomic structure, and a characteristic chemical composition and physical properties. The general term used to describe the physical makeup of a rock is its *lithology*.

There are three main categories of rocks, within which there are many subtypes, but we'll just deal with the main groups:

Igneous. This type is formed by cooling and consolidation (crystallization) of hot, molten rock (*magma*) into minerals. The type is either extruded *onto* the Earth's surface as "extrusives" or "volcanics," such as lava and ash flows, or intruded *into* older, pre-existing rocks of any type, usually deep below the Earth's surface, as "intrusives." The former typically form in layers on the Earth's surface, while intrusives, when exposed, typically form massive, unlayered bodies.

The igneous rock's grain size is indicative of the magma's cooling history: coarse-grain texture indicates slow cooling, whereas fine-grain indicates rapid cooling or chilling. If intrusive igneous rocks crop out today at the surface, it follows that they have been uplifted and that the rocks that once occurred above them have been eroded away. Usually, this type tends to form the cores of mountain ranges. One intrusive igneous rock, *granite*, is germane to our story of Sandia Mountain. (More below under Rocks of Sandia Mountain.)

Sedimentary. These rocks are consolidated *sediment* formed on top of, or very close to, the Earth's surface. One type of sediment is loose particulate material derived by the physical or chemical breakdown of pre-existing rocks, and then transported by flowing water, wind, or downward-creep due to gravity and then deposited in layers. Examples of these, from the original sediment to their consolidated forms and classified by grain size, are: 1) boulders, cobbles or pebbles →*conglomerate*; 2) sand →*sandstone*; and 3) silt/clay →*mudstone* (or *shale* if the rock is platy). With increasing distance of transport, the particles become rounded due to abrasion and are finer grained.

A second group of sedimentary rocks is *non*-fragmental, i.e., those formed in place (vs. transported), most often in layers on the ocean floor. The most common is *limestone* (calcium carbonate, $CaCO_3$), formed through the chemical action of living organisms Less common types are chemical precipitates such as *gypsum* (calcium sulfate, $CaSO_4$), and *travertine* (also $CaCO_3$ but deposited by fresh-water springs).

Metamorphic. This type is formed by alteration (metamorphism) of pre-existing rocks of any kind, and recrystallization by high pressure and/or high temperature. Like intrusive igneous rocks these were formed deep below the Earth's surface and, if exposed, have been uplifted and unroofed. They are therefore typically very old and usually hard, sometimes highly distorted, and like intrusive igneous rocks, tend to form the cores of mountains. They are subdivided by crystal size, composition, and texture. A few examples, from the original form to the altered product, are 1) mud →*slate*→*schist* →*gneiss*; 2) sand →*quartzite* (welded sandstone); and 3) limestone →*marble*.

Bodies of Rock. Rocks occur in two general modes: layered, and unlayered (massive). Sometimes the two can be confusing mixtures. Layers are formed one at a time, one atop the other. Almost always these were formed on the Earth's surface where the requisite upper free space was available. Most often these are sedimentary rocks, but also can be stacks of extrusive igneous, volcanic materials. In contrast, massive rocks are those sedimentary types that lack bedding planes, and intrusive igneous or metamorphic rocks that have been formed at one single time *en masse*. Geologists lump rocks together into mappable bodies and give them names.

Formations and Groups. A sequence of layered rock large enough to be mapped in the field at a reasonable scale, typically 1:24,000 (the scale of standard 7.5-minute U.S. Geological Survey topographic maps) is the called a *formation*. The term is of fundamental importance in geologic mapping. A formation has characteristic physical properties such as thickness, particle grain size, color, bedding shapes, etc. Each formation is given a proper name taken from a geographic locality or prominent geographic feature (a town, river, mountain, etc.) near where the unit was first officially described in the literature (see the lexicon of unit names in *APPENDIX IV*). That location becomes the formation's *type section*. This reference place is always consulted when questions arise about the nature and definition of the unit. Finer subdivisions of formations are called *members*, and larger clumpings of related formations are called *groups*.

Intrusive Igneous Rocks. These form as large irregular bodies called *plutons*. They can form horizontal, tablet-shaped bodies called *sills* (think of window sills), vertical wall-like bodies called *dikes* (think of Holland), or mushroom-shaped masses called *laccoliths*.

Structures. Two types of structure are most common: folds and faults.

Folds are simply mappable bends of layered rocks. They come mainly in three varieties: anticlines, synclines, and monoclines (*Figure 0-10A*).

Faults are fractures in bodies of rock where there has been relative movement along the plane of the fracture (*Figure 0-10B*). Now for a little gobbledygook: the block on top of an inclined fault plane is termed the *hanging wall block*, and that on bottom the *footwall block*.

There are three main types of faults. The first is the *normal fault* formed under tension (top of *Figure 0-10B*) where the hanging-wall block drops down along the fault surface with respect to the foot-wall block due to gravity. Typically, the fault angle is about 60° from the horizontal. On maps, bar-and-ball symbols indicate the down-side of the fault (top right of *Figure 0-10B*).

The second type, the *reverse fault*, is formed under compression (middle of *Figure 0-10B*) where the hanging wall block is thrown up with respect to the footwall block. On maps, black triangles indicate the up-side of the fault (middle right of *Figure 0-10B*).

The third type is the *strike-slip fault*, formed under a twisting, lateral compression (bottom of *Figure 0-10B*). Here, one block moves laterally with respect to another. On maps, arrows show the sense of movement. This type has an odd naming system. Imagine standing astride the fault, looking along its trace. If the blocks move as in the figure in the lower left, you will be turned to the left and the fault is called *left lateral*. Similarly, if you are astride the fault on the lower right, you will be turned to the right and the fault is *right-lateral*.

Unfolded Rock Layer

Anticline

Syncline

Monocline

A. Folds

Created by tension

Normal fault

Normal fault after erosion

Created by compression

Reverse fault

Reverse fault after erosion

Created by Compression at an Angle

Left-lateral, strike-slip fault

Right-lateral, strike-slip fault

B. Faults

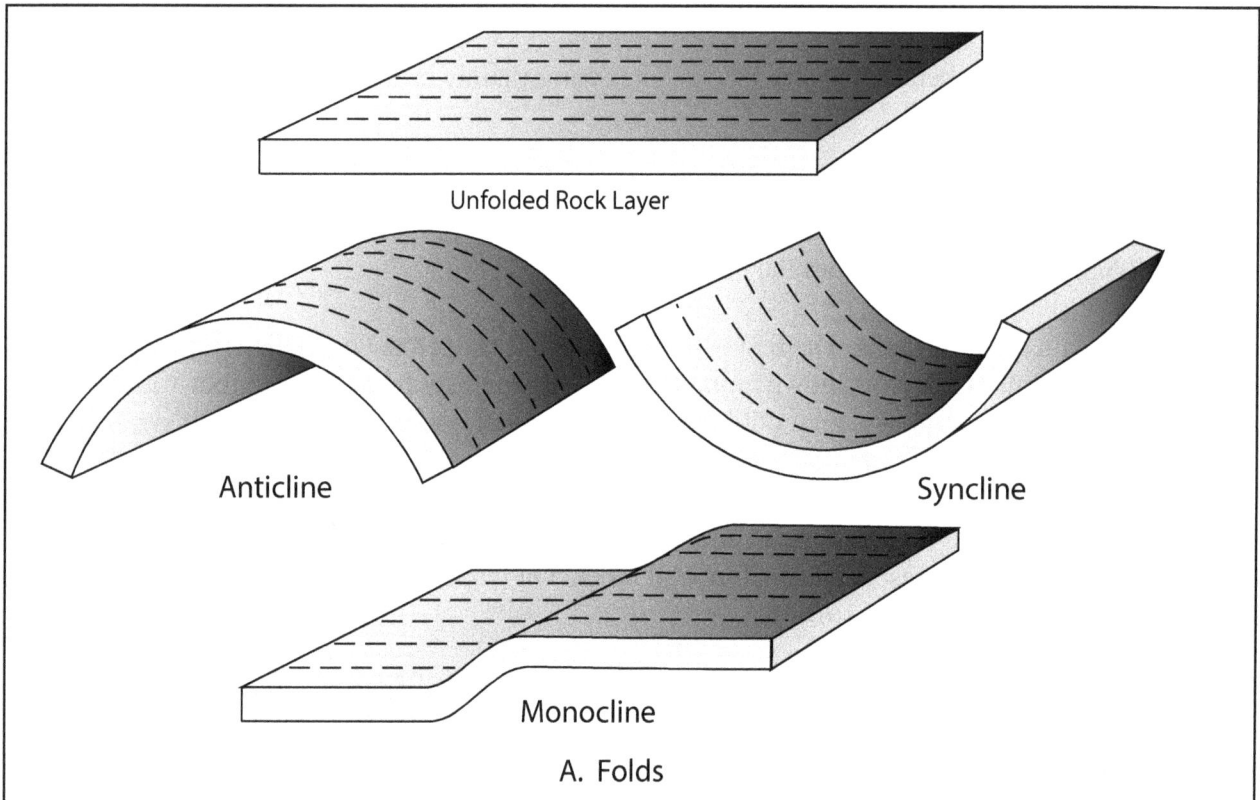

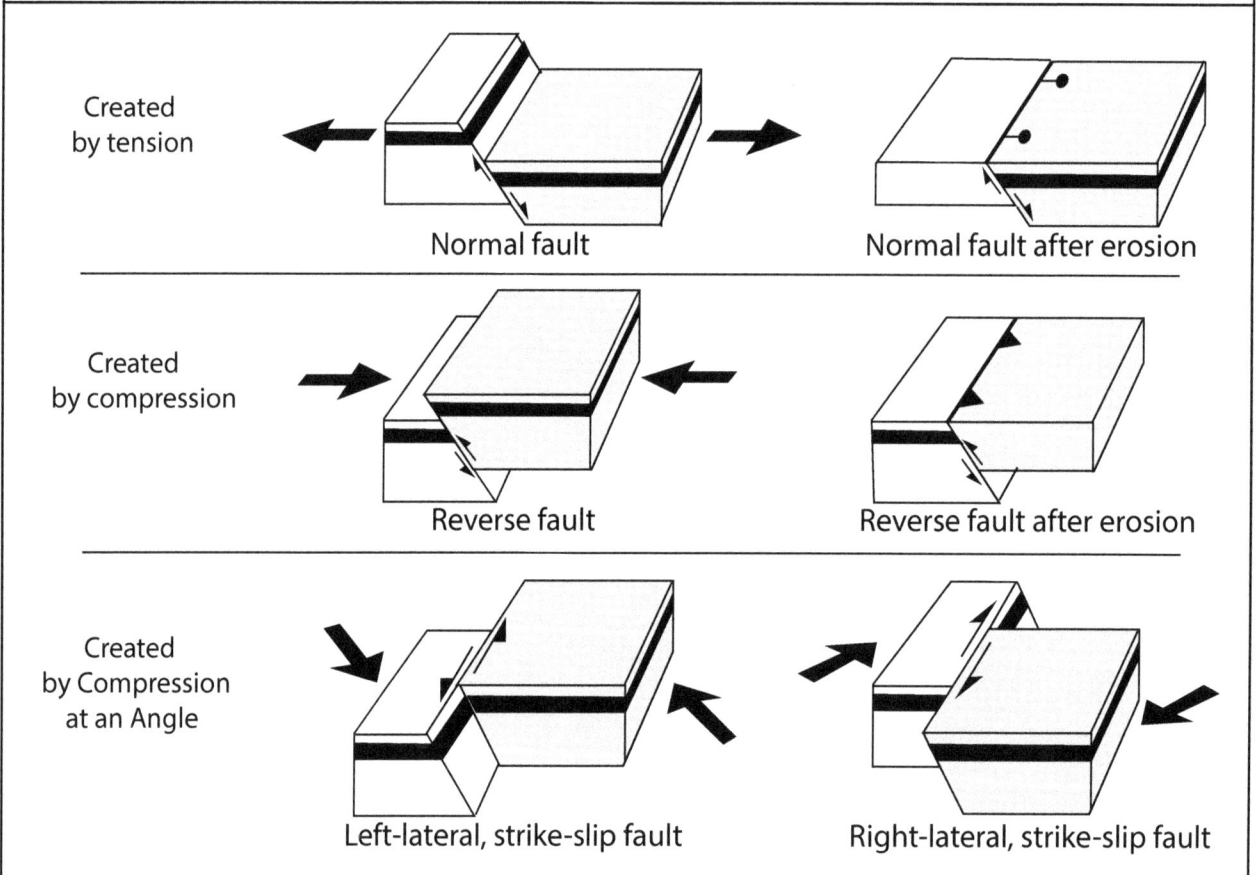

Figure 0-10. Types of Structures

Sinistral vs. Dextral. Leave it to geologists to invoke Latin. Left-lateral faults are sometimes called *sinistral*, and right-lateral *dextral*. These Latin words crop up in many places and contexts, and mean "to-the-left" and "to-the-right," respectively, but they've acquired deeper and more nuanced meanings. The left hand for centuries has been considered evil, unclean, or "sinister." A "left-handed" compliment is one that is less than welcome. Dextral shows up in biology and refers to coiling to the right, and in chemistry as in *dextrose*, a type of sugar in which certain optical properties turn to the right.

Geologic Illustrations. There are 12 types of geologic illustrations commonly used by geologists. These are listed below.

Topographic Maps. A topographic map illustrates the shape of the landscape by means of lines ("contours") connecting points of equal elevation. The contours are separated by a constant interval of elevation, the "contour interval" or "CI." The CI chosen depends on the amount of topographic relief present in an area and on the scale of the map. The U.S. Geological Survey has published a vast series of topographic maps, "quadrangles," typically measuring 7.5 x 7.5 minutes of latitude and longitude, respectively, with a scale of 1:24,000, or 2000 ft = 1 inch. The quadrangle maps are given names of some prominent feature or town on the map. (Older "quads" were usually 15 x 15 minutes on a side with a scale of 1:48,000.) Some modern topographic maps cover an even greater area and measure 30 x 60 minutes with a scale of 1:100,000, reflecting the modern tendency to go metric.

Geologic Maps. Such maps show the distribution of different bodies of rock, both layered and unlayered types, exposed on the Earth's surface. The rock units, usually formations, are assigned patterns and/or colors that differentiate them by age and type. The structure of the rocks (horizontal, inclined, folded, faulted) controls the distribution of the rocks on the surface, which in turn controls the relief of the landscape. Generally, a geologic map shows the "bedrock," i.e., the hard, solid rocks that lie below the younger unconsolidated units being currently distributed by streams. Some geologists focus their efforts solely on the younger, "surficial" material. Bedrock mappers jokingly consider the surficial material above bedrock to be "overburden," or stuff" in the way." In contrast, surficial mappers sometimes consider the bedrock as "underburden," a platform supporting the important stuff on top.

"Surface" vs. "surficial." This terminology, unfortunately, can be confusing. The noun "surface" means the top of a topographic or geologic unit, while the adjective "surface" means *at* the Earth's surface. The adjective "subsurface" means *below* the surface and the term "subsurface geology" refers to the bedrock geology hidden from view below the Earth's surface. On the other hand, the noun "surficial" indicates the youngest, usually Quaternary-age deposits (Q) that form a thin veneer over the underlying bedrock, and "surficial geology" deals with that thin veneer.

Structure Maps. Like topographic maps, structure maps describe the elevation of a select datum, such as the top of a rock formation, below the surface using contour lines with a constant interval, or they can even project a datum into thin air where it has been removed by erosion using dashed contour lines.

Block Diagrams. These 3-D figures are usually employed to better illustrate some sort of structure (e.g., *Figure 0-2*).

Cross Sections. A geologic cross section is drawn along a vertical plane parallel to a line drawn on a geologic map. A cross section shows the distribution of the rocks below the surface along that line and in that plane. To be accurate the line must be straight, but geologic trends sometimes do not cooperate. In this case the geologist will bend the line from place to place, losing some accuracy in the process. Cross sections are essential companions to geologic maps and generally use the same patterns and colors. A truly accurate cross section is drawn to true scale, i.e., the vertical ("V") and horizontal ("H") scales being the same. However, some of these sections will be overly wide and skinny because the vertical distance shown (usually 100s or 1,000s of feet) is often so much less that the horizontal distance (usually miles). To alleviate this inherent problem, cross sections are often drawn with a built-in vertical exaggeration of anywhere from two (V:H = 2) to as much as 50 (V:H = 50). Another type of cross section is the schematic variety. These are drawn casually without a scale to illustrate some general relationship. They are extremely useful but the schematic nature must be clearly stated.

Stratigraphic Columns. A column is an artificial construct, showing all the rocks of an area (almost always sedimentary) stacked vertically as if they were present in one place. Although rocks are instead usually spread out over a wide area, such a column serves as a reference tool to characterize an area, to show which rock units are present there and how thick they are.

Isopach Maps. The word "isopach" means equal thickness. An isopach map is one showing the varying thickness of a geologic unit, such as a formation, over an area by contour lines connecting points of equal thickness. These are extremely important in certain quarters, particularly in oil and gas exploration, but none appears in this book.

Paleogeographic Maps. These are synthesis maps showing an ancient geography. They show the ancient *paleo* distribution of land areas, bodies of water, and certain environments (deltas, coastal plains, floodplains, etc.) using different colors, shades of gray and/or patterns. They are extremely useful in reconstructing an area's geologic history. The non-geologist might have some difficulty feeling comfortable with a paleogeographic map (*Figure 0-11A*) and feel more at ease with a "bird's eye" map. The latter is a view of an area that might be seen from space (on an exceptionally clear day; *Figure 0-11B*). These are excellently illustrated by the beautiful art work of Dr. Ron Blakey in his several books on the western U.S. (Blakey and Ranney, 2008, 2018) and on his website. (Blakey is a Professor Emeritus of Geology from Northern Arizona University, and has been creating his wonderful images since 2009). Although a bird's-eye map provides more comfort to the non-geologist, the paleogeographic map provides much more information. We will revisit both maps in *Chapter 3*.

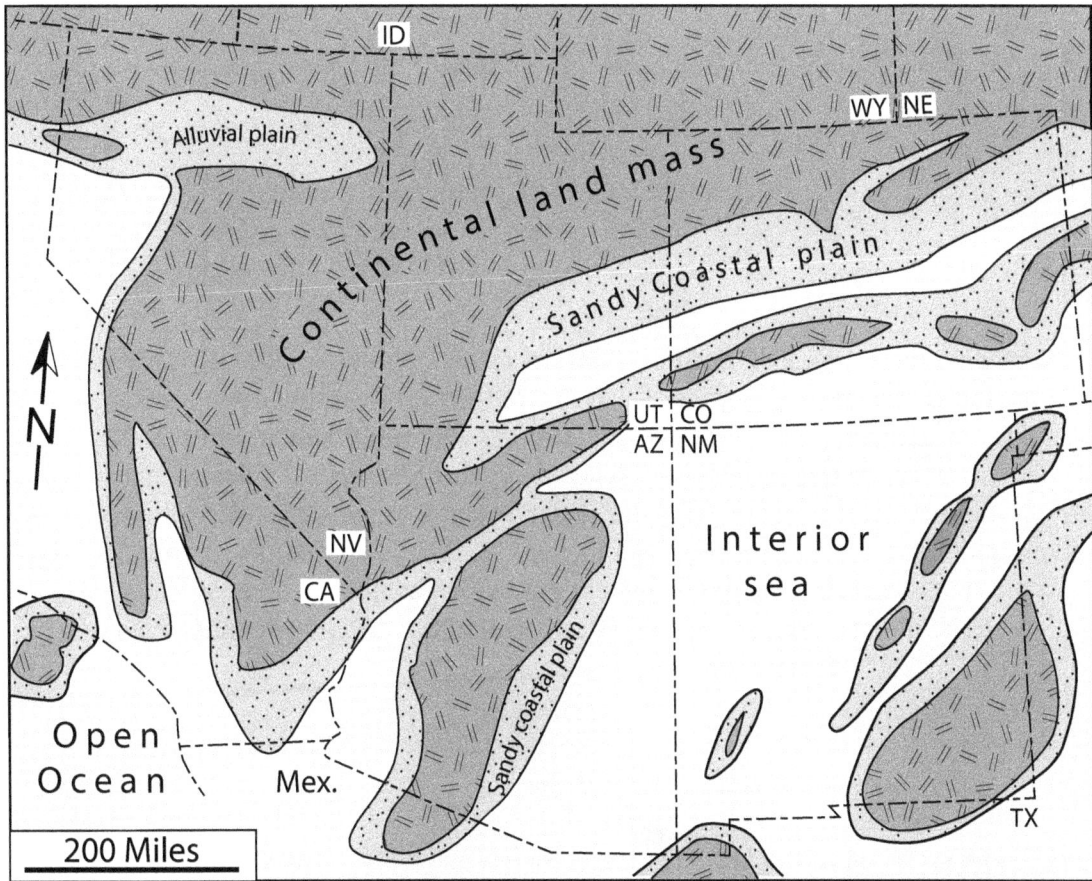

A Paleogeographic Map of Southwest North America, 1.7 Ga

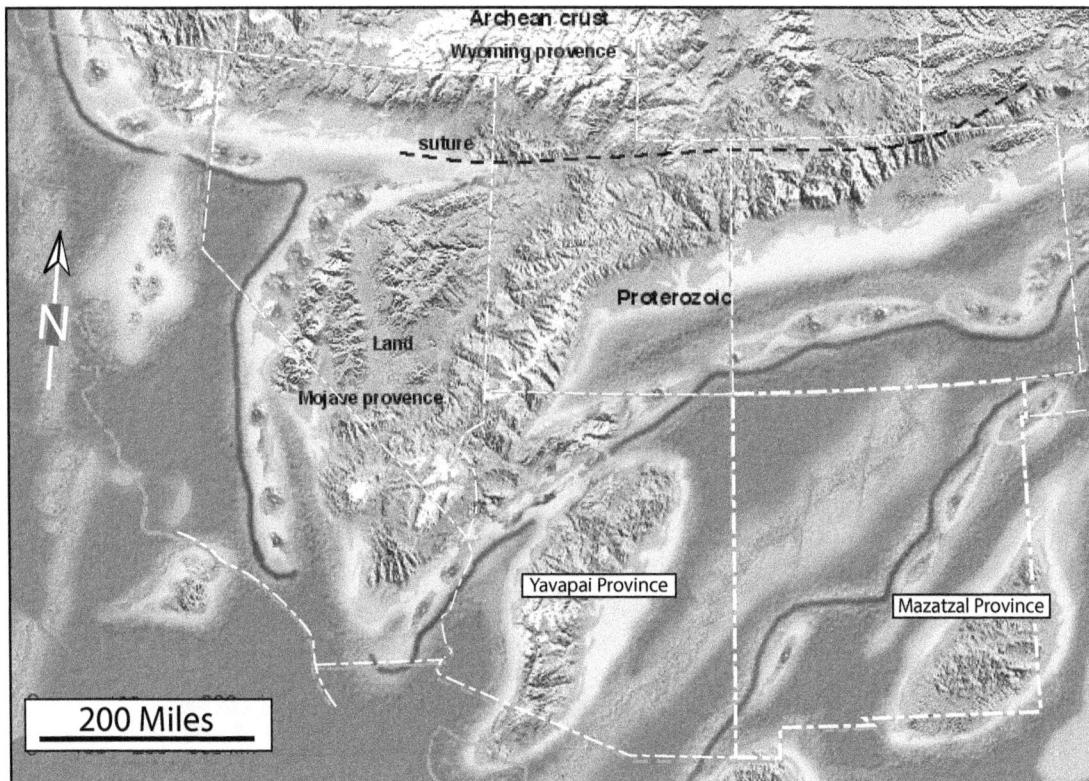

B. "Bird's Eye" View of Southwest North America, 1.7 Ga (Modified from R. Blakey, Used with Permission)

Figure 0-11. Contrast Between Paleogeographic Map and "Bird's Eye" View

Paleogeographic maps and accommodation space. The non-geologist may have trouble fully grasping the significance of a series of these maps that scrolls through time-successive geographies. Each ancient geography, in sequence, actually lies *on top of* the earlier ones. To explain we need to appreciate the important concept of *accommodation space*. It goes like this. As sediment is delivered to an area by water or wind, the land under it tends to subside due to the added weight. This creates a topographic low, a "basin" that allows the new sediment to accumulate in the same place rather than being swept across and deposited somewhere else downstream. In other words, subsidence creates accommodation space in which the new sediment can accumulate and thus be preserved to become part of the geologic record.

Cartoons. These figures, which tend to draw a smile, have no scale and are highly stylized. However, they can be extremely useful to illustrate a geologic principle.

Google Earth Images. *Google Earth* is a boon to geologists and geographic researchers alike via a personal computer. The program has been available for free on the Internet since 2004 when *Google* acquired the pioneering company *Keyhole Inc*. Since then, it has added *Street View* (2008), *Historical Imagery* (2009), and *3D Imagery* (2012). Its resolution is often superior to conventional aerial photography. The ability to tilt the images to view features at oblique angles is a wonderful asset, but there is some distortion where the relief is high, so some caution is warranted. *Google Earth* provides instant gratification though, avoiding the hassle of purchasing the aerial photos, waiting on delivery, somehow hustling access to the required (and rare) stereo viewer, and of course paying for them. I have used *Google Earth* extensively in this book.

Photographs. Photographs are indispensable, and Sandia Mountain is extremely photogenic. Roadcuts provide many rock exposures and some of the best ones are courtesy of the highway engineers. The engineers wear two hats: the good-guy hat where they helpfully provide fresh, unweathered outcrops, and the bad-guy hat where they may fiendishly feel it necessary to stabilize ("protect") outcrops by coating them with a layer of *gunnite* (gun-sprayed concrete slurry; *Figure 0-12*).

Figure 0-12. An "improved" outcrop of Permian Abo Formation, I-40 at Tijeras
(Photograph by Author)

Other. As soon as geologists generate numerical data that do not lend themselves to mapping, they need a way to present them. A host of esoteric charts, plots, graphs, etc., are thus employed, but we will not deal with any of these in this book.

Geologic (Deep) Time. An appreciation of geologic time requires a very different mindset. The evolution of this crucial concept is one of mankind's greatest intellectual achievements and led to the development of the *geologic time scale* (*Figure 0-13*).

Years	Eras	Periods		Rocks of Sandia Mt.	
				Sedimentary	Igneous/Metamorphic
Present					
2.6 Ma	Cenozoic (Cz) (tme of "modern life")	Quaternary (Q)			← Young alluvium
5 Ma		Tertiary (T)	Neogene	Santa Fe Group' Alluvial fill of Rio Grande rift and Hagan basin	
10 Ma					
15 Ma					Igneous dike at Placitas
20 Ma △					
30 Ma △			Paleogene		▬ 31 Ma ▬
40 Ma					
50 Ma					
60 Ma					
66 Ma △					
100 Ma	Mesozoic (Mz) (time of "middle life")	Cretaceous (K)			Rocks on back side of Sandia Mt. and in Placitas area
150 Ma		Jurassic (J)			
200 Ma					
		Triassic (TR)			
245 Ma △		Permian (P)			
300 Ma	Paleozoic (Pz) (time of "ancient life")	Pennsylvanian (IP)			Rocks on Sandia Crest
		Mississippian			
400 Ma		Devonian			1.1 Ga "Hiatus" (the *Great Uncomformity*
		Silurian			
		Ordovician			
500 Ma		Cambrian			
540 Ma					
600 Ma △	Proterozoic "basement" (X and Y)			Sandia Granite (Xg) Rocks of Rincon Ridge (Ym)	
1.0 Ga					
1.5 Ga					1.4 Ga / 1.7 Ga

△ = Scale change   Ma = million years   Ga = billion years

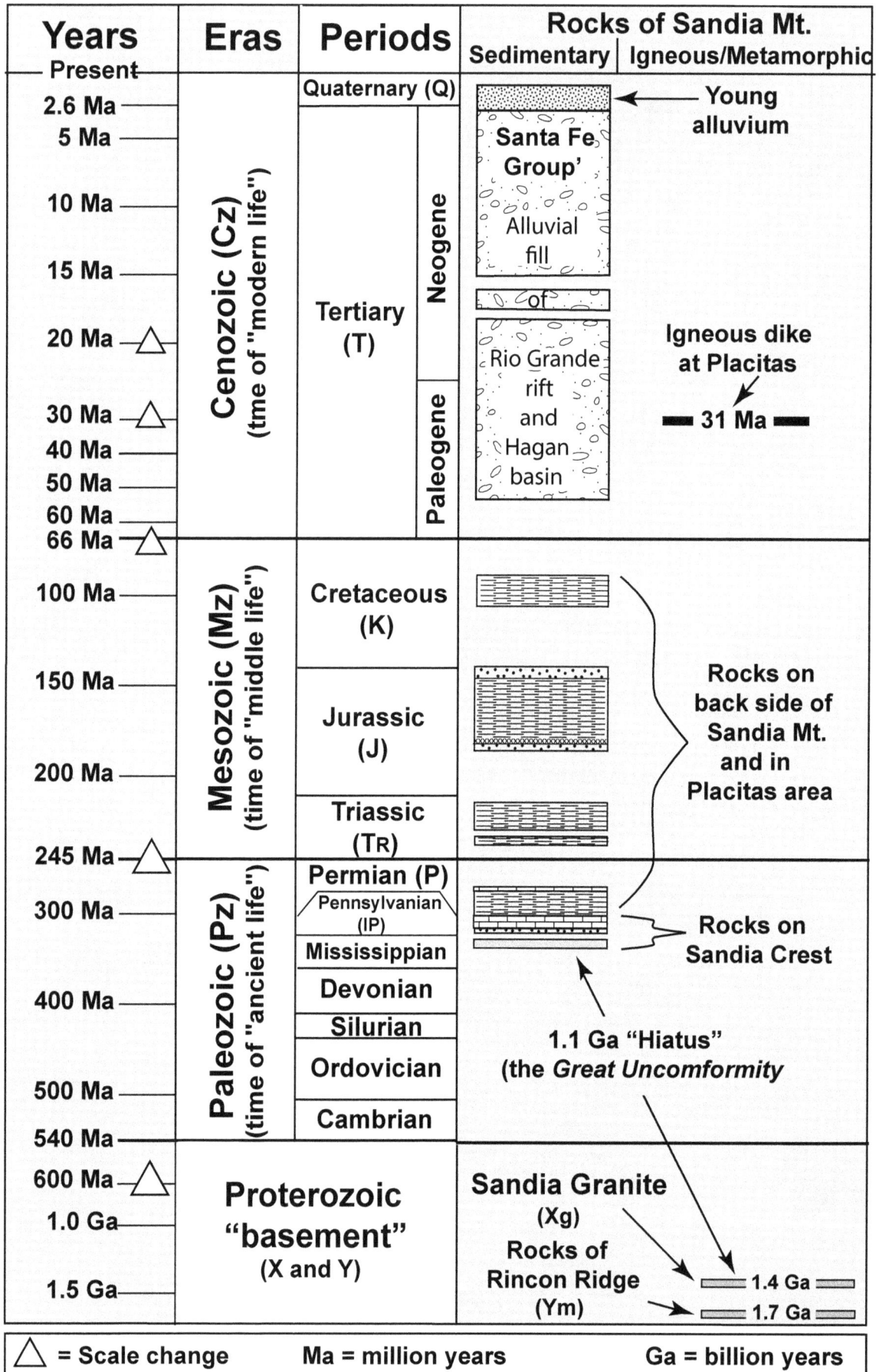

Figure 0-13. Geologic Time Scale and Rocks of Sandia Mountain

Problem of Scale. To somehow present the vast scope of geologic time on a single page requires downward compression of the time scale. The five changes of downward scale in the figure are marked by the Greek letter *alpha*, Δ, designating the point of change. To illustrate, if the 1.1 Ga "hiatus" shown near the base of the figure were shown at the same scale at that for the top 20 Ma, it alone would need about 8.5 ft of paper!

"Time" is a concept that exists mainly in the human mind. We structure our lives around it. We're obsessed by it. We humans think of "normal time" in terms of astronomical years, grouped into generations (20-30 years), lifetimes (70-90 years), historic eras (i.e., Middle Ages ~1000 years, the Renaissance ~300 years), and even pre-historic Ages (i.e., Stone Age ~6,000 years). In geologic terms these are blinks of an eye, and anything much longer is beyond our every-day comprehension. The total time span of the ages of the rocks of Sandia Mountain is about a whopping 1.7 billion years. This is *deep time*!

Billion. The number "billion" is used here in the American English sense, denoting one thousand million, 1,000,000,000 or 10^9, sometimes called the "short scale." Since about 1974, British English has used "billion" to mean one *million* million, or 10^{12}, the "long scale." This is our "trillion." Some American technical writers often try to avoid the confusion by using the term "thousand million" instead of "billion." Today the short scale is used both in this country and in the United Kingdom, as well as in this book, but France, being France, commonly still uses the long scale.

Several abbreviations are employed in the geologic literature to denote these large time numbers, with the help of Latin. These terms will crop up frequently in this book (the term "trillion" is not necessary!):

Ka: *Kilo anni* = thousands of years/thousands of years old/thousands of years ago (for some reason often written with a lower-case "k").
Ma: *Mega anni* = similar to above, for millions of years.
Ga: *Giga anni* = similar to above, for billions of years.

Geologic time is broken down in *Eras* and smaller *Periods* (*Figure 0-13*). The era names are based on the general type of fossil evidence the rocks contain. From oldest to youngest these are the Paleozoic (*Pz*), Mesozoic (*Mz*), and Cenozoic (*Cz*), which denote ancient, middle, and modern life, respectively. The period names all have interesting histories that we'll visit in APPENDIX II. The relevant abbreviations for all these units of time are shown on the time scale (*Figure 0-13*).

The gaps between the rock units shown on the time scale (*Figure 0-13*) warrant explanation. The rocks record discrete, discontinuous geologic *events*, but time itself flows steadily. A comparison to a tape-recording session of a rock-music combo might be instructive (*Figure 0-14*). The rock band in the studio will blast away for a certain time span as it is being recorded, and then take a coffee break while the recorder is turned off. The recorded sessions are analogous to the preserved geologic record, i.e., the formations, and the non-recorded coffee-breaks are the gaps between the formations. Geologists call time-gaps *unconformities*. For example, the time scale in *Figure 0-13* shows an enormous time gap between what is called the "basement" rocks and the first preserved rock formation above. This huge *hiatus* is called the *Great Unconformity* (to be discussed more in *Chapter 3*).

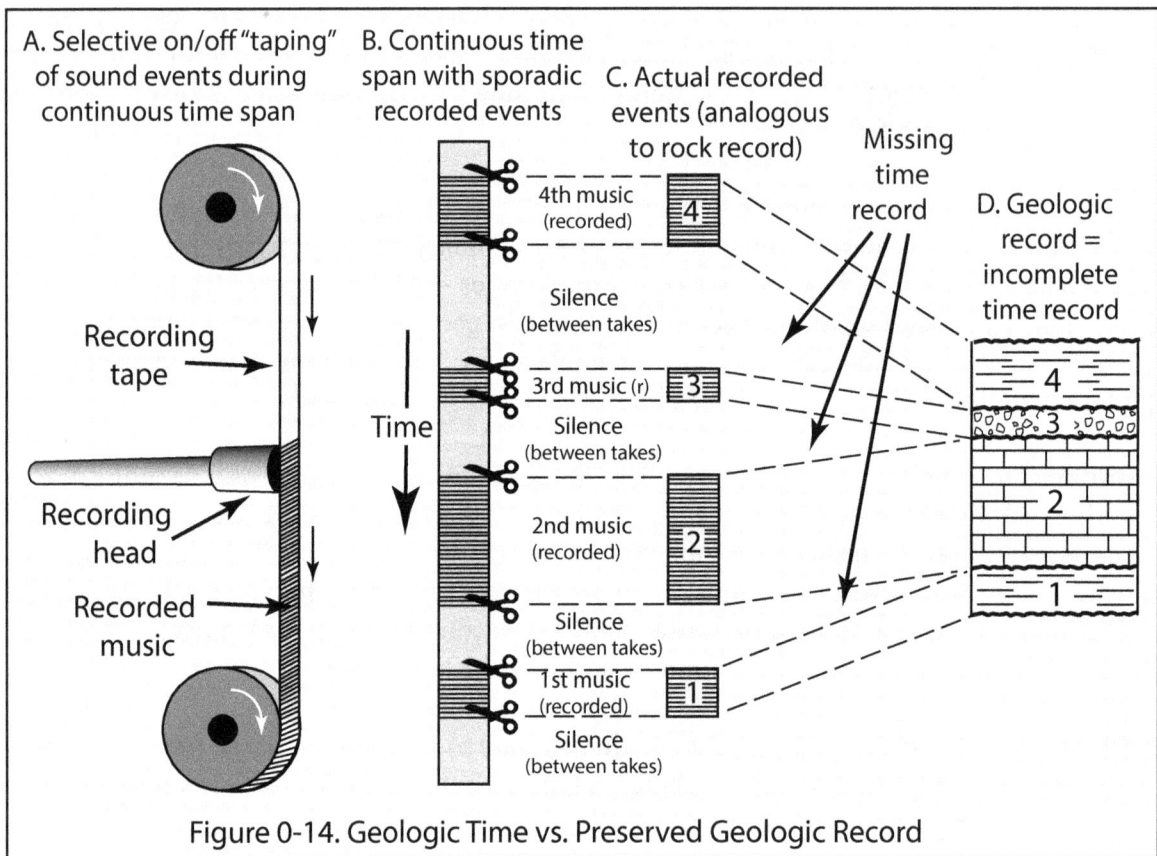

Figure 0-14. Geologic Time vs. Preserved Geologic Record

~~~

Age of Sandia Mountain

When one asks the age of Sandia Mountain, the person is unwittingly asking a trick question. To illustrate this conundrum, I point to the famous sculpture of *David* (*Figure 0-15*), carved by the Italian Renaissance master Michelangelo in 1504, and which presently resides in Florence, Italy. How old is David you may ask? Well, Michelangelo carved his masterpiece in 1504, so you might reply it's 516 years old (to 2020). No. David is carved from the exquisite *Carrera* marble that was quarried from near the northwest coast of Italy. The marble had been recrystallized from an original limestone to its present beautiful state between 27 and 15 Ma, so it must be 27-15 Ma old, right? Well, not quite. The precursor limestone for the *Carrera* marble had been formed even earlier at the bottom of a shallow sea about 200 Ma, so maybe *David* is 200 Ma, right? Wrong again!

The correct answer is all of the above. The sculpture *David* is a very young, 516-year-old piece of art composed of very old 200 Ma rock. The same idea applies to Sandia Mountain. It's a geologically young edifice, i.e., a mass of rock uplifted in stages since ca. 25 Ma to form the present landform, but one made of very old rock.

Rocks of Sandia Mountain

All three major rock types combine to make up Sandia Mountain. An igneous rock, granite, is supremely well-exposed on the western face and along the foothills. Sedimentary rocks cap the top of the mountain and its east flank. Metamorphic rocks make up the northwestern end and abut the southern end in Tijeras Canyon. Collectively the ages of these rocks (excluding the fill of the adjoining Albuquerque basin) range from 1.7 Ga to roughly 30 Ma.

Sandia Granite. The rugged west face of Sandia Mountain is dominated by the Sandia Granite, an intrusive igneous rock of Proterozoic age, radiometrically age-dated at 1.45 Ga (abbreviated *Xg* in illustrations; *Figure 0-16*).

Figure 0-15 . Michelangelo's *David*, 1504
(Photograph from Internet)

Figure 0-16. Rock Types Exposed on Sandia Mountain's West Face
(Photograph by Author)

By definition, granite contains mineral grains that are discernable to the naked eye and contains at least 20% quartz (glassy in appearance). The rock is characterized by some extra-course crystals, up to an inch in size, of a potassium feldspar mineral sometimes abbreviated as *K-spar*.

These coarsest grains range in color from off-white, light gray, to light pink, imparting the rock with its overall color. The pinkish variety is due to small amounts of iron contained within the K-spar's mineral matrix. Surrounding the K-spar is a finer-grained matrix or groundmass made up of a sodium-feldspar (*oligoclase*), fine quartz, and small amounts of the distinctive blackish mica, *biotite*.

Trace amounts of the black, magnetic, iron oxide mineral *magnetite* (Fe_2O_3) occur as sand-size grains, less than 1% of the total but occasionally as much as 4% (Shomaker 1965). The magnetite, being denser than the other minerals, tends to weather loose, lag behind and accumulate in arroyo bottoms as black streaks (more in *Chapter 11*).

Relatively fresh granite forms the sheer upper cliffs of the mountain's west face where it is somewhat protected from the weathering effect of penetrating water (*Figure 0-16*). Other occurrences of fresh granite are restricted to construction zones where the rock has been blasted, such as along the south side of I-40 in Tijeras Canyon.

When weathered the color of the granite is a drab brownish or brownish gray. It presents a knobby surface because the coarse K-spar grains weather more slowly than the matrix and so jut out. The weathering process tends to produce spheroidal bodies called "corestones"(see *Chapter 11*). Granite in its various stages of weathering, from fresh below to corestones above, is well displayed along NM-333 in Tijeras Canyon (*Figure 0-17A*).

A. Igneous Rock: Sandia Granite (NM-333, Tijeras Canyon; Roadbed for Scale)

B. Sedimentary Rock: Layered Madera Limestone
(Sandia Crest Road)

C. Metamorphic Rock: Contorted Rincon Quartzite
(Juan Tabo Canyon)

Figure 0-17. Types of Rock Forming Sandia Mountain
(Photographs by Author)

There are three relatively uncommon rock types associated with the Sandia Granite. Two of these, *enclaves* and *aplite dikes*, occur within the Sandia Granite, while the third type, *pegmatite*, generally occurs nearby but separate from the main granite body.

Enclaves. This type occurs as irregular gray to dark gray, often fist-sized or larger, oblate bodies within the granite. They are most likely older pieces of basaltic "underplating" derived from the very bottom of the fluid granite magma body as it intruded and filled its chamber (more in *Chapter 2;* Kelley and Northrop 1975).

Aplites. These are crisscrossing, light-colored veins (often confusingly called "dikes") of igneous rocks, usually a few inches wide within the granite. The minerals are similar to the granite but instead are very fine-grained. These were late intrusions into the already-crystallized and cooling magma and were accordingly chilled.

Pegmatites. This third type occurs as irregularly-shaped bodies near the margin of an igneous intrusion, often as dikes like the aplite but more commonly as irregular-shaped bodies. They consist of unusually very coarse grains up to several inches across of quartz, K-spar, and the white mica *muscovite*. Like aplite, these were also formed very late in the cooling process where the residual melt had been highly enriched in volatile ingredients (that resisted incorporation into the granite) such as water vapor.

The example shown in *Figure 0-18* is a "twofer" for enclaves and aplite dikes. Here, recently cooled, solidified granite, along with a large enclosed enclave, was fractured—enclave and all, and the fracture filled by a younger aplite dike.

A. Portion of Granite Boulder (Photograph by Author; Elena Gallegos Area)

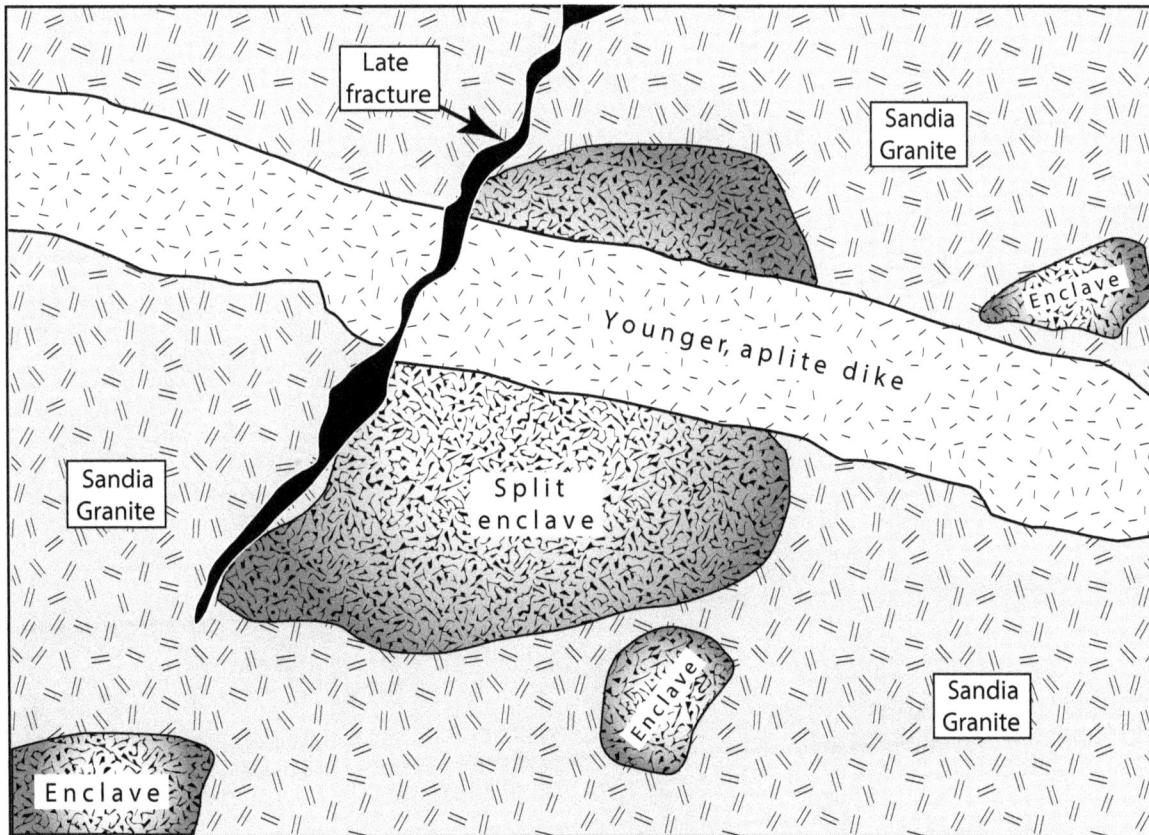

B. Interpretation of "A" Above

Figure 0-18. Sandia Granite with Enclosed Basaltic "Enclaves," Split by Younger Aplite Dike

Orbicular granite. This bizarre, and quite rare, rock type is characterized by numerous spheroid bodies, *orbicules*, within a granitic host rock (*Figure 0-19*). Each orbicule is several inches in diameter and consists of a core encased by one or more concentric rings or shells of crystals of differing mineralogies, all encased within a finer-grained matrix. The result is a suite of utterly fascinating and highly photogenic rocks.

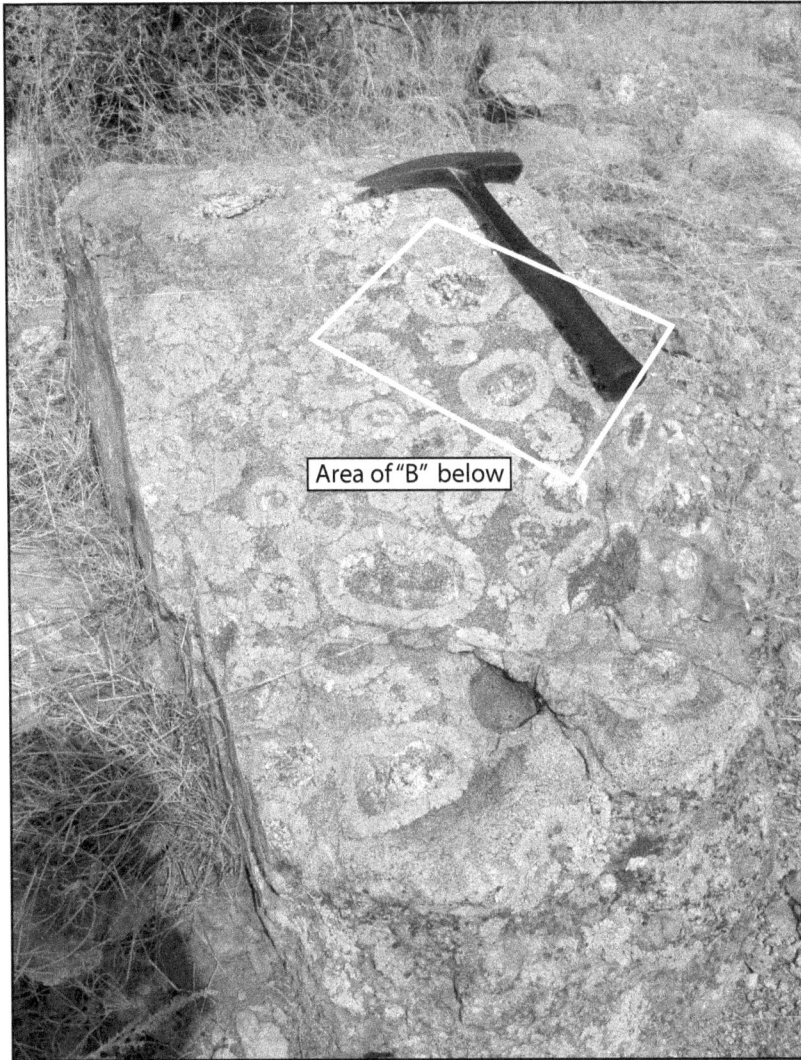

A. Boulder at North End of Outcrop (12-Inch Hammer for Scale)

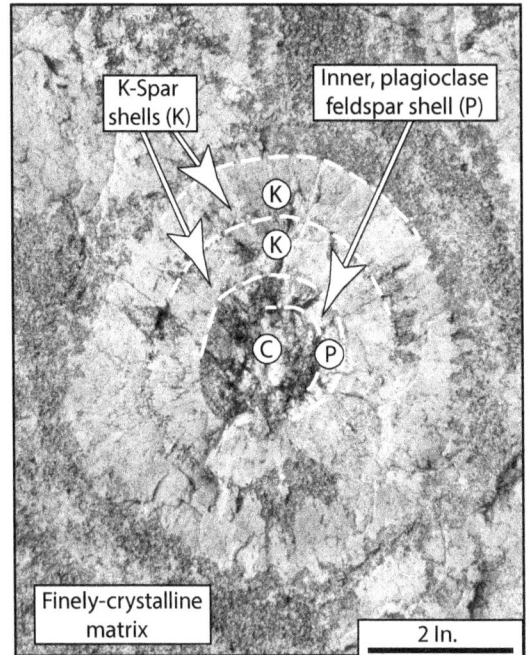

B. Detail of "A" Above

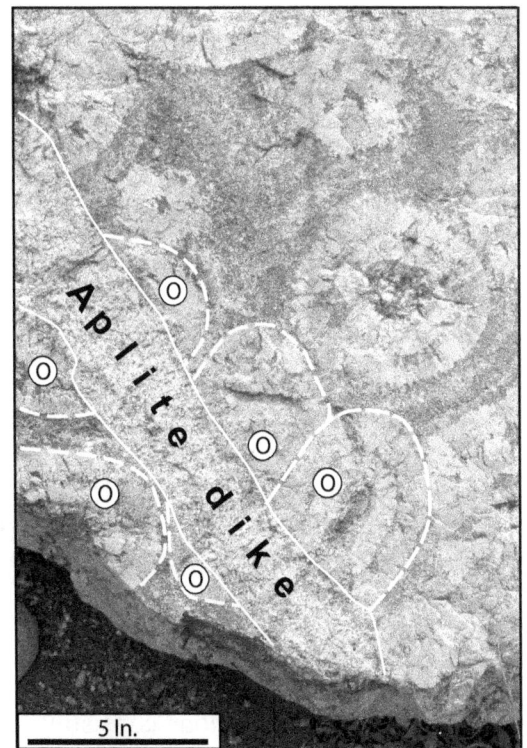

C. Detail of Triple-Shelled Orbicule in "D" Below

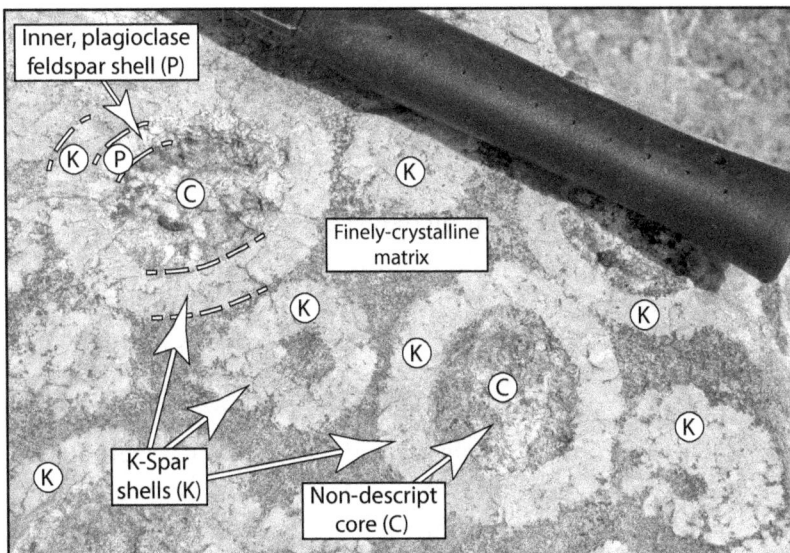

D. Boulder with Triple-Shelled Orbicule and Select Orbicules (O) Cut by Later Aplite Dike

Figure 0-19. Orbicular Granite (Photographs by Author)

There are two occurrences of such rocks on Sandia Mountain. The first consists of two small outcrops along the La Luz Trail in the Juan Tabó area at an elevation of about 7800 ft (Kelley and Nordstrom 1975). I have never been able to locate these and I suspect that they have been hacked away. The second is located north of I-40 in Tijeras Canyon, holding up a little hill or knoll northeast of the Montecello subdivision, (see *Chapter 12*).

Orbicules consist of a core, often of a dark-colored pre-existing rock fragment probably derived from the pluton's pre-existing wall rock, or of a nondescript nature. The core is surrounded by one or more radial shells, each of a single mineral. Inner shells are white, plagioclase feldspar crystals, often poorly developed, around the edges of the seed. This is followed by a shell of salmon-colored K-spar crystals (*Figure 0-19B*). Some have two K-spar shells (*Figure 0-19C*), and others demonstrate later igneous events (*Figure 0-19D*).

Orbicular rocks worldwide vary in their mineralogy and structure, but all are highly dependent on the composition of the original magma and on the prevailing pressure-temperature regime at the time of formation. These rocks typically occur as dike- or pod-like bodies of limited extent, possibly injected along the margins of partially crystallized magma and a body of slightly younger, more fluid magma. (It is important to point out that intrusion of a magma body the size of the Sandia Granite likely occurred in a series of pulses, building from the bottom up.) As earlier pulses partially crystallized to a "mush," later, more water-rich magma was somehow segregated from the earlier mush, chilled, and channeled to form dike-like bodies along or near the contact between the two.

Elsewhere in the world such rocks have been better exposed and studied in more detail. A fine example from northern Argentina in South America is instructive (Grosse et al. 2010). There a water-rich "pocket" got squished (extruded) out from a main magma body by fracturing of the surroundings. Rapid chilling of the segregated volatile melt occurred as the pressure dropped in the pocket. Nucleation on older rock fragments in the melt—the seeds"—progressed rapidly. The matrix between the orbicules crystallized next even before the orbicules had fully hardened. The short time scale for the entire process was only a few weeks or months.

Sedimentary Rock Cover. Starting east below Sandia Crest, and forming the back side of the mountain (the "East Mountain") is a sequence or "section" of layered formations of sedimentary rocks (*Figure 0-17B*). These rocks range in age from 315 Ma (Pennsylvanian) to about 30 Ma (Oligocene). The section thickens from about 400 ft at Sandia Crest to nearly 14,000 ft in the region to the northeast called the Hagan basin (*Figure 0-20*; Kelley and Northrop 1975). The entire section was once present above the position of today's Sandia Crest, but that's not at all to suggest that the elevation of Sandia Crest was at one time thousands of feet higher than the present 10,678 ft. As the mountain was uplifted, erosion of the sedimentary section kept pace so that the mountain probably was never much higher than it is today. The entire section consists of some 15 discrete formations consisting of mudstone, sandstone, limestone and gypsum, and in the Hagan basin at the very top is a formation of volcanic debris.

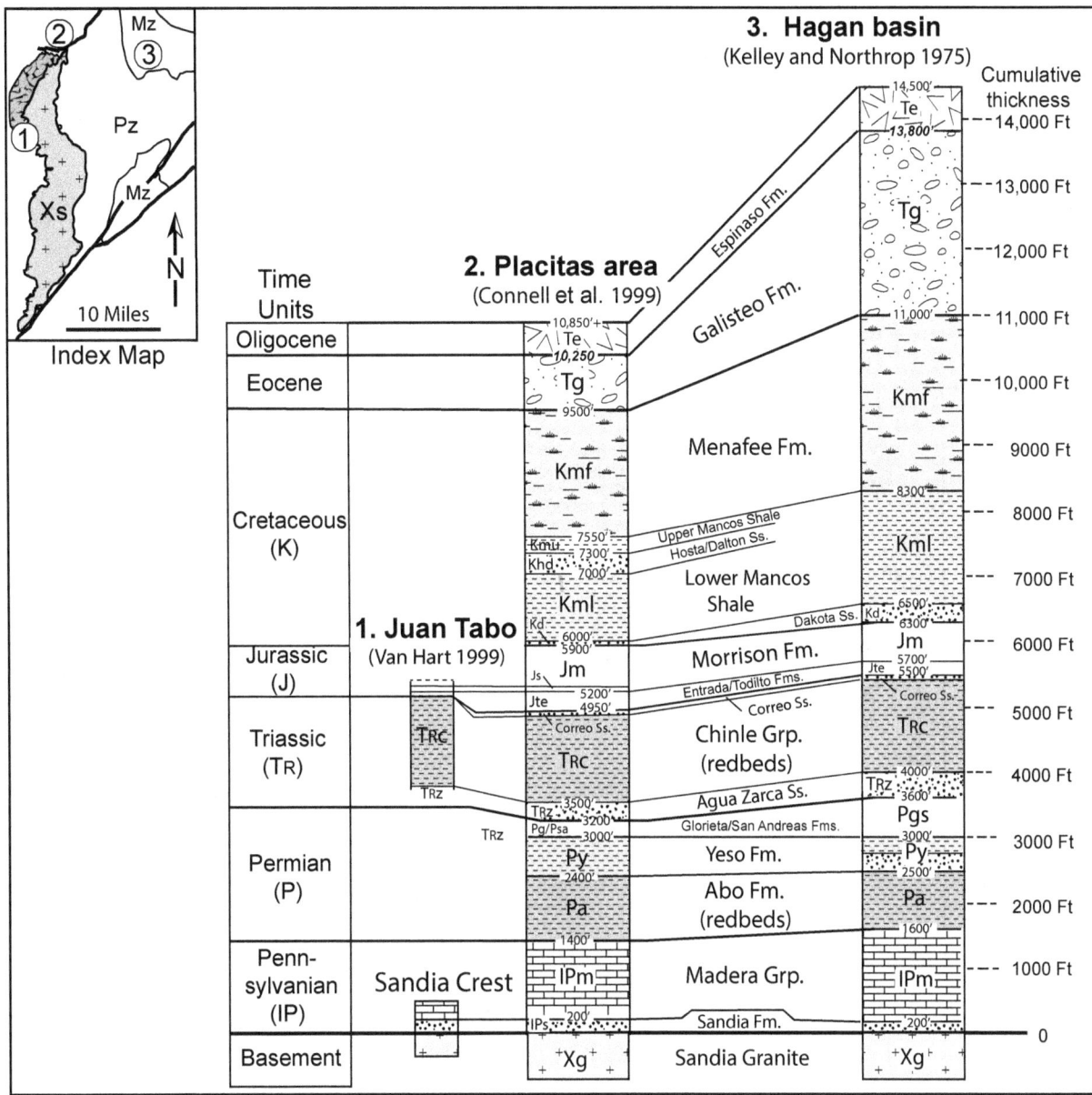

A. Key Stratigraphic Columns

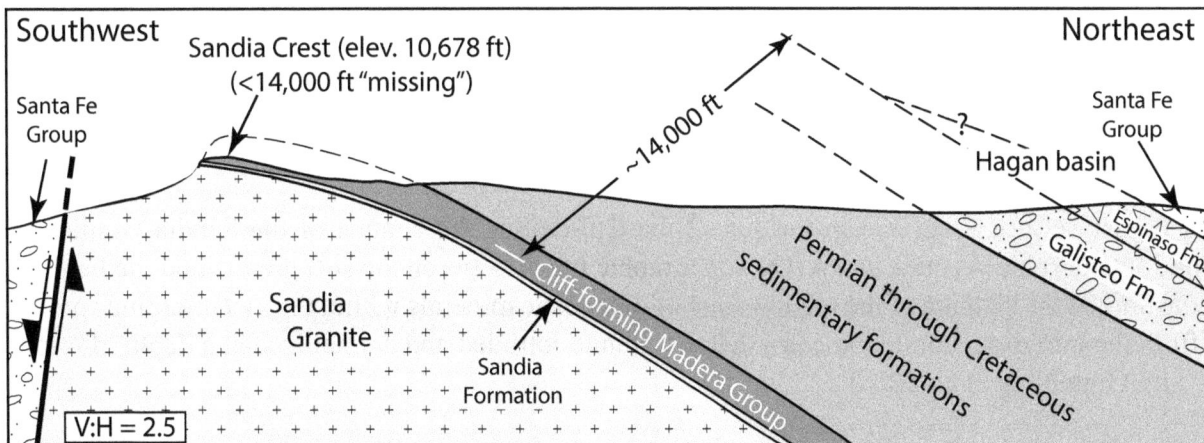

B. Schematic Cross Section from Sandia Crest to Hagan Basin, Showing Amount of "Missing" Rock Section (Based on Kelley and Northrop 1975, Cross Section B-B')

Figure 0-20. Sedimentary Rocks of Sandia Mountain

Rincón Metamorphics. Metamorphic rocks (abbreviated *Ym*), the oldest type at Sandia Mountain at 1.7 Ga in age, occur at the northwest end of Sandia Mountain (*Figure 0-21*). The rocks there consist of light gray, micaceous quartzite (a metamorphosed muddy sandstone (*Figure 0-17C*) grading to quartz-mica *schist* (Kelley and Northrop 1975).

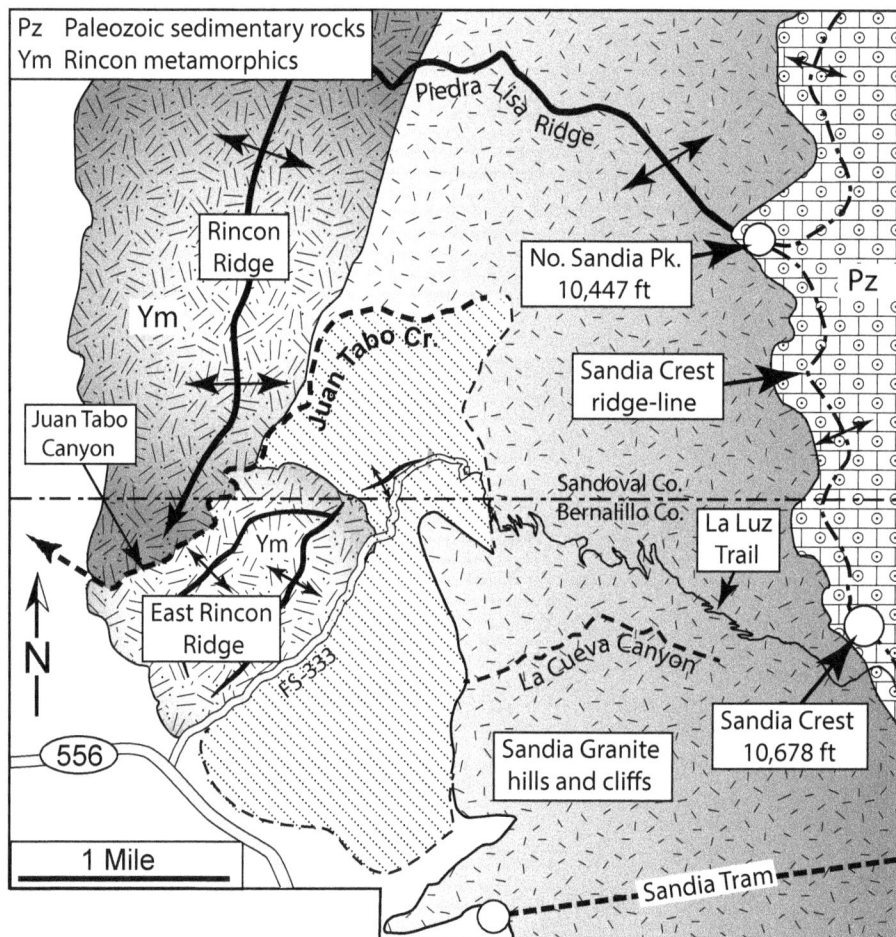

Figure 0-21. Location Map of Topographically-Low Juan Tabo Cul-de-Sac (Cross-Hatched), Northwestern Sandia Mountain

The main characteristic of this metamorphic rock is its relative hardness. It therefore rises above the more easily-weathered granite to its east, forming a south-plunging, south-southwest-trending feature called Rincón Ridge. To the southeast of the main ridge is a subsidiary, lower companion, here referred to as East Rincón Ridge (*Figure 0-21*).

Along Forest Service Road 333 in the Juan Tabó Cul-de-Sac the contact of the Sandia Granite with the Rincón Metamorphics is quite sharp. The topographic break between the soft granite and the hard quartzite has constrained the location of the road. A suite of diagnostic minerals within the metamorphics produced by heat from the intruding granite indicates that the granite intruded and crystallized at a depth of 4 to 6 miles (7-10 km; Grambling et al. 2016).

Regional Geologic Setting

It is now time to put the Sandia Mountain into a more regional context. A geologic map of the state of New Mexico does just that (*Figure 0-22A*). The map reveals that, in simple terms, the entire state is composed of three main provinces. To the west and northwest is the southeastern end of the vast *Colorado Plateau*, an area characterized by relatively undisturbed sedimentary rocks. The east half of the state consists of the western part of the even more vast *Stable Interior*, also characterized by relatively undisturbed sedimentary rocks. Between these two is a relatively low-lying zone, a basin, filled with relatively young unconsolidated sediment. This is the *Rio Grande rift* (*Figure 0-22B*), a feature that hosts the Rio Grande. The rift expands in breadth to the south and southwest where it loses its linear character, merging with and becoming the *Basin and Range Province*.

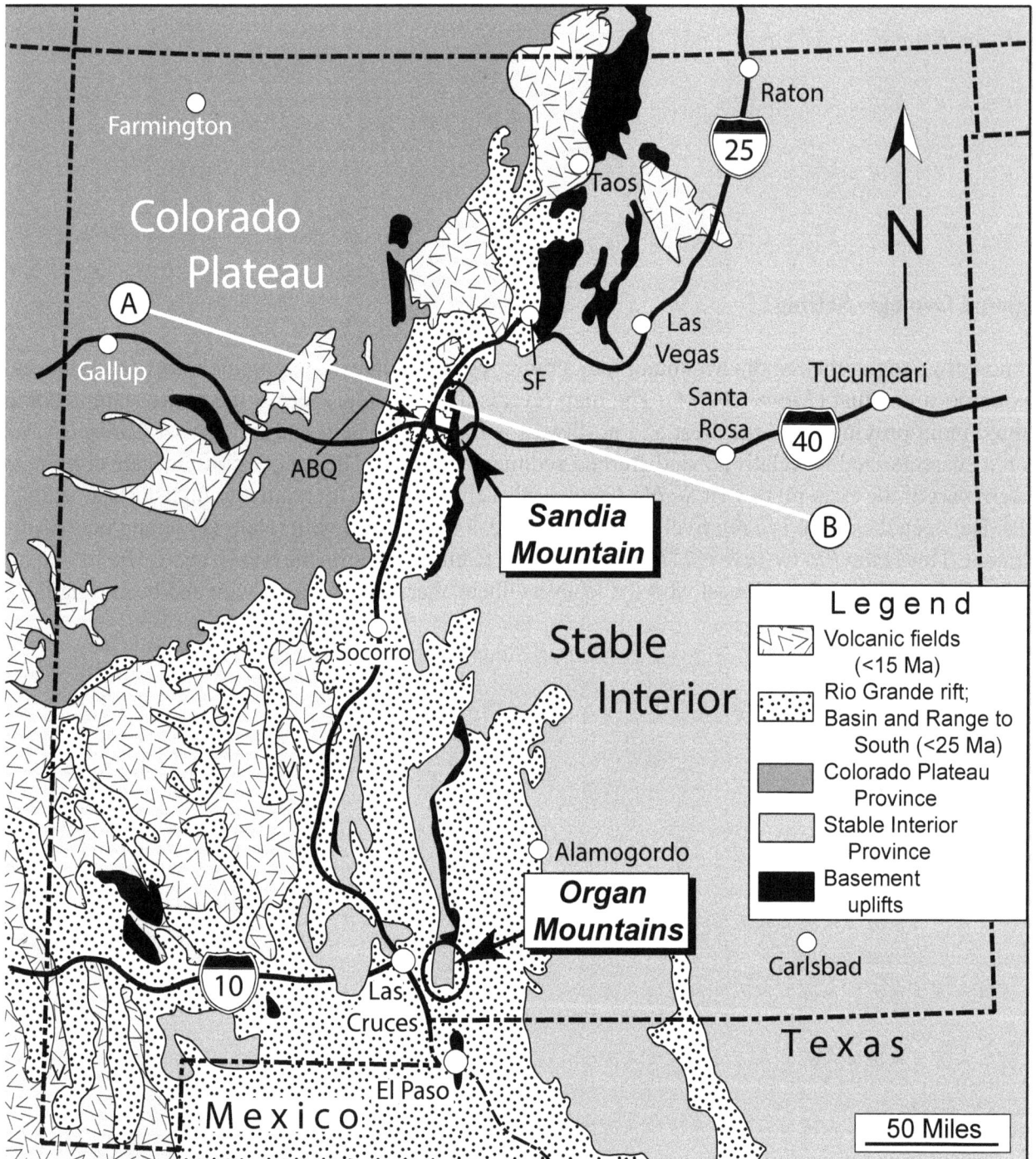

A. Generalized Geologic Map

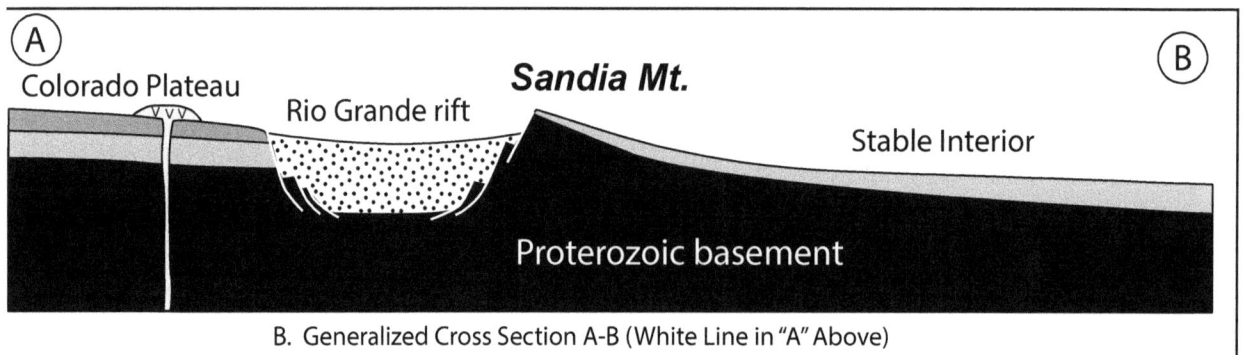

Legend

Volcanic fields (<15 Ma)

Rio Grande rift; Basin and Range to South (<25 Ma)

Colorado Plateau Province

Stable Interior Province

Basement uplifts

50 Miles

B. Generalized Cross Section A-B (White Line in "A" Above)

Figure 0-22. Geology of New Mexico.

Underlying the sedimentary rock cover of the Colorado Plateau and Stable Interior is a complex assemblage of metamorphic and intrusive igneous rocks of Proterozoic age, collectively called the "basement." Parts of this basement have been locally uplifted such that they peek through the sedimentary cover and form mountains (black in *Figures 0-22* and *0-23*). The basement structure map shows that one of these uplifts, Sandia Mountain, forms part of the eastern shoulder of the adjoining Rio Grande rift. The mountain is clearly affiliated with the rift.

Geologic Map of Sandia Mountain

Having prepared some background, we are now ready to construct some maps of Sandia Mountain itself. A structure map on top of the mountain's ancient basement reveals the broad, elongated domal character, the east-sloping ramp, the Tijeras fault zone, and the Hagan basin (*Figure 0-24*). Note that the structure west of Sandia Crest is projected using dashed contour lines, i.e., into where it is interpreted to be if it hadn't been eroded away.

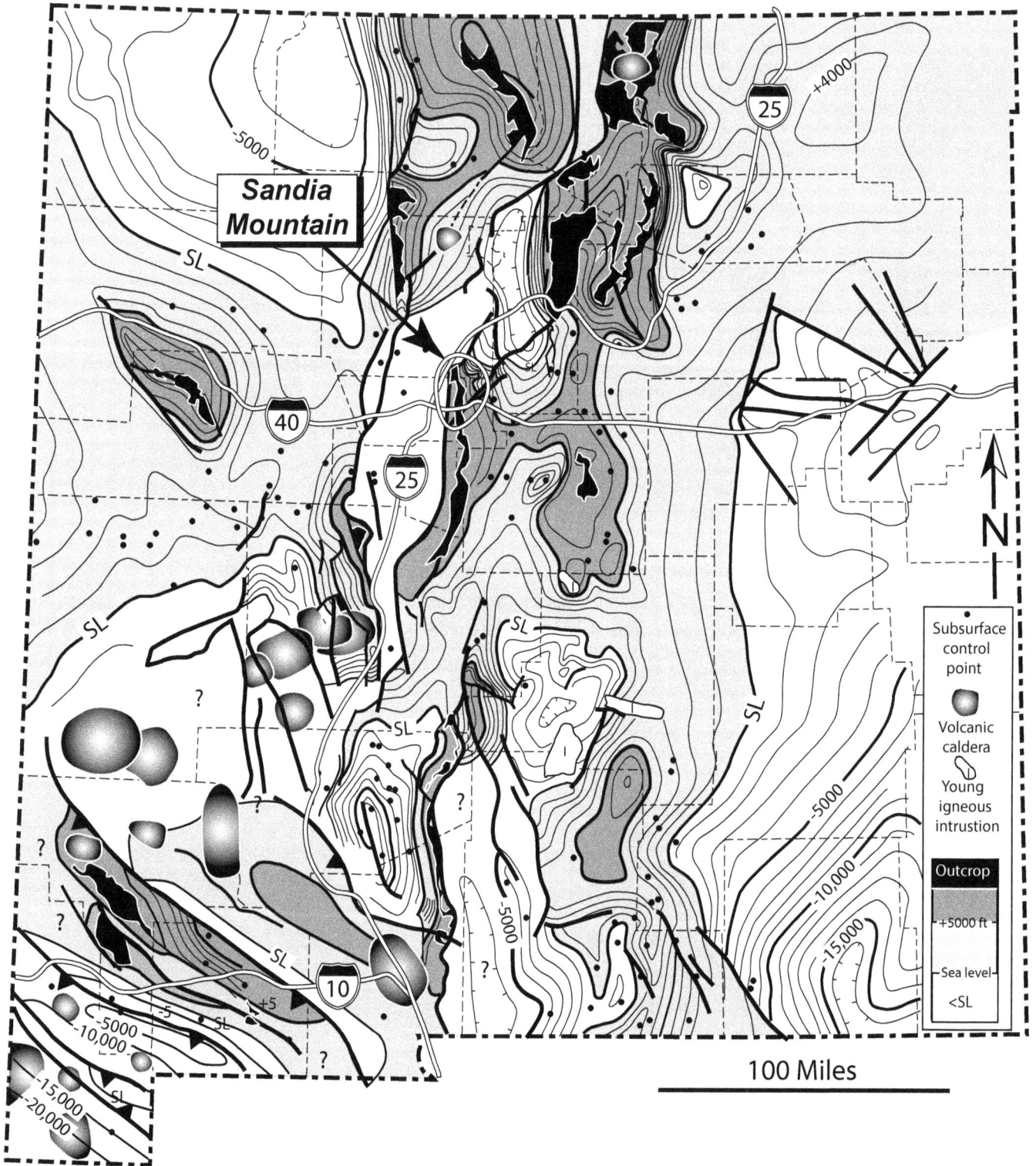

Figure 0-23. Generalized and Partly-Speculative Structure Map on New Mexico Basement

Legend:
- Subsurface control point
- Volcanic caldera
- Young igneous intrustion
- Outcrop
- +5000 ft
- Sea level
- <SL

100 Miles

Sandia Mountain

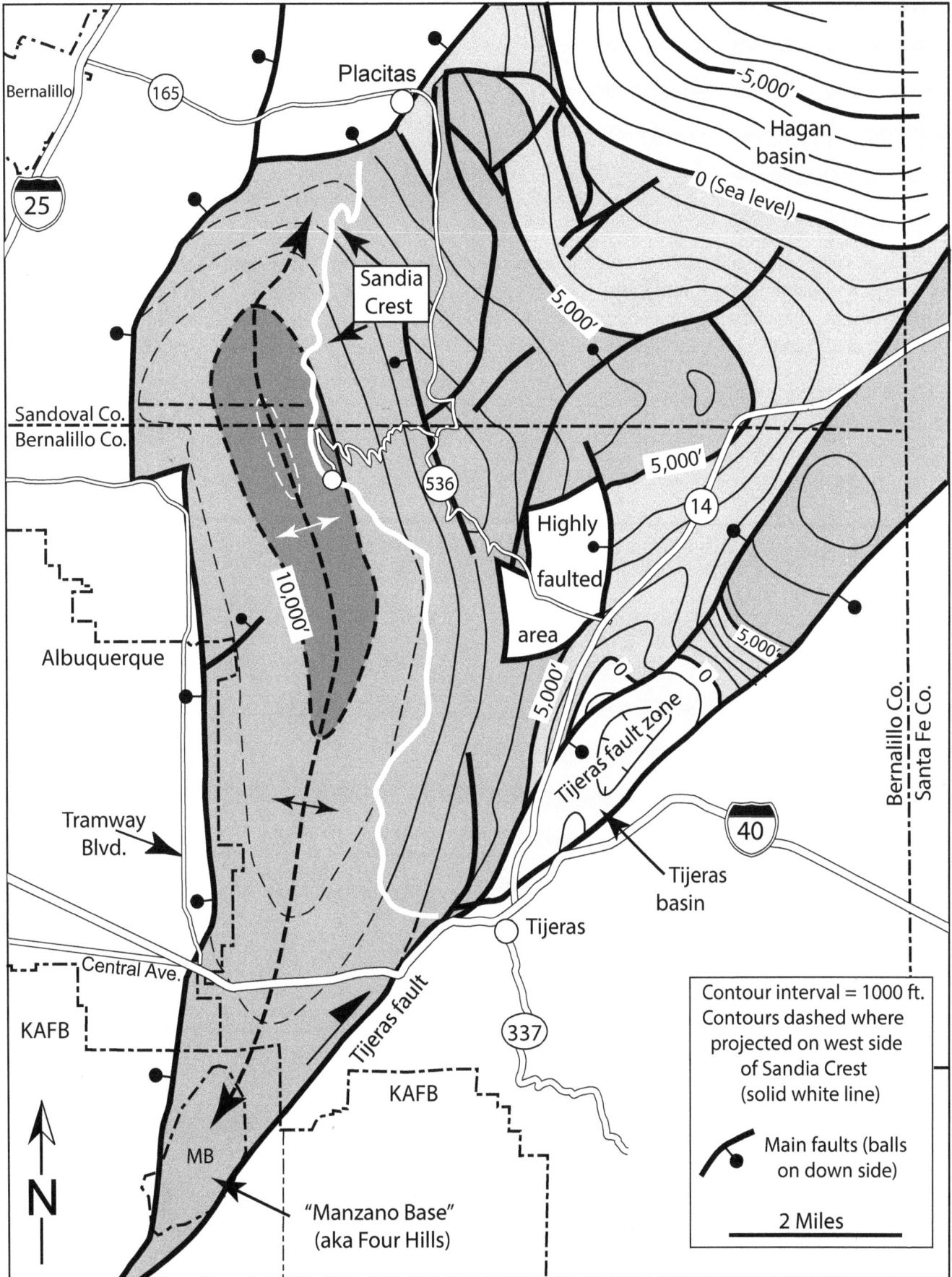

Figure 0-24. Structure Map on Sandia Mountain Basement
(Modified from Kelley and Northrop 1975)

Labels within the figure:

Bernalillo
165
Placitas
25
-5,000'
Hagan basin
0 (Sea level)
Sandia Crest
5,000'
Sandoval Co.
Bernalillo Co.
536
5,000'
14
Highly faulted area
10,000'
Albuquerque
5,000'
0
0
Tijeras fault zone
Tramway Blvd.
Bernalillo Co.
Santa Fe Co.
40
Tijeras basin
Tijeras
337
Central Ave.
KAFB
Tijeras fault
KAFB
N
MB
"Manzano Base" (aka Four Hills)

Contour interval = 1000 ft.
Contours dashed where projected on west side of Sandia Crest (solid white line)

Main faults (balls on down side)

2 Miles

A *geologic* map of the mountain shows the aerial distribution of the *bedrock* (the solid rock exclusive of the overlying unconsolidated materials) that is visible at the surface (*Figure 0-25*). This map is chock full of information, also showing the general ages of the rocks, the declination angles (*dips*) of the sedimentary rocks, the main faults, and some associated geography. We will refer back to this important map again.

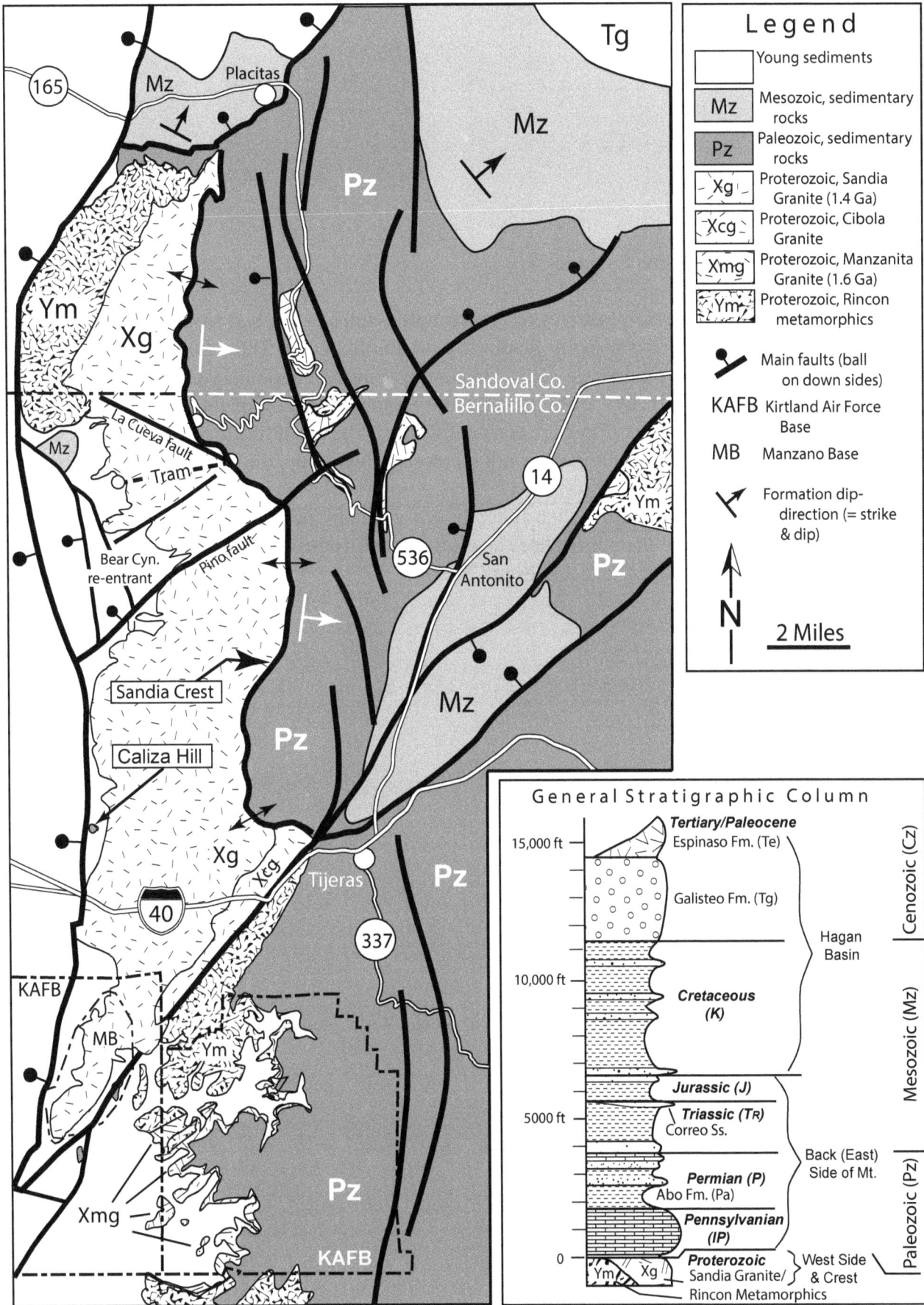

Legend

- [ ] Young sediments
- Mz — Mesozoic, sedimentary rocks
- Pz — Paleozoic, sedimentary rocks
- Xg — Proterozoic, Sandia Granite (1.4 Ga)
- Xcg — Proterozoic, Cibola Granite
- Xmg — Proterozoic, Manzanita Granite (1.6 Ga)
- Ym — Proterozoic, Rincon metamorphics

- Main faults (ball on down sides)
- KAFB — Kirtland Air Force Base
- MB — Manzano Base
- Formation dip-direction (= strike & dip)
- N
- 2 Miles

General Stratigraphic Column

15,000 ft	Espinaso Fm. (Te)	*Tertiary/Paleocene*
	Galisteo Fm. (Tg)	Cenozoic (Cz) / Hagan Basin
10,000 ft	*Cretaceous (K)*	Mesozoic (Mz)
	Jurassic (J)	
5000 ft	*Triassic (TR)* — Correo Ss.	Back (East) Side of Mt.
	Permian (P) — Abo Fm. (Pa)	Paleozoic (Pz)
	Pennsylvanian (IP)	West Side & Crest
0	*Proterozoic* — Sandia Granite/ Rincon Metamorphics	West Side & Crest

Figure 0-25. Geologic Map and Rock Column of Sandia Mountain
(Modified from Karlstrom et al. 1999; Read et al. 1999)

77

Introduction to the Term "Plate Tectonics"

Geologists refer to the large-scale processes of crustal uplift, subsidence, and shifting about, collectively as *tectonics*. Tectonic events are constantly modifying the Earth's crust. Then there is the larger concept of *plate tectonics*, the basis of modern geology. This overarching, revolutionary concept emerged only in the years 1960-1968. The term refers to the global-scale segments, *plates*, of the Earth's crust that move relative to each other. New, oceanic plates are constantly being formed at linear zones in the oceans called "spreading ridges." But because planet Earth is not increasing in volume, the forward edges of the oceanic crustal plates are constantly being consumed, i.e., shoved or *subducted* under other, pre-existing plates. Some of this subducted crust becomes melted at depth below the pre-existing plates, and the melt (magma) ascends and adds its material to the upper plates as chains of intrusions and volcanoes. This is the essence of plate tectonics (we will revisit plate tectonics in *Chapter 15*).

PART I:
BUILDING A MOUNTAIN

1: Back to the Beginning

Synopsis

Over the course of almost a billion years (1.8 – 0.9 Ga) the proto-North American continent, called *Laurentia*, was buffeted by three collisions of island-arc terranes on the south, and by one collision of continental terranes on the east and west. All of these terranes became welded onto the Laurentian core, vastly adding to its extent. When the dust had settled ca. 0.9 Ga, Laurentia had been greatly enlarged to become the supercontinent of *Rodinia*. From 825 to 550 Ma, Rodinia itself broke up, with continent-sized chunks on three sides separating and going on their way, creating something beginning to resemble modern North America. In what would become New Mexico, tensional faults as well as old sutures remaining from the continental assembly were jostled about, providing "scars" that would profoundly influence later tectonic events.

Introduction

With the very brief introduction to plate tectonics on hand from the previous chapter, we now are ready to build, piece by piece, the continent of today's North America.

Continental Assemblage (1.8 – 0.9 Ga)

North America was not always as it appears today. It has been assembled, piece by piece, from about 1.8 Ga to about 0.9 Ga, from enormous chunks of *island-arc* material (see below), crashing northward against, and welding ("docking") with, the more ancient continental core.

Island arcs. Island arcs, sometimes called volcanic arcs, are curved linear belts of active volcanic islands built up on the sea floor. Modern examples are mainly associated with the "Ring of Fire," the linear belt of volcanoes that surrounds the Pacific Ocean. Modern examples include the Aleutian Islands of western Alaska, the Sunda-Banda arc (including the islands of Java and Bali) of Indonesia, the Kuril-Kamchatka arc of eastern Russia, and the islands of the West Indies.

Now let's take a detour to define some terms, and then we'll move on. As touched on above, what we today call "North America" is the product of extreme modification of an ancient continental "core" over almost a billion years There is a name for each sequential stage of construction: *Laurentia*→*Rodinia*→*North America*. The first two of these terms will come up only in *Chapters 1* and *3*, but it will eliminate confusion by defining them all now (*Figure 1-1*).

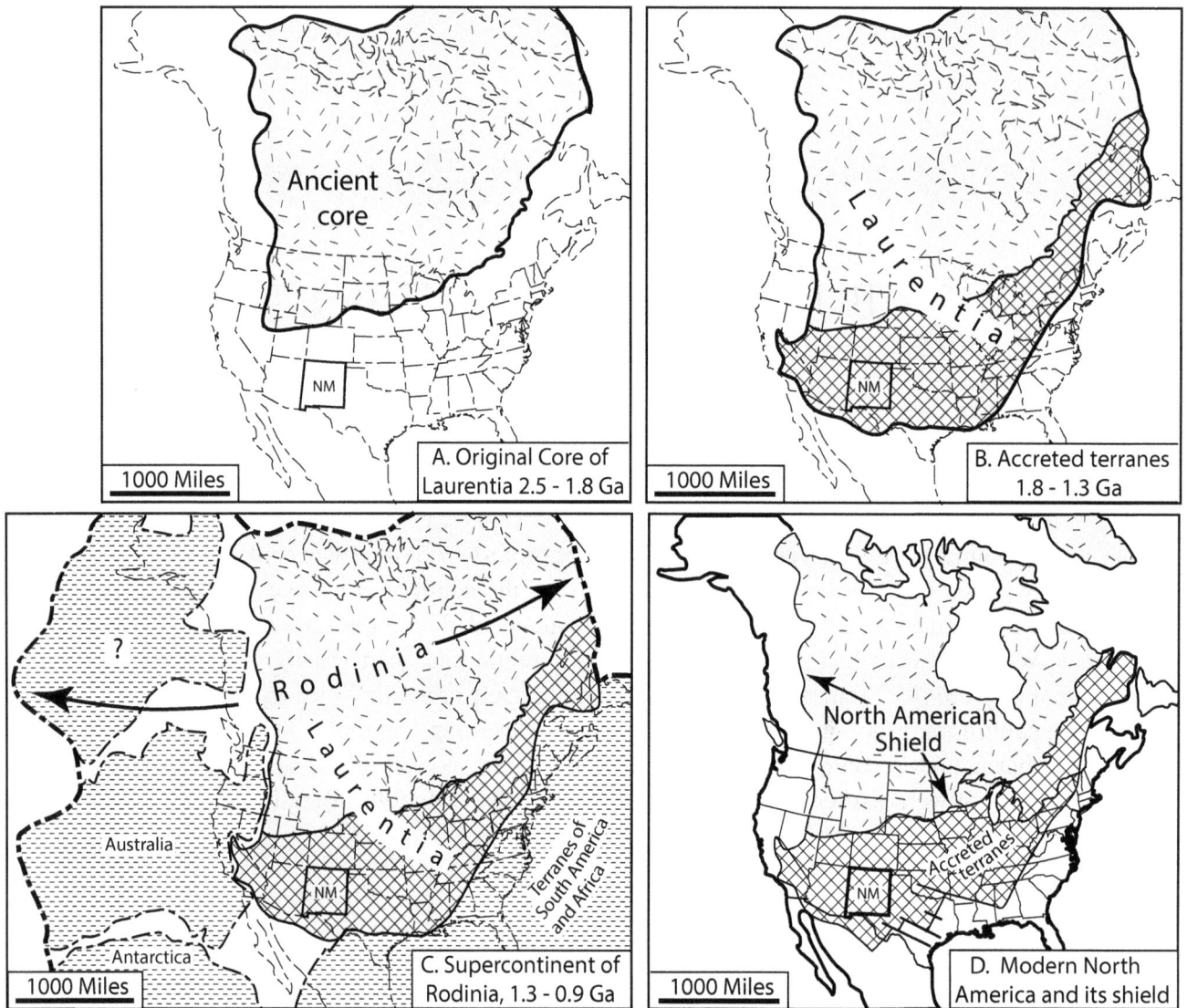

Figure 1-1. Definition and Comparison of Terms "Laurentia," "Rodinia," and "Modern North America"
(Outlines of Modern North America and New Mexico Shown for Reference)

Terms *Laurentia*, *Rodinia*, and *North America*. The continent of "North America" is what we recognize as home today. Less known is that it was built up over about a billion years piece by piece by continental-scale collisions, and then later partly plucked apart to form the present entity. The sequence began with a core of ancient rocks older than 2.5 Ga, itself the product of even earlier collisions from 2.5 to 1.8 Ga (*Figure 1-1A*). The southern margin of the core next became the site of three sequential, ultra-slow-motion collisions ("smash-ups") of island-arc crustal material onto it to form the landmass, *Laurentia* (*Figure 1-1B)*. These collisions, from 1.8 to 1.3 Ga, were followed by a fourth event of continental collisions and dockings on three sides from 1.3 to 0.9 Ga to form the enormous new supercontinent of *Rodinia* (*Figure 1-1C*). Geologists call these crustal additions *terranes*—a term not to be confused with "terrains," which are areas of surface landscape. What would become New Mexico consists of terranes from the first and second island-arc additions. From about 825 to 550 Ma most of the continental additions from the final collision rifted away, leaving behind something close to today's *North America* (*Figure 1-1D*).

Now let's take a brief look at each of the four collisions onto Laurentia's core. These collided terranes collectively form the largely-buried basement of New Mexico and therefore exert a powerful influence on all future events.

Collisions #1 and #2. The first collision on Laurentia's southern margin, the *Yavapai* (name from a Native American tribe in Arizona), 1.8 Ga (*Figures 1-2A* and *1-2B*), was quickly followed, again on the south by the *Mazatzal* (name from a mountain range northeast of Phoenix AZ), 1.65-1.6 Ga (*Figure 1-2C*). These two additions make up the basement of almost all of New Mexico, except its southeasternmost corner.

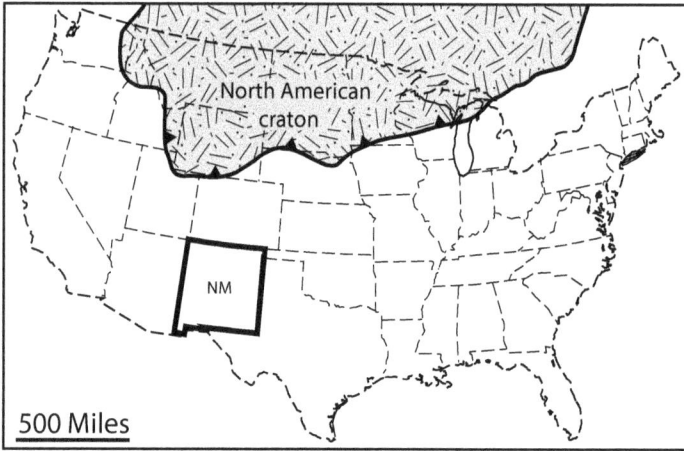

A. North American Craton, 1.8 Ga

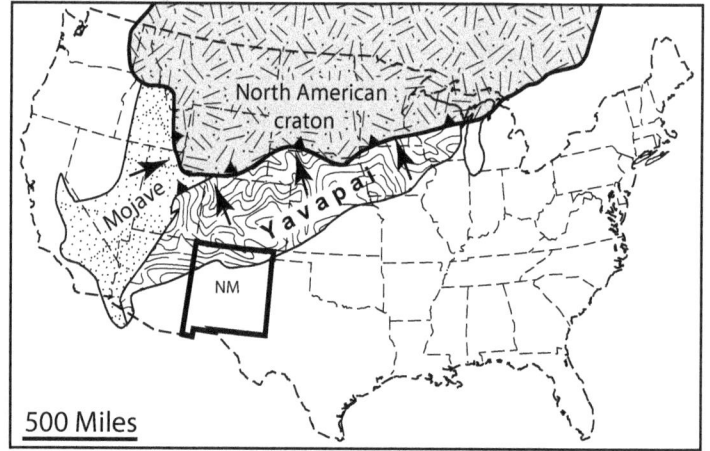

B. Collision #1: Addition of Mojave and Yavapai Terranes, 1.8 - 1.7 Ga

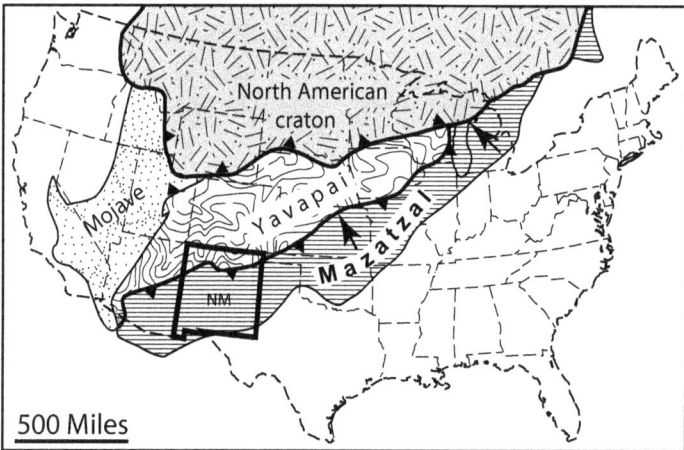

C. Collision #2: Addition of Mazatzal Terrane, 1.65-1.6 Ga

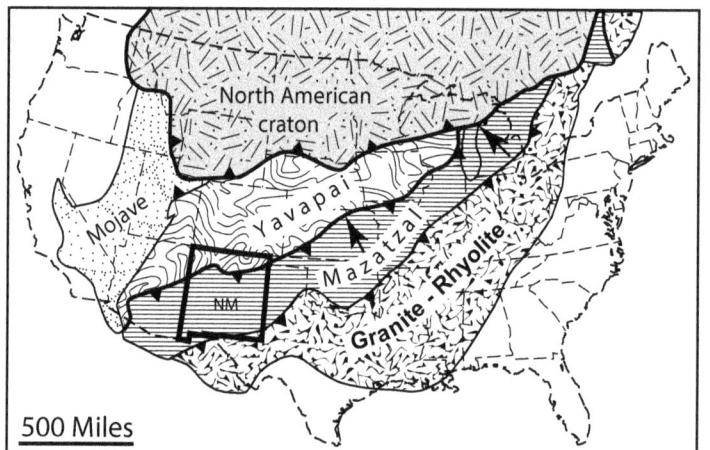

D. Collision #3: Addition of Granite-Rhyolite Terrane, 1.45-1.3 Ga

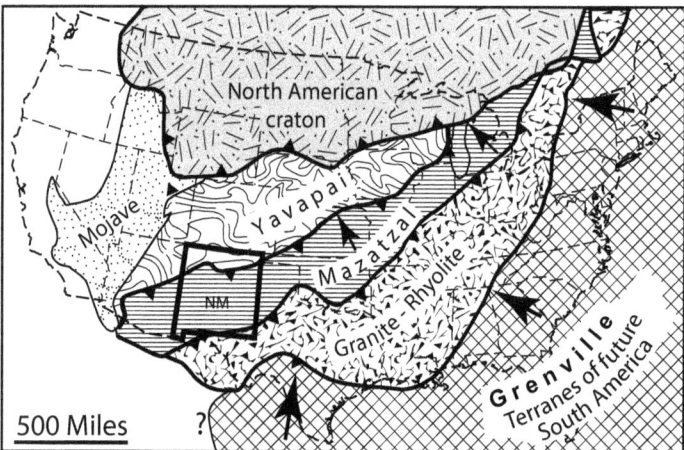

E. Collision #4: Addition of Grenville Terranes, 1.3-0.9 Ga

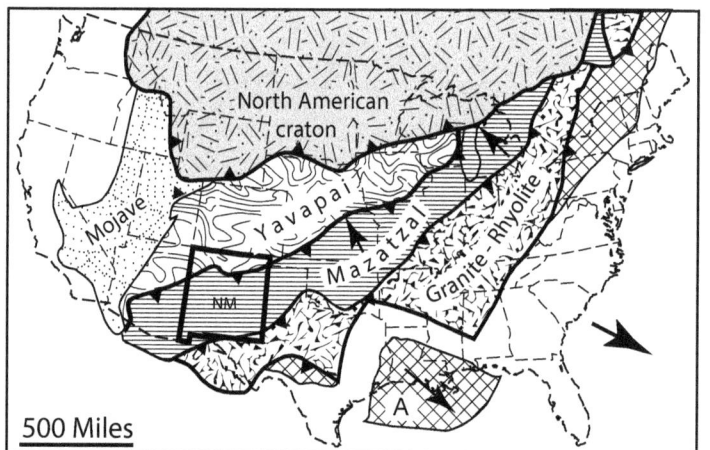

F. Rifting and Separation of Laurentia and Argentine Precordillera (A), 550 Ma

Figure 1-2. Accretion and Assembly of Laurentia
(Present USA shown for reference; modified from Whitmeyer, and Karlstrom 2007)

Collision #3. After a lull from 1.6 to 1.55 Ga, a third collision, the *Granite-Rhyolite*, 1.45-1.3 Ga, added another piece to the southern margin of Laurentia (*Figure 1-2D*). This third collision so thickened the crust over a vast area that the lower part of that crust was shoved down to such great depths and heated that it reached the temperature of partial melting. This resulted in an enormous and very widespread thermal event (*Figure 1-3A*), during which buoyant blobs of magma ascended from the lower crust, cooled, and embedded themselves up in the middle crust. One of these blobs became the Sandia Granite. (I postpone further discussion of this seminal thermal event to *Chapter 2* and ask the reader to hold on.) An imperfect modern analog to this event is the collision of India with Eurasia (the name for an ancient supercontinent comprised of Europe and Asia) roughly 50 Ma that buckled up the Tibetan Plateau (*Figure 1-3B*).

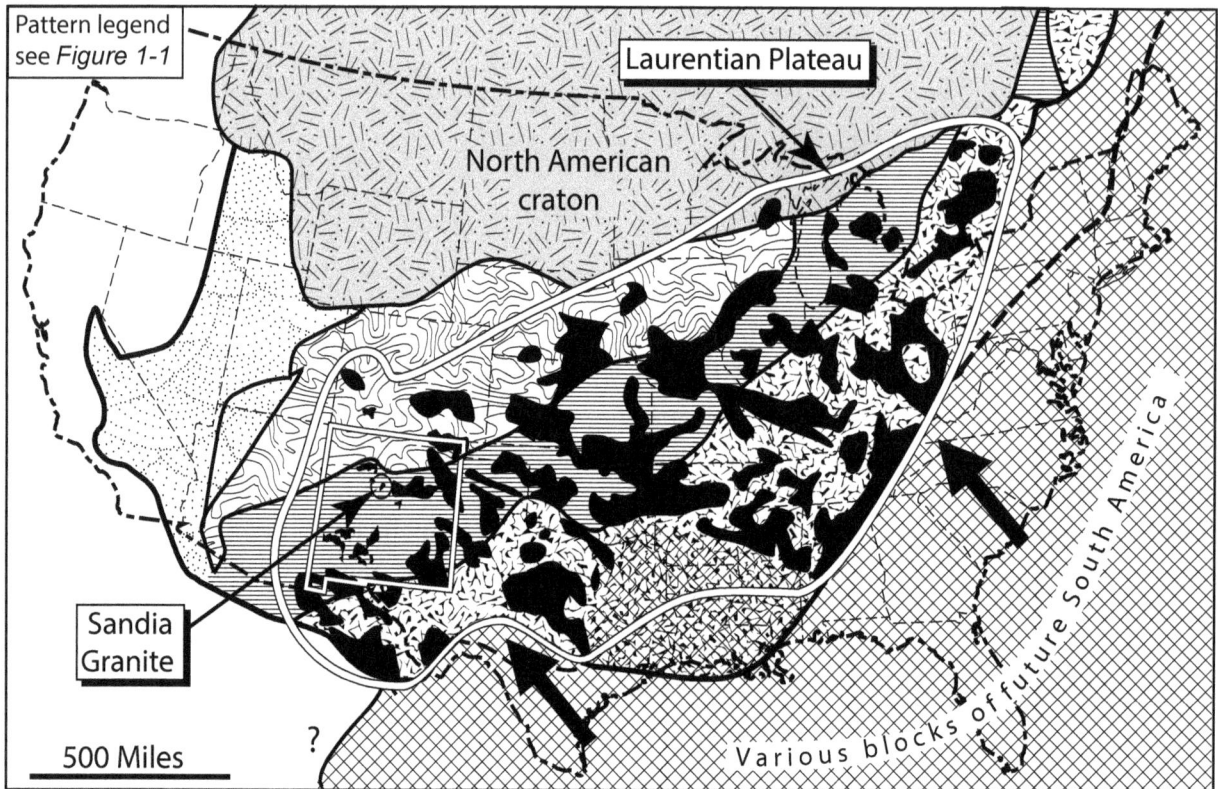

A. Distribution of 1.45-1.35-Ga Granitic Rocks (Black) in Accreted Terranes of Laurentia, and Postulated Plateau (from Karlstrom et al. 2004; Whitmeyer and Karlstrom, 2007; Modern Geography Added for Reference)

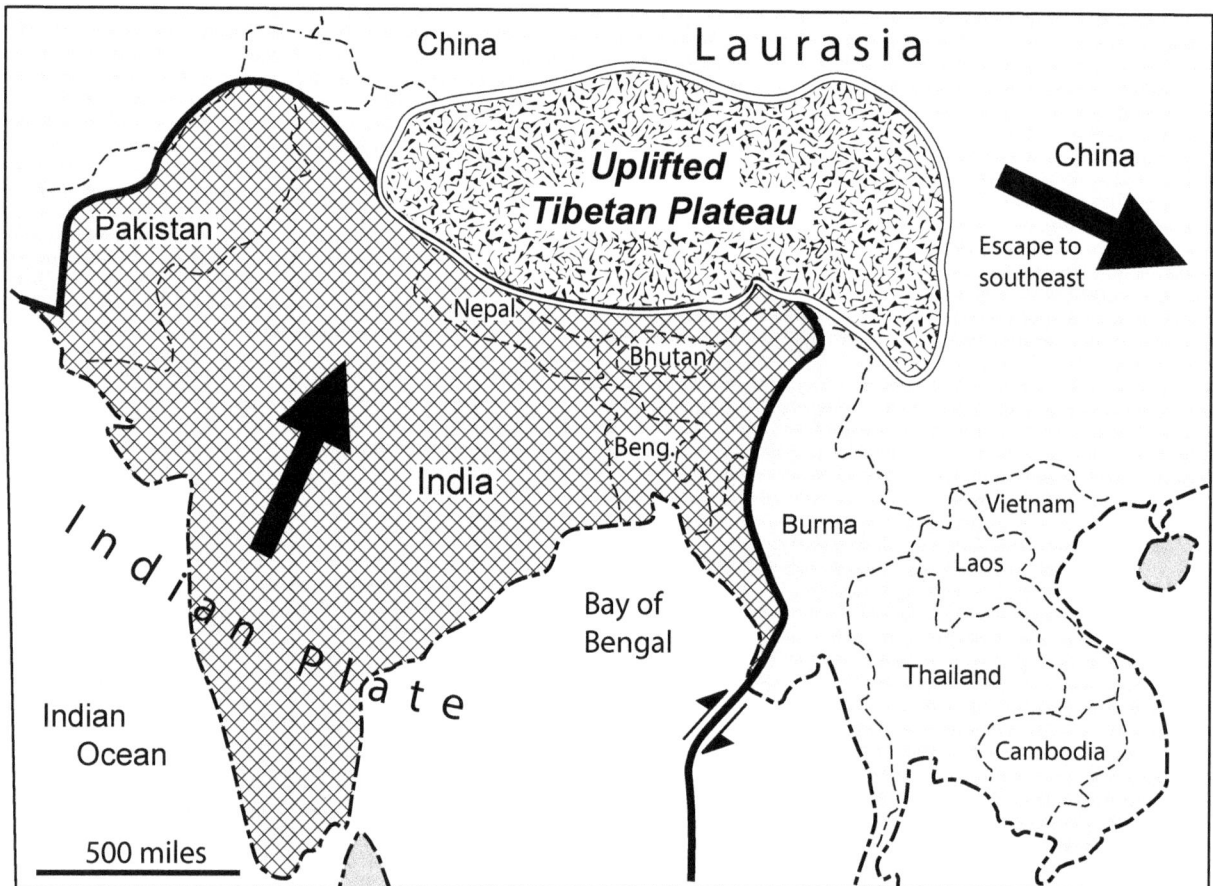

B. Modern Day Tibetan Plateau

Figure 1-3. Comparison of Elevated Laurentian Plateau (1.45-1.35 Ga) and Modern Tibetan Plateau

To visualize how these collisional events appeared, it helps to compare a typical "bird's-eye" map of Eurasia (with its realistic imagery) during the first of the collisions, (*Figure 1-4A*), to a tectonic map of modern Indonesia (*Figure 1-4B*). The latter depicts the continent of Australia moving north to collide with the Eurasian plate, producing a tortuous *melange* of twisted masses and volcanic arcs. This region is a mess but it does have a somewhat obscure sense of order, i.e., fragments of oceanic plates and island arcs colliding with and welding onto a segmented continental plate. The bird's eye view gives a "feel" for how the collision appeared in real life.

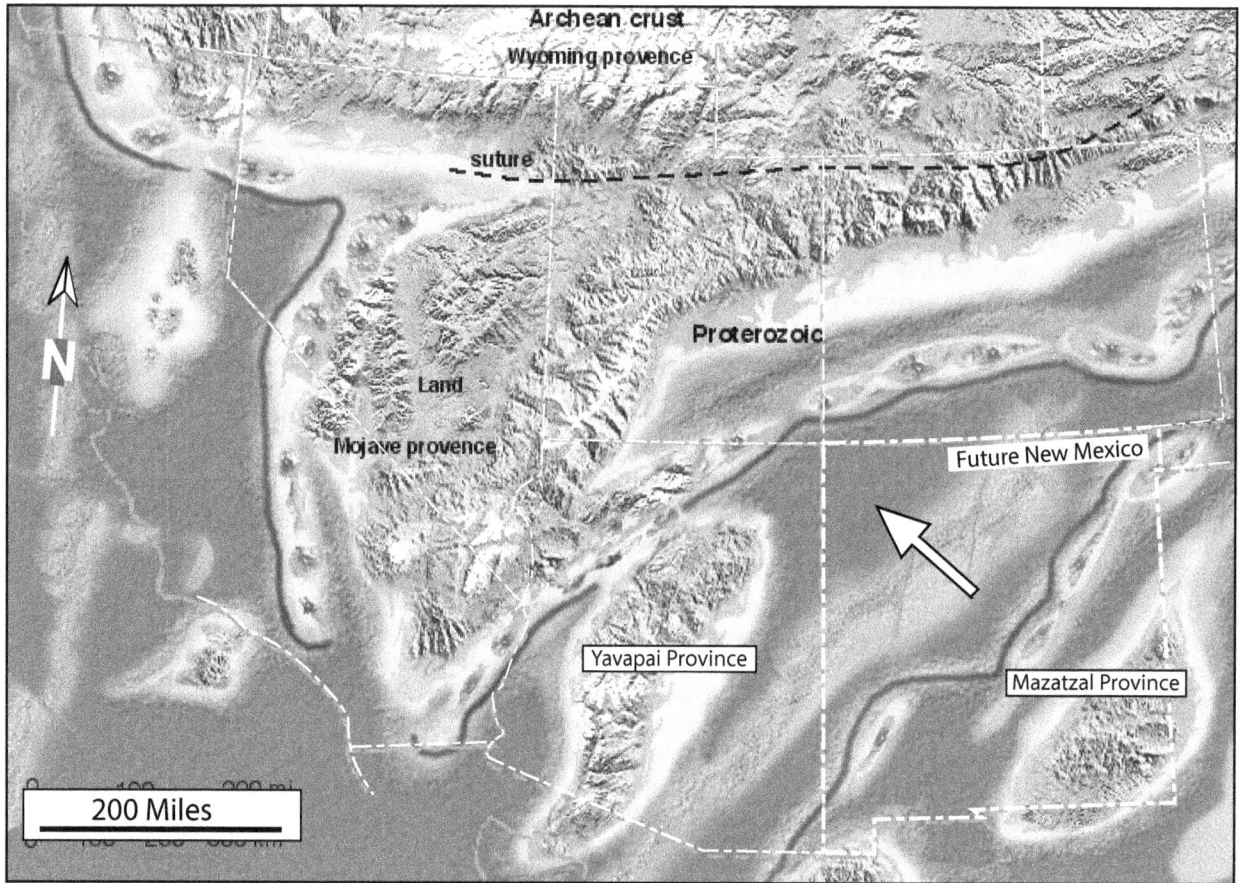

A. Paleogeographic "Bird's Eye" Snapshot of Laurentian Assembly, 1.7 Ga
(Modified from R. Blakey; Modern Geography Added for Reference)

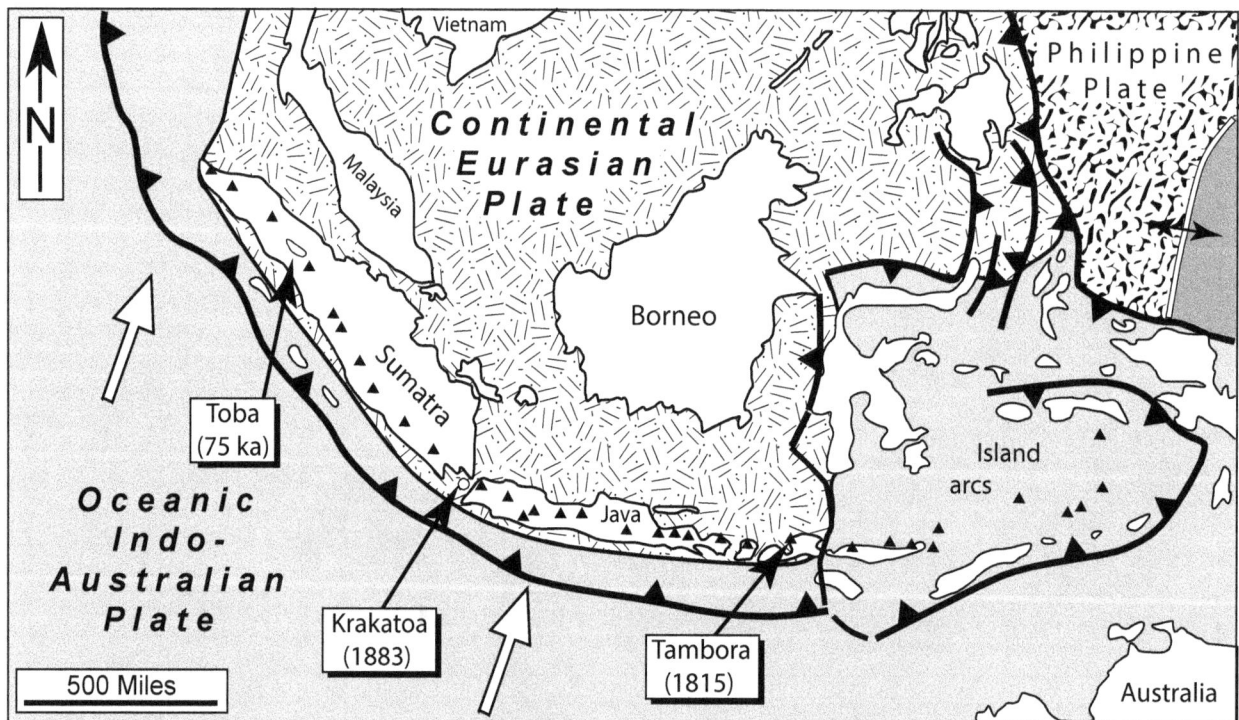

B. Indonesia: Modern Plate-Tectonic Analog for Assembly of Laurentia
(Triangles = Volcanoes; Heavy Lines = Subduction Zones, Barbs on Up-Side; (Major Volcanic Eruptions Indicated)

Figure 1-4. Comparison of Paleogeographic Map of Laurentia
and Tectonic Map of Modern Indonesia (Note Different Scales)

Collision #4. A fourth and final collision, a continent-to-continent one coming on the heels of the third, was the *Grenville* (1.3 – 0.9 Ga) that tacked continental masses onto the margins of Laurentia, creating the supercontinent of Rodinia (*Figure 1-2E*). This concluded almost a billion years (1.8- 0.9 Ga) of continental accretion. But the Grenville was not quite finished. Parts of the new continent, particularly what would later become New Mexico, were ripped through during the closing stages of the collision by northwest- and north-trending extensional (tensional) and strike-slip faults (*Figure 1-5*).

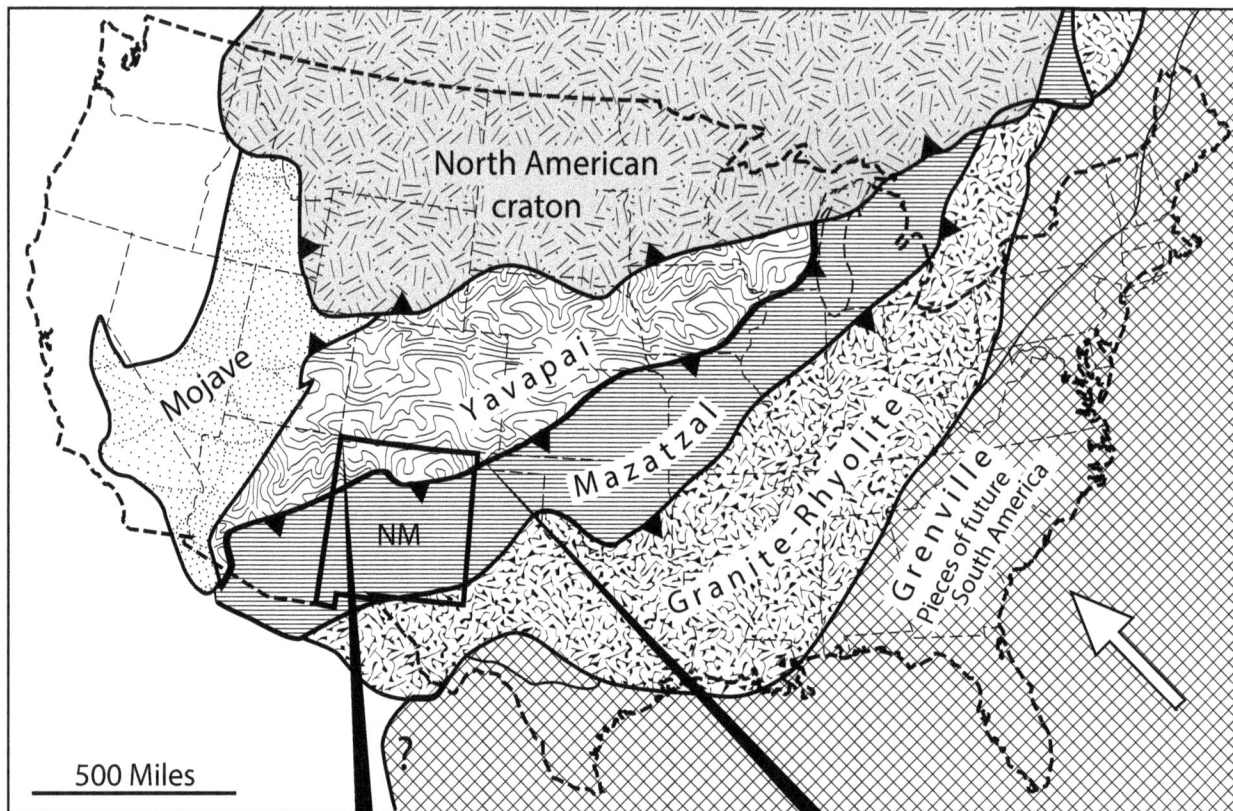

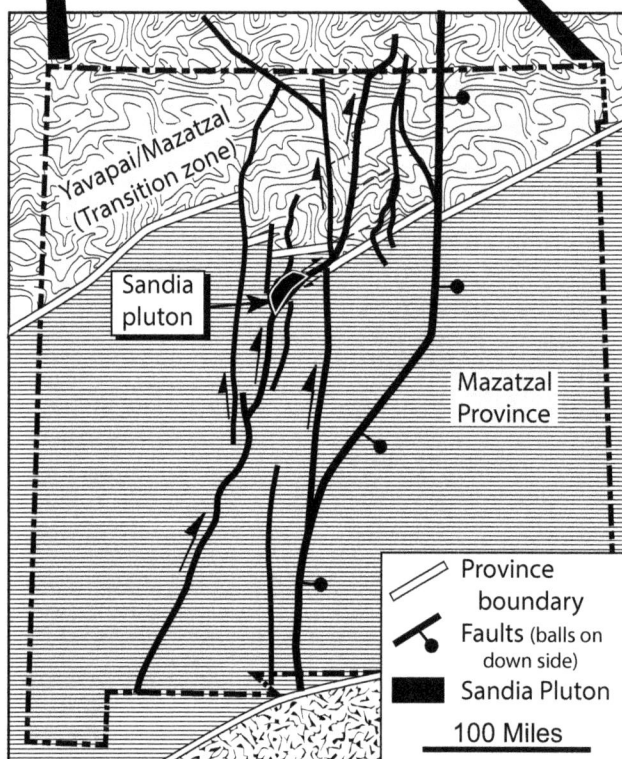

Figure 1-5. Grenville Shearing and Fracturing of New Mexico , 1.3 - 0.9 Ga
(Present USA Shown for Reference; Modified from Whitmeyer and Karlstrom 2007)

Supercontinent of Rodinia and its Breakup (825-550 Ma)

Now we gradually transition to a finer time scale, leaving the term *Ga* behind and scaling down to *Ma*. By about 900 Ma (0.9 Ga) Rodinia had been completely constructed (*Figure 1-6*). Its enormous extent acted as an effective thermal insulator for the Earth's internal heat, like a pressure lid on a pot of boiling water, preventing heat from escaping to the surface from the hot mantle below. At ca. 825 Ma, an enormous reservoir of heat had built up, "pluming" from the mantle under the supercontinent. This destabilized Rodinia such that it began to fracture and break apart. During an interval of only 75 Ma (825-750 Ma), the entire supercontinent swiveled around 90° counterclockwise to something like its present orientation (Li et al. 2008). Then, ca. 750 Ma, Rodinia's two huge appendages on the west side—Australia and Antarctica—rifted off and went on their way. Next, ca. 600 Ma, massive chunks of the eastern margin—pieces of the future South America and Africa—did the same. Finally, ca. 550 Ma, a piece on the south popped off and began its long journey to what is now Argentina in South America (*Figures 1-2F* and *1-5*). *Voilà*, modern North American was beginning to take shape. (However, not until the Mesozoic, about 300 Ma later, would the modern west coast begin to experience its own tectonic collisions and terrane additions to assume its final contour, and would most of the terrane-boundary offsets begin to form.)

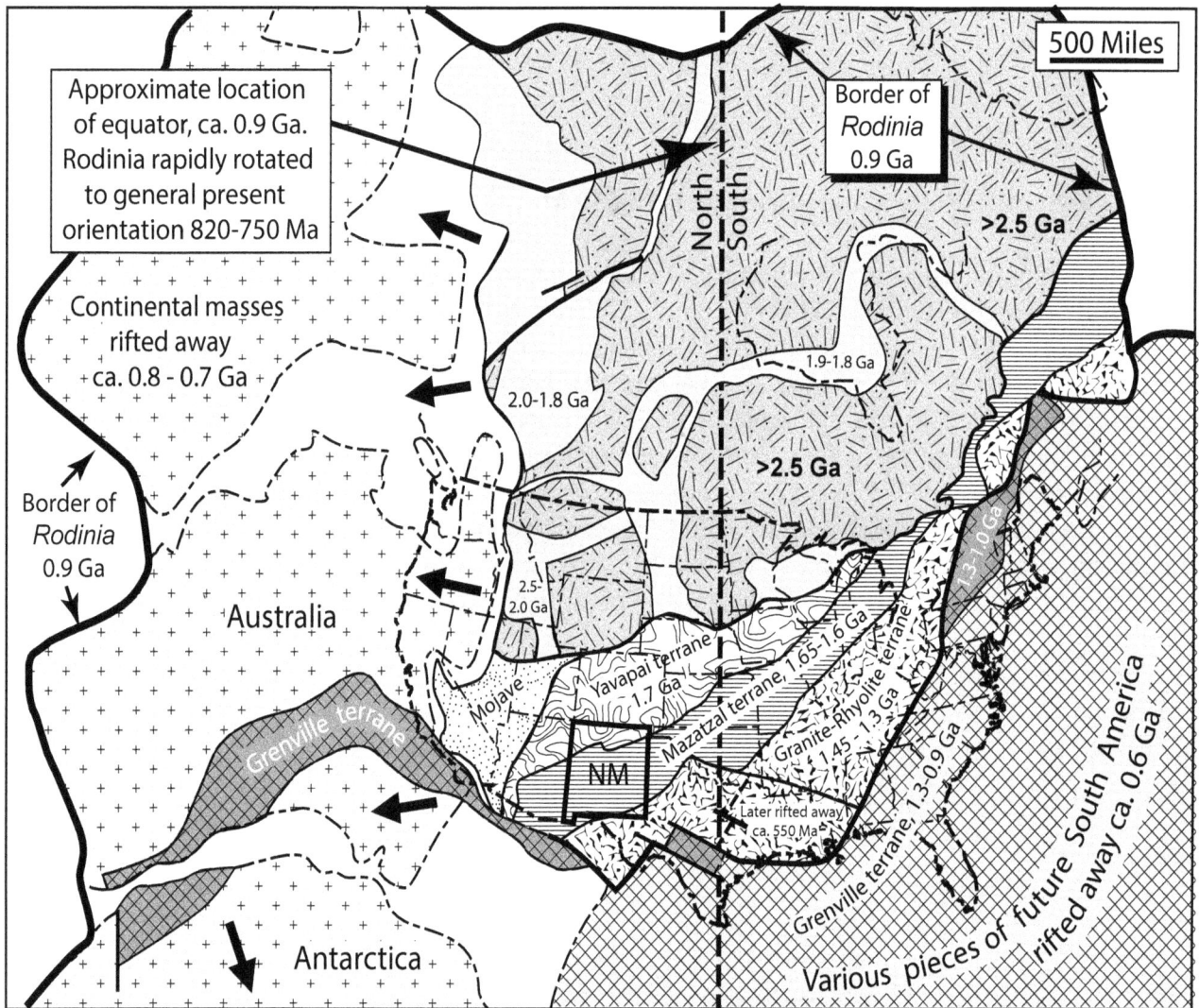

Approximate location of equator, ca. 0.9 Ga. Rodinia rapidly rotated to general present orientation 820-750 Ma

Continental masses rifted away ca. 0.8 - 0.7 Ga

Border of *Rodinia* 0.9 Ga

Australia

Grenville terrane

Antarctica

North South

Border of *Rodinia* 0.9 Ga

500 Miles

>2.5 Ga

1.9-1.8 Ga

2.0-1.8 Ga

>2.5 Ga

2.5-2.0 Ga

Mojave

Yavapai terrane 1.7 Ga

Mazatzal terrane, 1.65-1.6 Ga

Granite-Rhyolite terrane, 1.45-1.3 Ga

1.3-1.0 Ga

NM

Later rifted away ca. 550 Ma

Grenville terrane, 1.3-0.9 Ga

Various pieces of future South America rifted away ca. 0.6 Ga

Figure 1-6. Supercontinent of Rodinia 0.9 Ga
(Modern Geographic Boundaries Shown for Reference;
Modified from Karlstrom et al. 2001; Whitmeyer and Karlstrom 2007)

2:
EMPLACEMENT OF THE SANDIA GRANITE (1.45 GA)

Synopsis

At 1.45 Ga, a blob of granitic magma, part of a vast volume of melting in the lower crust, rose via buoyancy to a depth of 4 to 6 miles into older rock in the middle crust. Irregularities in the "host" rock of the middle crust likely played a role in controlling the shape of the intruding magma, especially its top in the Juan Tabó Cul-de-Sac. The magma solidified to form a body of rock, a pluton, later to be called the Sandia Granite. Field data suggest that the pluton has roughly a "mushroom" shape, inclined about 45° to the northwest, with a base exposed in Tijeras Canyon and an irregular top exposed in the Juan Tabó area.

~~~

Introduction

In the INTRODUCTION I promised the reader that I would make this narrative intelligible to the non-geologist. There will be no graphs or chemical formulas, but I do want gently take the reader outside his/her comfort zone somewhat and delve into an essential geologic process. As much as possible I will draw upon familiar analogies for geologic phenomena to ease the pain.

Intrusion

First, we need to focus on the intrusion of the Sandia Granite—the backbone of Sandia Mountain. This event was just one of a large number of granitic intrusions spread over a vast area at this time (*Figure 1-3A*). Much ink has been spilled to explain this continental-scale phenomenon. To recap, a compelling hypothesis is that the pile-up of terranes, shoved into and, in places, onto the southeast rim of the growing continent, so thickened the crust that the top was raised to form a high, vast plateau, such as today's Tibetan Plateau in Asia (*Figure 1-3B*). At the same time the base of this uplifted mass was pressed down to deep zones of high heat such that melting of this lower crust occurred over an enormous area (*Figure 1-3A*)

The resulting melt—*granitic magma*, being less dense (because of both its composition and thermal expansion) and therefore more buoyant than the encasing rock, gradually worked its way upward much like a blob in a *lava lamp*—that odd device so popular in the 1960s (*Figure 2-1*). And remember, all this happened in super slow-motion! As the magma bodies ascended, they gradually lost their heat and thus partially crystallized, forming a "mush." When the melt finally became so viscous (and encountered overlying rock in the middle crust that was cool and brittle, and thus could not be shouldered aside), it could rise no farther (*Figure 2-1*). There it ponded, crystallized fully, and became bodies, or plutons, of igneous rock. To clarify, the Sandia pluton is a *body* of Sandia Granite, not the granite itself.

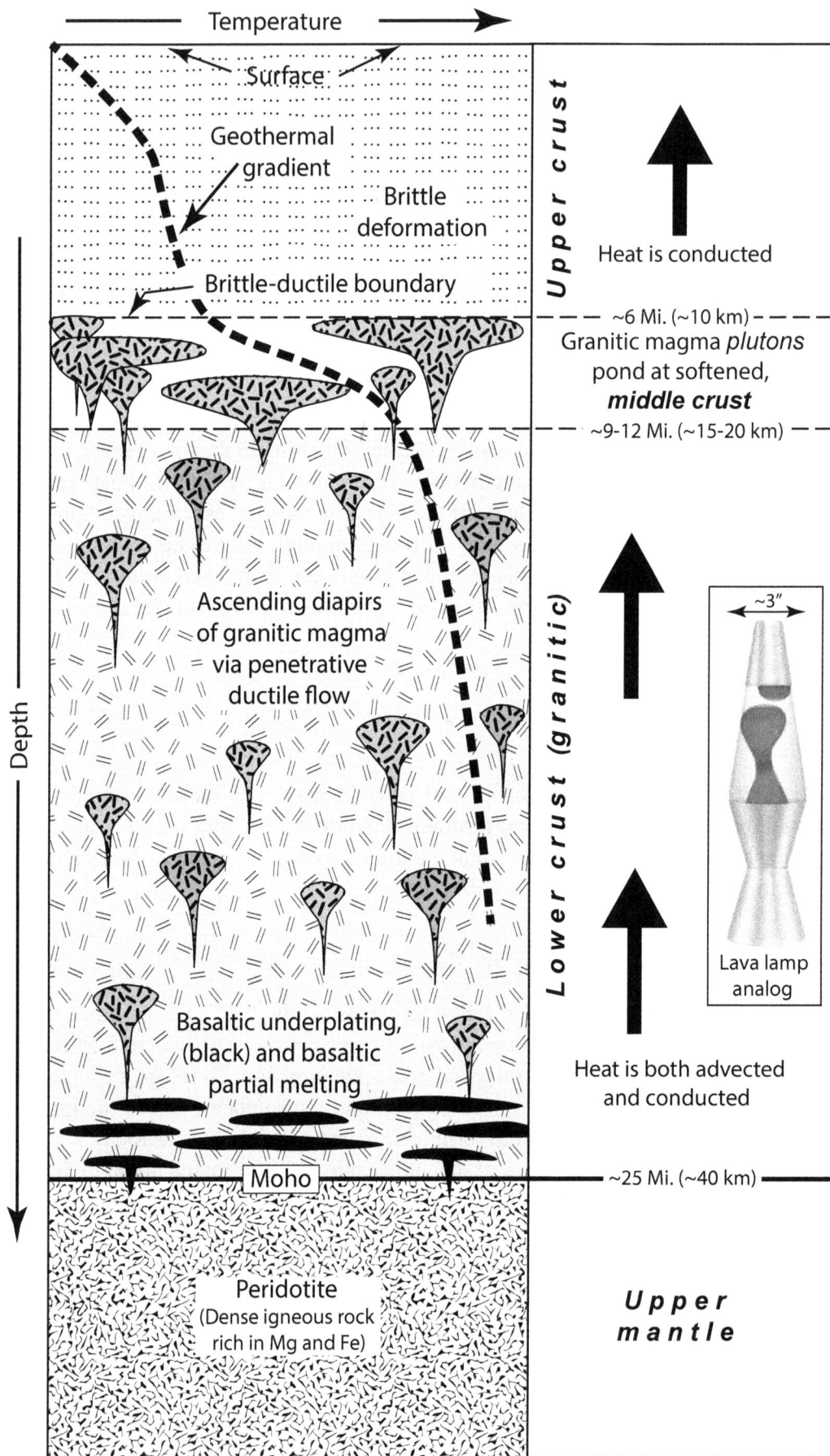

Figure 2-1. Sketch Showing Origin, Rise, and Ponding of Diapiric Magma Bodies in Middle Crust (Modified from Shaw et al. 2005; Keller et al. 2005)

The final depth of crystallization of the granite in the middle crust was roughly 4 to 6 miles (7 to 10 km; Grambling et al. 2016). Much later, ca. 315 Ma, when the first sedimentary rock formed over its top, the granite was at sea level. Today the Sandia Granite resides at an elevation of over 10,000 ft. It has been uplifted some 6 to 8 miles! The Sandia Granite has had quite a ride!

Size and Shape of the Sandia Granite Pluton

Only a small portion of the Sandia Granite pluton is exposed today, but this exposure is the beginning point to determine the pluton's overall size and shape (*Figure 2-2*). For the pluton's aerial extent to the northeast, some writers envision a huge body extending from the Sandia Mountain to well beyond and including the granite of Hermit Peak near Las Vegas NM, for a total length of about 125 miles (Karlstrom et al. 2004). As to the southwestern limit, it has been proposed that a western chunk of the pluton has been ripped off by a right-lateral fault and now resides below Fenton Hill in the Jémez Mountains, 60 miles to the north (*Figure 2-3*; Laughlin 1991). There a granite about 8500 ft below the surface was encountered by the drilling of geothermal test wells. This granite is very similar in age, chemistry, and mineralogy to the Sandia Granite, making for a compelling and intriguing argument. The responsible fault, running parallel to the later Rincón fault (on the west side of Sandia Mountain), must reside somewhere west out in the Rio Grande rift. This view is supported by Daniel et al. (1995) who postulate the same right-lateral fault, but placed it in a Laramide-age context. This is consistent with the enigmatic sawtooth offset of large-scale geologic features seen in *Figure 1-5* (more about this in *Chapter 4*).

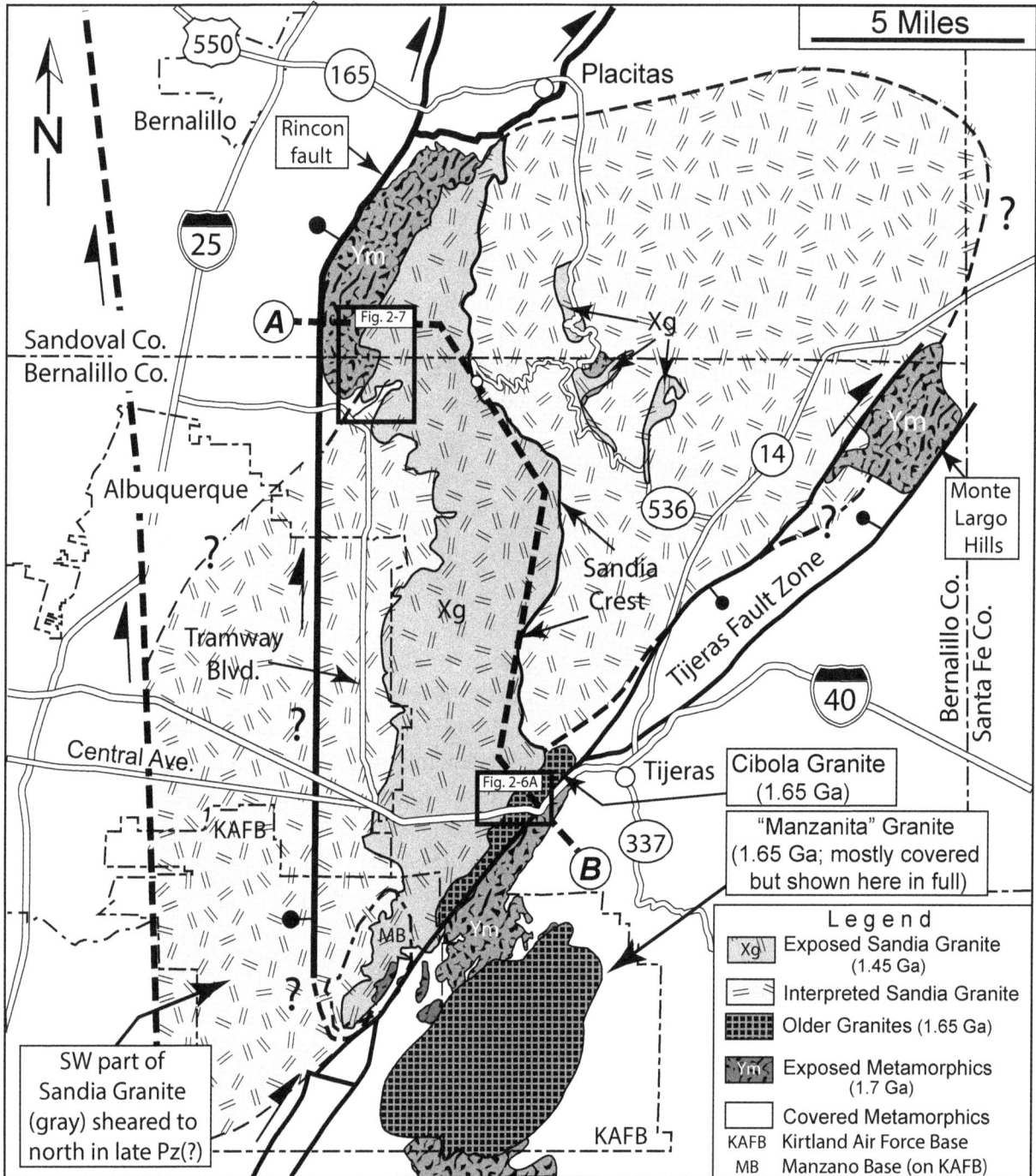

Figure 2-2. Surface and Interpreted Subsurface Distribution of Sandia Granite Pluton
(Schematic Cross Section A-B in Figure 2-5A)

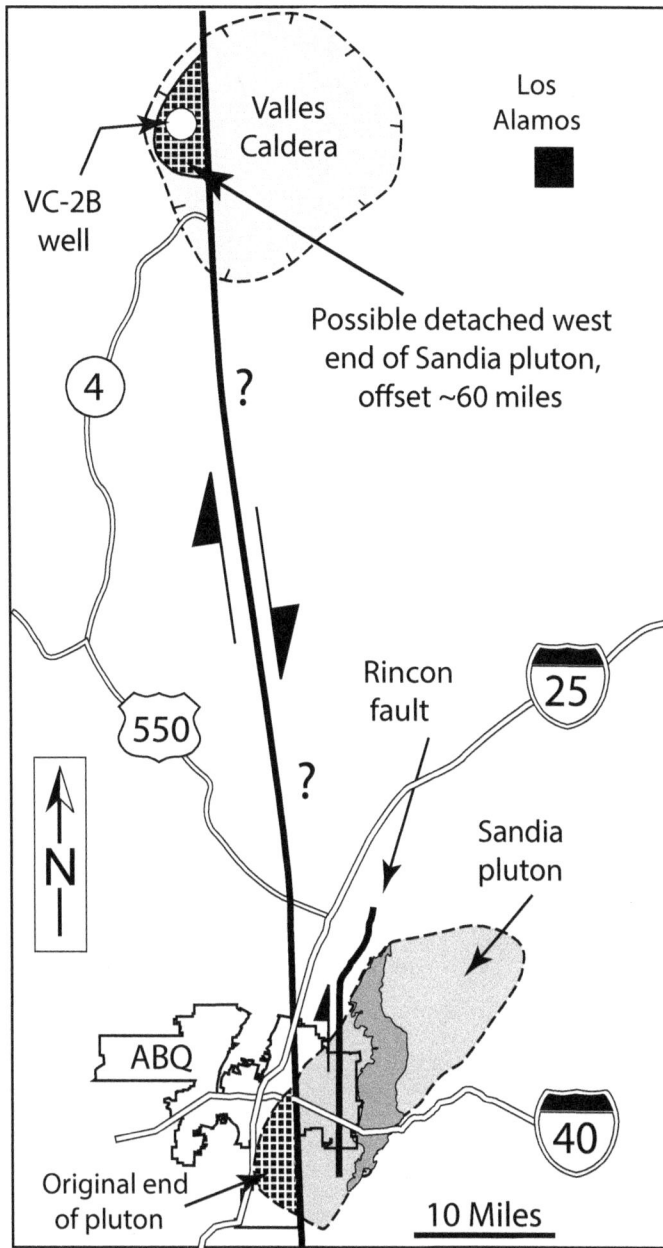

Figure 2-3. Possible Sandia-Valles Caldera Connection
(Modified from Laughlin 1991)

Thickness. A graduate student at UNM (Kirby 1994) interpreted the Sandia Granite pluton as having a complicated, sheared bottom along old US-66 (NM-333) in Tijeras Canyon, and he determined that the base of the pluton dips about 45° down to the northwest (*Figure 2-4*). This all-important dip angle constrains the interpreted thickness of the Sandia pluton. Simple trigonometry, using the outcrop-width (measured normal to the contacts) of about 9 miles and the 45° NW dip suggests that the maximum *vertical* thickness is also about 9 miles. This important number will come up later in *Chapter 6*. Most of what we interpret about the size and shape of the pluton comes from observations at these two main contacts: the southeastern one in the I-40/NM-333 area of Tijeras Canyon, and the northwestern one in the Juan Tabó Cul-de-Sac (*Figure 2-5A*).

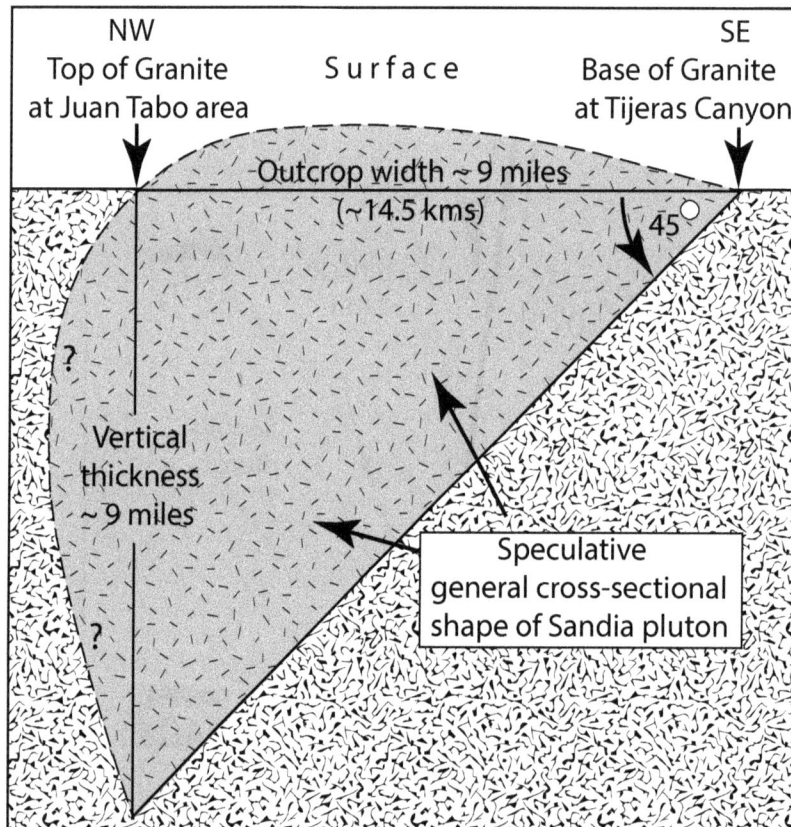

Figure 2-4. Estimated Thickness of Sandia Granite Pluton
(Based on Kirby 1994)

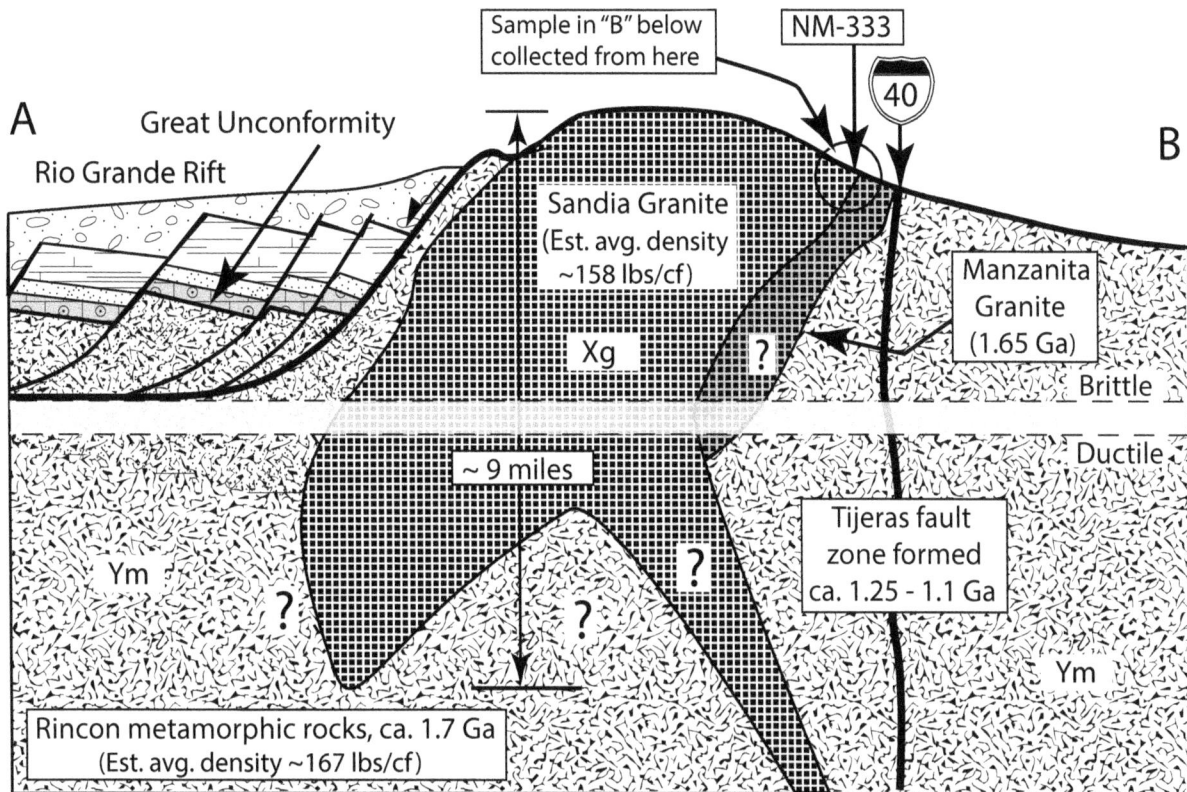

A. Schematic Cross Section (Location in Figure 2-2) Showing Present Setting of Sandia Granite Pluton

Labels in part A:

Great Unconformity

Rio Grande Rift

Sample in "B" below collected from here

NM-333

40

Sandia Granite (Est. avg. density ~158 lbs/cf)

Xg

?

Manzanita Granite (1.65 Ga)

Brittle

Ductile

~ 9 miles

Ym

?

?

?

Tijeras fault zone formed ca. 1.25 - 1.1 Ga

Ym

Rincon metamorphic rocks, ca. 1.7 Ga (Est. avg. density ~167 lbs/cf)

A

B

B. Sample of Basal Part of Sandia Granite Showing Flowage Fabric (Photograph by Author)

Figure 2-5. Sandia Granite Pluton

Southeastern Contact. This contact preserves evidence of lateral flowage of the intruding granitic magma as it filled the expanding, accommodating magma chamber. The rock with this streaky fabric at the very bottom can no longer be called granite in the strict sense (*Figure 2-5B*). At one particularly fine outcrop along NM-333 (thanks to the Highway Department) the contact zone is known as the Seven Springs shear zone (*Figure 2-6).*

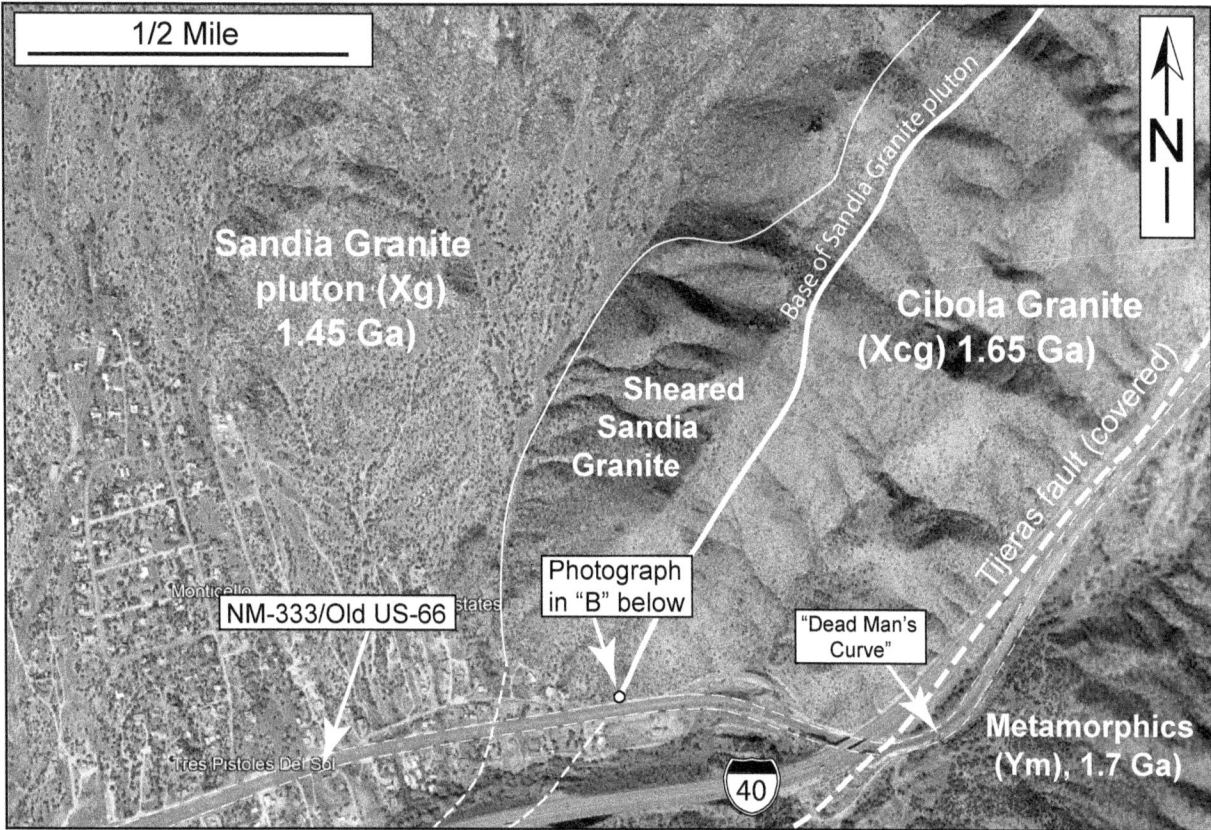

A. *Google Earth* Image of Base of Sandia Granite Pluton (Location in *Figure 2-2*)

B. North View of Part of Seven Springs Shear Zone, Mile 3.8 on NM-333 (Photograph by Author)

Figure 2-6. Seven Springs Shear Zone, Tijeras Canyon

Seven Springs. A specific name such as "Seven Springs" warrants an explanation. For centuries, springs from the mountain along Tijeras Canyon provided a resting spot for travelers. In the early 1920s a little store was established at the site of several springs and by 1926 it was named the "Springs Store." By the early 1930s the building was gone and today the Cañon de Carnuel Land Grant Building occupies the site, 1.1 miles northeast of the Deadman's Curve on the south side of NM-333 (East Mountain Historical Society 2020). Geologists have borrowed the name.

Northwestern Contact. The contact in the Juan Tabó Cul-de-Sac is both highly visible and accessible, and provides a superb laboratory for observing intrusive details of the Sandia Granite pluton. The bedrock geology here consists almost entirely of Proterozoic-age basement rocks. They are the 1.45 Ga intrusive Sandia Granite (*Xg*) and the older, enclosing "country rock," the 1.7 Ga Rincón Metamorphics (*Ym*). The contact between the two (*Xg/Ym*) trends northeast along Forest Service Road (FS) 333 at the eastern margin of "East Rincón Ridge" (*Figures 2-7* and *2-8*). The topographic break here is caused by differential erosion at the contact between the harder, metamorphic rock on the northwest and the less resistant Sandia Granite on the southeast. The break determined the best grade for the road.

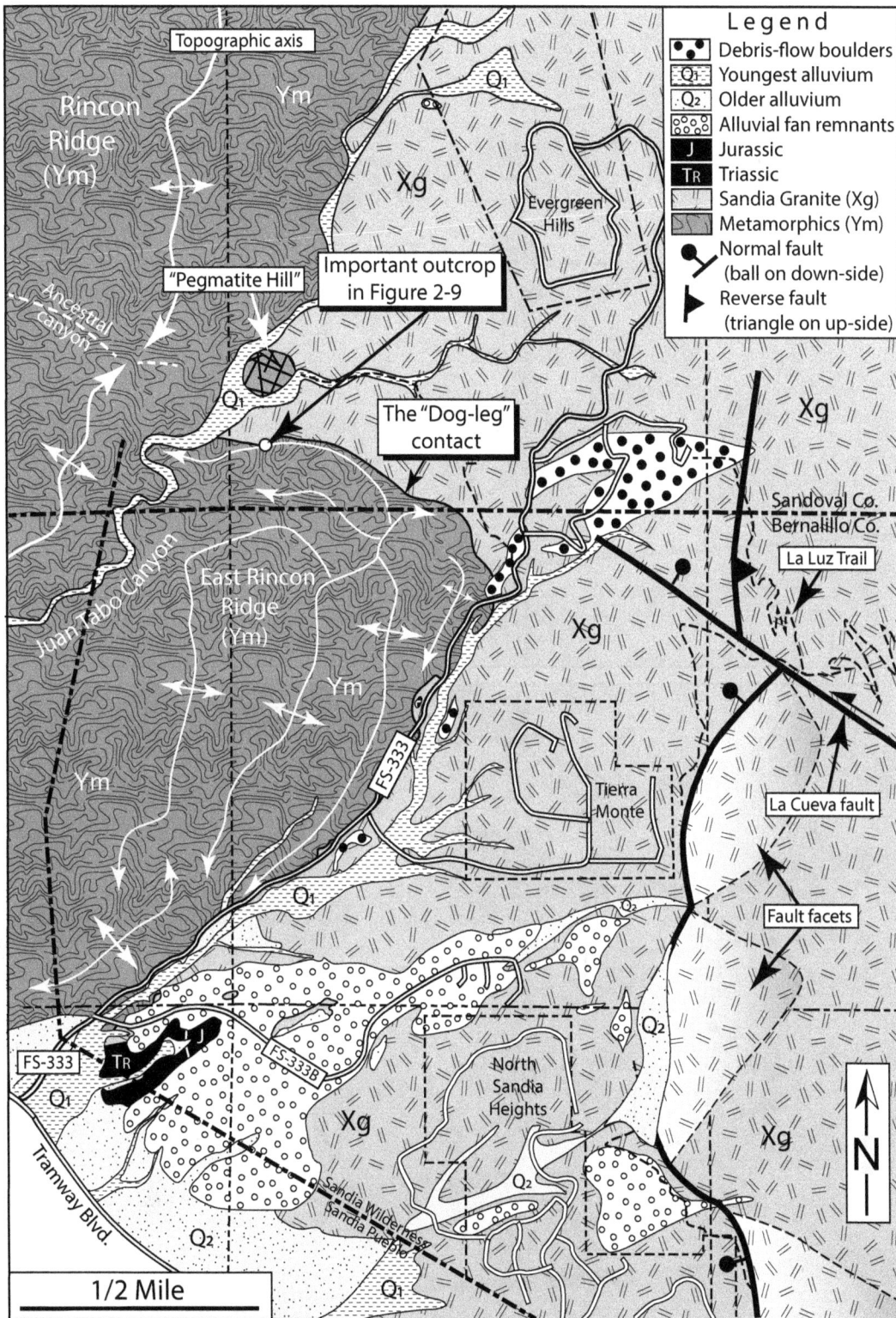

Legend

- **Debris-flow boulders**
- **Q₁** Youngest alluvium
- **Q₂** Older alluvium
- **Alluvial fan remnants**
- **J** Jurassic
- **TR** Triassic
- Sandia Granite (Xg)
- Metamorphics (Ym)
- **Normal fault** (ball on down-side)
- **Reverse fault** (triangle on up-side)

Rincon Ridge (Ym)

Topographic axis

Ym

Xg

Evergreen Hills

"Pegmatite Hill"

Important outcrop in Figure 2-9

Ancestral canyon

Q₁

The "Dog-leg" contact

Xg

Sandoval Co.
Bernalillo Co.

La Luz Trail

Juan Tabo Canyon

East Rincon Ridge (Ym)

Ym

Xg

La Cueva fault

FS-333

Ym

Tierra Monte

Fault facets

Q₁

Q₂

Q₂

FS-333

TR

J

FS-333B

North Sandia Heights

Xg

Q₂

Xg

Q₁

Tramway Blvd.

Sandia Wilderness
Sandia Pueblo

Q₂

Q₁

N

1/2 Mile

Figure 2-7. Geologic Map of Southern End of Juan Tabo Cul-de-Sac
(Modified from Read et al. 1999)

101

Figure 2-8. Outcrops of "Smeared" Contact Between 1.45 Ga Sandia Granite (Xg)
and 1.7 Ga Rincon Metamorphics (Ym; FS-333, Juan Tabo Area)
(Photographs by Author)

At the northeast end of East Rincón Ridge the *Xg/Ym* contact makes a sharp 90° dog-leg zig to the northwest for about ¾ mile, and then a 90° zag back to the northeast along the eastern base of the main Rincón Ridge (*Figure 2-7*). Most published maps (e.g., Kelley and Northrop 1975; Read et al. 1999; Williams and Cole 2007) posit that a down-to-the-north feature called the *La Cueva fault* extends across from the east, creating this offset (*Figure 2-7*). I humbly suggest an alternative interpretation. (The reason I offer this in such detail is that this lineation probably plays a crucial role in the origin of Juan Tabó Canyon, to be discussed later in *Chapter 9*.)

"Dog-leg" contact. This little aside is for the reader who likes to check out things in the field. The *Xg/Ym* contact forming the northern side of East Rincón Ridge can be studied at a single location (*Figures 2-7* and *2-9*). At this spot I can find no evidence of a fault contact, but rather I see an *igneous* contact similar to that along FS-333 (*Figure 2-8)*.

A. East-Northeast View of Key Outcrop

B.. North Close-Up View of Contact in "A"

Figure 2-9. Contact Between Sandia Granite and Rincon Metamorphics
(Location in *Figure 2-7*; Photographs by Author)

To investigate this lineation in the other direction we have to take a 2.5 to 3-mile hike up part of the La Luz Trail to the east. A spectacular display of this northwest-southeast lineation can be seen looking west down from the trail (*Figure 2-10*). In 1969 a strong-legged UNM graduate student named H. Feinberg studied this part of the La Cueva lineament. He found no trace of a fault there, nor can I. Rather, and importantly, Feinberg (1969) notes a massive zone of what is called *mylonite* (ground-up and sheared rock) along the trace, rather than a fault. The La Cueva lineation continues southeast up the mountain and terminates just below Sandia Crest at the Great Unconformity, indicating that the lineation is older than then the 315-Ma rocks above the unconformity.

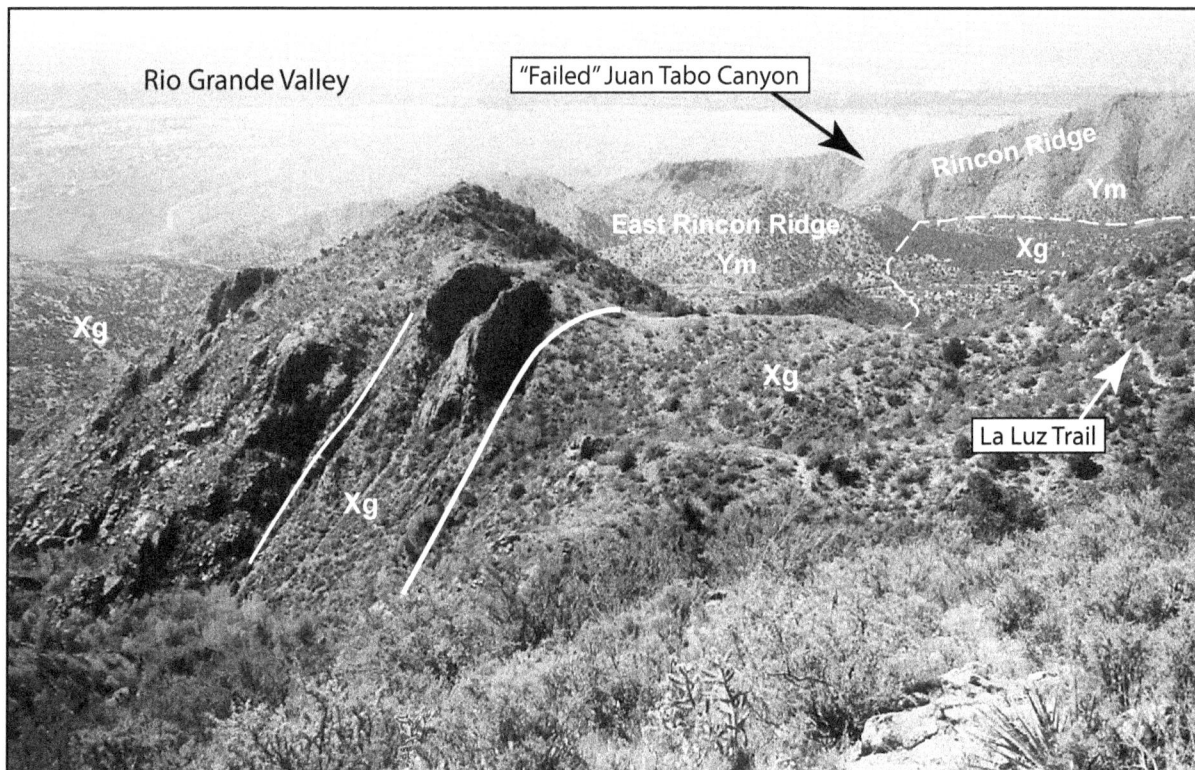

A. West Northwest View across Trace of Ancient La Cueva Lineament (Inside Two White Lines)

B. Spectacular Closer View of Sheared and Mylonitized, Ancient La Cueva Lineament

Figure 2-10. La Cueva Lineament from La Luz Trail
(Photographs by Author)

Meanwhile, down below at the far western end of the zig-zag contact is a little circular hill that I have informally dubbed "Pegmatite Hill" (*Figure 2-7*). Its rock is not granite (which is defined by composition and *texture*), but rather an extremely coarse-grained type consisting of whitish K-spar, glassy quartz, and some transparent mica called *muscovite.*(*Figure 2-11*).This texture is typical of late crystallization of a gas and water-rich melt in the upper reaches of a pluton, forming a rock type called *pegmatite*. In map view the hill does in fact seem to occupy the ancient topmost position of the northwest-dipping Sandia Granite pluton (*Figure 2-12*).

A. Photograph (By Author)

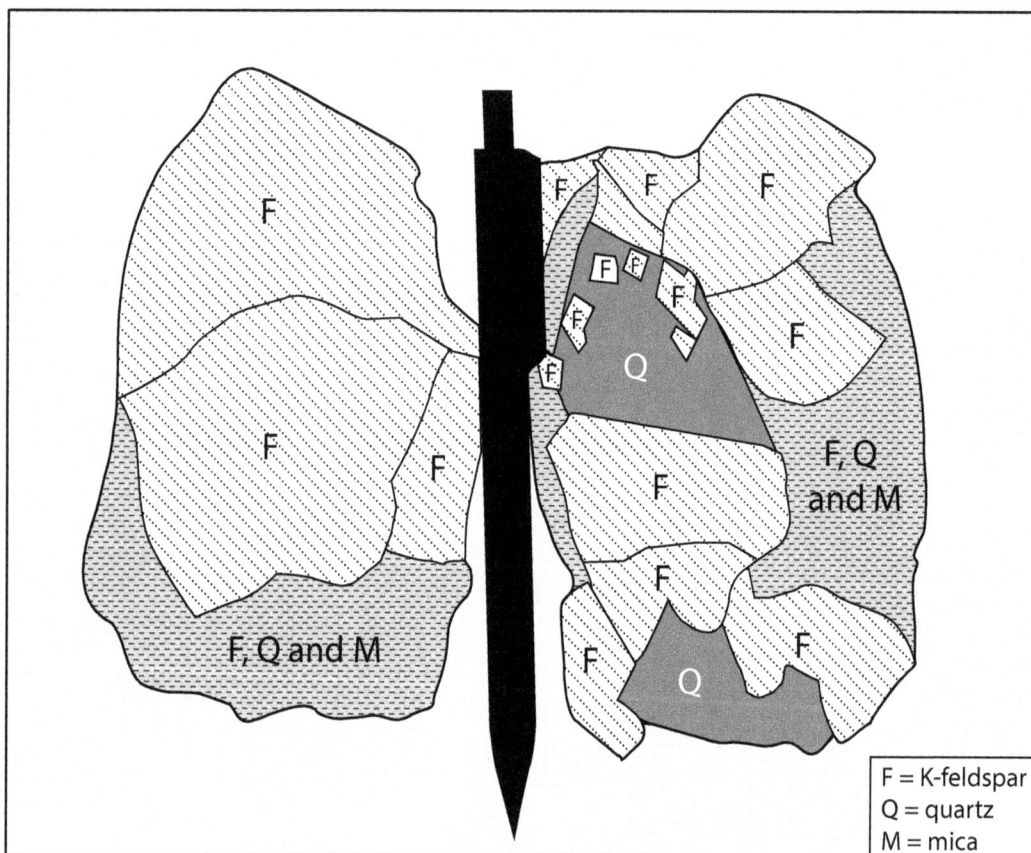

F = K-feldspar
Q = quartz
M = mica

B. Interpretative Sketch of "A" Above

Figure 2-11. Rock Samples from "Pegmatite Hill"

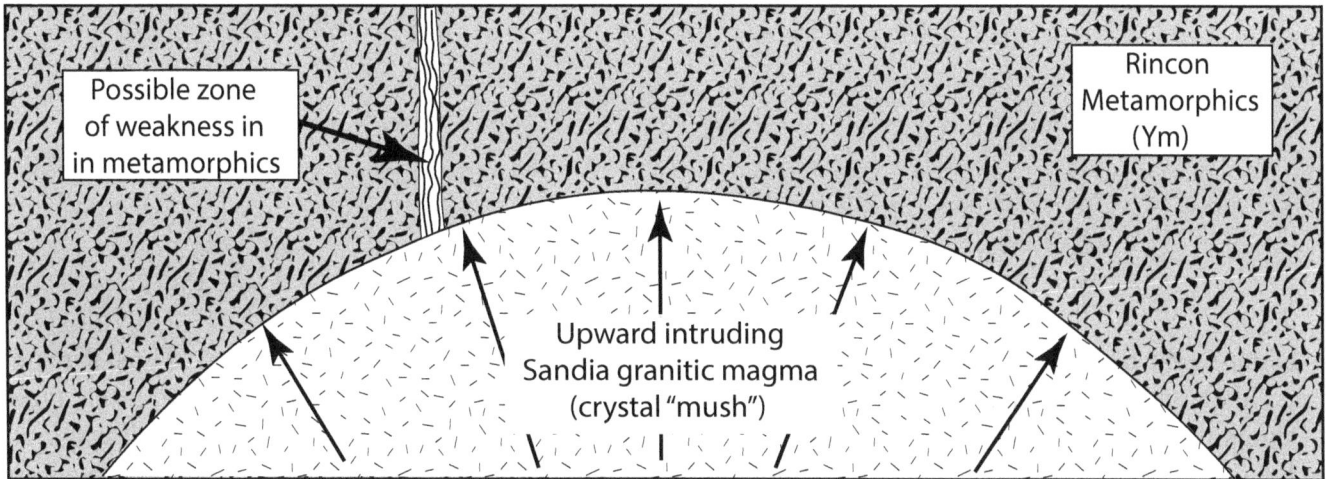

A. Initial Intrusion of 1.45 Ga Sandia Granitic Magma into 1.7 Ga Rincon Metamorphics

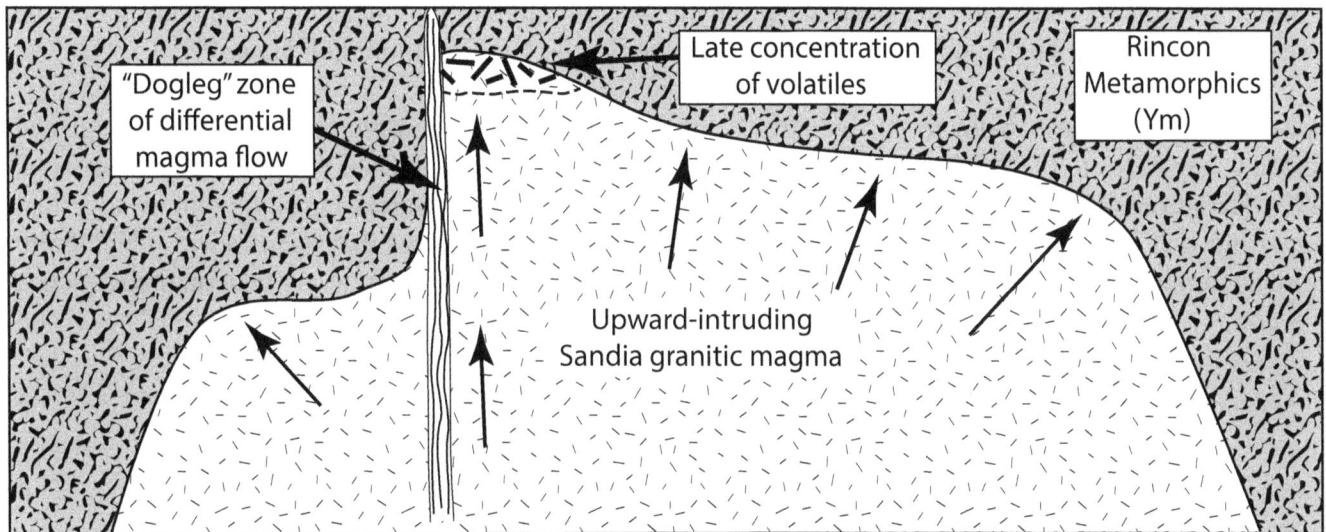

B. Late-Stage Distortion of Sandia Magma Along Zone of Differential Flow (Perhaps Taking Advantage of Ancient Zone of Weakness in Older Rincon Metamorphic Country Rock)

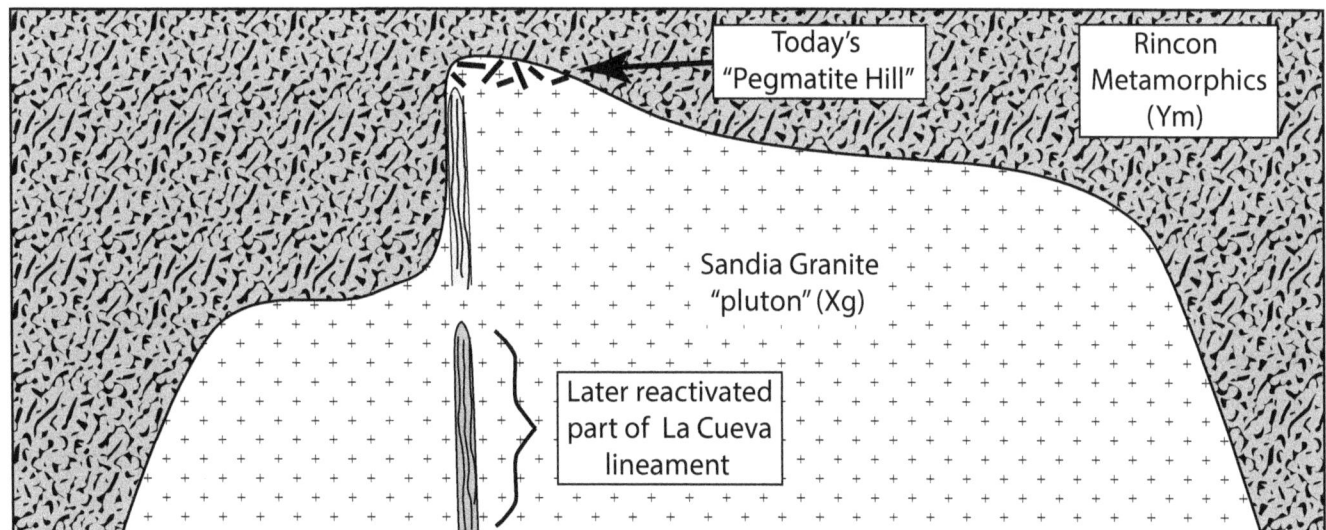

C. Final Concentration of Volatiles in Highest Part of Sandia Pluton, Producing Coarsely-Crystalline, Feldspar-Quartz-Muscovite Rock of Future "Pegmatite Hill"

Figure 2-12. Cartoon Cross Sections Showing Postulated Irregular Intrusion of Sandia Granitic Magma in Juan Tabo Cul-de-Sac (Compare to Figure 2-7)

Considering the lack of fault evidence at the *Xg/Ym* contact at the key outcrop on the north end of East Rincón Ridge, the mylonite along the La Cueva lineament high up along the La Luz Trail, and the bizarre rock type at Pegmatite Hill, I conclude that the lineation is truly ancient, caused perhaps by differential flowage of the Sandia Granite magma during its forceable intrusion 1.45 Ga into the older rock. The intruding magma perhaps exploited a zone of weakness within the older, encasing Rincón Metamorphics. Importantly, down-to-the-north faulting much later occurred along only a short segment of the lineament, reactivating that old zone of weakness, but had no effect on the outcrop of the Sandia Granite (more in *Chapter 9*).

A possible, more ancient analog to this scenario is another pluton, the *Manzanita* pluton, 1.65 Ga in age and located a few miles to the south of Sandia Mountain on Kirtland Air Force Base (*Figure 2-2*; Brown et al. 1999). Karlstrom et al. (2016) constructed a cross section tying the Sandia and Manzanita plutons together, suggesting that both, though at different times, were intruded into terranes that were being subjected to regional compression (*Figure 2-13*). I have modified the cross section to illustrate my take on the sequence of events.

Northwest　**A n c i e n t　l a n d　s u r f a c e, >1.65 Ga**　**Southeast**

Brittle, upper crust

300° C

Depth-controlled, brittle/ductile transition zone (middle crust)

Hot, ductile lower crust

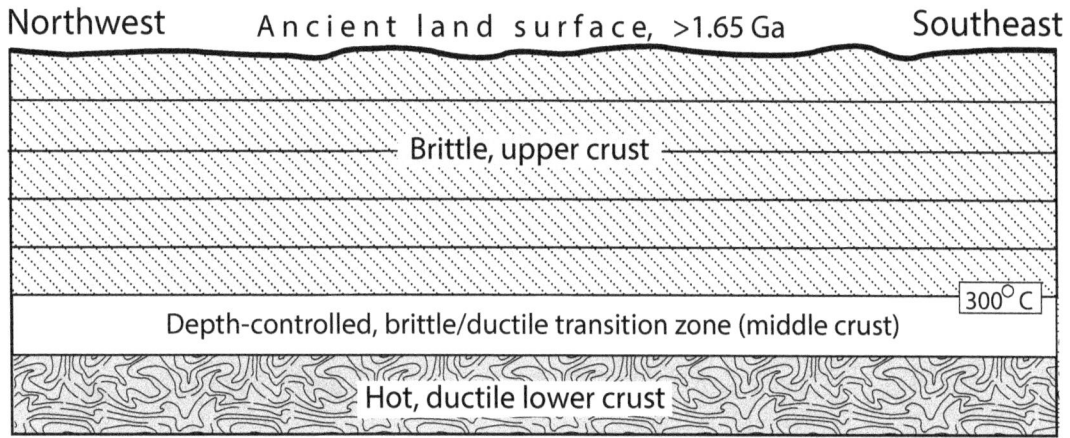

A. Crust Prior to Deformation, and Emplacement of Granitic Plutons, >1.65 Ga

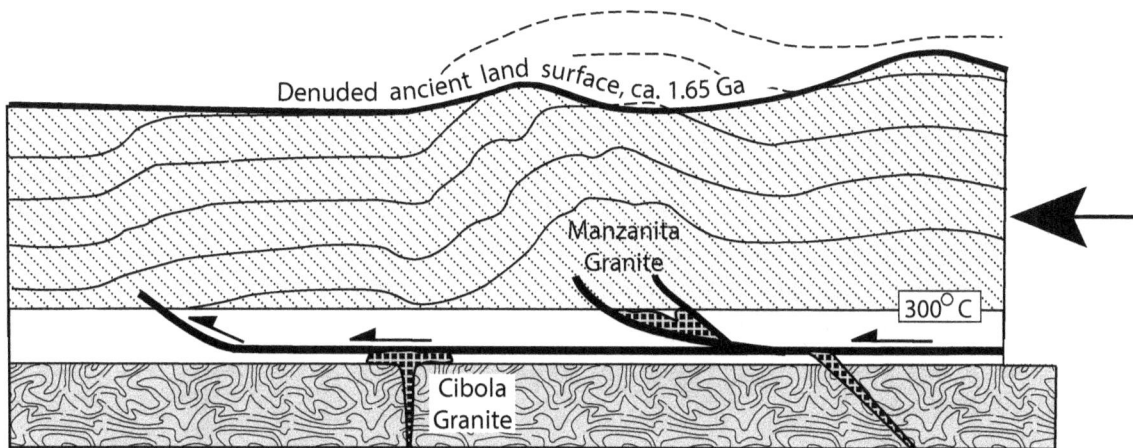

Denuded ancient land surface, ca. 1.65 Ga

Manzanita Granite

300° C

Cibola Granite

B. Emplacement of Cibola and Manzanita Granitic Plutons (Checkered Pattern), 1.65 Ga

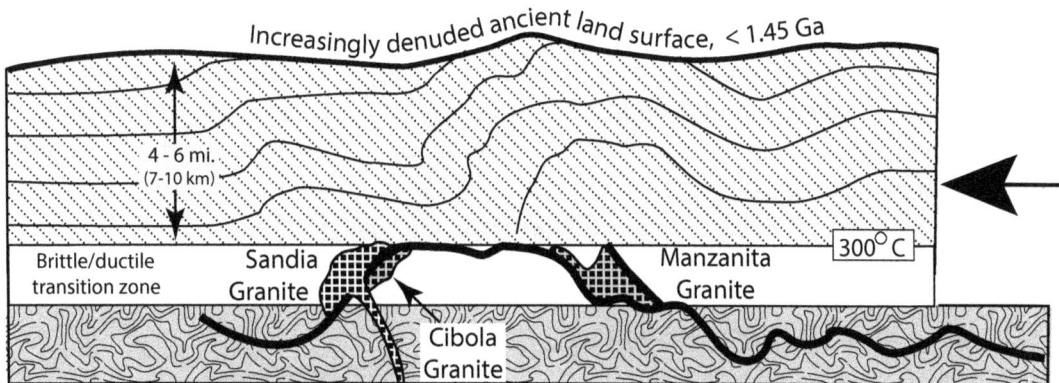

Increasingly denuded ancient land surface, < 1.45 Ga

4 - 6 mi.
(7-10 km)

300° C

Brittle/ductile transition zone

Sandia Granite

Cibola Granite

Manzanita Granite

C. Emplacement of Sandia Granite and Deformation of Older Plutons (Checkered Patterns), 1.45 Ga

Figure 2-13. Schematic Cross Sections Showing Postulated Emplacement Sequence of Granite Plutons
(Modifed from Brown et al.1999; Karlstrom et al. 2016)

111

3:
UNROOFING OF THE SANDIA GRANITE
AND ITS SUBSEQUENT REBURIAL
(1.45 GA-315 MA)

Synopsis

During the 1.1 Ga period following the intrusion of the Sandia Granite (1.45 Ga—315 Ma) the ancient land surface rose, causing the removal of 4 to 6 miles of cover and exposing the top of pluton at the new, lowered land surface. Then began a 240 Ma (315-75 Ma) period characterized by pulses of subsidence and deposition that buried the pluton under nearly 14,000 ft of sedimentary rock. During this time, ca. 290—270 Ma, a wave of continental collisions created the Ancestral Rocky Mountains, a chain likely rivaling the much later, modern Rocky Mountains in grandeur. This earlier range was then beveled smooth, and by ca. 75 Ma the land surface in New Mexico, i.e., the top of the Cretaceous sediments, was a vast alluvial plain lying just barely above sea level.

~~~

Introduction: Setting the Stage

Between the story of the Sandia Granite (*Chapter* 2) and the rocks that cover it, three important and extremely interesting events occurred. 1) the deep erosion of the Laurentian continent to form the *Great Unconformity* (GU), 2) the subsequent subsidence of the GU and its reburial by about 14,000 ft of sedimentary rock, and 3) the creation of the *Ancestral Rocky Mountains* (a range that has nothing to do with our present Rocky Mountains).

The Great *Unconformity* (GU, 1.45 Ga – 315 Ma)

What is an *unconformity*? A clinical definition is that it is a contact within a sequence of rocks that marks a significant time gap (a *hiatus*) between rocks on either side of it. It can be simply one between parallel layered rocks of significantly different ages (a *disconformity*), between rocks of both different ages and types (e.g., sedimentary atop igneous or metamorphic, a *nonconformity*), or between rocks of different orientation (e.g., horizontal atop inclined, an *angular unconformity*). The three types are indicated by the wavy lines in *Figure 3-1A*.

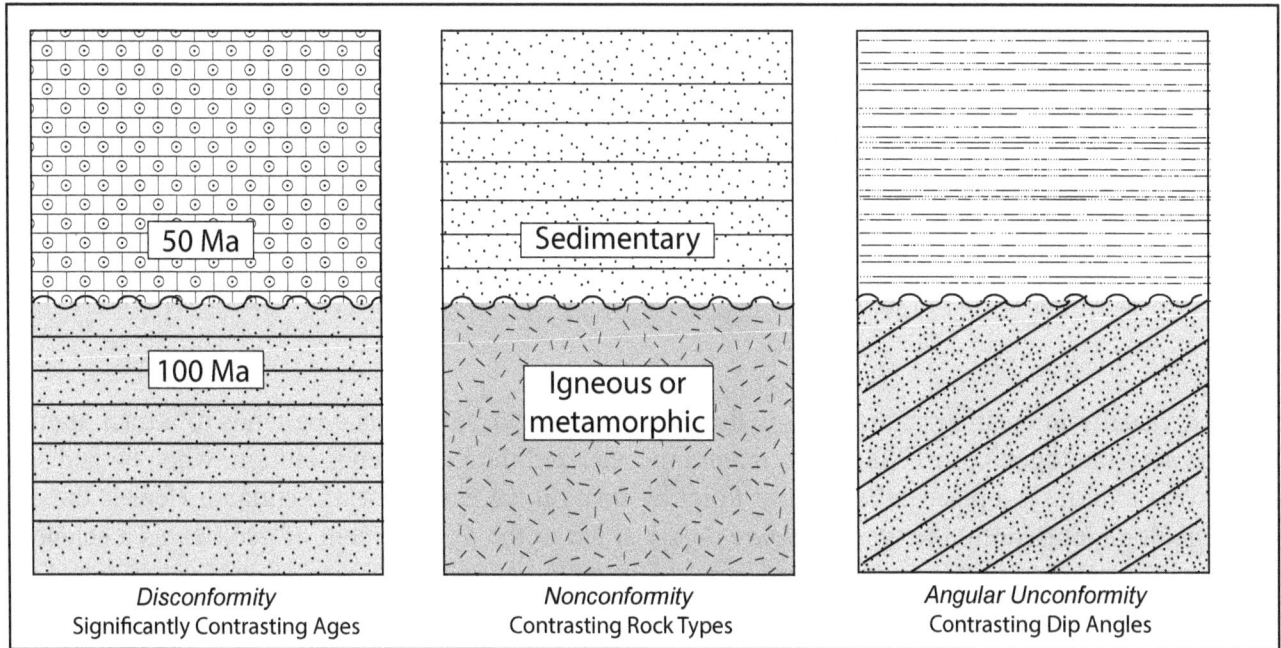

Disconformity
Significantly Contrasting Ages

Nonconformity
Contrasting Rock Types

Angular Unconformity
Contrasting Dip Angles

A. Types of Unconformities

B. Angular Unconformity at Siccar Point, Scotland, World's Most Famous Geologic Outcrop (Photograph from Internet)

Figure 3-1. Unconformities

The most visually interesting of the three is the *angular unconformity*. This type tends to attract attention and to be very photogenic. In 1788 such a feature mesmerized the Scot, James Hutton, when he and some friends were exploring a place called Siccar Point by boat on the east coast of Scotland. Here is the most famous geologic outcrop in the world (*Figure 3-1B*). It led to the development of geology as a science, it introduced the concept of "deep time," and earned Hutton the sobriquet "father of modern geology." The angularity displayed by the rocks at Siccar Point is striking, but the time gap at the contact is modest—only ca. 65 Ma.

What is generally called the Great Unconformity, or the GU, is one of vast extent and an enormous hiatus, and depending on location either an angular unconformity or a nonconformity. There is some conflict concerning who first used the term. One opinion is that it came from John Wesley Powell after his 1869 pioneering trip down the Colorado River through the Grand Canyon. A second opinion is that it was first coined by Clarence Dutton in his 1882 book, *Tertiary History of the Grand Cañon District* (Dutten 1882). Since Dutton's time, the GU has been revealed to be a global-scale surface spanning much of North America as well as other continents.

At Sandia Mountain the Great Unconformity is a *nonconformity* (center of *Figure 3-1A*), with sedimentary rocks lying atop mainly ancient igneous rocks. It is on full display along the shear west face (*Figure 3-2A*). It is not as visually spectacular as Siccar Point but its time gap is roughly 20 times larger (1.1 Ga vs. 65 Ma). The actual contact on the west face is obscured by soft sediment sloughing down from the overlying Sandia Formation (*Figure 3-2B*).

A. Sandia Crest and West Face from Sandia Tram

B. Interpretation of "A" Above

Figure 3-2. Rocks of Upper West Face of Sandia Mountain
(Photograph by Author 1986)

North America's GU. About 540 Ma, at the beginning of the Cambrian Period, the western edge of Laurentia began to subside below sea level. As subsidence progressed, the sea/land contact advanced ("transgressed" in geologic parlance) from west to east across the ancient landscape. Therefore, the age gap between the new rocks being deposited and the underlying old rocks became greater with the eastward advance. For example, in the Grand Canyon the GU superimposes rocks of ca. 525 Ma above those no younger than ca. 740 Ma, for an age gap of at least 215 Ma. In what later became New Mexico, the superposition is between rocks ca. 315 Ma overlying rocks no younger than ca. 1450 Ma (the Sandia Granite) below, for a hiatus of 1135 Ma (1450–315 Ma), an enormous, more impressive span.

Missing record. The absence of a record in the rocks is itself a type of information. Its temporal vastness (1.1 Ga!) taxes the imagination. According to the famous words of the French music composer Claude Debussy: "Music is the space between the notes." The Great Unconformity also enthralled the American poet, Reg Saner (b. 1931), who penned a wonderful, evocative essay on it entitled, "The Ideal Particle and the Great Unconformity" (Saner 1991), a fascinating read available on the internet.

Those first sedimentary rocks atop the GU at the western edge of Laurentia are notable in that for the first time they contain abundant preserved fossils. Something profoundly important happened at this time: the preserved record of abundant life started here, at the beginning of the Cambrian Period, ca. 540 Ma. Two important questions thus raise their heads: 1) Why did sedimentary rocks suddenly begin to accumulate after such a long time-gap, and 2) what change in ocean chemistry allowed life for the first time to produce preservable hard parts (e.g., shells) at the beginning of the Cambrian?

As to the first question (sediment accumulation), two intense continental glaciations of global scale between 717 and 635 Ma deeply scoured the continental surfaces, lowering their elevation. This created "accommodation space" (see below) into which sediment could accumulate and be preserved, rather than flushed away to the deep sea (Keller et al. 2019). Second, global sea level rose at this time due in part to the breakup of Rodinia (*Chapter 1*), and led to the slow landward sweep of the shorelines over the new continental edges.

Accommodation space at Sandia Mountain. With a few interruptions, Sandia Mountain was the site of nearly 14,000 feet of sediment accumulation stacked atop the Great Unconformity. But how is it possible to pile stuff up so high? We don't. It is important to realize that sedimentary formations do not simply plop atop one another and build upward, because gravity won't allow that. Rather, the land subsides in tandem with deposition, providing the requisite "accommodation space" for the new sediment to accumulate. As sediment is delivered to an area, its heavy mass creates a huge downward pressure. The deep crust below (being ductile; more in *Chapter 6*) very slowly and plastically flows laterally away from the area of higher pressure, allowing the surface to sink to provide that accommodation space (i.e., a *basin*) to catch the arriving sediment. If there were no such space, the streams, laden with sediment, would simply continue on their way, dragging along their sediment load and dumping it somewhere further downstream wherever accommodation space existed, perhaps as far away as the sea. If the basin floor continues to subside as sediment is entrapped, thousands of feet can accumulate to form a thick column of sedimentary rock, as is the case here.

As to the second question (ocean chemistry), the advancing, abrasive shoreline (sometimes called the "wave-base razor" because of the way this energetic environment carves up the land's edge) reworked and stripped the thick blanketing soils that had developed over millions of years, exposing enormous areas of fresh rock to the corrosive effects of chemical weathering. The fresh surge of ions (such as Ca^+, Na^+, Mg^+, and PO_4^{-3})

in solution into the sea was toxic to the existing, very primitive marine life that until then lacked hard parts (Peters and Gaines 2012). So, what's a soft-bodied animal to do when exposed to such offensive chemistry? Well, get rid of it by precipitating it as mineral matter (think of kidney stones in humans)!

In sum, a veritable perfect storm was produced by the breakup of Rodinia, by global glaciations that deeply scraped away the tops of the continents creating accommodation space for sediment, and by the surge of unpleasant chemicals eroded from the continents and dumped into the oceans. In concert these three caused a pivotal event in Earth history: an explosion of life in the seas that left behind a lasting record in the rocks.

Reburial (315–70 Ma)

The eastward-advancing ("transgressing") sea, slowly, but with fits and starts, finally reached the area of the future Sandia Mountain ca. 315 Ma. (The edge of the sea had briefly reached New Mexico ca. 360 Ma, but soon withdrew, leaving little sediment behind.) For the next 245 Ma nearly 14,000 ft of sediments buried the Sandia Granite in four distinct phases with a mountain-building interruption between the first and second (see *Figure 0-20A*).

Phase 1: Pennsylvania Sandia Formation and Madera Group (315-290 Ma). An excellent visual introduction to the sedimentary-rock cover is provided by the mountain's shear upper west face (*Figure 3-2*). Here the first sediment deposited above the Great Unconformity is on full, but veiled display. The non-descript slope above the GU is formed by the shaley Sandia Formation, which unfortunately sloughs down and obscures the GU itself. Above the soft Sandia Formation is the more robust, cliff-forming Madera Group that creates the mountain's ridge line called Sandia Crest. Limestone is a non-fragmental sedimentary rock of calcium carbonate ($CaCO_3$) precipitated from sea water via the action of marine animals. (We will revisit the Madera Group rocks in *Chapter 12*.)

Sandia Crest. This name is confusingly used in two senses. First, it is the north-south topographic divide separating the mountain's east-dipping backslope from the precipitous west side. Second, it is the highest point along the crest line, 10,678 ft, at the upper end of the Sandia Crest Byway, a place where the Crest House café/gift shop is located.

Interuption: The Ancestral Rocky Mountains, and Antarctica. As the sea was sweeping across the land from west to east and creating the GU, the other continents were performing a "continental dance" (*Figures 3-3A* to *3-3D*). From ca. 315 to 245 Ma, during and following the deposition of the Madera Group, a huge southern continent called *Gondwana* crashed into Laurentia from the east and southeast, suturing together to form another, new supercontinent called *Pangaea* (*Figures 3-3E* and *3-3F*). This collision between Laurentia and Gondwana created the *Ancestral Rocky Mountains*.

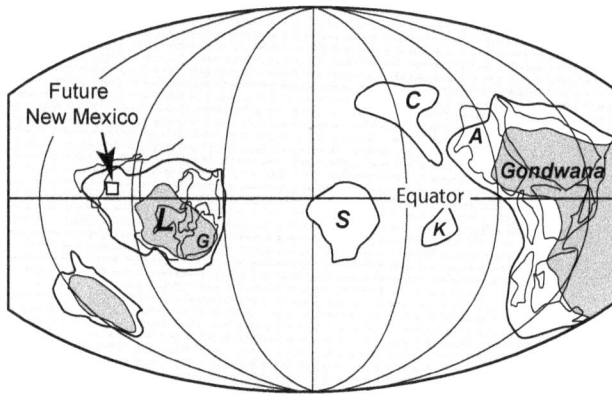

A. 510 Ma

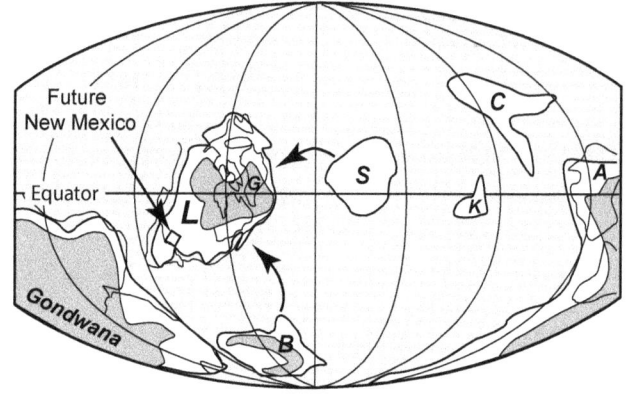

B. 475 Ma

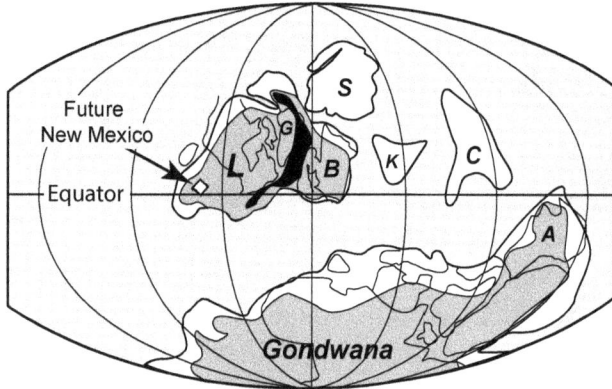

C. 400 Ma

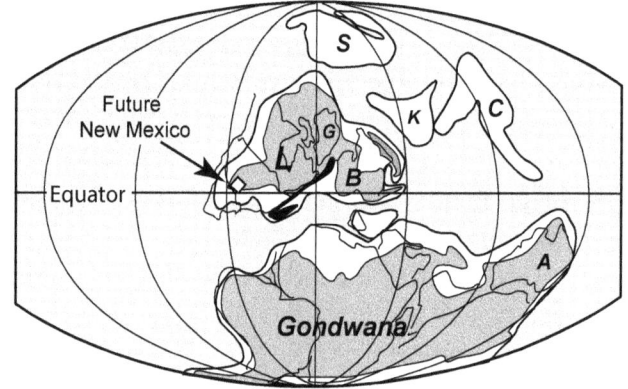

D. 340 Ma

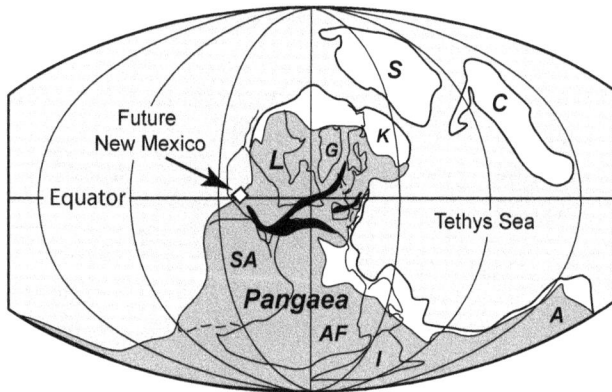

E. 315 Ma, Time of Pennsylvanian Sandia Formation

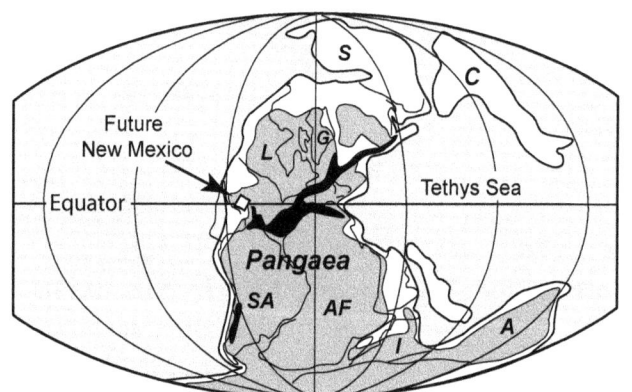

F. 280 Ma, Time of Permian Abo Formation

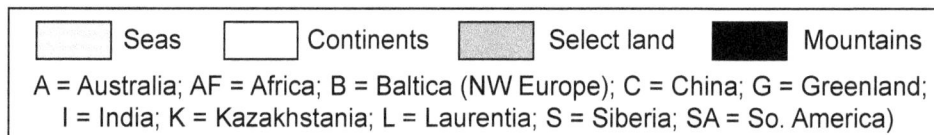

Figure 3-3. Assembly of Supercontinent of Pangaea
(Modified from Stanley 1986)

Ancestral Rocky Mountains. Laurentia at this time was already damaged goods following the Grenville continental-continent collision (1.3-0.9 Ga). Old faults in Laurentia—the "scars"—formed earlier far from the collision zone were jostled and reactivated. Crustal blocks on the west side of these north-trending faults were moved north along those faults. This complicated movement resulted in the saw-tooth offset of the Proterozoic province boundary (*Figure 3-4*). Unfortunately, the term "Ancestral Rocky Mountains" can become confused with what we know as our present "Rocky Mountains." However, the former range is much older than the latter, and the two ranges had totally different origins. By 75 Ma, the old, *Ancestral* Rockies had been worn down to nubs, and the modern Rockies began to form and overprint the former (*Figure 3-4*; more in *Chapter 4*).

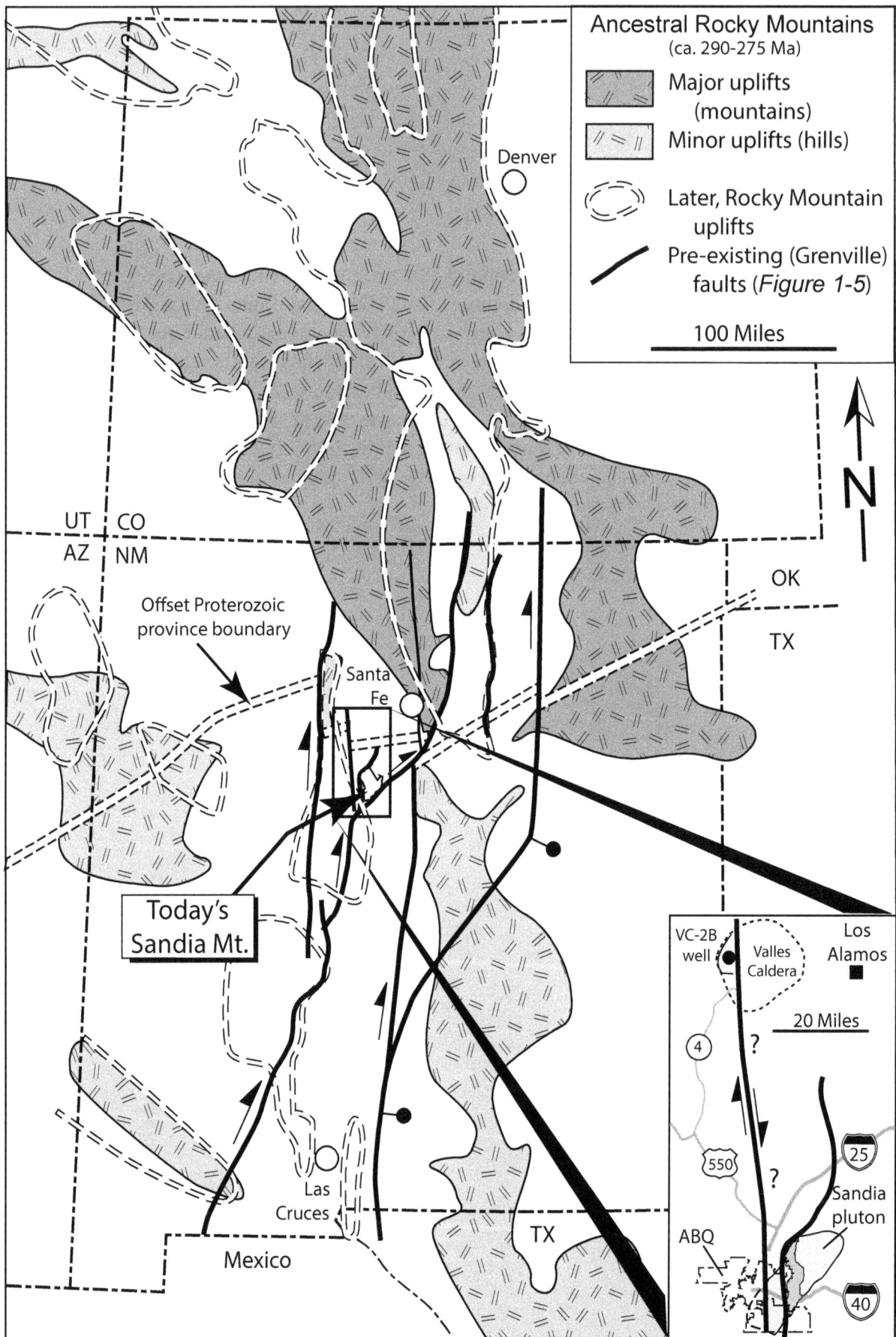

Figure 3-4. Ancestral Rocky Mountains and Future Sandia Mountain
(Modified from Chapin and Seager 1975; Laughlin 1991; Chapin et al. 2014)

The labels within the figure include:

Ancestral Rocky Mountains (ca. 290-275 Ma)
- Major uplifts (mountains)
- Minor uplifts (hills)
- Later, Rocky Mountain uplifts
- Pre-existing (Grenville) faults (*Figure 1-5*)
- 100 Miles

N

Denver

UT | CO
AZ | NM

OK
TX

Offset Proterozoic province boundary

Santa Fe

Today's Sandia Mt.

Las Cruces

Mexico

TX

Inset:
VC-2B well
Valles Caldera
Los Alamos
20 Miles
(4)
?
550
?
25
Sandia pluton
ABQ
40

The entire composite continent of Pangaea with its little New Mexico part then slowly migrated northward from the equatorial region to the northern temperate zone (*Figure 3-5*). The change of *paleolatitude* had an enormous effect on the nature of the sedimentary rocks that were then deposited.

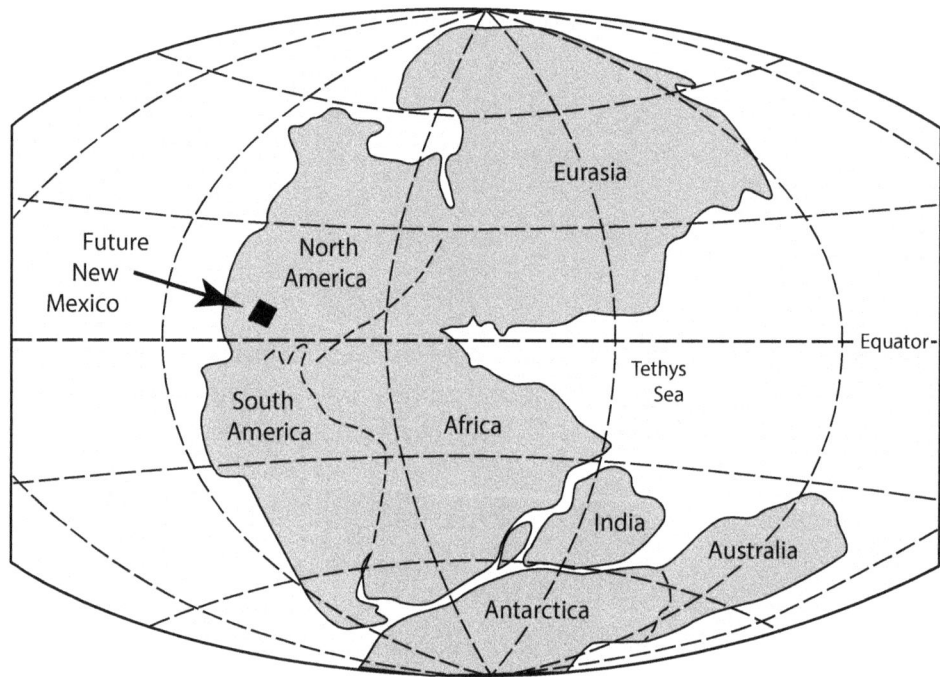

A. *Pangaea*, 250 Ma

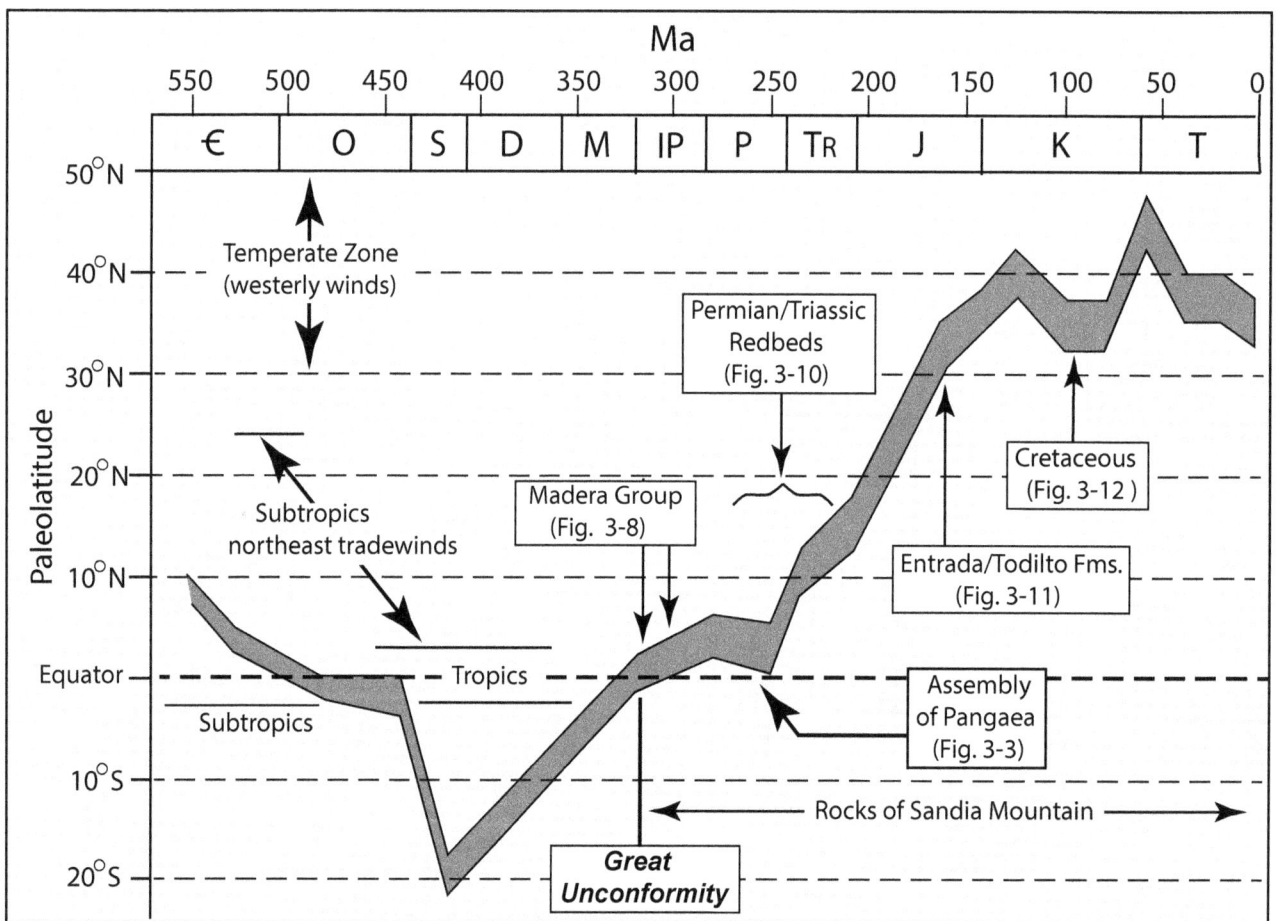

B. Plot of Paleolatitude vs. Time (for New Mexico's Four Corners Area (Modified from Dickenson 1989)

Figure 3-5. Supercontinent of *Pangaea* and New Mexico

Let's now focus on a critical, southern extremity of Pangaea—what would later separate and become the continent of Antarctica (*Figure 3-5A*). During this time this segment was dancing around the South Pole, and ca. 300 Ma it actually centered on the pole, allowing a huge continental ice sheet to accumulate there (*Figures 3-6* and *3-7*). Once the ice sheet was firmly in place, astronomical variations in the shape of Earth's orbit caused variation in the intensity of sunlight, and as a result, the volume of the ice sheet to wax and wane. This in turn caused the global sea level to rise and fall, greatly controlling the depositional character of the Madera Group in far-away New Mexico (*Figure 3-8*).

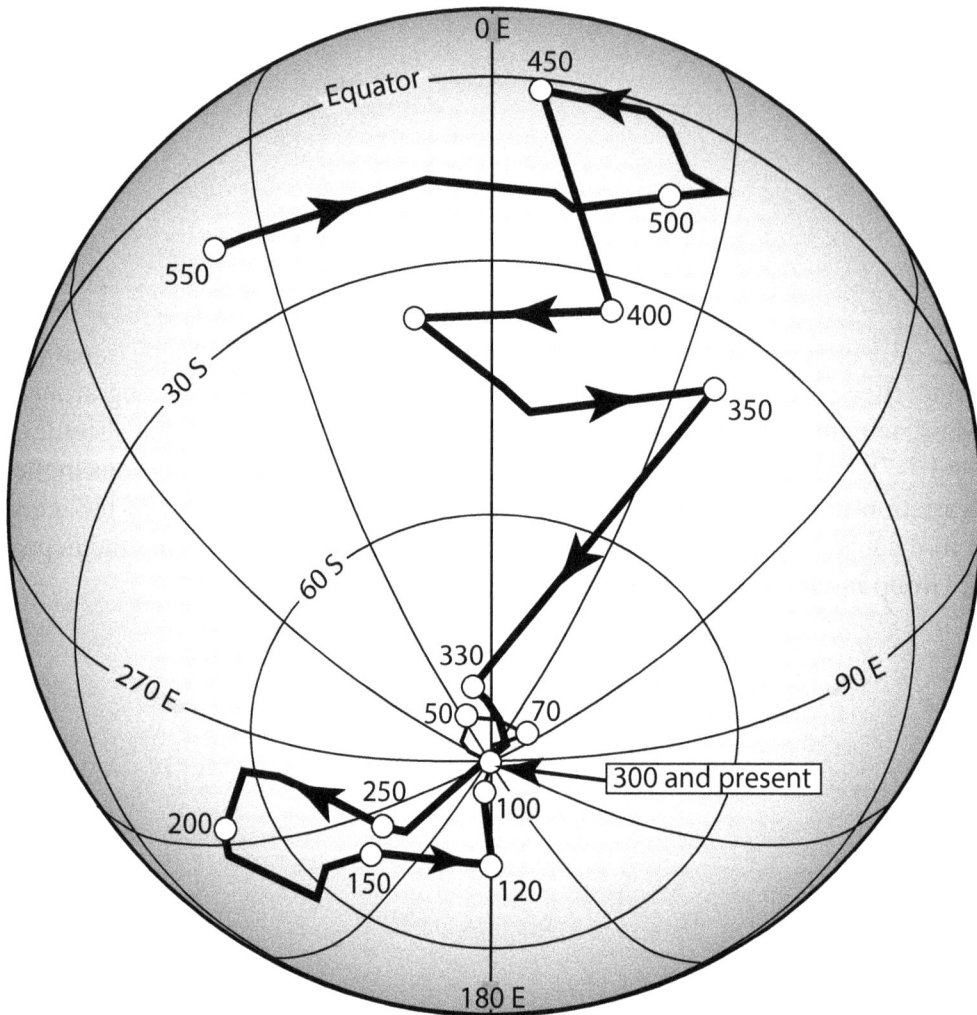

A. Migration Path of East Antarctica Through Time (in Ma)

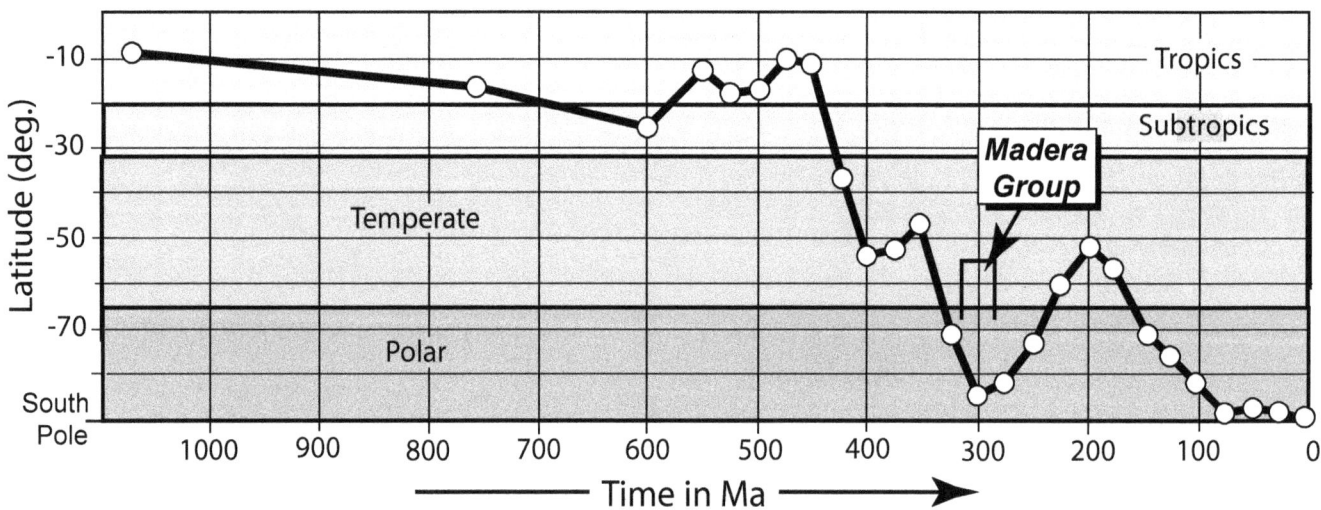

B. Migration Path of East Antarctica Through Latitude

Figure 3-6. Link between the Madera Group, Glaciation, and Migration of East Antarctica
(Modified from Torsvik et al., 2008)

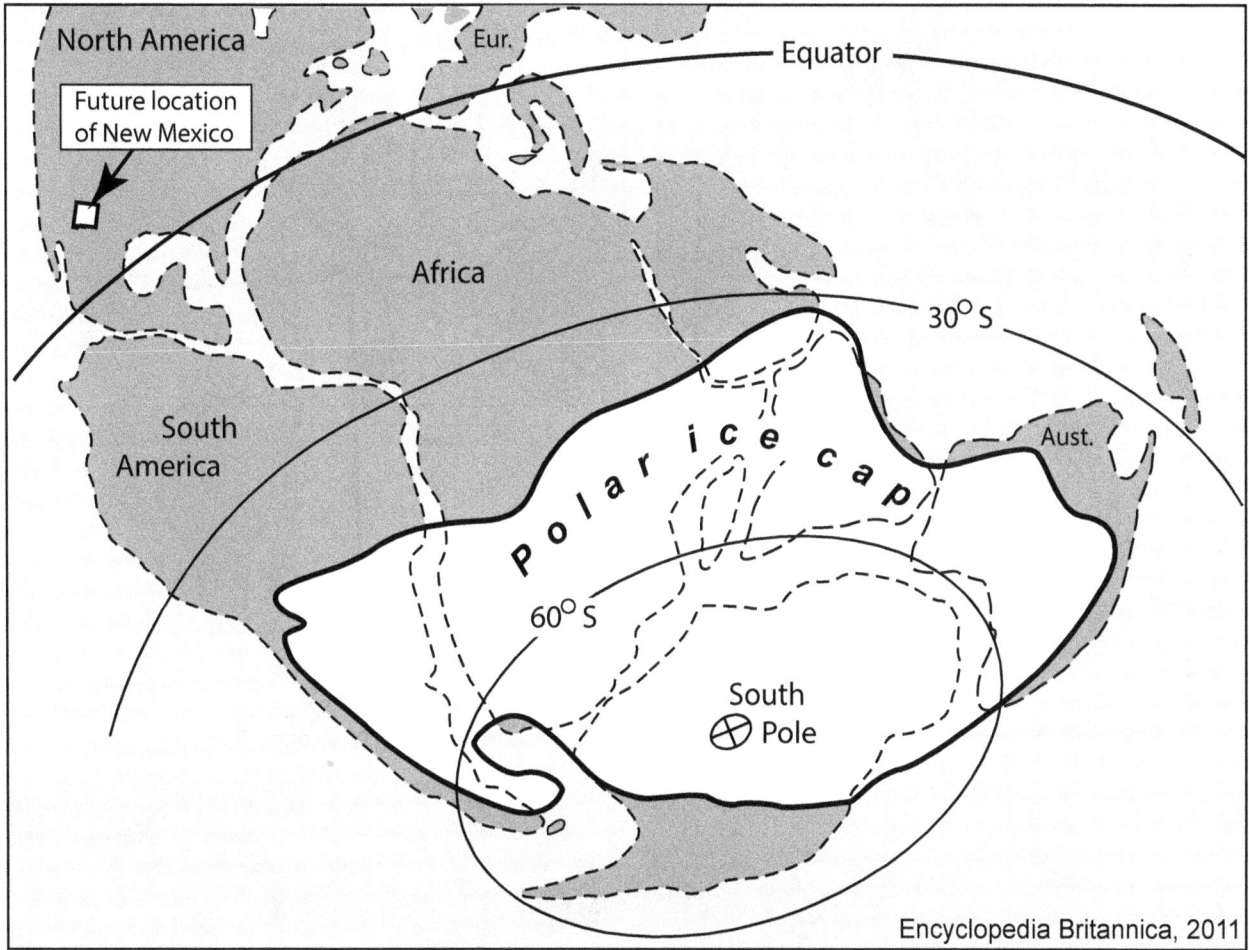

Figure 3-7. Global Paleogeography During Time of Madera Group (ca. 300 Ma)

125

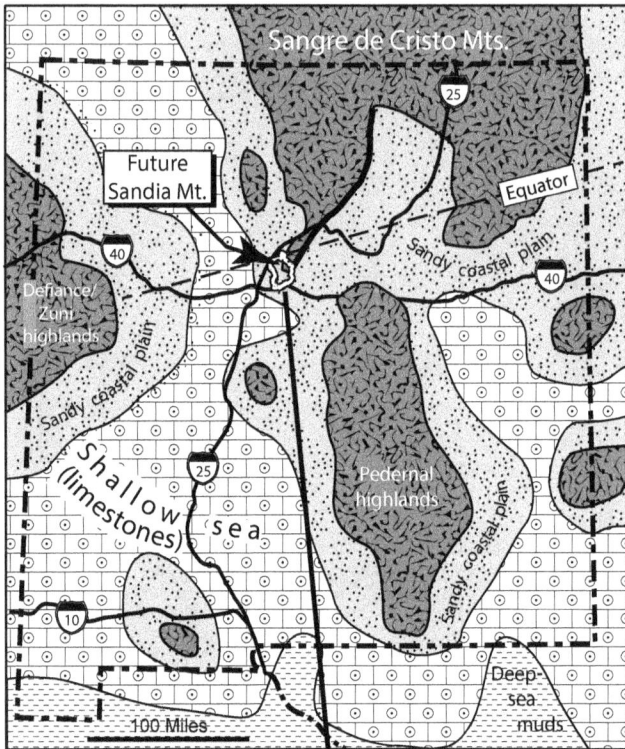

A. Relatively Low Sea-Level

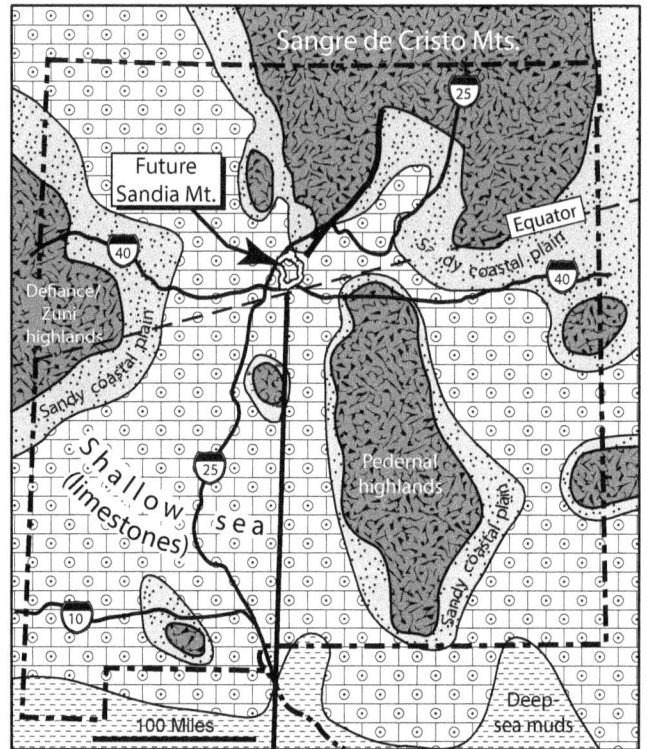

B. Relatively High Sea-Level (Modified from Kues 2004)

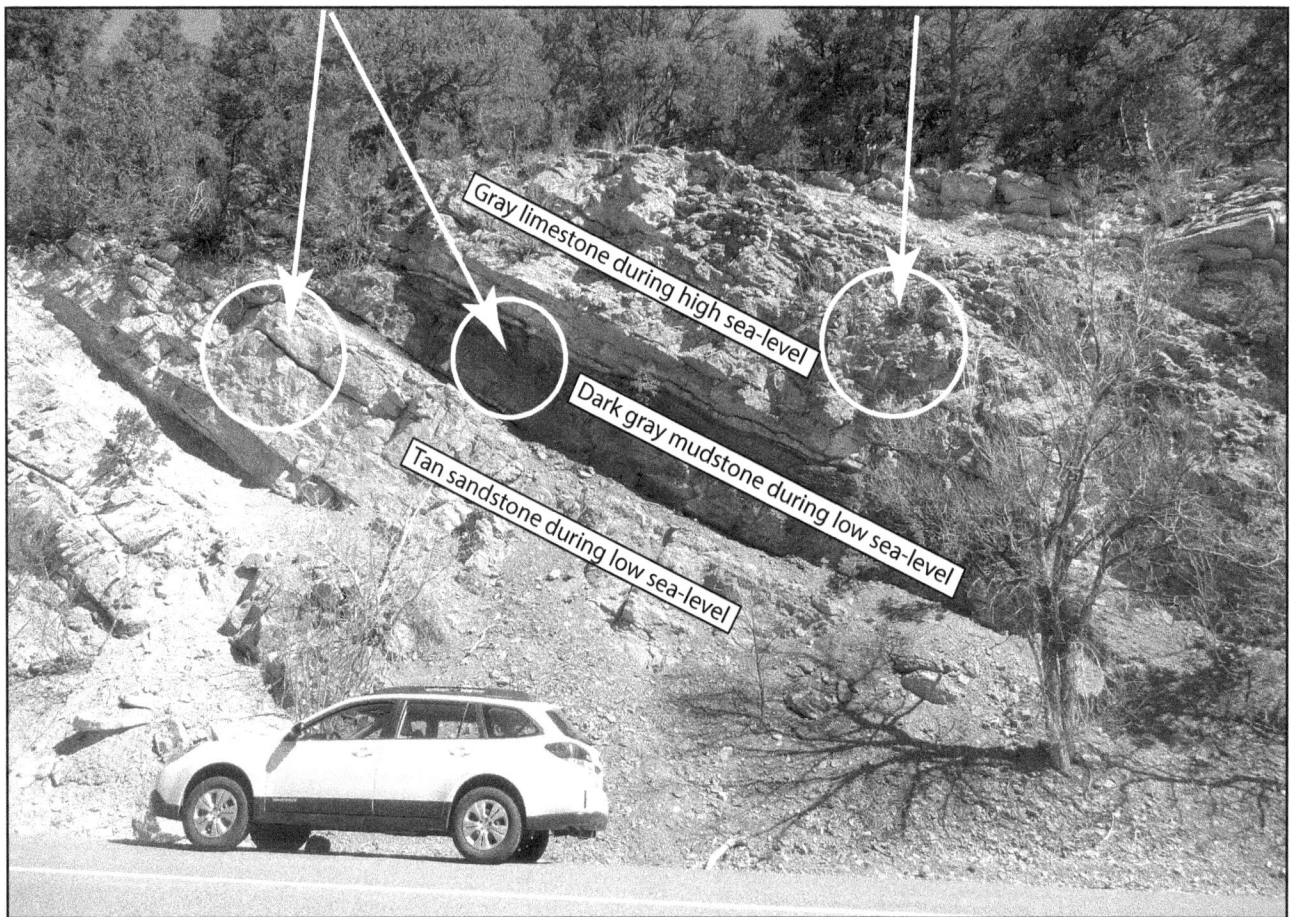

C. Sandia Crest Byway (NM-536), Mile 2.9, Madera Group Showing Influence of Changing Sea-Level on Rocks
(Photograph by Author)

Figure 3-8. Paleogeography-I: Madera Group (ca. 310-300 Ma)

(Select Modern Geography in "A" and "B" Added for Reference)

Breakup of Pangaea

Like Rodinia before it, the massive Pangaean supercontinent behaved like a lid on a kettle of boiling water. Eventually, ca. 175 Ma, Pangaea cracked open and the pieces separated and formed the continents that we recognize today. At this point we're ready to resume with the second of the four phases of burial.

Phase 2: Permian and Triassic Redbeds (290–205 Ma). From ca. 290 Ma the area of the future New Mexico was dry land and characterized by a flood of *redbeds*, i.e., reddish sands and muds deposited by rivers. Where did all the sand and mud come from, and more importantly, why are they dominantly reddish in color? The source of all that sediment was the eroding Ancestral Rockies (*Figure 3-9A*). The answer to the second question—why is all this sediment reddish—is more complicated, and much more interesting.

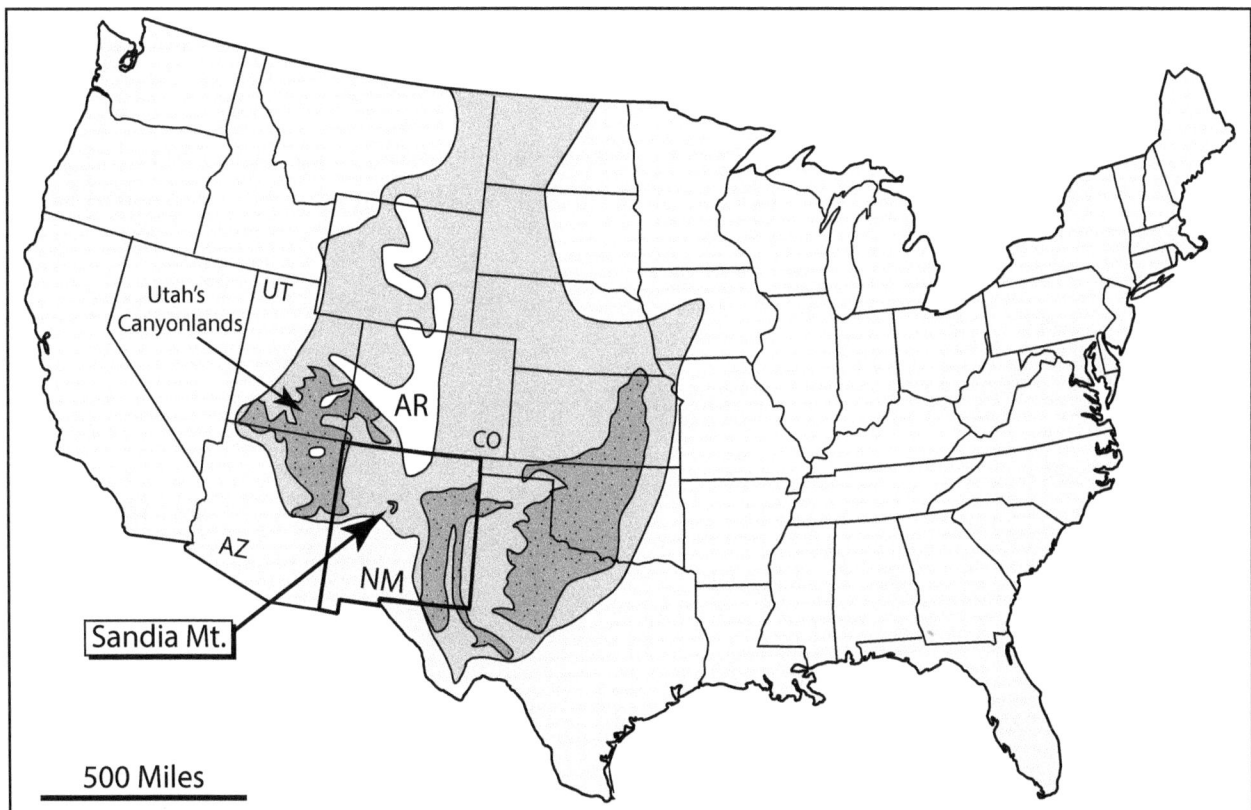

Figure 3-9. Generalized Extent of Permian and Triassic Redbeds of Western North America
(Light Gray = Original Extent, Dark Gray = Present Extent; AR = Ancestral Rockies)

Redbeds. The sudden appearance of reddish sediments in the geologic record begs an explanation. As mentioned above, by ca. 250 Ma the world's continents had finished crashing together and had completely merged into one gigantic supercontinent called *Pangaea* that stretched from pole to pole (*Figure 3-5A*). But even before the final amalgamation, by ca. 280 Ma, the region of today's West had become part of a vast continental interior located near the equator and far from the coasts (*Figure 3-5B*). The climate was hot and arid to semi-arid, and therefore oxidizing ("rusting"). A huge volume of sediments was shed from the uplifted Ancestral Rockies (AR in *Figure 3-9*) and deposited as a vast apron. The sediments, especially the muddy ones, turned reddish, perhaps tens of millions of years *after* they were deposited due to mixing with oxidizing ground-water. Finely diseminated iron held within the clayey films coating the individual sand grains became oxidized to the red mineral *hematite*, converting the sediments to "redbeds."

The first redbed unit deposited in the Sandia Mountain area was the Permian-age Abo Formation (*Figure 3-10A*). It is best observed at one convenient place along the lower reach (mile 0.9 to 1) of the Sandia Crest Byway (more in *Chapter 12*).

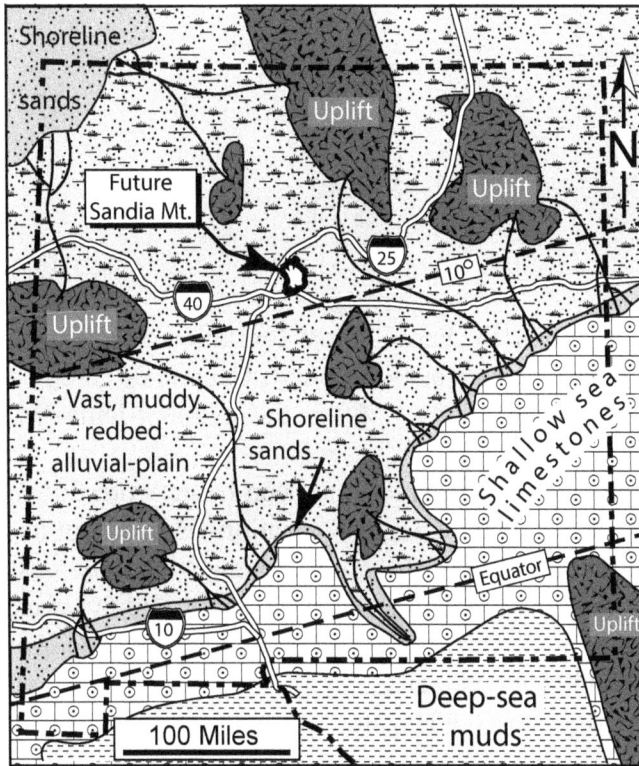

A. Permian, Abo Formation, ca. 280 Ma
(Modified from Kues 2004)

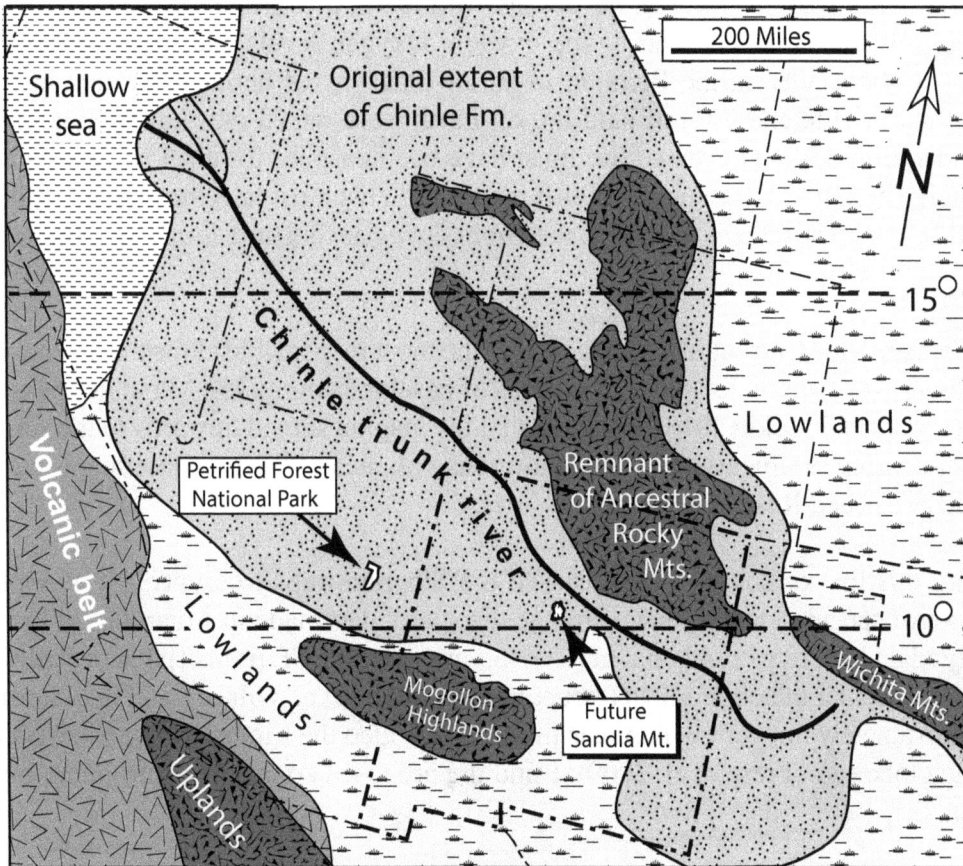

B. Regional Late Triassic Paleogeography, Chinle Formation, ca. 230 Ma
(Showing Paleo-Latitude; Modified from Baldridge 2004)

Figure 3-10. Paleogeography - II: Permian and Triassic (ca. 280-230 Ma)
(Select Modern Geography Added for Reference)

129

To the west of Sandia Mountain, the Permian saw formation of extensive wind-blown sand seas, or *ergs*, that have been stained red, such as in Arizona's Canyon de Chelly and in New Mexico's "Red Rock" area on Jémez Pueblo land along NM-4 in the southern Jémez Mountains.

By late in the Triassic Period the Ancestral Rockies had been being worn down to low hills. Their eroded debris—the brick-red muds and sands of the Chinle Formation—spread out as a vast sheet on top of the Permian (*Figure 3-10B*). (Note that the area of the future Sandia Mountain was then at a latitude of about 10° N, or about that of today's Panama in Central America). The Chinle Formation is the unit that hosts the fabulous petrified wood of Arizona's Petrified Forest National Park. The group also forms much of the lower, reddish slopes west of the village of Placitas, a part of that enigmatic outlier in the Juan Tabó area, and exposures far afield to the northeast in the Hagan basin.

Climate change. By early in the Triassic, ca. 250-245 Ma, the climate crashed. Atmospheric levels of carbon dioxide soared upward, the oceans warmed and turned mildly acidic, killing off the microorganisms that produce most of the oxygen. Oxygen levels had dropped from an all-time maximum of about 30% in the Permian to an all-time low of about 12% in a span of only 20 Ma. About 96% of marine life and 70% of land life on Earth were wiped out. Land animal habitats were restricted to very low altitudes, and even so life must have been gasping for air. Living with 12% oxygen is like trying to catch a breath today atop a 14,000-ft mountain. This must have been a miserable time to be alive on the planet! In the great scheme of things, the *Great Dying* as this time is called was a very close call for life on Earth.

What caused this catastrophe? Current ideas suggest that the most responsible culprit is an enormous, short-lived (< 1 Ma) outpouring of about 720,000 cubic miles of basaltic lavas at 252 Ma in northeastern Siberia, the so-called "Siberian traps" (Oskin 2013). Diabolically, the magma surged upwards to the surface through limestones and coal-bearing sediments, releasing enormous volumes of the greenhouse gas carbon dioxide into the atmosphere (Ross and Ross, undated).

Delving deeper into this line of enquiry, what caused the Siberian traps to erupt? We now place our toe daintily into the pool of speculation. In 2008 a strange discovery was made in Antarctica, a location almost (but not quite) antipodal to the traps. An aerial gravity survey (the GRACE project) discovered what appears to be a 300-mile-wide circular meteoric impact structure below the ice of Wilkes Land in eastern Antarctica. The hypothesis, unfortunately not yet verifiable, is that this provided the trigger for the Siberian traps on the opposite side of the globe (von Frese et al. 2009). I find the proposed connection intriguing, but the jury is still very much out.

Phase 3: Jurassic Entrada, Todilto, and Morrison Formations (165-145 Ma). After a 40-Ma time gap the area of the future New Mexico once again began to accumulate non-marine sediment. This triplet of formations is responsible for some of northern New Mexico's more interesting scenery—particularly in the so-called "Georgia O'Keeffe country" northwest of Abiquiu, 75 miles to the north (*Figure 3-11*).

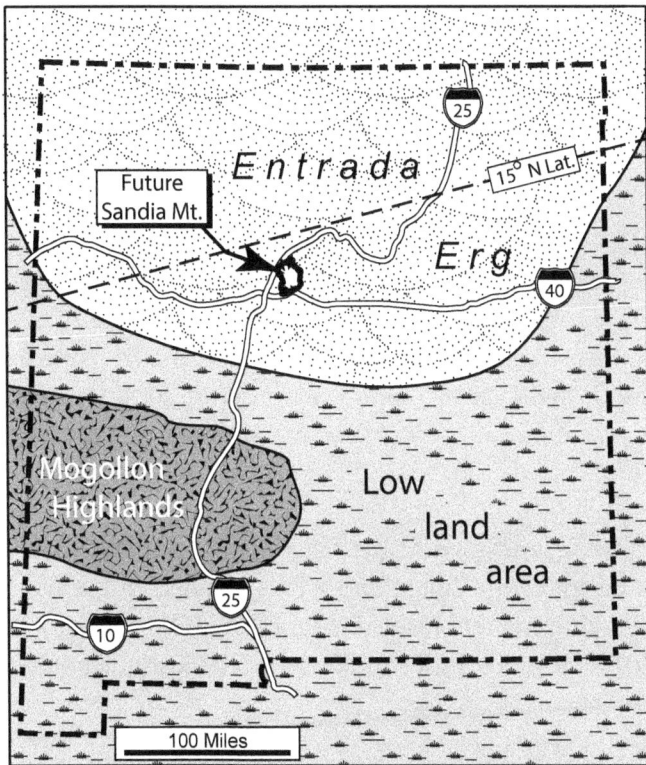

A. Entrada Sand Sea ("Erg"), ca. 165 Ma
(Modified from Lucas and Anderson 1997)

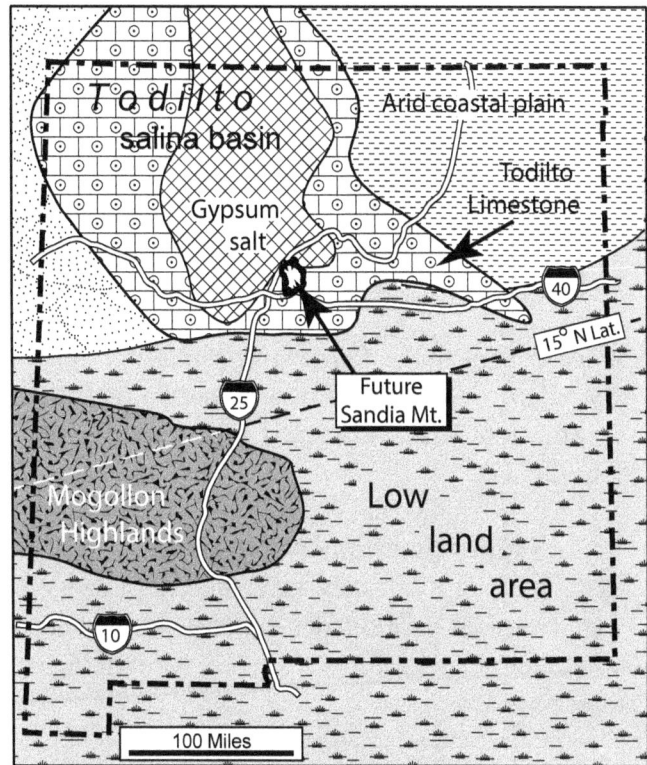

B.. Todilto Formation, ca. 160 Ma
(Modified from Lucas and Heckert 2003; Lucas 2004)

C. North View of Entrada Formation and Limestone Member of Todilto Formation
(The "Alcove" Area West of Placitas; Photographs by Author)

Figure 3-11. Paleogeography-III: Middle Jurassic (ca. 165-160 Ma
(Select Modern Geography in Maps for Reference)

The lower unit is the *Entrada Formation*, a massive, bulky light-colored sandstone that usually forms a cliffy outcrop. Above the Entrada are two units that together make up the *Todilto Formation*. First is a thin, gray limestone unit, succeeded by a massive, resistive, whitish unit of *gypsum* (a calcium sulphate salt, $CaSO_4$-$2H_2O$).

Todilto limestone and gypsum members. The thin limestone at the base of the Todilto is truly unique. It is finely laminated to platey, crenulated into little folds, and when crushed in the hand releases a fetid smell because it's loaded with organic matter (inset in *Figure 3-11C*). This unit is the "Rosetta Stone," critical for making sense of a strange outlier in the Juan Tabó area (more in *Chapter 6*).

An excellent display of these units is located 25 miles northwest of Sandia Mountain, just west of the village of San Ysidro at the appropriately-named White Mesa. Here is a 400-ft north-facing cliff of the Entrada and the two Todilto members, with the gypsum on top. The brilliant whiteness of the gypsum is a result of the ongoing quarrying operation by Zia Pueblo. The product is trucked to the American Gypsum Company's plants in Bernalillo and Albuquerque for the manufacture of wallboard.

At Sandia Mountain the best exposure of this triplet is on the north end, west of Placitas in an area I call the "Alcove" (*Chapter 8*). At the Juan Tabó area to the south only the limestone member is present (*Chapter 10*). On the East Mountain, behind the Mountain Christian Church along NM-14 is an abandoned Todilto gypsum mine (in the early days, gypsum was used to white the walls of houses).

I will say little about the two formations lying above the Todilto because they provide few notable exposures at Sandia Mountain. One of these, the Morrison, is extremely important elsewhere due to its important uranium deposits and its famous dinosaur fossils.

Phase 4: Cretaceous (100-75 Ma). During this time (from late Jurassic to the late Cretaceous) a new range of mountains, what would become the modern *Rocky Mountains*, began to progressively rise to the west and southwest. The debris from those mountains washed down to the east and northeast via a number of pulses to form a vast alluvial plain ending at the edge of a shallow sea (*Figure 3-12A*).

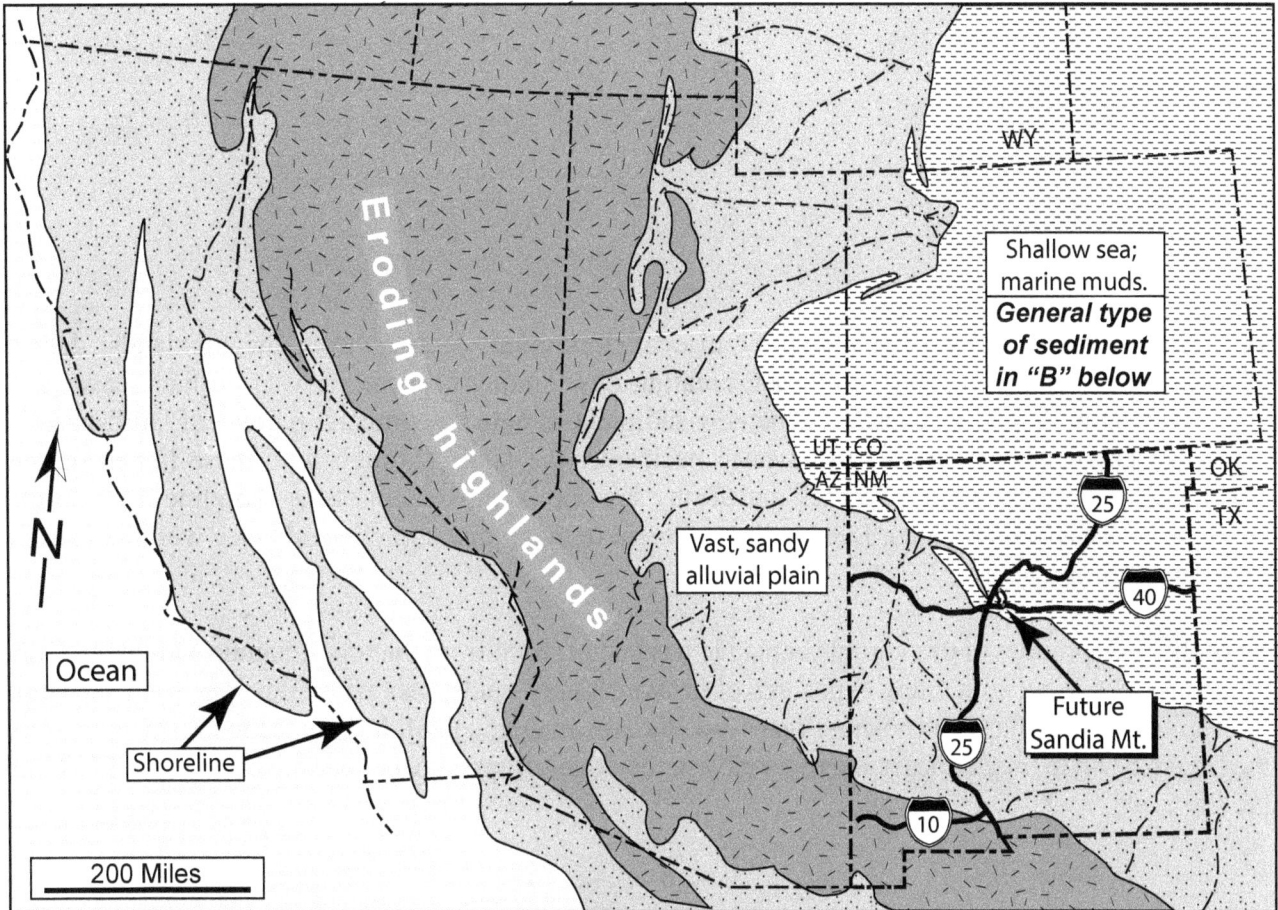

A. Paleogeographic Map, ca. 75 Ma (Select Modern Geography Added for Reference)

B. North View of Typical Cretaceous Mudstone (Mancos Shale, *Km*; NM-165 Mile 3.9 West of Placitas)

Figure 3-12. Paleogeography-IV: Late Cretaceous (ca. 75 Ma)

Depositional Pulses. Each time a pulse of deposition, sourced from the highlands, delivered sediment to the east and northeast, creating new land, the shoreline was pushed seaward. The gravitational loading caused by the new sediment initiated a companion pulse of land subsidence that enabled the shoreline to advance back over the land. Through time these linked pulses of deposition and subsidence caused a back-and-forth "saw-tooth" movement of the shoreline from northeast to southwest across New Mexico. Eventually, sediment delivery and shoreline advance won the battle and filled in the shallow sea.

Most of the rock product of this complex episode has been eroded from the top of Sandia Mountain. Only in the Placitas area do we find a few exposures of brownish gray mudstones with a few enclosed sandstones, representing offshore marine deposits (*Figure 3-12B*). However, much of the Cretaceous has been preserved in the Hagan basin to the northeast and especially in the down-dropped Albuquerque basin in the Rio Grande rift to the west.

~~~

In sum, over an enormous time-span, from ca. 315 to 75 Ma, a series of depositional pulses sequentially buried the ancient basements rocks with an accumulation of nearly 14,000 ft of sedimentary cover. These discrete units, today tilted and partially exposed at the surface, create a tapestry of variable colors and erosional characteristics of the Sandia Mountain landscape (*Figure 3-13*).

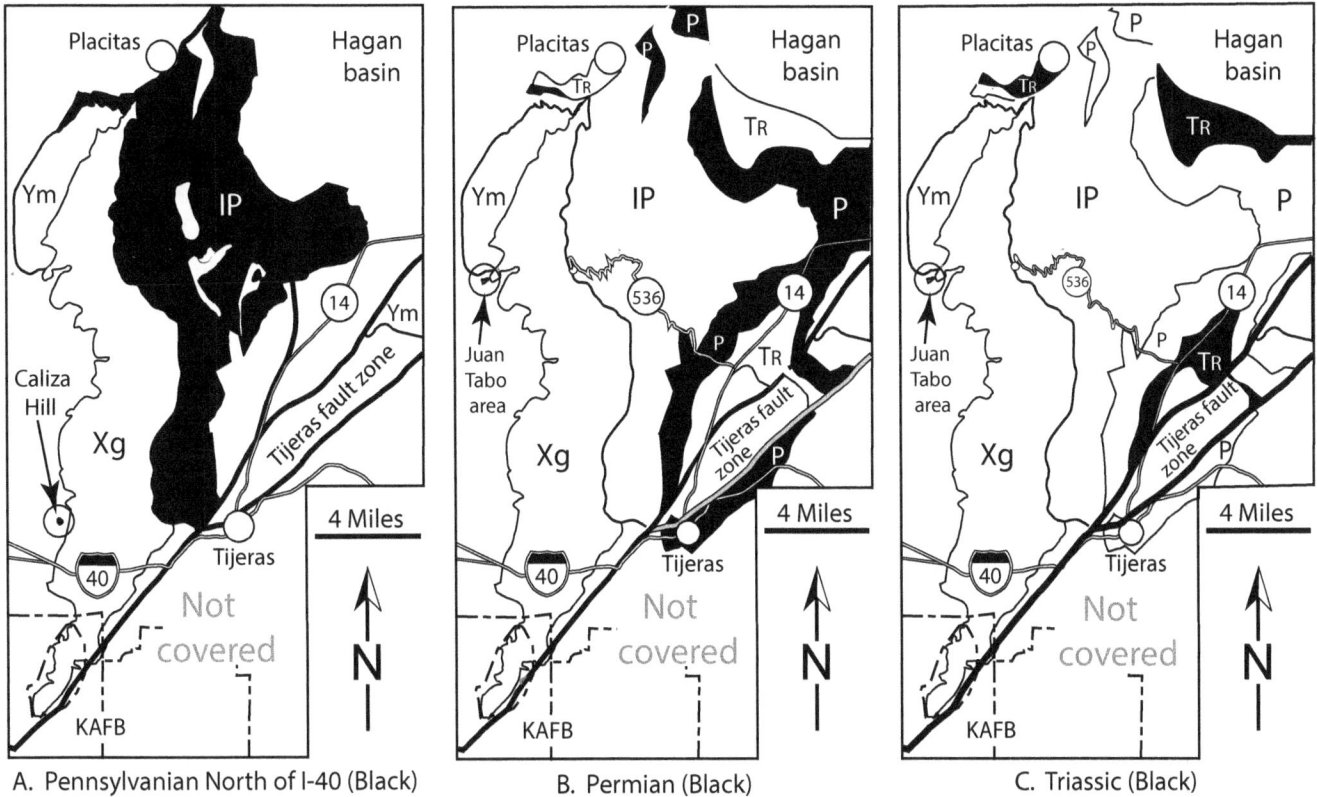

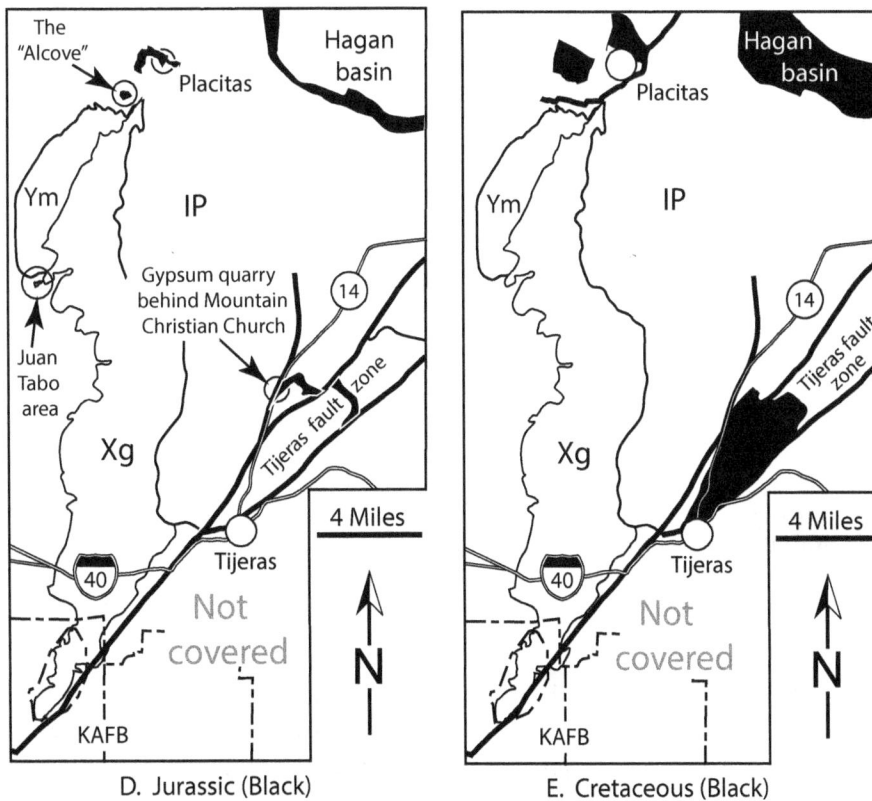

Figure 3-13. Distrubution of Rocks by Age on Sandia Mountain
(Karlstrom et al. 1999; Read et al. 1999)

4:
LARAMIDE COMPRESSION
AND THE ROCKY MOUNTAINS

Synopsis

The mountain-building event called the Laramide, from ca. 75 to 40 Ma, caused by North America's relentless, westward advance across the globe, forever changed the western half of the continent. Low-angle subduction of the oceanic Pacific (Farallon) Plate under the western edge of North America projected compressive force far into the western continental interior. This rippled the overlying basement rock, forcing large blocks to "pop up," forming what are called "Laramide type" mountains. Surprisingly, the area that would become Sandia Mountain, particularly the part anchored by the Sandia Granite, was only partly involved in the event. East-northeast, followed by northeast compression pushed the northern portion (near present Placitas), upward and probably slightly to the north. The eastern side of the Sandia uplift relieved the stress by an array of minor, generally north-trending up-to-the-west reverse faults. The Tijeras fault, an old northeast-trending zone of weakness, experienced some right-lateral wrenching. Later, some of these Laramide structural "wounds" would be reactivated and reversed in their direction of movement when the Rio Grande rift developed.

~~~

Introduction

At this point we must broaden our context in order to appreciate the next stage of development, namely the *Laramide* event that produced the modern Rocky Mountains.

Just What is the "Laramide?"

The name comes from Jacques LaRamie (1784-1821), a French-Canadian fur trapper. In Wyoming his name has been used for the city of Laramie, Fort Laramie, the Laramie River, and the Laramie Mountains. The regional compressive event that created the Rocky Mountains was first recognized in the Laramie Mountains and thus given the name.

Sometimes I suggest that Sandia Mountain is not really part of the Rocky Mountains, but this is splitting hairs. The mountain differs in *origin* from "normal" Rocky Mountains, but is indeed *surrounded* by them. The Laramide compressive event certainly did jostle what became the Sandia Mountain, but it didn't *create* the mountain itself. That happened much later (*Chapter 6*).

As a prelude to the origin of the Rocky Mountains it is instructive to make a detour here to focus on the two fundamental sources of energy that provided the engine for the overarching paradigm of geology, namely *plate tectonics*. We will revisit some of this in *Chapter 15*, but to understand the Laramide we need to introduce these issues now.

The two energies. The energies that govern large-scale events in geology are: 1) the potential energy of gravity, and 2) the molecular, kinetic energy of heat. Together these two, in concert with the force of friction, result in the motion of crustal blocks, and thus mountain formation.

For gravity, consider a heavy, rigid slab of material perched atop a surface of another, hot, slippery, plastic material (top of *Figure 4-1A*). If somehow this plastic surface were uplifted on one side, the semi-rigid slab would slide laterally, downslope on the slick surface due to gravity's pull (center of *Figure 4-1A*). Take it another step and suppose the plastic surface were bowed up in the middle, below the rigid block, forcing the block to break in half and allowing the two halves to slide outward from the bulge (bottom of *Figure 4-1A*). Gravity causes motion.

① Surface — Cool, rigid block — Surface

Hot, soft plastic

② Gravity-driven movement

Uplift

Hot, soft plastic

③ Tension and fracture

Gravity-driven movement — Gravity-driven movement

D r a g — D r a g

U p l i f t

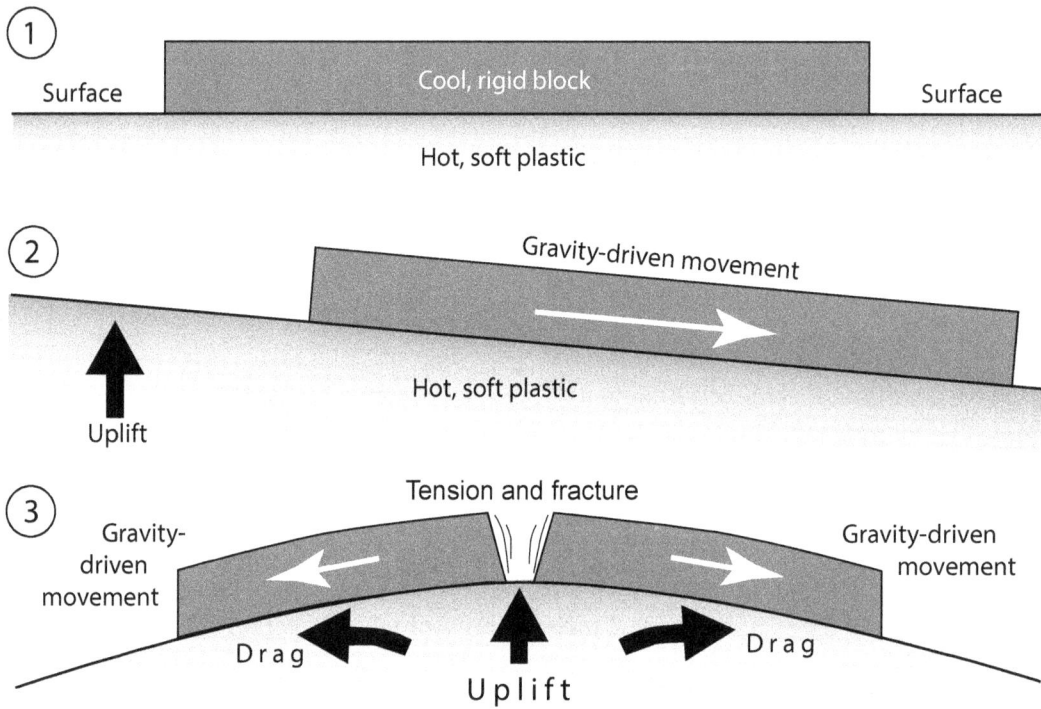

A. Gravity Driving Plate Movement

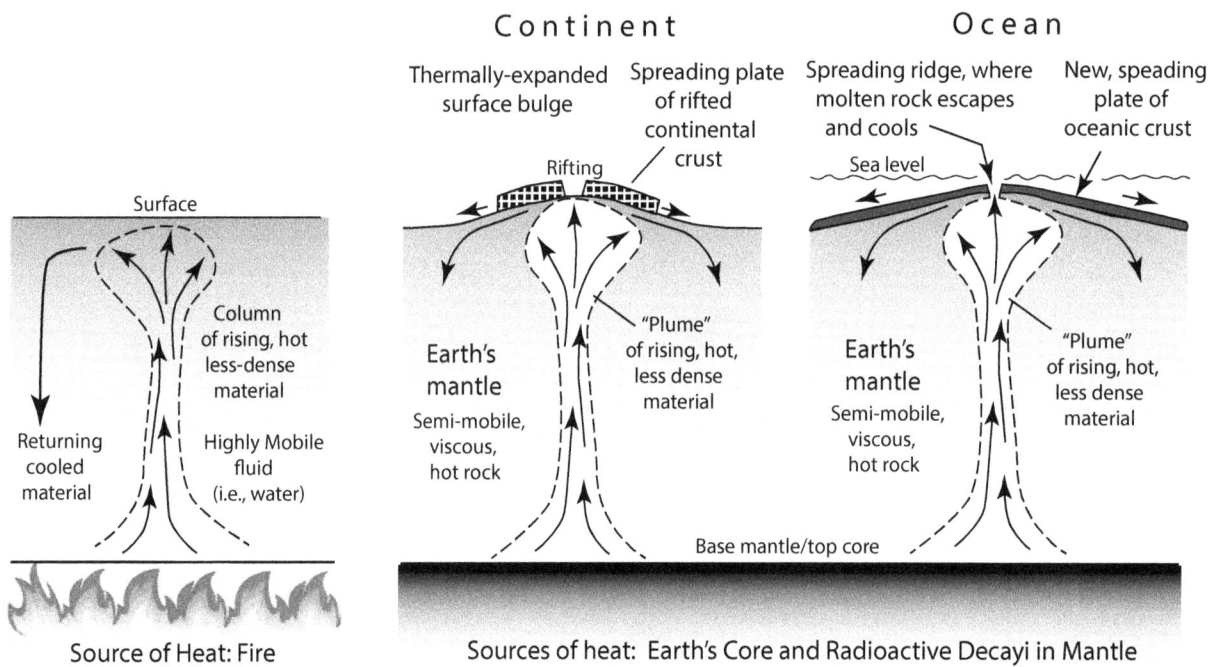

Continent

Thermally-expanded surface bulge

Spreading plate of rifted continental crust

Rifting

Earth's mantle

Semi-mobile, viscous, hot rock

"Plume" of rising, hot, less dense material

Ocean

Spreading ridge, where molten rock escapes and cools

New, speading plate of oceanic crust

Sea level

Earth's mantle

Semi-mobile, viscous, hot rock

"Plume" of rising, hot, less dense material

Base mantle/top core

Surface

Column of rising, hot less-dense material

Returning cooled material

Highly Mobile fluid (i.e., water)

Source of Heat: Fire

Sources of heat: Earth's Core and Radioactive Decayi in Mantle

B. Heat Producing Rising Columns in Mobile Fluid (e.g., Water; Vertical Scale = Inches)

C. Earth's Internal Heat Producing Rising "Plumes" in Viscous Medium (Earth's Mantle) in Ocean (Left) Vs. on Continent (Right) (Vertical Scale = 100s of Miles)

Figure 4-1. The Two Energy Sources that Move Rock Masses: Gravity and Heat

Heat causes rock (and much else) to expand. The interior of Earth is constantly losing heat from the interior moving upward to the surface by *convection*. This process contrasts with the much less efficient *conduction* (illustrated by an iron poker stuck in a fire that slowly becomes too hot to hold). Convection is the physical movement of hot material from a hot place to a cooler one. Think of a mobile fluid such as water in a pot with a source of heat at its base (left of *Figure 4-1B*). The water closest to the source of heat becomes hottest, expands as it becomes less dense, and rises in columns or "plumes" to the surface, a cooler region of lower density. As the rising fluid loses its heat near the surface it becomes denser and thus begins to sink.

With a more viscous fluid though, such as the hot plastic rock in Earth's mantle, plumes of hot, less dense material rise through the mantle (in super slow motion) and transport heat toward the surface (*Figure 4-1C*). As the plumes near the surface, they shed some of their heat to the surrounding rock. The resulting thermal expansion of rock near the surface causes upward bulge, or uplift, creating a gravitational "head."

On the continents (left of *Figure 4-1C*), uplifted bulges of the crust are subject to tension and tend to fracture into smaller blocks. These then can slide laterally to lower regions via gravity. In the oceans (right of *Figure 4-1C*), things are somewhat different. Without a capping, thick continental block, the hot plumes reach almost to the surface and thermally bulge the ocean floor, forming a splitting ridge. Hot, liquid rock (magma) from the plume extrudes into the ridge and out to the sides onto the ocean floor, cooling and crystallizing, continually forming the dense igneous rock called *basalt*. This new basaltic material forms two plates of separating oceanic crust. Because they are elevated, gravity pulls at them and sets them in motion in opposing directions.

To understand the Laramide effects on western North America we must keep in mind the past history—the "pre-existing conditions." When the Laramide compression began in central New Mexico, the area was already damaged goods. First there was the "grain" of the basement rocks inherited from the assembly of the continent (*Chapter 1*). Then there was the wreckage of the worn-down Ancestral Rocky Mountains with the inherited "scars" of their development (*Chapter 3*). During the Laramide these flaws were the first things to snap.

As we look back to the geologic time scale (*Figure 4-2*), but with *tectonic* events added on the right column, we see that the Laramide event followed a long, 170-Ma (245-75 Ma) period of relative stability. The onset of the Laramide event coincides with the change of the movement of the North American plate (*Figure 4-3A*). This plate (today's North America plus the western half of the Atlantic Ocean) migrated some 5500 miles across the globe between about 75 and 40 Ma, all the while colliding with and overriding the oceanic Pacific plate on its west side. The journey was not uniform, but variable in both its direction and velocity (extremely slow of course). In the following aside I touch on some of the nuts and bolts of Laramide tectonic activity.

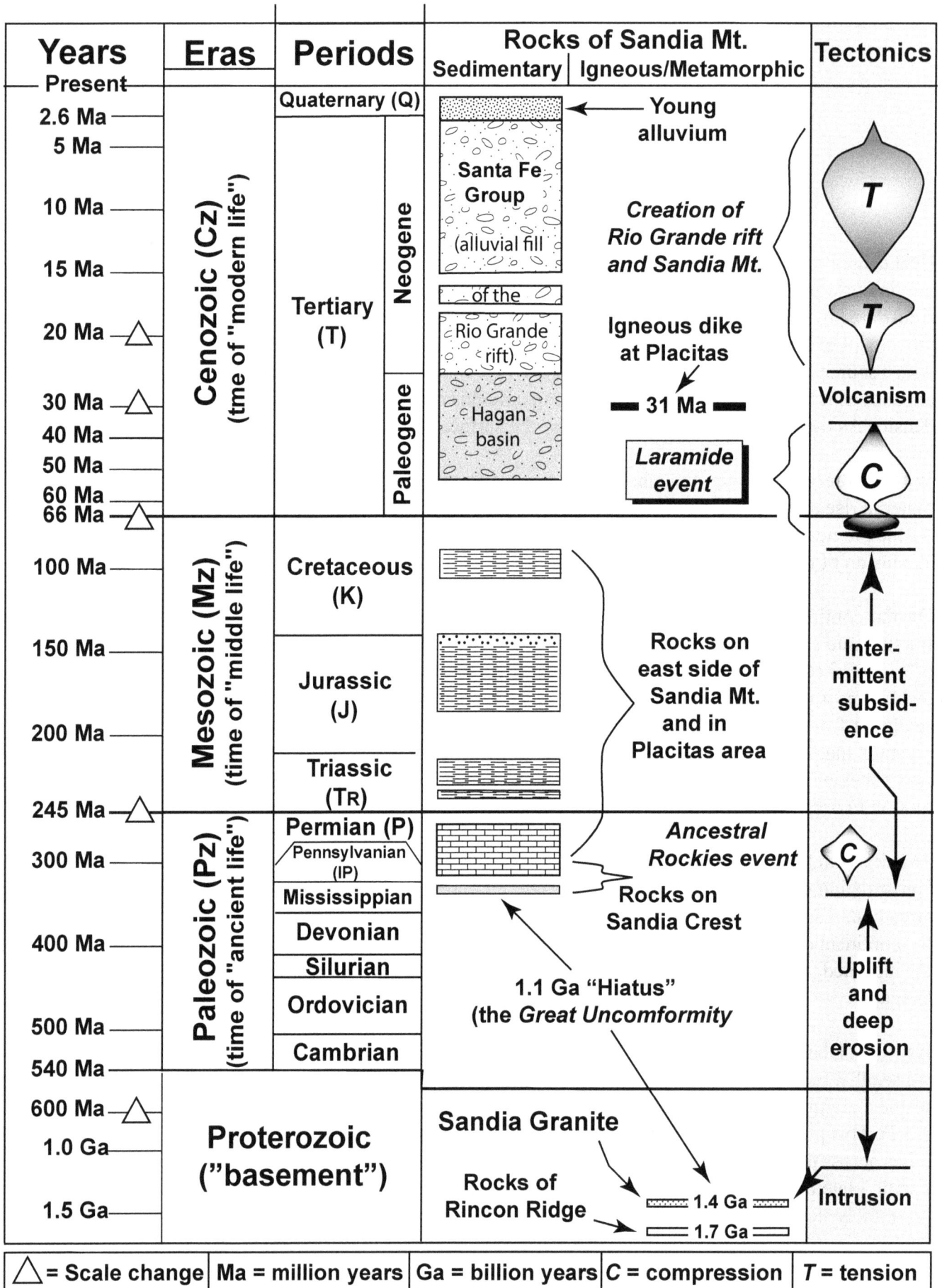

Figure 4-2. Geologic Time Scale, Rocks and Events Affecting Sandia Mountain

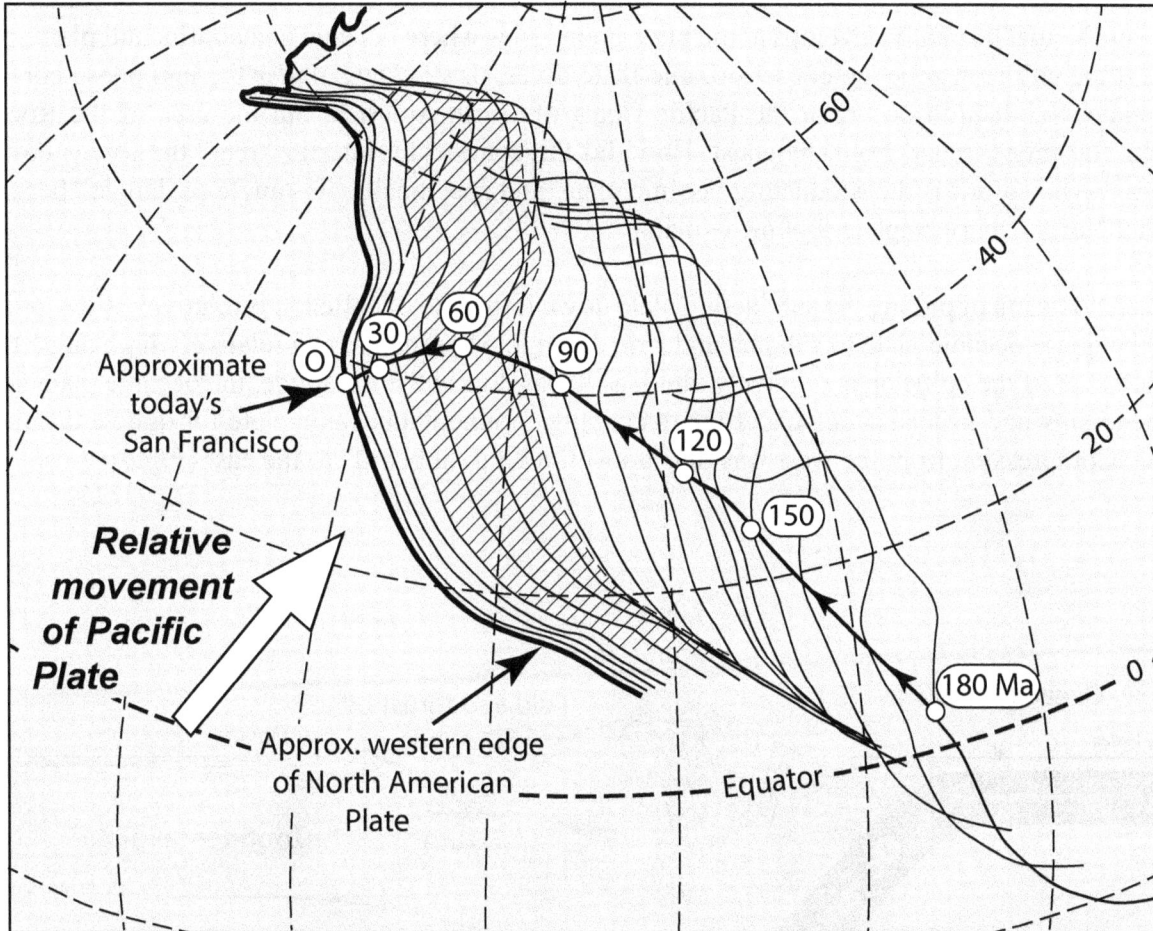

A. North American Plate's 5500-Mile Journey During Past 180 Ma
(Modified from Chapin 1987; Laramide Period of Rapid Movement, 75-40 Ma, Hatchured)

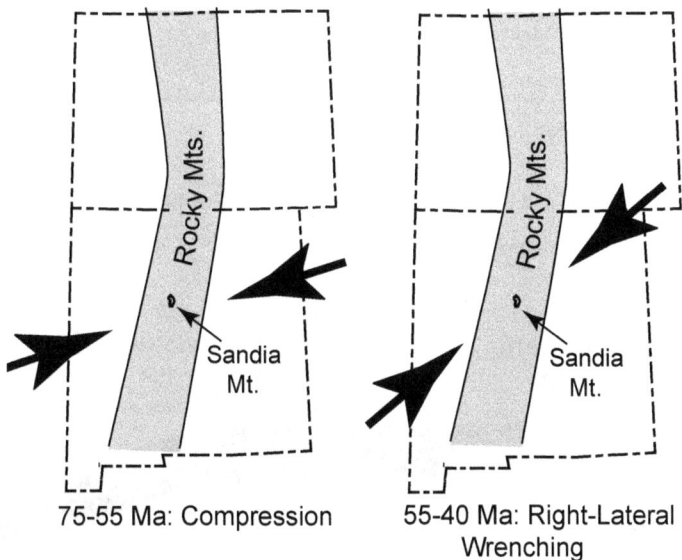

75-55 Ma: Compression

55-40 Ma: Right-Lateral Wrenching

B. Forces Affecting Southern Rocky Mountain Area
(From Chapin and Cather 1981)

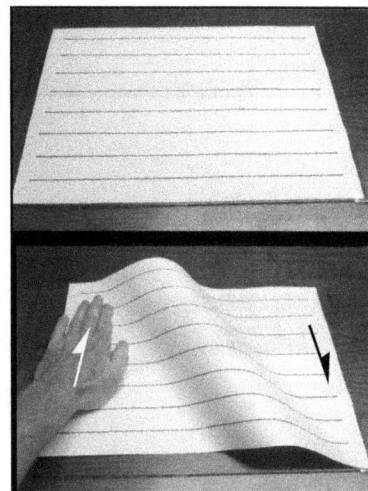

C. Example of Anticlinal Fold Developed by Right-Lateral Wrenching

Figure 4-3. Forces Affecting Western North America During Laramide

Laramide nuts and bolts. To simplify, the main action in mountain building is the compressional "push" that normally develops at the plate boundaries where oceanic and continental plates collide. At the collision zone the denser oceanic plate dives, or subducts, *under* the less dense continental plate. Subduction occurs at the Pacific Coast of North America, but the area of the Rockies is more than 500 miles inland from the coast. How did the continental interior "feel" the forces developed so far away? The culprit is the subducting oceanic plate, and specifically, its "angle of dangle" beneath the continental North American plate. Let me explain.

Gravity can do its job of pulling the subducting plate down if the rate of collision is relatively slow, providing time for the dense oceanic plate to subduct and sink down into the mantle at a relatively high angle (*Figure 4-4A*). However, if the collision is relatively rapid, as it was during the Laramide, the oceanic plate does not have time to sink and instead slides under the overriding continental oceanic plate for quite a distance at a low angle. Compression from the west was thus transmitted far "inboard" to the east (*Figure 4-5*).

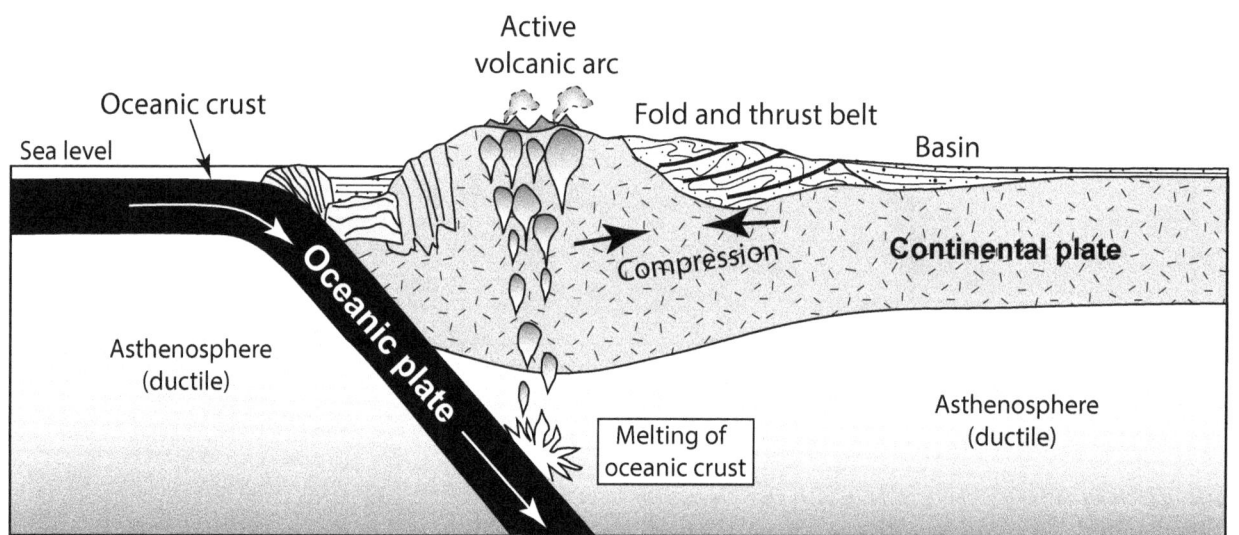

A. Relatively-Slow, High-Angle Subduction

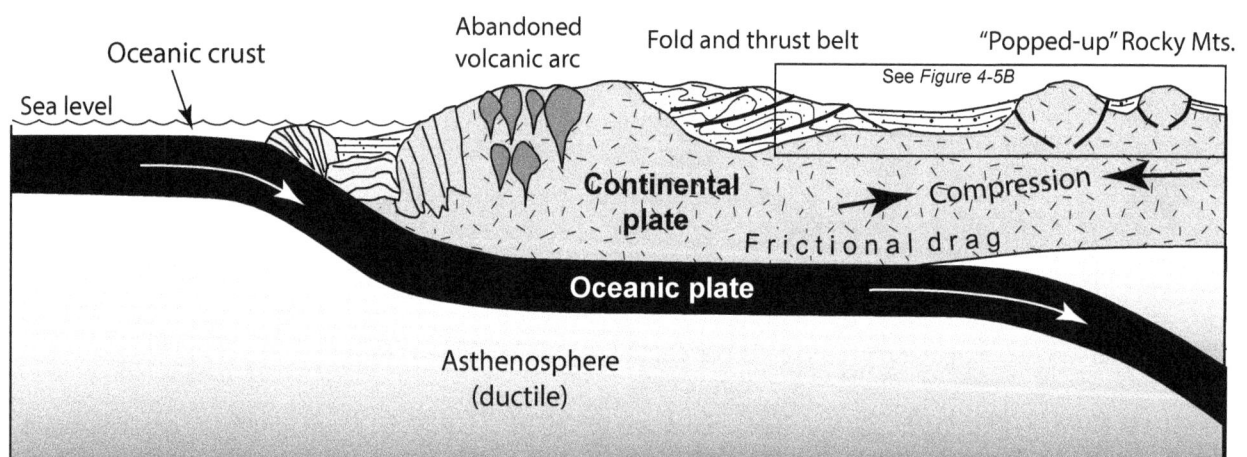

B. Relatively-Rapid, Low-Angle (Laramide-Like) Subduction

Figure 4-4. Schematic Cross Sections Showing Importance of Subduction Angle
(Modified from Meldahl 2013)

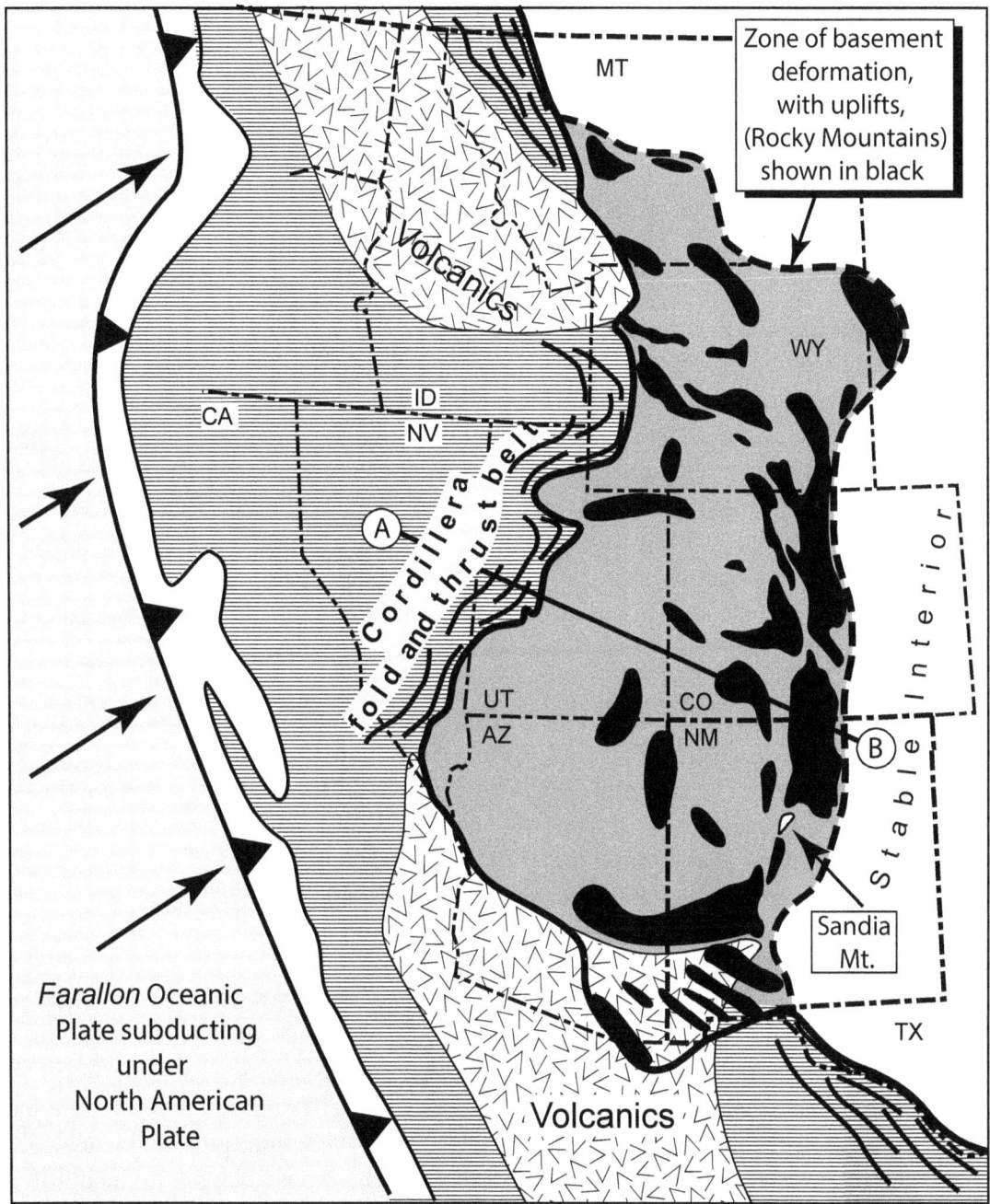

A. Schematic Map Showing Region Affected by Laramide Compression

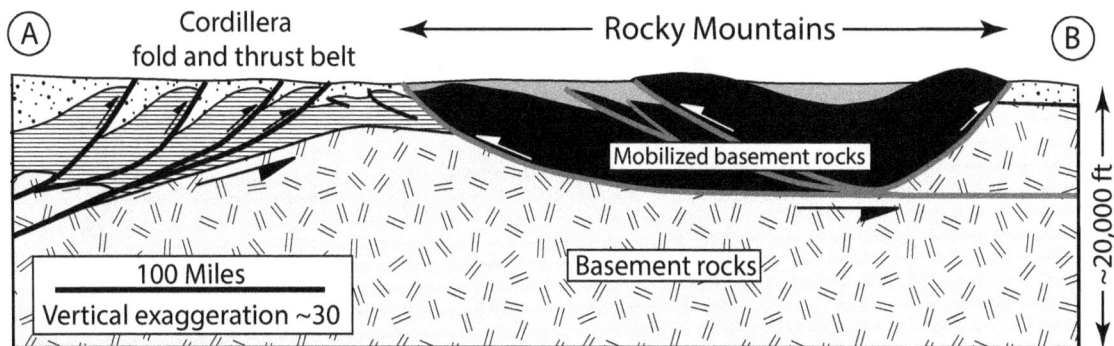

B. Schematic Cross Section A-B in "A" Above (Also See Figure 4-4B; Modified from Lowell 1983)

Figure 4-5. Deformation of Western North America During Laramide

143

Starting about 75 Ma overall, but mostly from ca. 55 to 40 Ma in New Mexico, the western half of North America was subjected to compression from the west (the time indicated by the cross-hatchured area in *Figure 4-3A*). At about 55 Ma the movement of North America changed and the compressive force shifted from east-northeast to northeast, which then encouraged right-lateral wrenching along old faults (*Figure 4-3B*). This multi-directional collision and the shift of force direction produced our complex Rocky Mountains. The individual segments of the Rockies (e.g., Sangre de Cristo Mountains) are so-called "basement-cored" uplifts, meaning that the underlying, rigid basement rocks are integral parts of the mountains. Blocks of basement rock, when subject to lateral compression, do not bend easily, so something has to give. Segments of the continental crust "popped up" along one or more faults like watermelon pits from pinching fingers (*Figure 4-4B*).

Much of New Mexico, situated in the southern part of the affected area, was caught up in this pervasive event (*Figure 4-5*). Strangely, and amidst all this, Sandia Mountain is not a Laramide *uplift*, but it was indeed shoved around at the time. The Laramide *influence* on Sandia Mountain though is another thing. The effects are underappreciated because most of them are subtle and some have since been confused by later reversal of movement. The easing of compression ca. 40 Ma introduced a period of transition leading to one of regional tension 10-15 Ma later that created our modern Sandia Mountain (*Chapter 5*).

Laramide Effects on Sandia Block

As mentioned above, the Laramide (75-40 Ma) did indeed influence what later became Sandia Mountain (*Figure 4-6*) but did not create it. An important argument against any *major* Laramide uplift is the lack of sediment washed off it and deposited somewhere nearby. Geologists have searched high and low for such material, to no avail. This sedimentary debris would be the "smoking-gun" evidence for a nearby uplift. Importantly, though, perhaps only soft Cretaceous shales were exposed on top of the uplifted regions. Erosion had less than 1000 ft of this soft stuff to work on, and this perhaps is the reason no accumulation of sediment is found (Karlstrom et al. 1999).

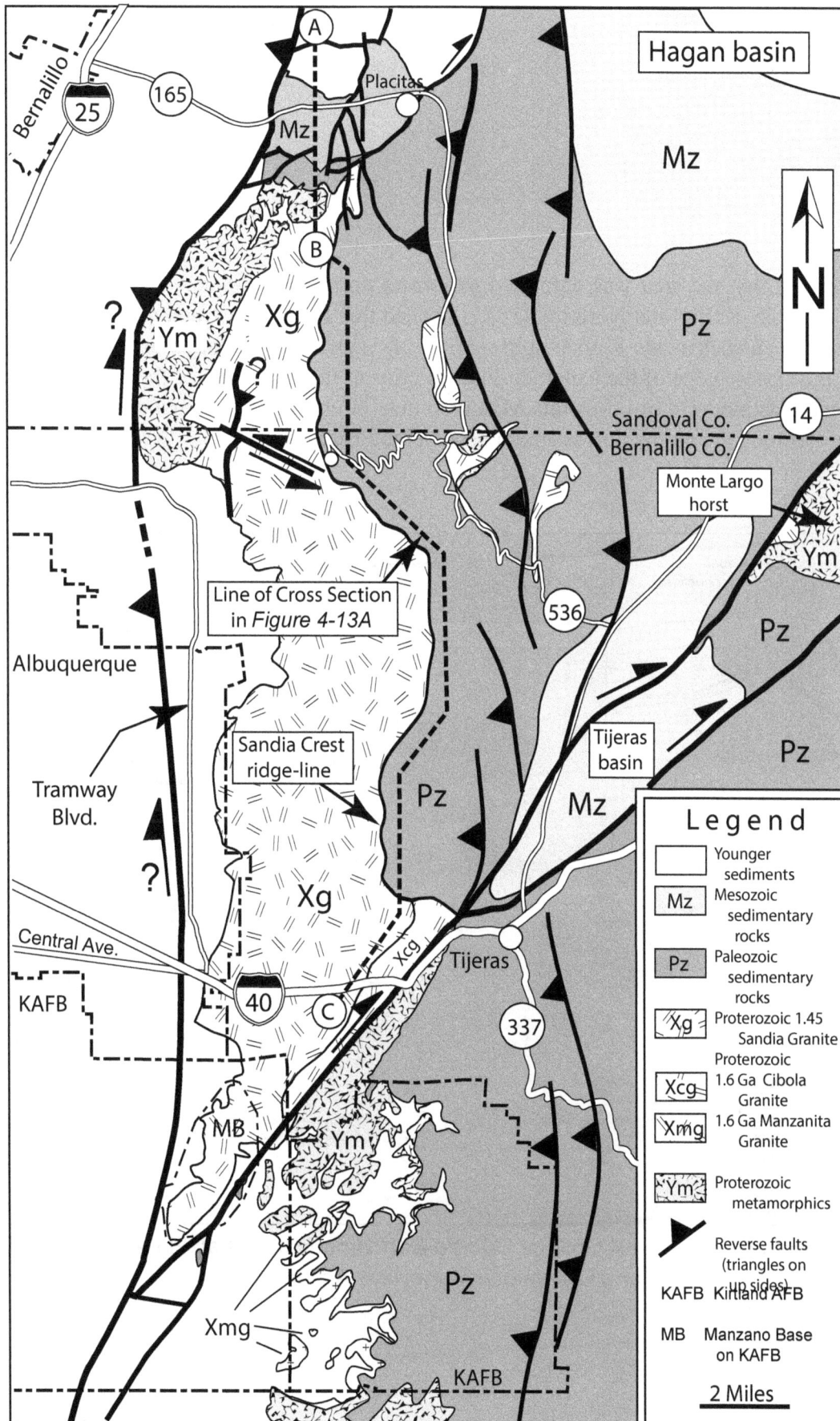

Figure 4-6. Geologic Map of Laramide Structures (Bold Lines) of Sandia Mountain
(Modified from Karlstrom et al. 1999)

Map labels:

Hagan basin

Bernalillo — 25 — 165

A — Placitas

Mz

B

? — Ym — Xg — ?

Mz

Pz

N

Sandoval Co.
Bernalillo Co. — 14

Monte Largo horst

Ym

Line of Cross Section in *Figure 4-13A*

536

Albuquerque

Pz

Pz

Sandia Crest ridge-line

Tramway Blvd.

Tijeras basin

?

Xg

Mz

Central Ave.

KAFB

40 — C — Xcg — Tijeras

337

MB — Ym

Pz

Xmg

KAFB

Legend

	Younger sediments
Mz	Mesozoic sedimentary rocks
Pz	Paleozoic sedimentary rocks
Xg	Proterozoic 1.45 Sandia Granite
Xcg	Proterozoic 1.6 Ga Cibola Granite
Xmg	1.6 Ga Manzanita Granite
Ym	Proterozoic metamorphics

Reverse faults (triangles on up sides)

KAFB Kirtland AFB

MB Manzano Base on KAFB

2 Miles

145

Some have proposed that the area was deformed to form a north-facing *monocline*, i.e., a *proto*-Sandia Mountain. For example, Kelley and Northrop (1975) likened this suspected structure to monoclinal uplifts of the Colorado Plateau such as the Kaibab uplift (*Figure 4-7*) and went as far as to suggest that this area may have been the eastern edge of the Colorado Plateau prior to the formation of the Rio Grande rift. The idea of a monocline as part of a proto-Sandia Mountain may be the key to understand the creation of the strange structure forming the mountain's northern nose in the Placitas area.

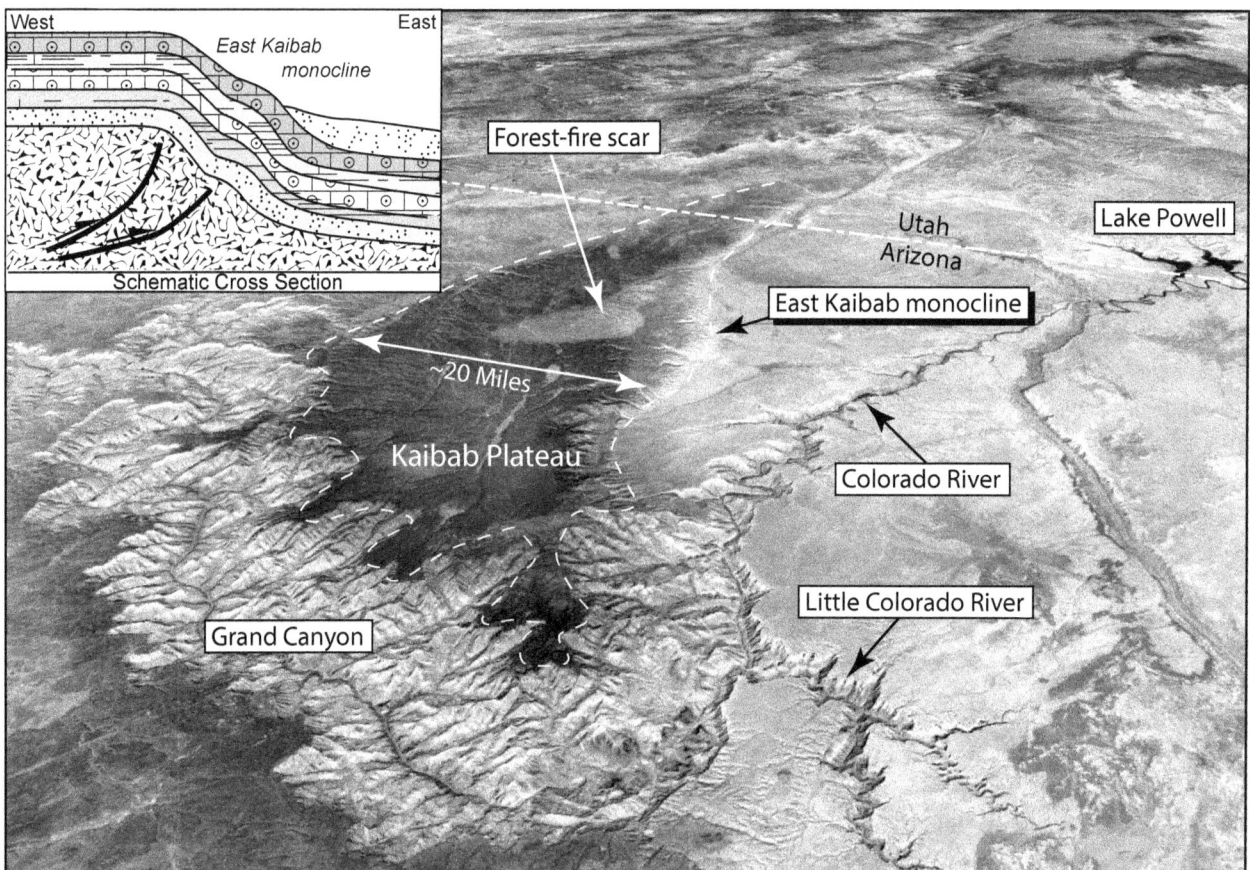

Figure 4-7. North-Northwest Oblique *Google Earth* Image of Kaibab Plateau (Forested) and East Kaibab Monocline, Arizona

Placitas Area

The mountain's faulted northern nose is a glaring anomaly that was formed during the Laramide. For years the area was not at all understood. This was partly corrected by a detailed geologic mapping project that produced a wonderful cross section of the area (Connell et al. 1999; *Figures 4-8* and *4-9*). The structure shown is that of a monocline, with a single steepened section of the normally horizontal layers.

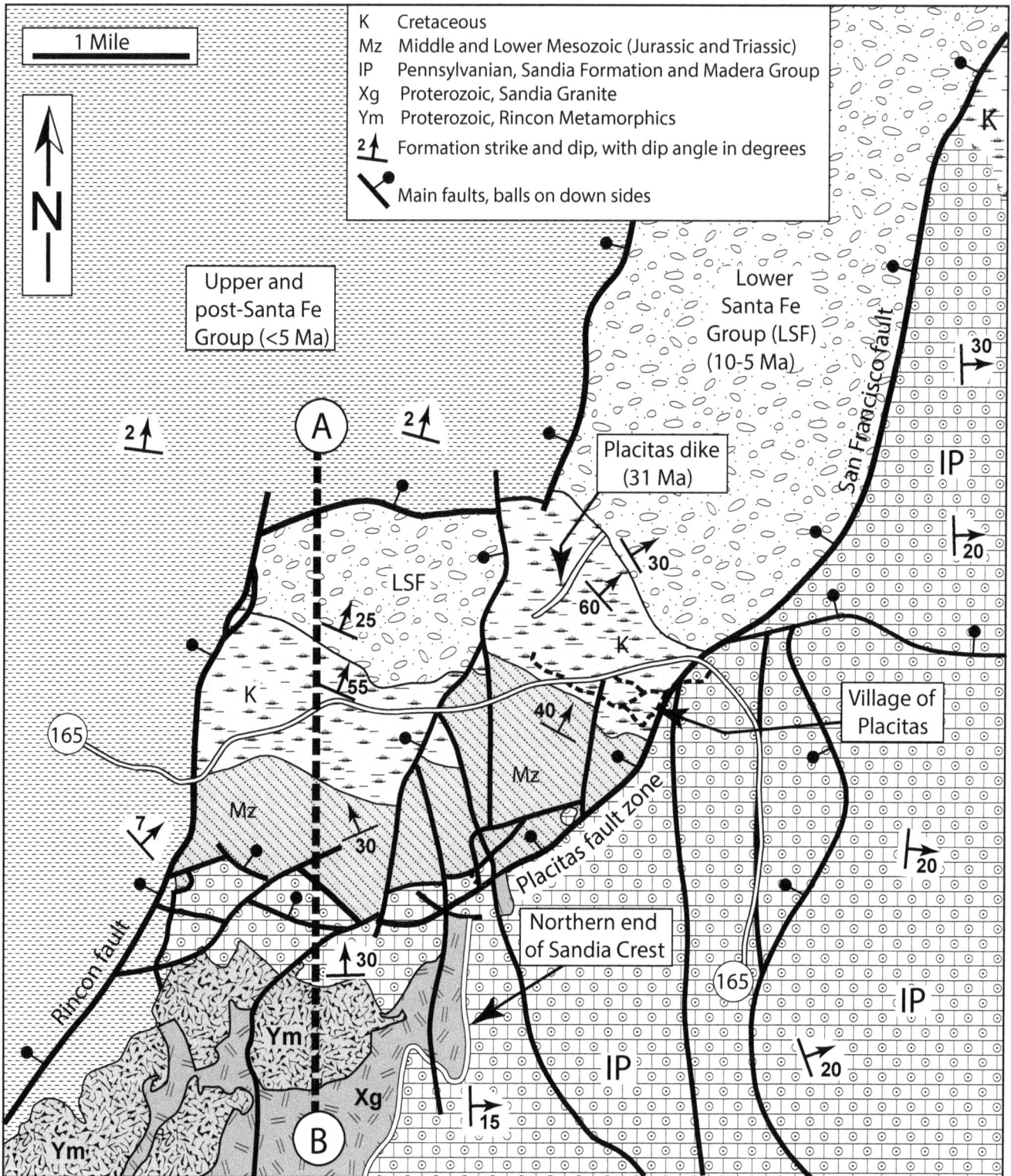

Figure 4-8. Simplified Bedrock Geologic Map of Northern Nose (Placitas Area) of Sandia Mountain (From Connell et al., 1999; Cross Section A-B Shown in Figure 4-9)

Legend:
USF Upper Santa Fe Group
LSF Lower Santa Fe Group
LT Lower Tertiary
K Cretaceous
J-TR Jurassic and Triassic
IPm Pennsylvanian, Madera Group

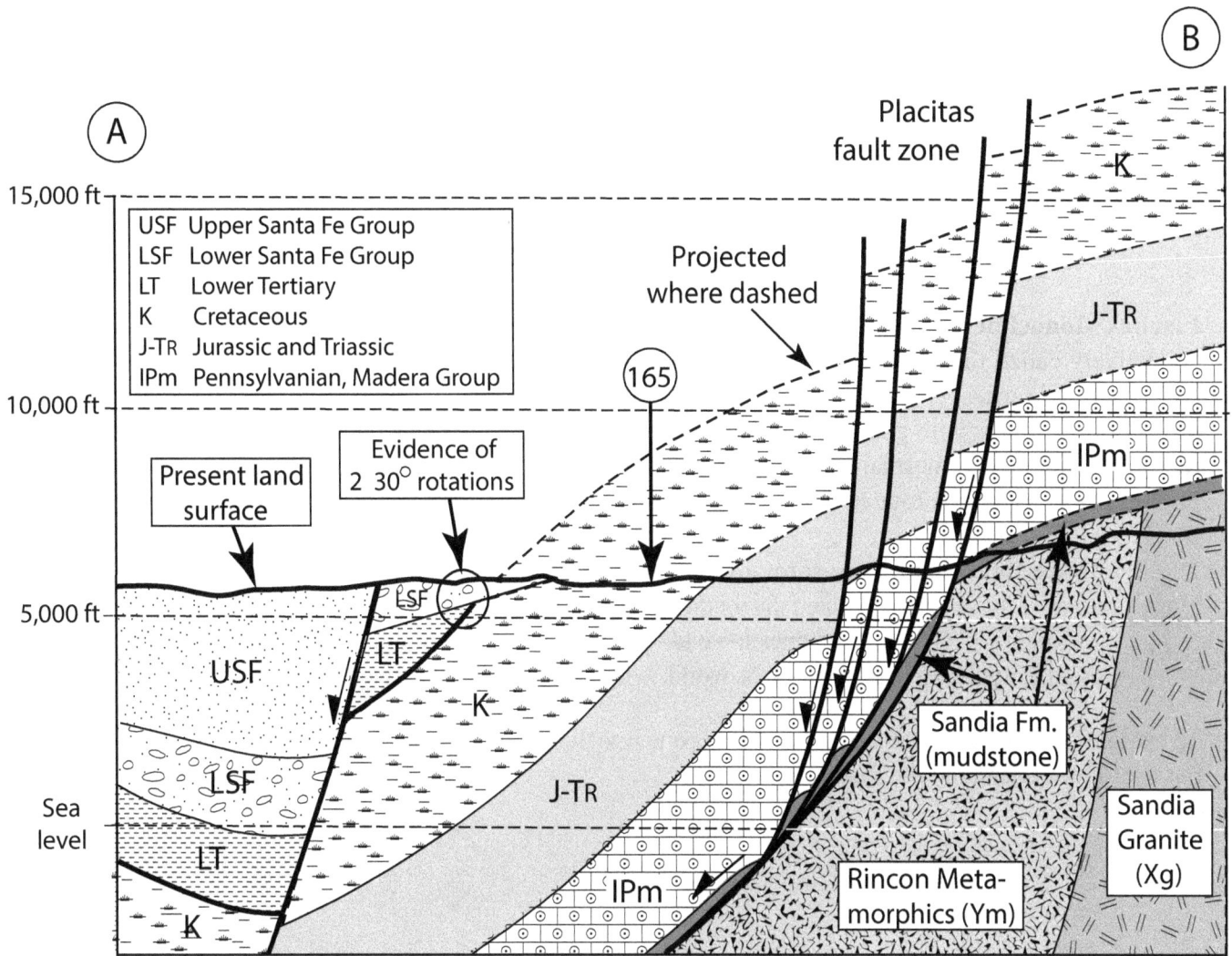

A. Present Structure (Modified from Connell et al. 1995; No Vertical Exaggeration)

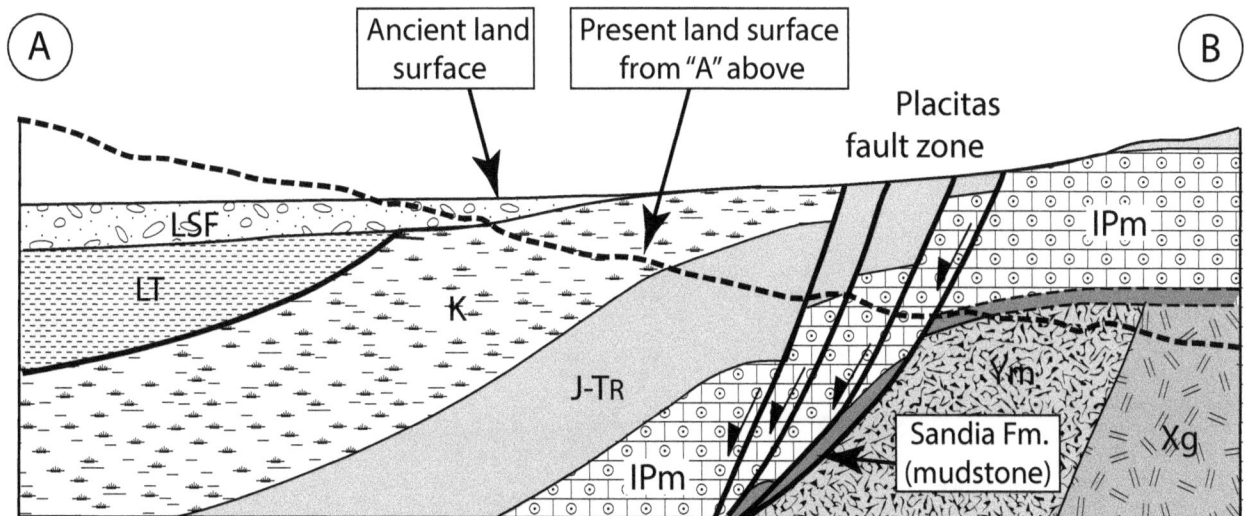

B. Cross Section Restored (Rotated 30° South) to Earlier (15-10 Ma) Orientation

Figure 4-9. North-to-South Cross Section A-B Across Placitas Area of Sandia Mountain
(Location in *Figure 4-8*)

149

Placitas Monocline. This structure is bordered on the south by a cluster of down-to-the-north normal faults, collectively called the Placitas fault zone (*Figures 4-8* and *4-9A*). This fault "chop-up" makes the Placitas area a geologic mess. (Folks living with poor water wells find themselves in a different world from their next-door neighbors with good wells.) The main normal faults are nearly vertical, indicating steepening by rotation from the more normal 60°. A close look at the cross section reveals that the stack of formations was pushed twice, each time rotated down to the north or northeast about 30° (*Figure 4-9B*).

Monoclines. In order to form a monocline, the layers of sedimentary rock must slip across each other. Layers are separated by bedding planes, which are zones of weakness. In order to fold sedimentary rocks, these zones have to "give." Compare bending a telephone book (*Figure 4-10A*) vs. attempting to bend a block of wood, which has no zones of weakness.

To explain how this structure was formed leads one into the tall weeds. The following paragraph is "skippable" but useful to understand the Placitas anomaly.

Legend

K = Cretaceous
J/TR = Jurassic and Triassic
P = Permian
IPm = Pennsylvanian Madera Group
IPs = Pennsylvanian Sandia Formation

Basement Rock = Rincon Metamorphics
and Sandia Granite

Post-Cretaceous Rocks (Lower Tertiary
and Santa Fe Group) not Shown

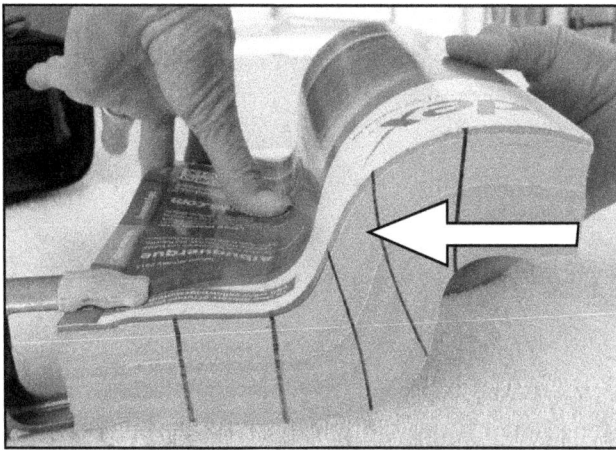

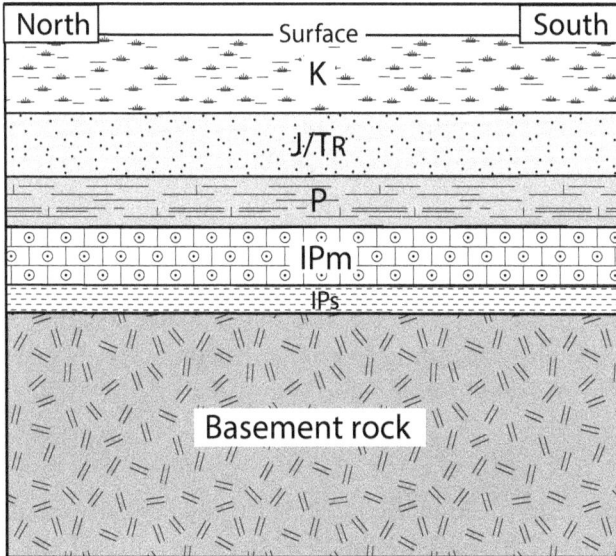

A. Simple Demonstration of Formation of Monocline

B. Pre-Laramide Compression

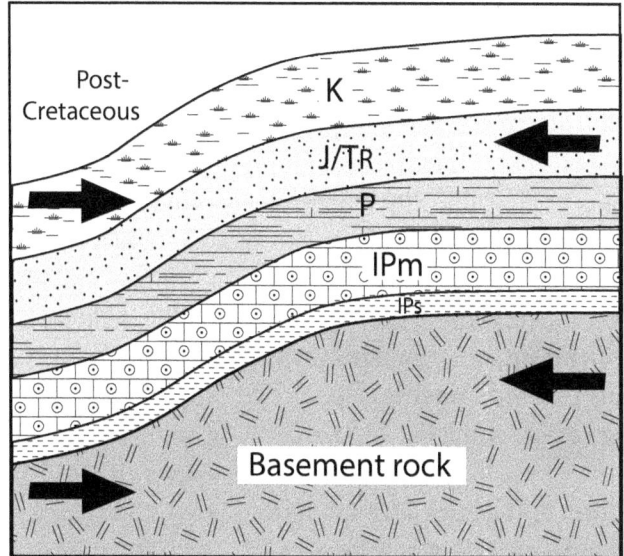

C. Beginning of Laramide Compression and Folding

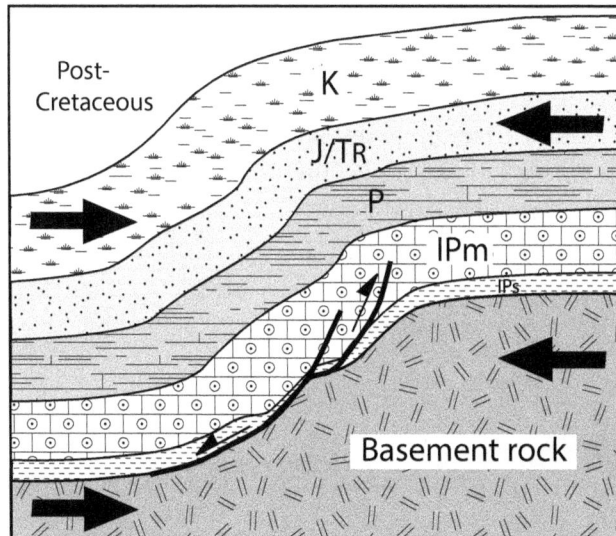

D. As Monocline Forms During Late Laramide Compressive Shortening, Fold's Axis Creates Local Compression and Faulting near Top of Basement

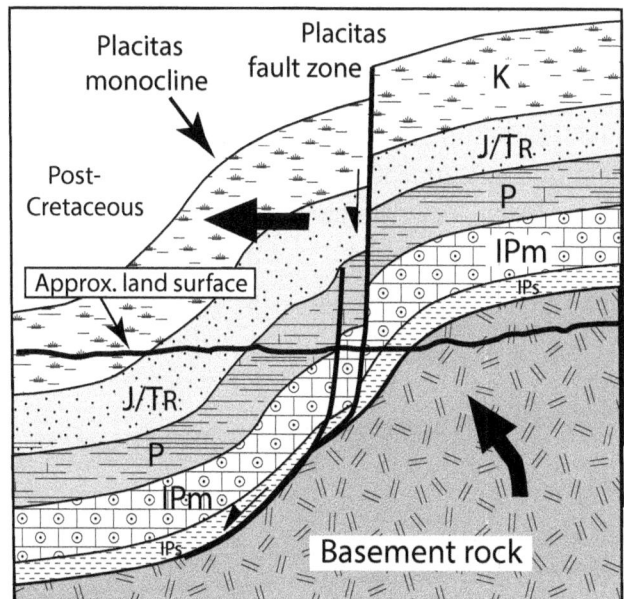

E. Rio Grande Rift Tension and Uplift of Sandia Block Rotates Formations down to North and Reverses Movement Along Faults

Figure 4-10. Schematic North-to-South Cross Sections Showing Possible Sequential Development of Placitas Monocline

Let's postulate what happens as a monocline like the one at Placitas develops. First, as compression from the south forms a flexural bend out in front, the bend experiences internal compression, which is relieved by compensating slippage along the bedding planes (*Figures 4-10B – 4-10E*; Brown 1983). If development of the monocline ceases at this early stage, these internal zones become inactive but they provide sites of weakness for later use. The published cross section of the Placitas area (Connell et al. 1999) projects the high-angle normal faults of the Placitas fault zone down ("out of sight") into the subsurface. Without altering the surface interpretation, those faults could be reinterpreted to "sole out" in the mechanically-weak shales of the Sandia Formation and become bedding-plane faults, lying just above the rigid basement. It follows that the Placitas fault zone may represent a set of Laramide bedding-plane thrusts reactivated as normal faults during later Rio Grande rift (RGR) extension.

The Placitas monocline seems to be the eastern, broken-off part of a larger, buried structure. Geophysical indications exist for a northwest-trending structure of some sort extending from the Rincón Ridge northwest into the Albuquerque basin. The feature has a name: the *Ziana anticline*, combining the names of Zia and Santa Ana Pueblos.

Ziana Anticline. This enigmatic feature may play an important role in the style of the later uplift of Sandia Mountain. The structure was first recognized as a structural "high" back in 1974 by Shell Oil (Black and Hiss 1974), which drilled its well, the Shell Oil No. 1 Santa Fe Pacific (unsuccessfully) on it. The structure was later delineated via a regional gravity survey and seen to extend northwest to north-northwest for about 20 miles (*Figure 4-11*; Grauch et al. 1999). ("High" gravity, the light shade on the map, indicates dense, basement rock closer to the surface, whereas "low" gravity, the dark shades, indicates deeper basement.) Whether the influence of the structure extends east across to Placitas is unknown, but it's quite feasible. Most of the sedimentary formations in the Placitas area (more below) dip to the northeast, suggesting that they perhaps once occupied the northeast flank of an old monocline—the Ziana?

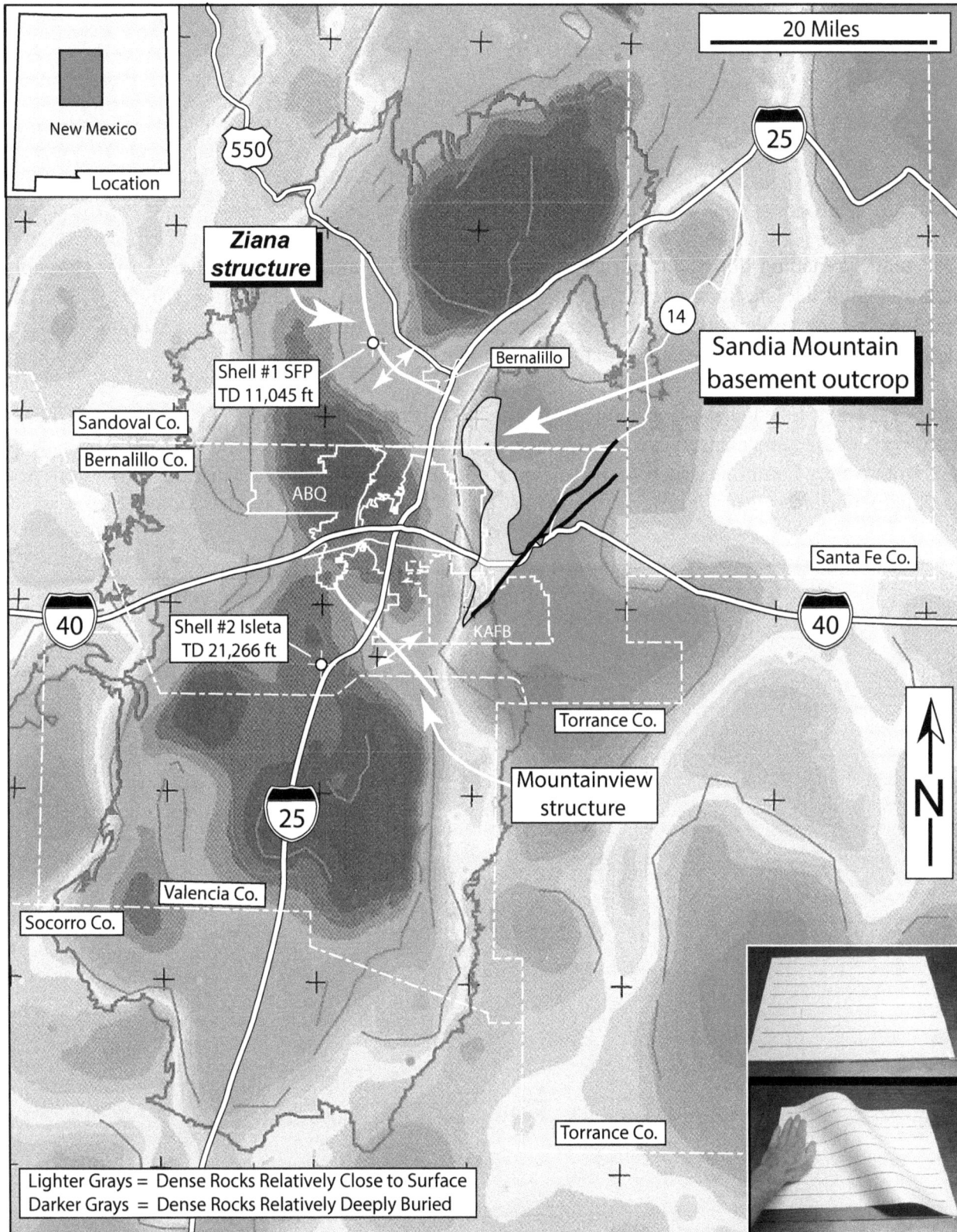

Figure 4-11. Gravity Map of Central New Mexico and Sandia Mountain Area
(Modified from Grauch *et al.*, 1999; Select Exploration Wells and Total Depths Shown;
Light Areas = Granity "Highs"; Dark Areas = Gravity "Lows")

The right-lateral wrenching during the earlier Ancestral Rocky Mountain event (315-275 Ma), possibly augmented by that of the late Laramide (55-40 Ma), produced northwest-trending folds (*Figure 4-11*). The Ziana structure is compatible with this style and may provide a glaring "pre-existing condition" for later Laramide events. (More about the Ziana structure in *Chapter 8*.)

One of the best ways to understand a geologic feature is to find a "look-alike," i.e., one that is better understood, very visible, and can serve as a teaching aid. There is no better place to look than in the state of Wyoming. Of the many Laramide structures there, one notorious example is located in the northwestern part of the state, namely *Rattlesnake Mountain* (*Figure 4-12*).

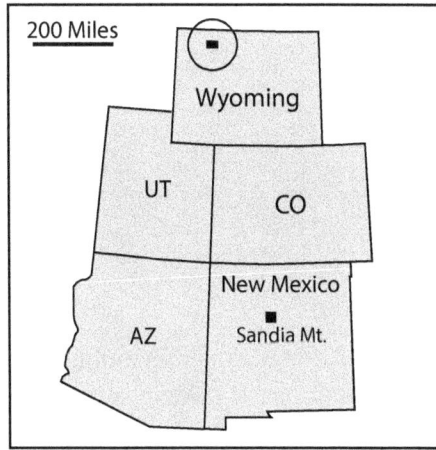

A. Location Map

B. Northwest Oblique *Google Earth* Image (Left) and Interpretation (Right)

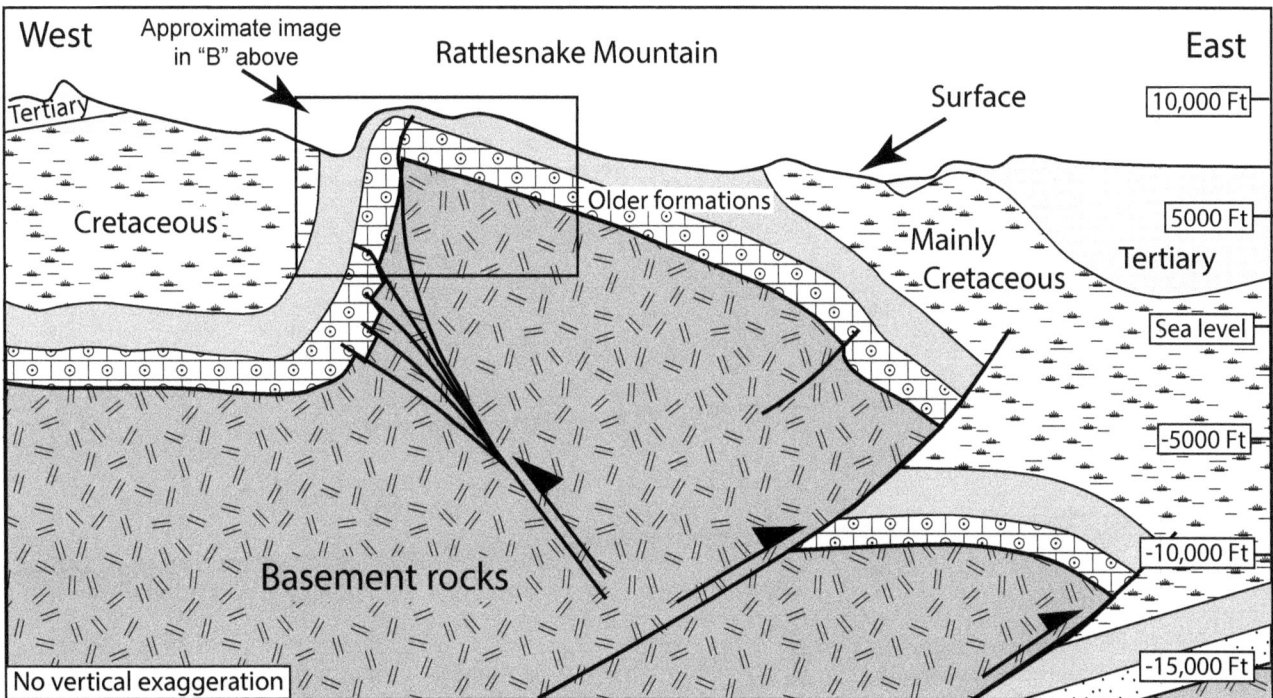

C. West-to-East Cross Section (Modified from Lageson and Spearing, 1988)

Figure 4-12. Rattlesnake Mountain, Wyoming: "Look-Alike" Monocline Analog for Northern Nose of Sandia Mountain (Placitas Area)

155

A Placitas Monocline look-alike. At Rattlesnake Mountain, as at Placitas, the basement is faulted and the overlying sedimentary formations are folded as well as faulted. The basement rock evidently deformed to accommodate compression, but in a very different way. The basement has no bedding planes to provide internal slip, so deformation is mainly by faulting. However, rigid basement rock, such as the Sandia Granite, can indeed exhibit some *apparent* folding when its main faults are accompanied by a myriad of tiny fractures, many with a small amount of slippage (*Figure 4-13*).

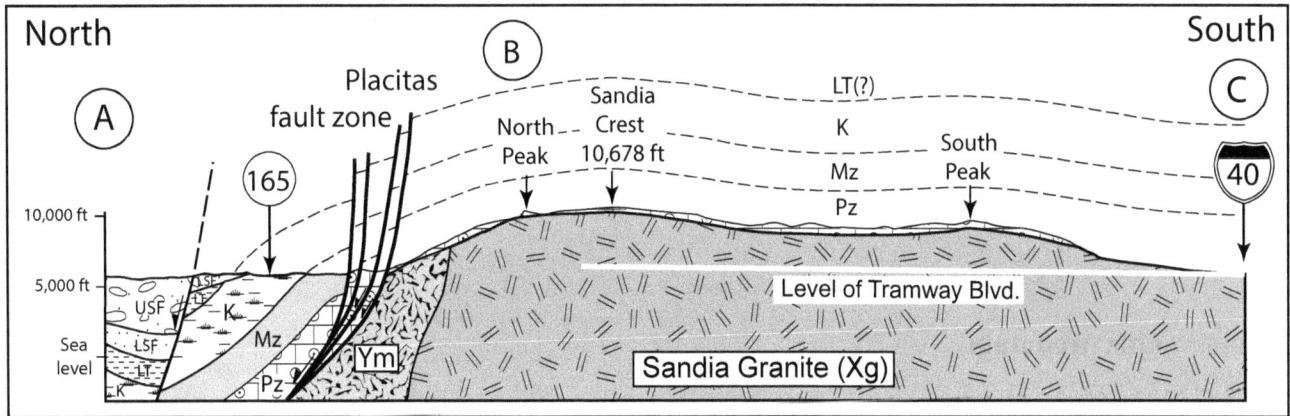

A. Cross Section A-B-C Across Sandia Mountain (Schematic South of Placitas Fault Zone; Location in Figure 4-6)

B. Intense Fracturing in Sandia Granite, Making it Possible to Deform a Mountain
(Photograph by Author from Sandia Tram)

Figure 4-13. Relationship of Northern Nose to Main Sandia Mountain

Uplift of Sandia Mountain

To review, the rates and directions of North America's general westward drift together produced a varying compressive force on the western continental margin. These forces, transmitted far "inboard" because of the low subduction angle, affected the area of the future Sandia Mountain during two distinct episodes.

Early Laramide (75-55 Ma). The first phase of Laramide compression, directed east-northeast (left of *Figure 4-3B*), caused the Sandia Mountain area to shorten (like pushing a flexible carpet, causing the ends to become closer to each other), producing the series of north-trending, up-to-the-west reverse faults that are now evident on the east side.

Late Laramide (55-40 Ma).The second phase of compression was directed northeast rather than east-northeast (right of *Figure 4-3B*). This created a right-lateral wrenching force, able to exploit the north-to-south "scars" inherited from the event that created the Ancestral Rockies. The resulting right-lateral movement may have rippled the overlying sedimentary package, forming the north-facing, northwest-trending Placitas monocline and its associated Ziana structure mentioned above (*Figure 4-11*). When the Laramide dust had settled, what would later become Sandia Mountain was likely a large low-relief, east- to northeast-facing monocline.

Forming part of the southern boundary of Sandia Mountain is the 60-mile, northeast-trending Tijeras fault system (*Figure 4-14*). In general, it consists of two main parallel wrench-fault strands. During the Laramide it experienced right-lateral motion and then left-lateral motion later during the development of the Rio Grande rift (*Chapter 6*).

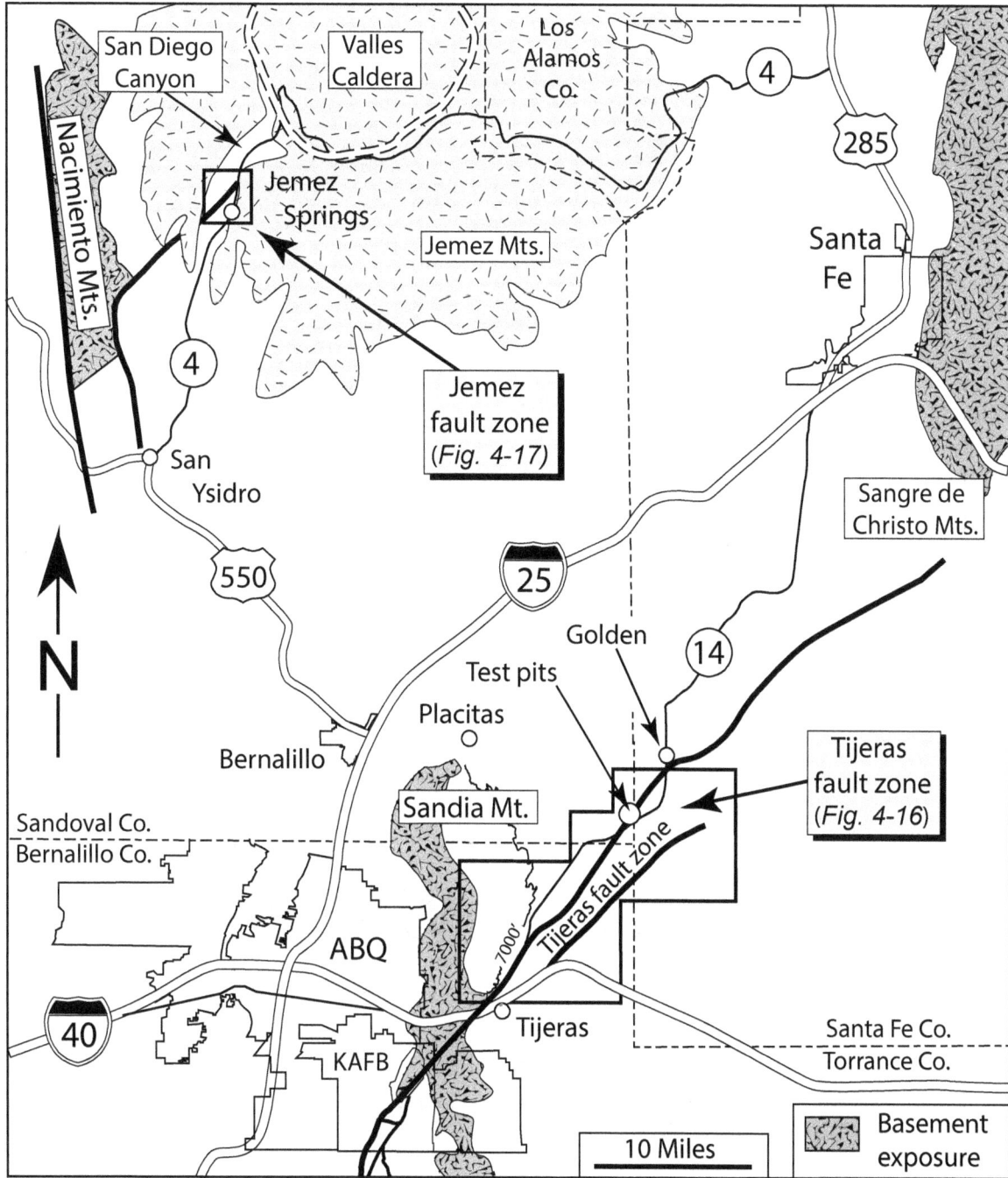

Figure 4-14. Two Wrench-Fault Systems in Central New Mexico

It is worth a diversion here to expand a little on wrench-fault systems. Such systems have been intensely studied by the petroleum industry because the associated structures often house significant accumulations of oil and/or gas. Probably the world's most studied is the modern San Andreas fault of California that rips through San Francisco.

Strange things can happen when crustal blocks crunch laterally against one another. Slices of rock can get caught up in the middle of the zone and get either popped up, like watermelon pits, or popped down (*Figure 4-15*).

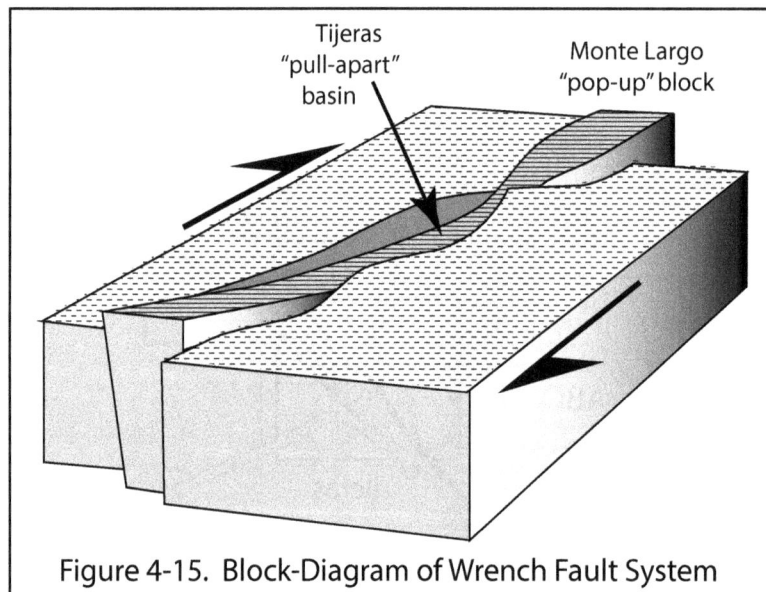

Figure 4-15. Block-Diagram of Wrench Fault System

Tijeras Fault System. This complex zone has both kinds of structures: a popped-down piece just to the north of the town of Tijeras called the Tijeras basin, and a popped-up segment farther to the northeast called the Monte Largo horst (*horst* = German for an upthrown block; *Figures 4-6* and *4-16*). The Tijeras fault zone is truly unlike anything else on the mountain and makes for a very confused, contorted place, and the bane of field geologists.

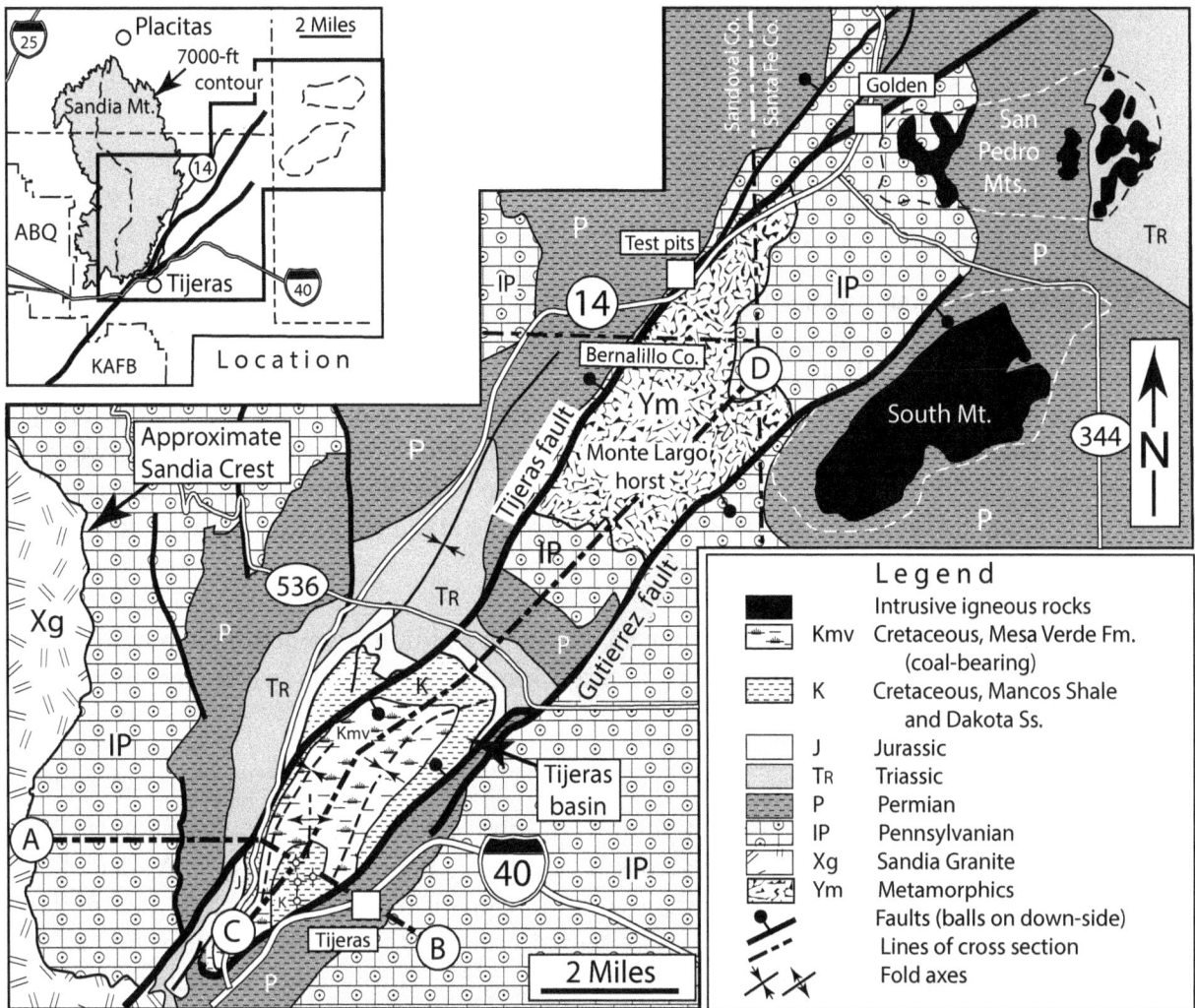

A. Geologic Map (Modified from Williams and Cole, 2007)

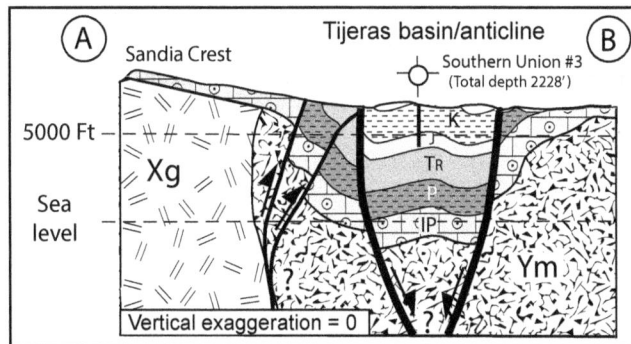

B. Cross Section A-B (Modified from Kelley and Northrop 1975; Read et al. 1998)

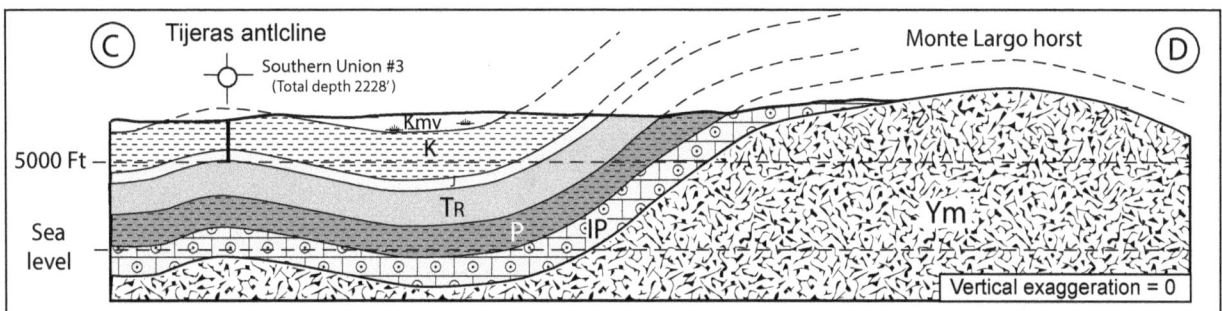

C. Cross Section C-D

Figure 4-16. Tijeras Wrench-Fault System

161

Test pits. About four miles southwest of Golden and just north of NM-14, geologists in 1997 dug three test pits across the fault (*Figure 4-16A*). Their findings revealed that the fault here had ruptured twice between 130 and 11 ka. If this event had involved the entire 25-mile fault sector from there to the southwest, and they suspect it had, it would have been about a 7 on the "moment magnitude" (or Mw) scale—a measure of earthquake intensity (Kelson et al. 1999). This is at the border between "strong" and "major." To get a feel for this cryptic number, the 1994 Northridge earthquake near Los Angeles, CA had a Mw of 6.7, and a Hiroshima-size atomic bomb would be about a 6.2. Fortunately, there has been no historical recurrence.

To demonstrate that the Tijeras fault system is not an oddity, an additional wrench-fault system is the nearby Jémez system, 1 ½ mile north of the village of Jémez Springs and better known as the site of Soda Dam (*Figure 4-14*).

Jémez fault system. For many decades Soda Dam has been the subject of postcards and weekend outings in San Diego Canyon—that southwest entry corridor to the Valles Grandes. A pair of northeast-trending wrench faults rip across Jémez Canyon here (*Figure 4-17*). A slice of basement rock pops up between the two faults, giving the canyon its narrow width at that point. As the Jémez River cut its canyon down to the top of the buried basement block, mainly since 800 ka, the river superimposed itself into and through the resistive rock, forming the narrow defile. As incision of the bedrock continued, leakage along the faults, which were conduits for hot, mineral-bearing waters from the Valles Caldara to the north, oozed to the surface and deposited the substance called *travertine* (see glossary in *APPENDIX III*). The lowest and newest travertine deposit is called Soda Dam, which is a mere 5 ka. These hot waters also carried a significant amount of arsenic in solution (and still do).

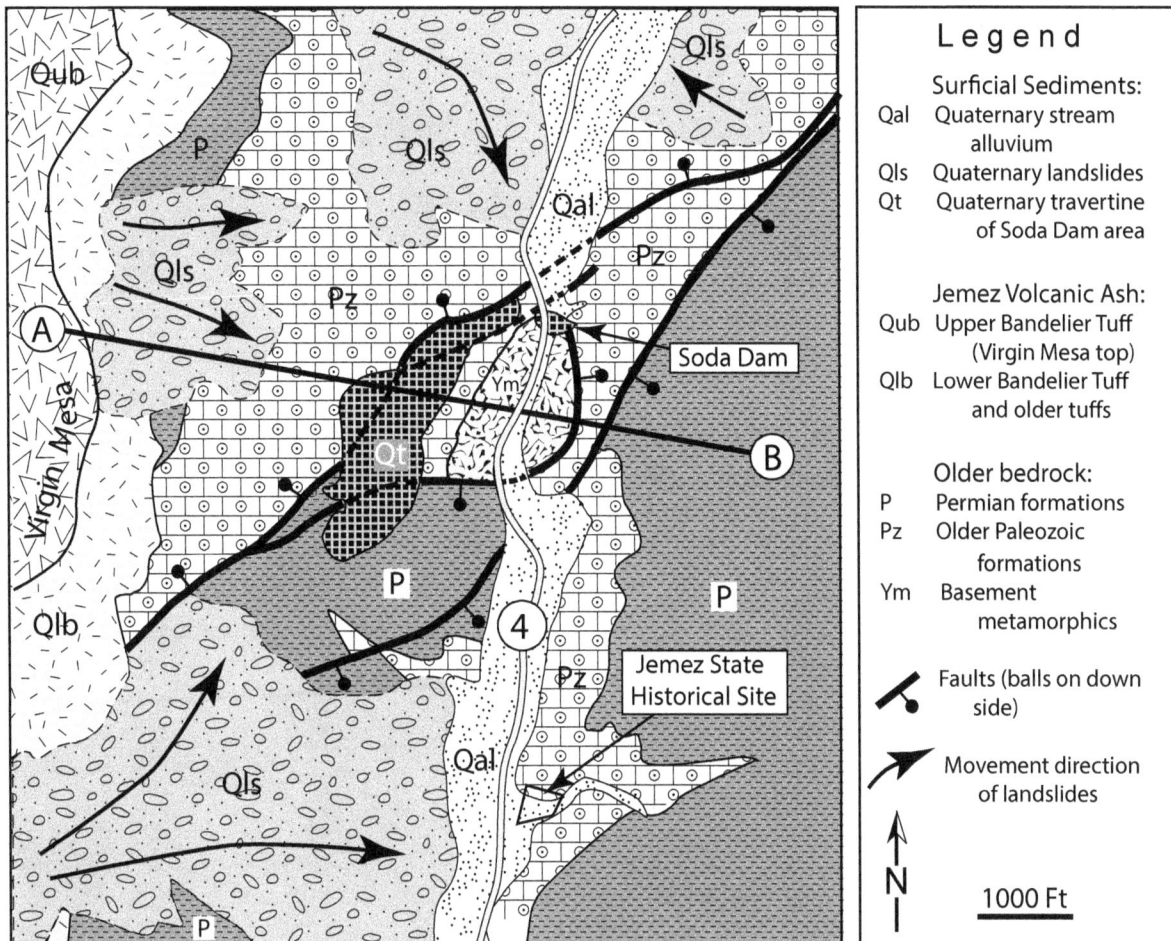

A. Geologic Map (Modified from Kelley et al. 2003)

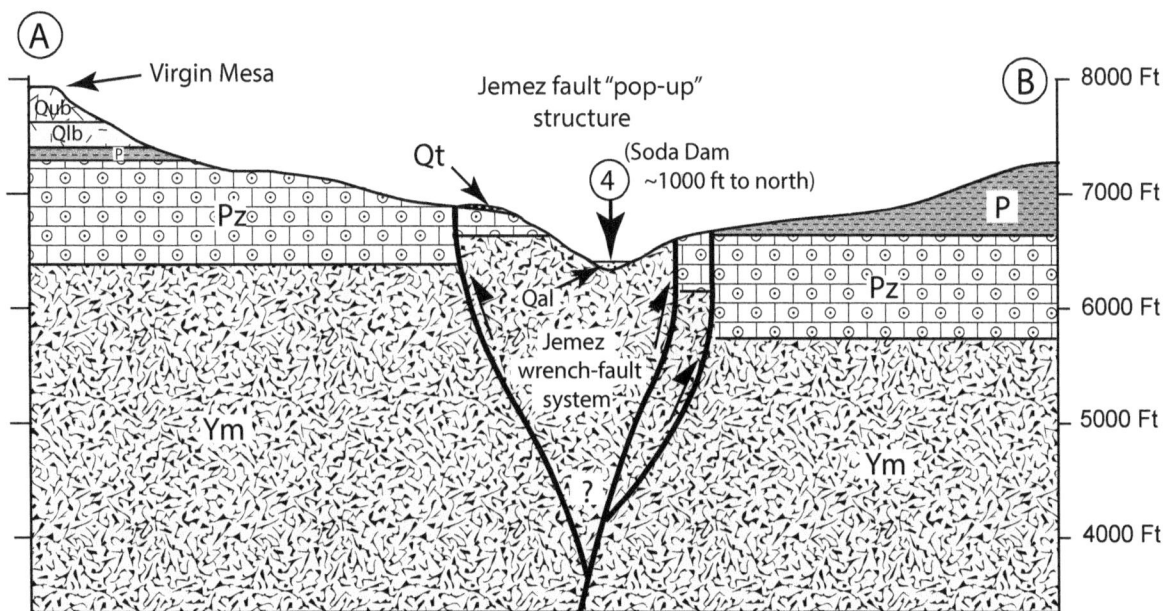

Legend

Surficial Sediments:
Qal   Quaternary stream alluvium
Qls   Quaternary landslides
Qt   Quaternary travertine of Soda Dam area

Jemez Volcanic Ash:
Qub   Upper Bandelier Tuff (Virgin Mesa top)
Qlb   Lower Bandelier Tuff and older tuffs

Older bedrock:
P   Permian formations
Pz   Older Paleozoic formations
Ym   Basement metamorphics

Faults (balls on down side)

Movement direction of landslides

N

1000 Ft

B. Cross Section A-B (Showing Bedrock Only; No Vertical Exaggeration)

Figure 4-17. Jemez Wrench-Fault System, San Diego Canyon
(Location in *Figure 4-14*)

163

Laramide Paleogeography

The Laramide event profoundly changed the landscape of New Mexico. By the end of the early Laramide, ca. 75 Ma, the landscape of New Mexico was an extensive alluvial plain lying just above sea level, with a few hilly bumps rising above it. Streams worked their way across this landscape to the east and northeast and emptied into a shrinking, shallow sea (*Figure 4-18A*). The Sandia Mountain area was merely a portion of this upland. By the late Laramide, ca. 50 Ma, the sea had been filled in and driven out of the interior of North America for good, and the rivers then delivered their loads directly to the Gulf of Mexico (*Figure 4-18B*).

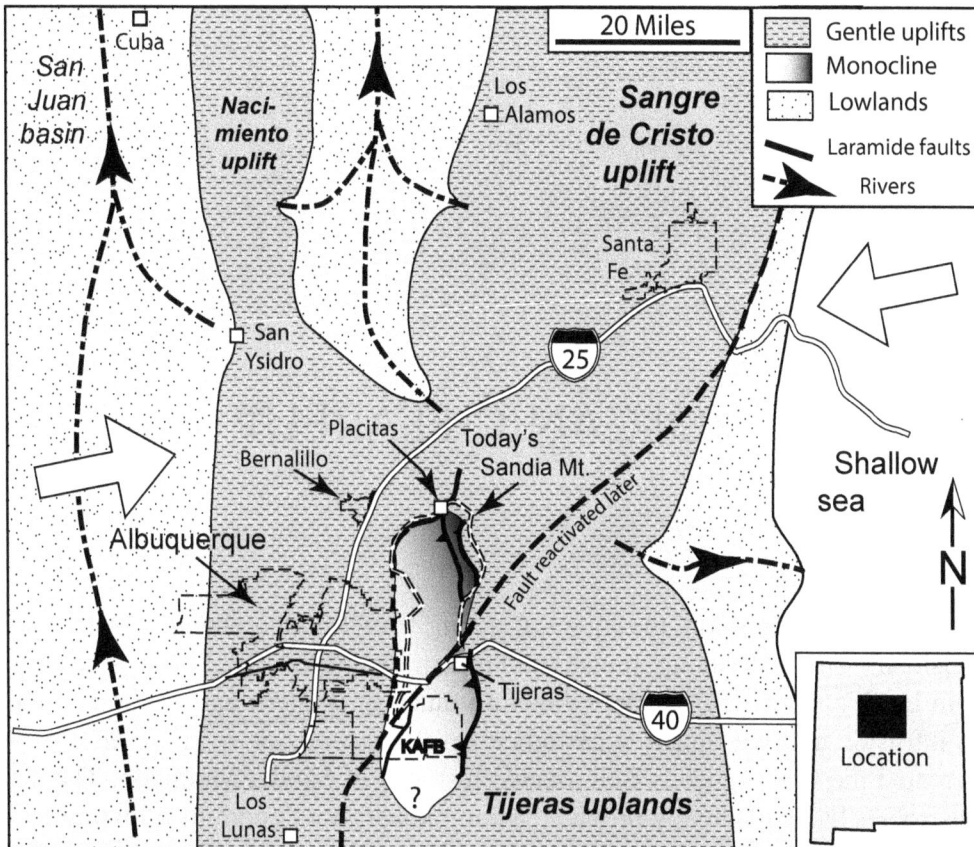

A. Early Laramide, 75 Ma (with Drainage; Modified from Cather 2004)

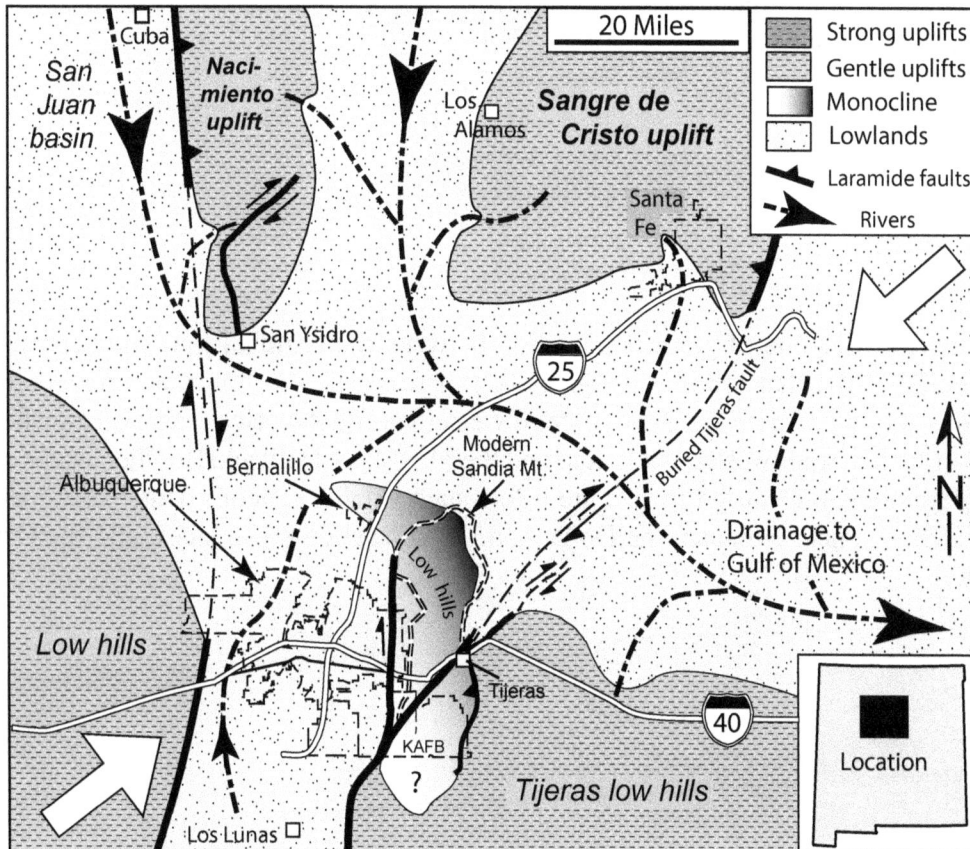

B. Late Laramide, 50 Ma (with Drainage; Modified from Cather 2004)

Figure 4-18. Paleogeography, and Compressive Force-Field Directions (Large White Arrows) During Laramide Event (75-40 Ma) (Select Modern Geography Added for Reference)

5:
VOLCANIC FLAREUP

Synopsis

Following the Laramide compression, the forces affecting New Mexico changed from compressional (the Laramide) to tensional. The intervening, transitional relaxation period, 36-27 Ma, saw an upward surge of hot material from Earth's mantle. A vast area covering much of western half of the continent experienced intense igneous intrusion and volcanism, a period called the *igneous* or *volcanic flareup*. A small portion of this flareup created the Ortiz Igneous Belt located just east of Sandia Mountain. The Tijeras fault zone and its numerous splays had collectively provided a complex array of conduits allowing leakage of magma toward and to the surface from four centers. The cores of these centers have been exposed via deep erosion and today form a string of low mountains. Three of these igneous centers, Los Cerrillos, Ortiz, and Dolores, are known for their mineral deposits.

~~~

Introduction

The period 55-40 Ma was one of gradual transition between the regional Laramide compression and the regional tension, 26-3 Ma, that followed (*Figure 4-2*). It was a time of elevated crustal temperatures and extensive igneous activity over an enormous area of western North America. Vast volcanic fields developed, including the San Juan in southwestern Colorado and the Mogollon-Datil and Bootheel areas of southwestern New Mexico (*Figure 5-1*). Dwarfed in size are a few such smaller igneous centers such as the Latir near Taos, the Sierra Blanca near Ruidoso, the Doña Ana-Organ east of Las Cruces, and the Ortiz belt east of Sandia Mountain.

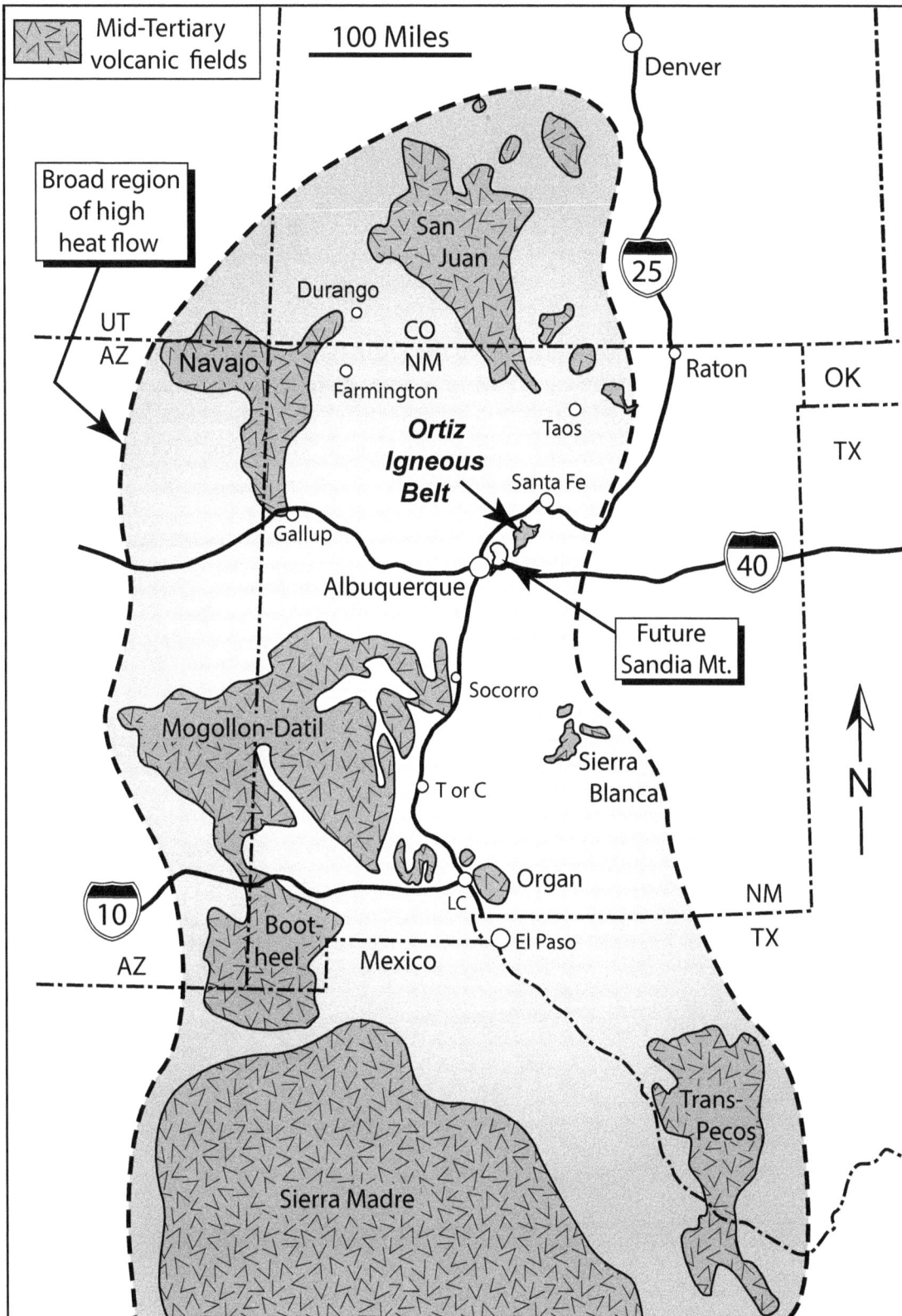

Legend:
- Mid-Tertiary volcanic fields

100 Miles

Broad region of high heat flow

Ortiz Igneous Belt

Future Sandia Mt.

Labels on map:
Denver, 25, Raton, OK, TX, UT, AZ, San Juan, Durango, CO, NM, Navajo, Farmington, Taos, Santa Fe, 40, Gallup, Albuquerque, Socorro, Mogollon-Datil, Future Sandia Mt., Sierra Blanca, N, T or C, Organ, LC, 10, Boot-heel, AZ, Mexico, El Paso, NM, TX, Sierra Madre, Trans-Pecos

Figure 5-1 Middle Cenozoic Volcanic Flareup

Ortiz Igneous Belt

This area east of Sandia Mountain is variously called the Ortiz Igneous Belt, the Ortiz Porphyry Belt, and the Ortiz Volcanic Belt (*Figure 5-2*). The name "Ortiz" comes from the Ortiz Mine Land Grant of 1833, a huge, square, almost 70,000-acre parcel covering what was then called the Placer Mountains. It is easy to assume that these hills are somehow related to the nearby Sandia Mountain, but this 25-mile-long belt is geologically quite distinct, both genetically and temporally (*Figure 5-3*).

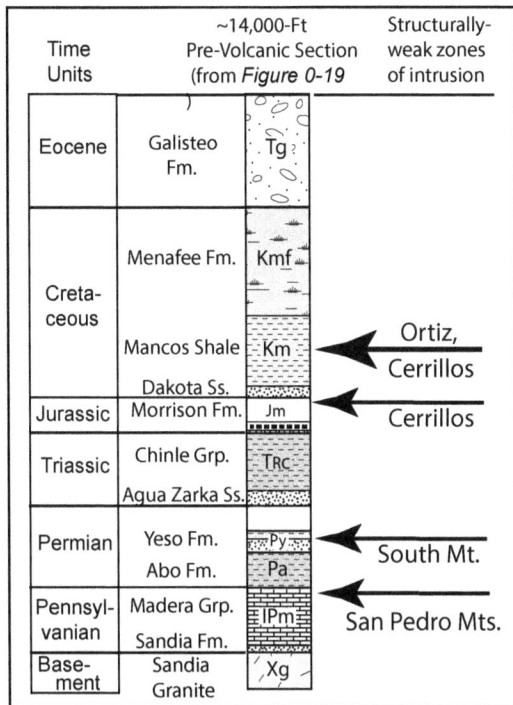

A. Rock Column and Horizons of Intrusion

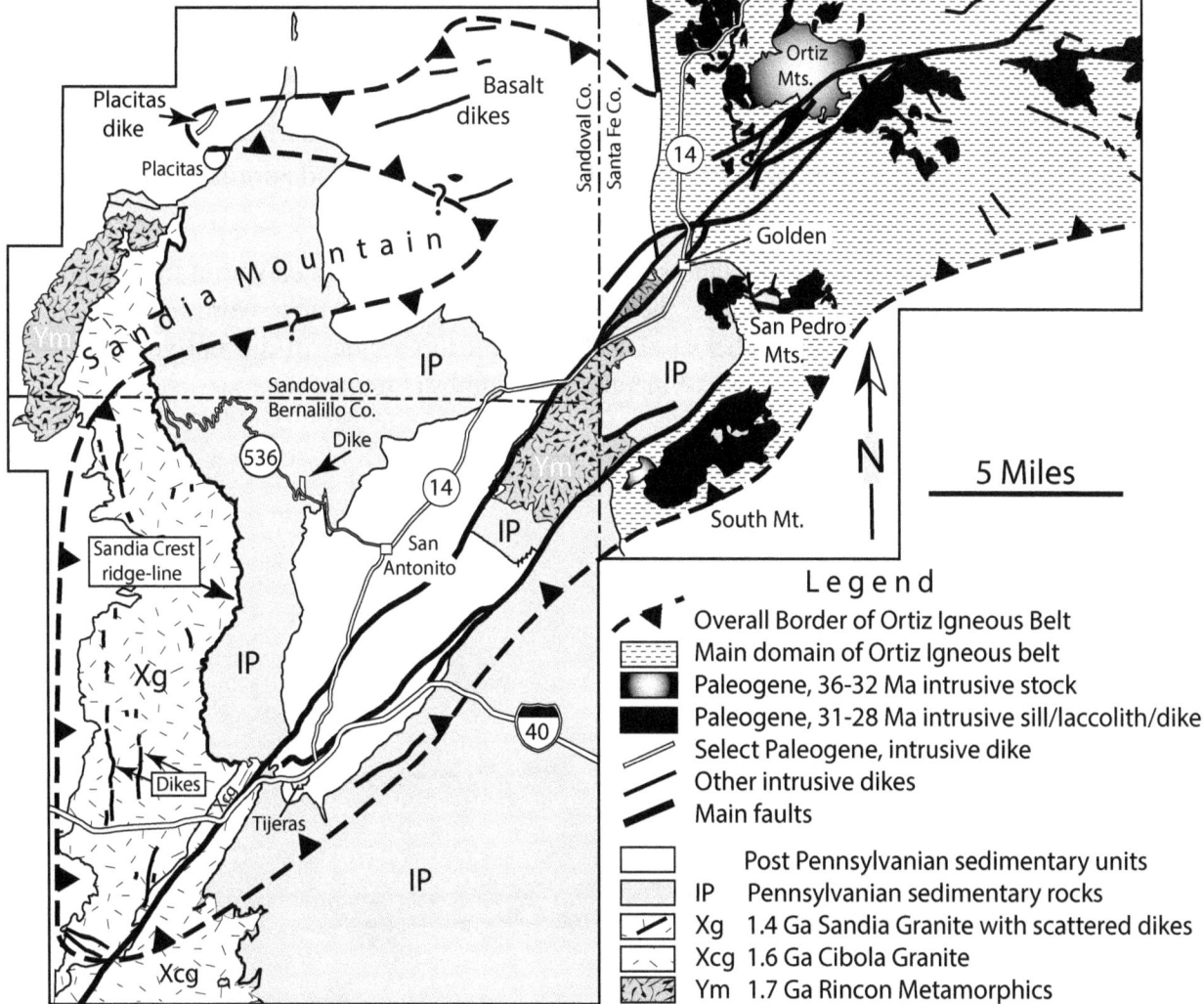

Time Units	~14,000-Ft Pre-Volcanic Section (from *Figure 0-19*)		Structurally-weak zones of intrusion
Eocene	Galisteo Fm.	Tg	
Cretaceous	Menafee Fm.	Kmf	
	Mancos Shale	Km	← Ortiz, Cerrillos
	Dakota Ss.		
Jurassic	Morrison Fm.	Jm	← Cerrillos
Triassic	Chinle Grp.	TRc	
	Agua Zarka Ss.		
Permian	Yeso Fm.	Py	← South Mt.
	Abo Fm.	Pa	
Pennsylvanian	Madera Grp.	IPm	← San Pedro Mts.
	Sandia Fm.		
Basement	Sandia Granite	Xg	

B. Geologic Map

Legend

- ◄┄┄ Overall Border of Ortiz Igneous Belt
- Main domain of Ortiz Igneous belt
- Paleogene, 36-32 Ma intrusive stock
- Paleogene, 31-28 Ma intrusive sill/laccolith/dike
- Select Paleogene, intrusive dike
- Other intrusive dikes
- Main faults

	Post Pennsylvanian sedimentary units
IP	Pennsylvanian sedimentary rocks
Xg	1.4 Ga Sandia Granite with scattered dikes
Xcg	1.6 Ga Cibola Granite
Ym	1.7 Ga Rincon Metamorphics

Figure 5-2. Sandia Mountain and Ortiz Igenous Belt

169

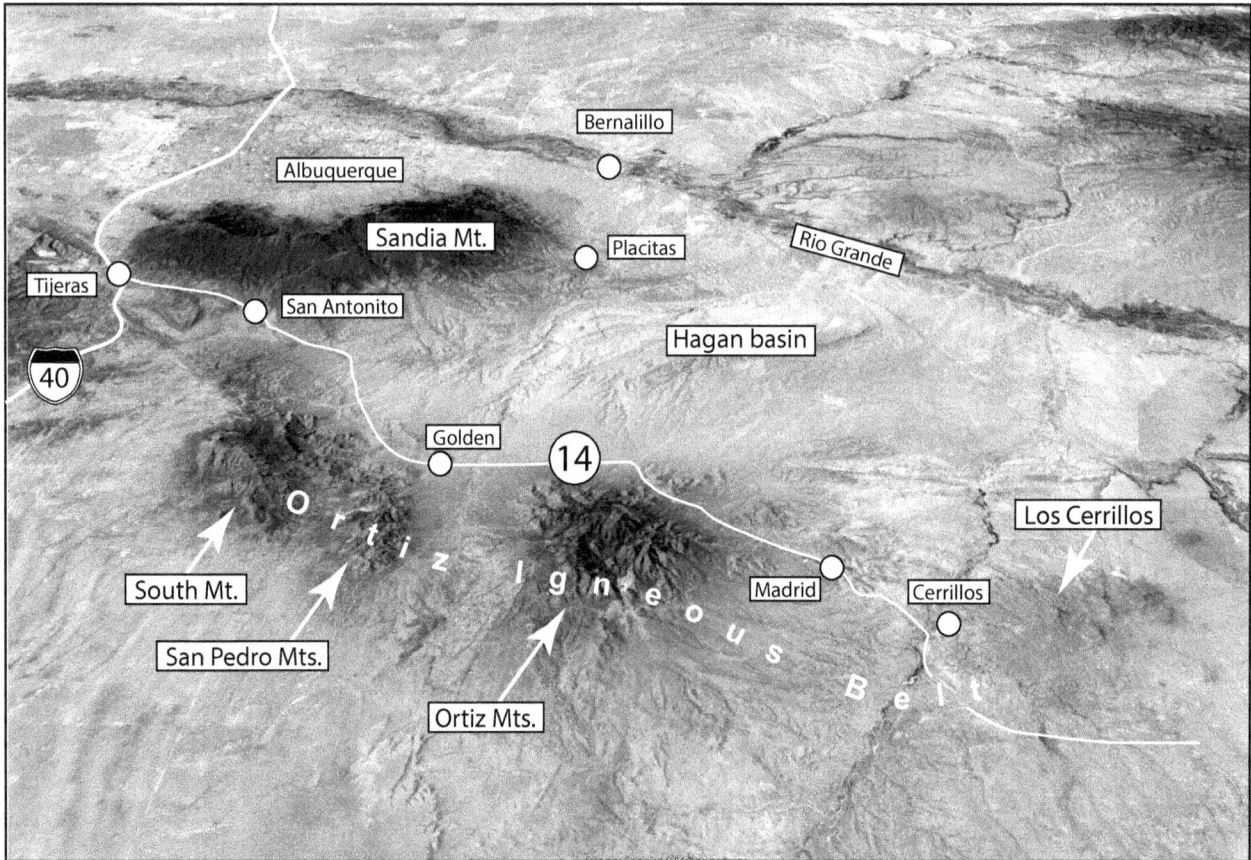

Figure 5-3. West Oblique *Google Earth* View of Ortiz Igneous Belt and Sandia Mountain

During the middle of the Cenozoic Era the land surface was at a much higher elevation than today. To the west was a vast Sahara-like field of sand dunes, the *Chuska Erg* (*erg* is the Arabic word for sand sea), with a surface elevation topping 10,000 ft (Cather et al. 2008). The Rio Grande rift did not yet exist, and the Laramide mountains were largely worn down to inconspicuous nubs (*Figure 5-4*).

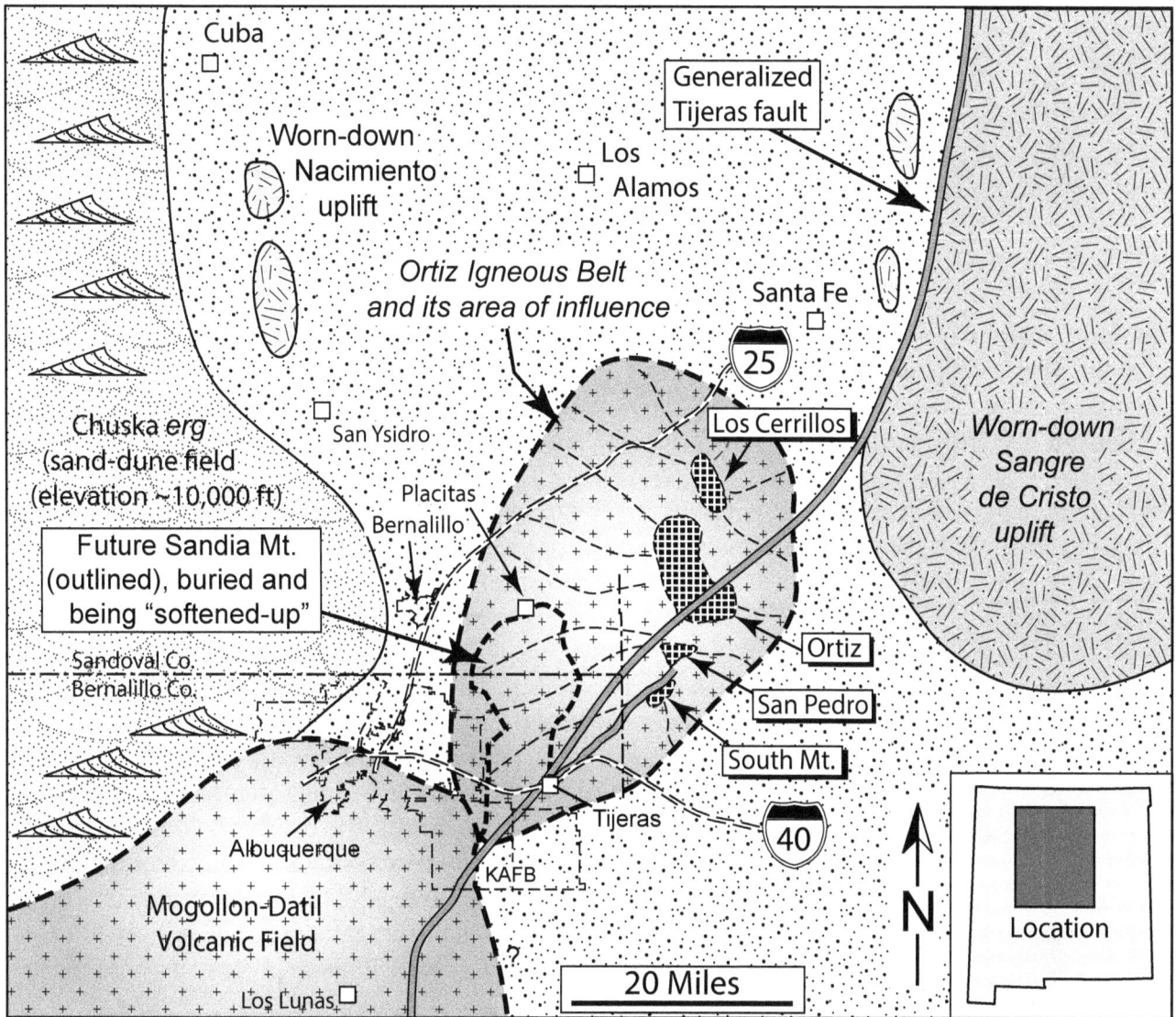

A. Paleogeography (Modified from Cather et al. 2008)

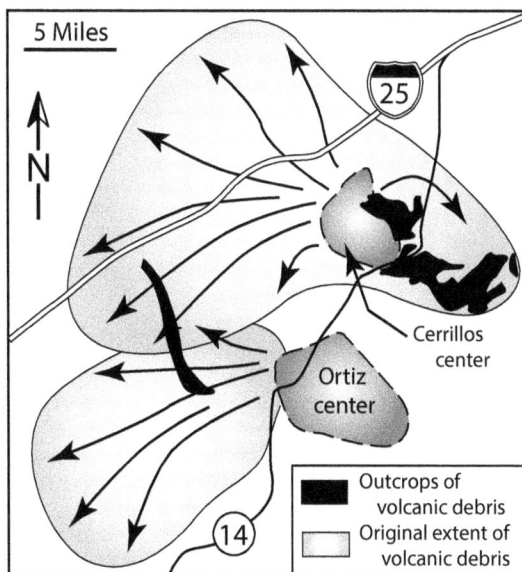

B. "Aprons" of Volcanic Debris from Two Northern Igneous Centers (Modified from Kautz et al. 1981)

Figure 5-4. Middle Cenozoic Geologic Setting of Ortiz Igneous Belt
(Select Modern Geography Added for Reference)

171

The slow transition from Laramide compression to Rio-Grande-rift tension reversed the sense of movement along the Tijeras fault system, from right lateral during the former to left lateral during the latter, further scrambling the geologic picture. Splays of this complex zone of structural weakness provided escape conduits for the Ortiz igneous activity (Woodward 1984).

This activity centered in four areas (*Figure 5-4*). From north to south they are: 1) Los Cerrillos near the village of Cerrillos, 2) the Ortiz Mountains just to the south of the coal-mining town of Madrid, 3) the San Pedro Mountains just to the east of the gold-mining town of Golden, and 4) the unimaginatively-named South Mountain that is mostly on private land north of I-40. (More about these places in *Chapter 13*.)

The Ortiz Belt igneous bodies were intruded in two waves. The first, 36-33 Ma, consisted of mushroom-shaped *laccoliths*, plus sheet-like sills injected below the surface in a sub-horizontal, accordant manner (*Figures 5-5A* and *5-5B*). As the magma began to work its way upward toward lower pressure near the surface it preferentially sought out paths of least resistance such as bedding planes in the soft beds of the Cretaceous-age Mancos Shale. Once the weak zones were found, the overlying rock was jacked up to provide space for the infilling magma. An adjustment to the stress field, extending from the final gasp of compression to the beginning of tension, led to the second wave, 32-27 Ma, producing sub-vertical to vertical magma bodies called *plugs* (not shown in *Figure 5-5*) and vertical, tabular bodies called *dikes* (27 Ma; more below), all of which cut *across* rock formations. It was some of these plugs that carried important ore-bearing fluids (e.g., gold, silver, lead and zinc; Maynard 2005).

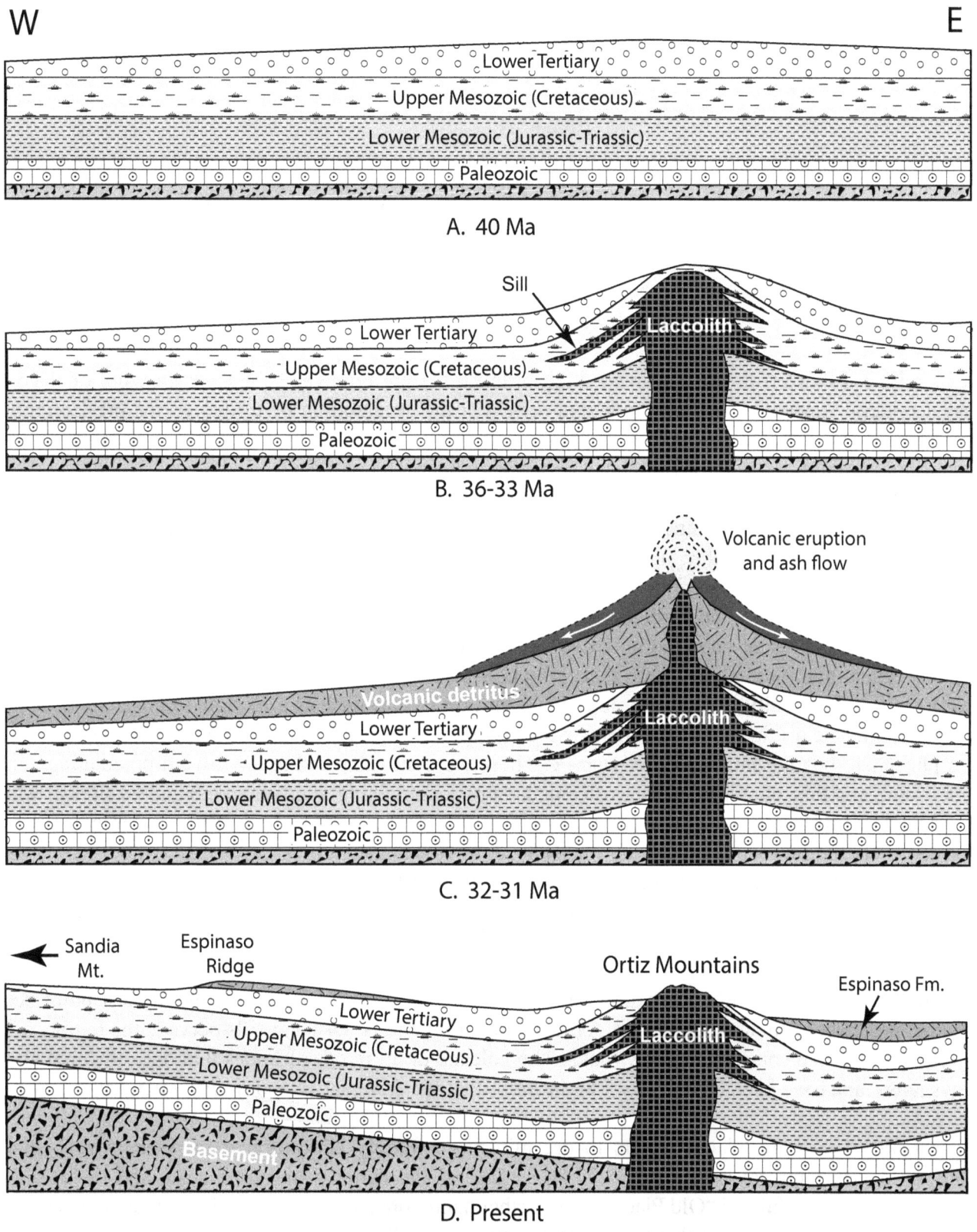

Figure 5-5. Schematic Cross Sections Showing Development of Northern Part of Ortiz Igneous Belt
(Modified from Disbrow and Stoll 1957)

Late during the intrusion of the laccoliths, 32-31 Ma, some of the magma found escape routes all the way to the surface and gushed out as volcanoes, spewing sheets of ash and debris outward in a broad apron. East tilting and erosion have since caused removal of the volcanoes and most of their debris, and beveled the landscape down to the core laccoliths themselves (*Figure 5-5C*). Only remnants of the volcanic-debris apron, known as the *Espinaso Formation*, remain (*Figure 5-5D*).

Los Cerrillos. Los Cerrillos (Spanish for "the little hills") are renowned for their deposits of the very unusual mineral *turquoise*. For centuries this area was the sole source of turquoise for pre-Columbian Native Americans of the Southwest, notably the Anasazi centered at Chaco Canyon. It is difficult to imagine transporting heavy stuff like rock for all that distance on the backs of men.

Turquoise. When the mineral first became known in Europe, it was generally thought that it came from the country Turkey, and the French word for the country was "turquoise." The Aztec word for the area was *chalchihuitl* (pronounced CHAL-chi-WHEETLE), and this awesome name has cruelly persisted for the northern part of the Cerrillos hills to this day. The question naturally arises, why the mineral is present here and nowhere else? Turquoise is a mineral with a thoroughly-captivating blue to greenish hue. Something about that rich color awakes a deep-seated emotion in the human brain that translates to irresistible beauty. To put a fine point on it, turquoise as a hydrated phosphate of copper and aluminum. The original raw ingredients were copper-bearing ore minerals, aluminum-bearing K-spar, and a phosphate-bearing mineral called *apatite*. These were all subjected to downward-percolating acidic groundwater, and this, combined with a volcanic host rock containing vugs (cavities) and fissures in which the turquoise could precipitate, provided the "perfect storm" that occurred at only this one place. Later, especially after the railroad came through in 1880, firmly establishing the town of Cerrillos, silver, gold, lead and zinc became the ores of choice. Prior to 1920s Los Cerrillos was the nation's highest producer of these metals. Today the mines are depleted and closed, but the Cerrillos Hills State Park north of town provides day use and plenty of open space.

Ortiz Mountains. These low mountains were once called the Placer Mountains, the highest point being the namesake Placer Mountain at 8897 ft. A menagerie of igneous stocks and sills pervade the region, all of them intruding Cretaceous sedimentary rocks. The volcano or array of volcanoes that once towered above today's landscape have been eroded down to the stubs that we drive across today (*Figure 5-5D*).

Gold. The term "placer" refers to ore minerals naturally concentrated in river gravels. Placer gold was found in these hills in 1928, setting off the nation's first gold rush, and the area was appropriately dubbed "Old Placers." To further hand-concentrate the gravels, abundant, dependable water was needed, and that was in short supply. By 1832 several gold-bearing veins were being worked, although on a small scale. From 1973 to 1986 *Consolidated Gold Fields* worked the Cunningham Hill deposit by the leach-and-heap method. In 1989 a joint venture resumed at the Carache Canyon deposit, but it soon shut down due to regulatory issues. In 2003 the 1350-acre *Ortiz Mountains Educational Preserve* (OMEP) was established by Santa Fe County as a settlement in a lawsuit against the joint venture, but it remains closed to the public.

San Pedro Mountains. This west-to-east mountain has two summits: San Pedro Mountain on the west (8242 ft) and Oro Quay Peak on the east (8226 ft). The intrusions here, unlike those of the Ortiz, are more sill-like and penetrate the upper part of the sedimentary rocks of the Pennsylvanian Madera Group.

More gold. Placer gold was discovered in 1839 and the San Pedros and the area soon became known as New Placers. Also, important deposits of copper and zinc were present. The little town of *El Real de San Francisco* popped up, and placer mining limped along until about 1880 when the town changed its name to the more palatable *Golden*. The town as well as the mines fizzled out after the mid-1880s.

South Mountain. This feature gets little press, but significant notice from motorists gazing north from I-40 east of Albuquerque. Not only does the mountain not host mineral deposits, but it lies mostly on private land and is therefore generally inaccessible (except from the north via a distant "finger" of BLM territory jutting south from NM-344). Its high point is at an appreciable elevation of 8748 ft. The mountain is formed by an igneous laccolith, now exhumed, that had been injected into a "weak spot" in Permian-age rocks, between a hard sandstone below and a soft shaley formation above (*Figure 5-2A*; more in *Chapter 13*).

Dikes. By definition dikes are near-vertical, tabular rock bodies that were intruded *below* the surface into rocks undergoing tension, and that often fill fractures or faults (*Figure 5-6A*). Two well-displayed and very accessible dikes occur in Sandia Mountain. The first is at mile 1.1 along NM-333 (old US-66) in Tijeras Canyon (*Figure 5-6B*). The second is at mile 3.2 along the Sandia Crest Byway (*Figure 5-6C*).

A. Classic North View of Dike Extending South
from Shiprock, Red Rock Highway,
San Juan County, New Mexico

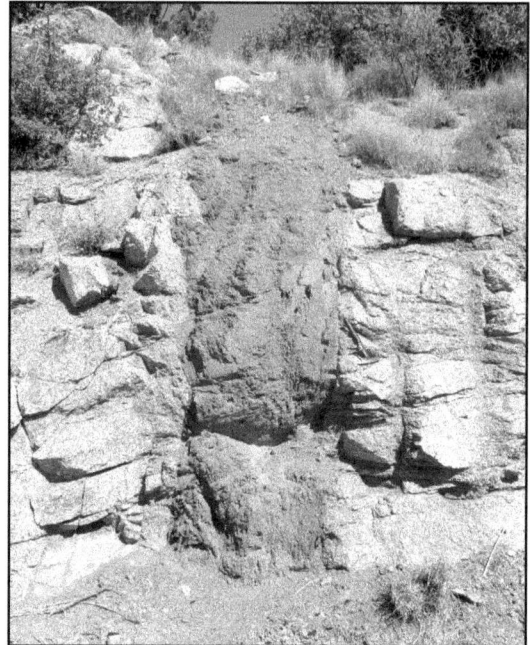

B. North View of Dike Along NM-333, Mile 1.1
(Opposite Elephant Rock in Tijeras Canyon;
Note Vegetation for Scale)

C. North View of Dike Occupying a Fault Along Sandia Crest Byway, Mile 3.2

Figure 5-6. Examples of Igneous Dikes (Photographs by Author)

Although Sandia Mountain and the Ortiz belt are quite unlike, the two shared an elevated-temperature regime during this time, and most of the dikes in both areas are interpreted to be about the same age (Shomaker 1965). In Sandia Mountain most of the dikes occur in small clusters, extending from I-40 in Tijeras Canyon on the south to as far north as the Juan Tabó Cul-de-Sac. The fractures that provided the necessary accommodation space suggest an environment of tension caused probably by thermal expansion—a phenomenon familiar in everyday life (*Figure 5-7*). At this time the Sandia Granite was still buried below more than 12,000 ft of sediment (even after about 1000 ft of Cretaceous shale had been eroded during the Laramide) and was probably at a temperature of nearly 300° F (Karlstrom et al. 1999).

Figure 5-7. Tensional (Expansion) Fractures in Rising Bread

Burial Again

Erosion has taken its toll on the volcanic centers. They are now largely buried in a blanket of Pleistocene-age alluvium and only the tops of the centers protrude through it. Today the alluvium itself has been partial eroded away, a process linked to the entrenchment of the Rio Grande after about 800 ka. This will be dealt with in *Chapter 7*.

6:
NEOGENE TENSION
AND CREATION OF A MOUNTAIN

Synopsis

Following the Laramide compression (75-40 Ma) and the volcanic flareup (36-27 Ma), the area of today's Sandia Mountain was subjected to a period of regional tension (ca. 26-3 Ma). Some of the Laramide reverse faults were re-used and converted to normal faults, the Tijeras fault zone experienced some left-lateral (opposite of Laramide) movement, and the crust stretched and fractured, creating the Rio Grande rift. The Sandia Granite pluton, long held in place by its total fusion to the encasing 1.7 Ga metamorphic country rock, was partially "freed" by major break-off faulting on its west, Rio Grande rift side. Opening of the rift thinned the crust under it, allowing "rebound" to uplift the rift's east flank and rotate the mountain back to the east. Once so freed, the pluton's relative-buoyancy was able to kick in and play a role. The greatest uplift was accordingly near where the pluton had its maximum vertical thickness. The uplift, i.e., the actual creation of Sandia Mountain, progressed via a serious of at least two pulses, each shedding aprons of eroded debris (sand and gravel) and filling the accommodation space produced by the growing rift to the west.

~~~

Introduction

Up to now we've been setting the stage by 1) assembling the supercontinent of Rodinia (*Chapter 1*), 2) intruding the Sandia Granite (*Chapter 2*), 3) unroofing the Sandia Granite to create the Great Unconformity (*Chapter 3*), 4) reburying the granite with about 14,000 ft of sedimentary rock and creating the Ancestral Rocky Mountains (also *Chapter 3*), 5) shoving the Sandia Mountain area around and uplifting it a little during the Laramide (*Chapter 4*), and 5) heating up the entire area and intruding a plethora of igneous rocks in the Ortiz Igneous Belt just to the east (*Chapter 5*). This chapter is where the rubber meets the road. Like a detective investigating a homicide, the first order of business is to establish a time line for the mountain's uplift history.

The Sandia Mountain/Rio Grande Rift Couple

As a segment of the Earth's crust rose to form Sandia Mountain after ca, 26 Ma it was already damaged goods. As the mountain was rising, erosion immediately kicked in and began to remove the highest, exposed rock, but at a slower rate than that of uplift. The eroded rock removed was transported downslope by streams and deposited as sediment on the mountain's flanks, especially in the adjoining Albuquerque basin (a segment of the larger Rio Grande rift). It is that sequence of sediment, coupled with some physical features of the northern nose of the mountain, that provide much of the forensic evidence for the mountain's uplift history.

After the regional thermal flareup of 36-27 Ma, an entirely new stress field developed in the area of central New Mexico. The regional compression that had created the Rocky Mountains gradually eased and regional tension slowly developed. The opening of the Rio Grande rift was accompanied by and genetically tied to the uplift of Sandia Mountain. At first glance such a link is counterintuitive, but the two go together like hand and glove (*Figure 6-1*).

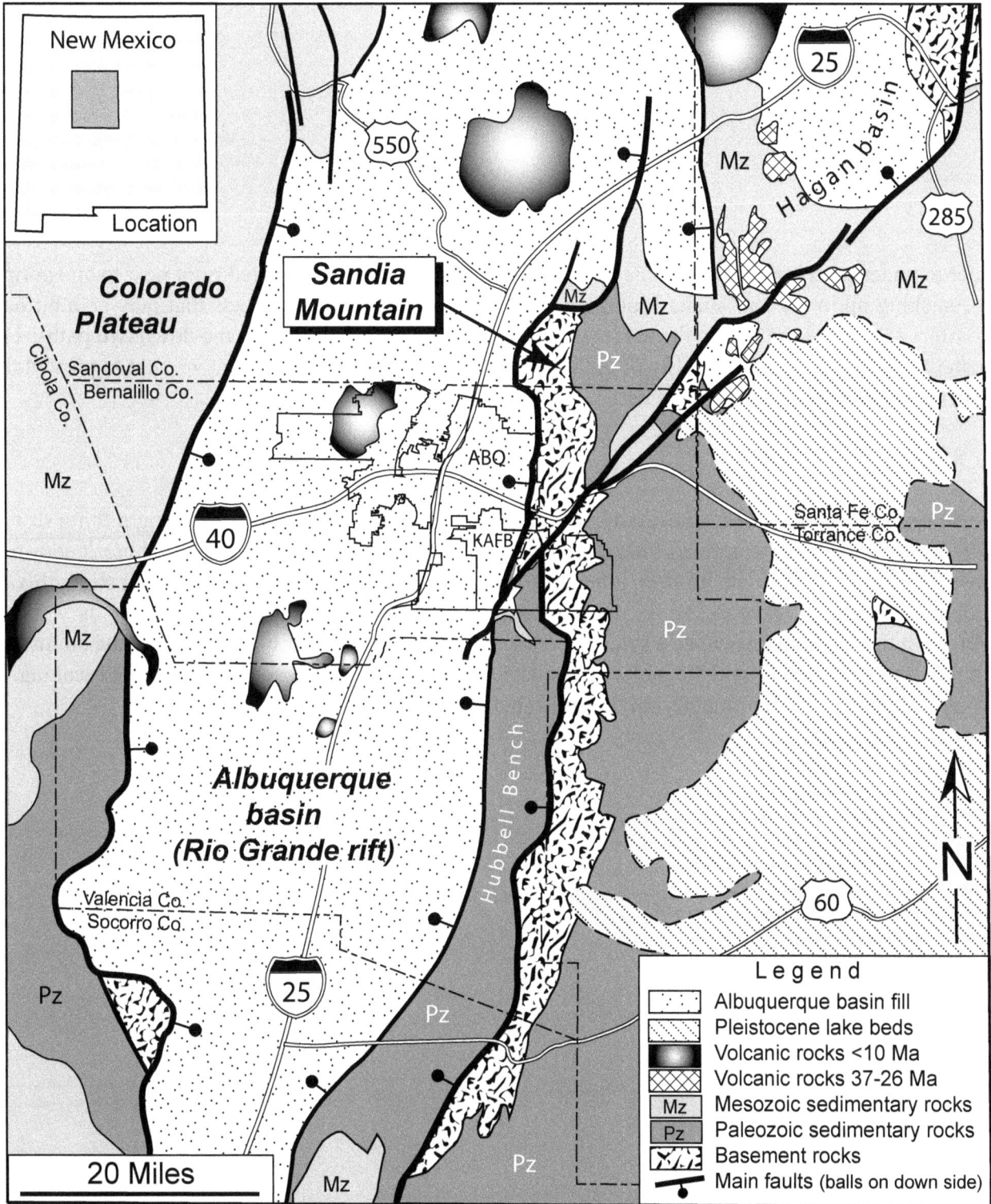

Figure 6-1. Generalized Geologic Map of Sandia Mountain and the Contiguous Albuquerque Basin Portion of the Rio Grande Rift (Modified from NMBG&MR 2003)

In order to understand how a "hole in the ground," the Rio Grande rift, is located right next to an enormous edifice sticking up in the air, Sandia Mountain, we need to touch on two subjects that may be a bit out of the comfort zone for some. I urge the reader to sit back, relax and briefly follow me down two paths: 1) the strength of rocks, and 2) the flexibility of the Earth's crust. In the passages below I attempt to present these ideas in hopefully easy-to-understand terms.

Rock Strength and Flexibility

Rock is hard and tough. Rarely, if ever, do we think of a solid rock like granite as being a thing that will bend. If we were to greatly compress a hand-sample of granite in a bench-top vice it would break and shatter because it is brittle. Let's scale up from a hand sample to a slab of the Earth's crust. Rock strength varies with depth of burial (*Figure 6-2A*). As depth increases, pressure and temperature increase as well. At a certain depth these conditions create a state where the strength of molecular bonds between the atoms in the rock's minerals is overcome and the molecular cells deform and glide past each other without breaking. The material thus becomes a deformable, *ductile* solid (but not a liquid).

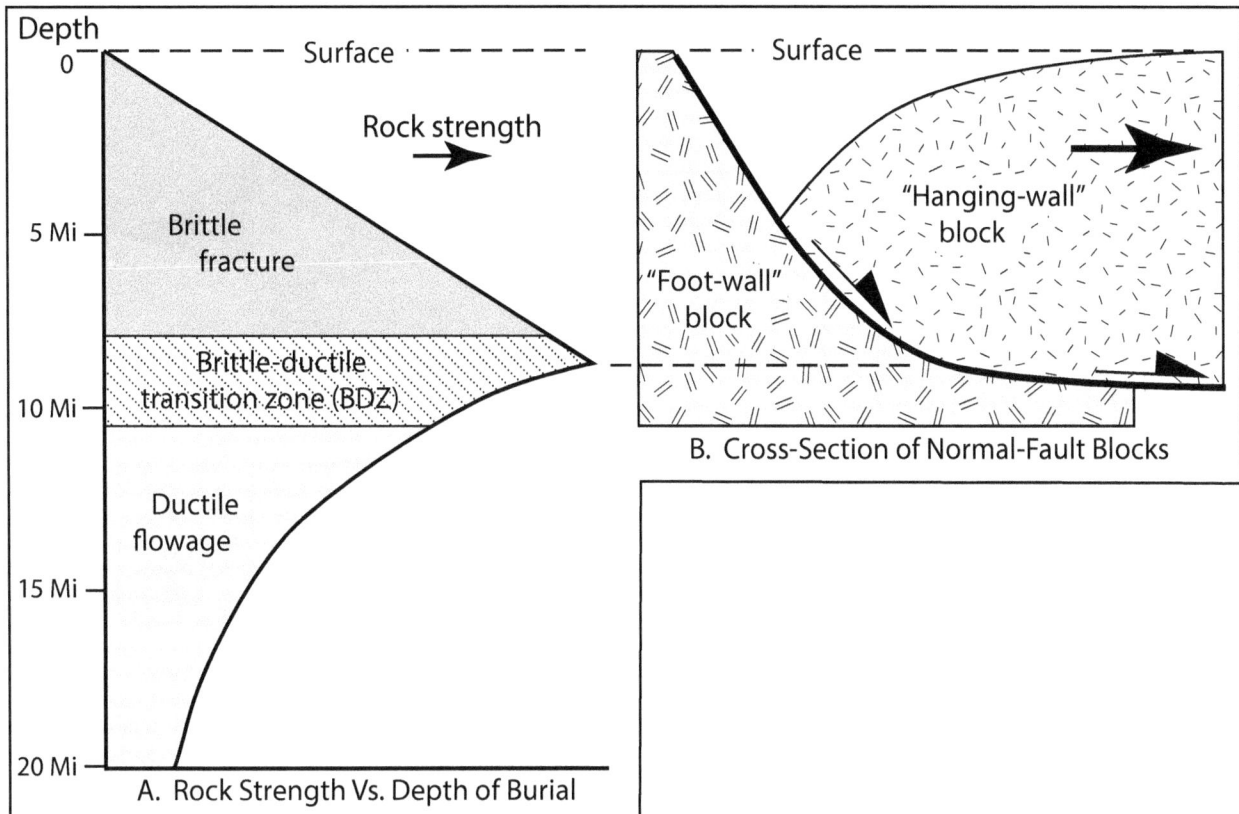

Figure 6-2, Depth-Controlled Strength of Continental-Crust
and Deep Geometry of Large Normal Faults

A critical depth is the *brittle-ductile transition* zone (BDZ), where stressed rock breaks above and flows like plastic below. Notice in *Figure 6-2B* how the heavy-lined fault (which is a break or fracture) in the shallow, brittle domain gradually "soles" (flattens) out at the BDZ, generally around 25,000 to 30,000 ft. Below that zone a fault cannot form because the rock will not break. This is how large segments of the Earth's crust stretch: by the combination of shallow brittle shattering and deep ductile flowage, with the critical BDZ separating the two.

Contrast this to a bar of copper metal (*Figure 6-3A*). If the bar were put under tension it would deform without breaking because it is *ductile*. Pulling the bar apart from its ends causes "necking" or flowage in the middle. The Earth's crust under tension, being more complicated than a copper bar, will break above the BDZ and plastically flow below it (*Figures 6-3B* and *6-3C*). This nature of the two behaviors is vital to understand the origin of Sandia Mountain and the Rio Grande rift (*Figure 6-3D*).

Unstretched ductile metal bar

Ductile "necking"

A. Stretched Ductile Metal Bar

Reference volume

Brittle-ductile transition zone

Earth's surface

Cool, brittle crust

BDZ

Hot, ductile crust

~20-30,000+ ft

B. Schematic Cross Section of Unstretched Slab of Crust

Fracture of brittle zone and formation of topographic sag, which fills up with alluvial material

Alluvial fill

BDZ

Ductile flowage

C. Schematic Cross Section of Stretched Slab of Crust

West

East

Colorado Plateau

Rio Grande rift

Sandia Mt.

Stable Interior

Alluvial fill

Basement

BDZ

Ductile flowage

Vertical exaggeration = 0

D. Schematic Cross Section Showing Principle of Stretched Crust Appied to Sandia Mt./Rio Grande Rift Couple

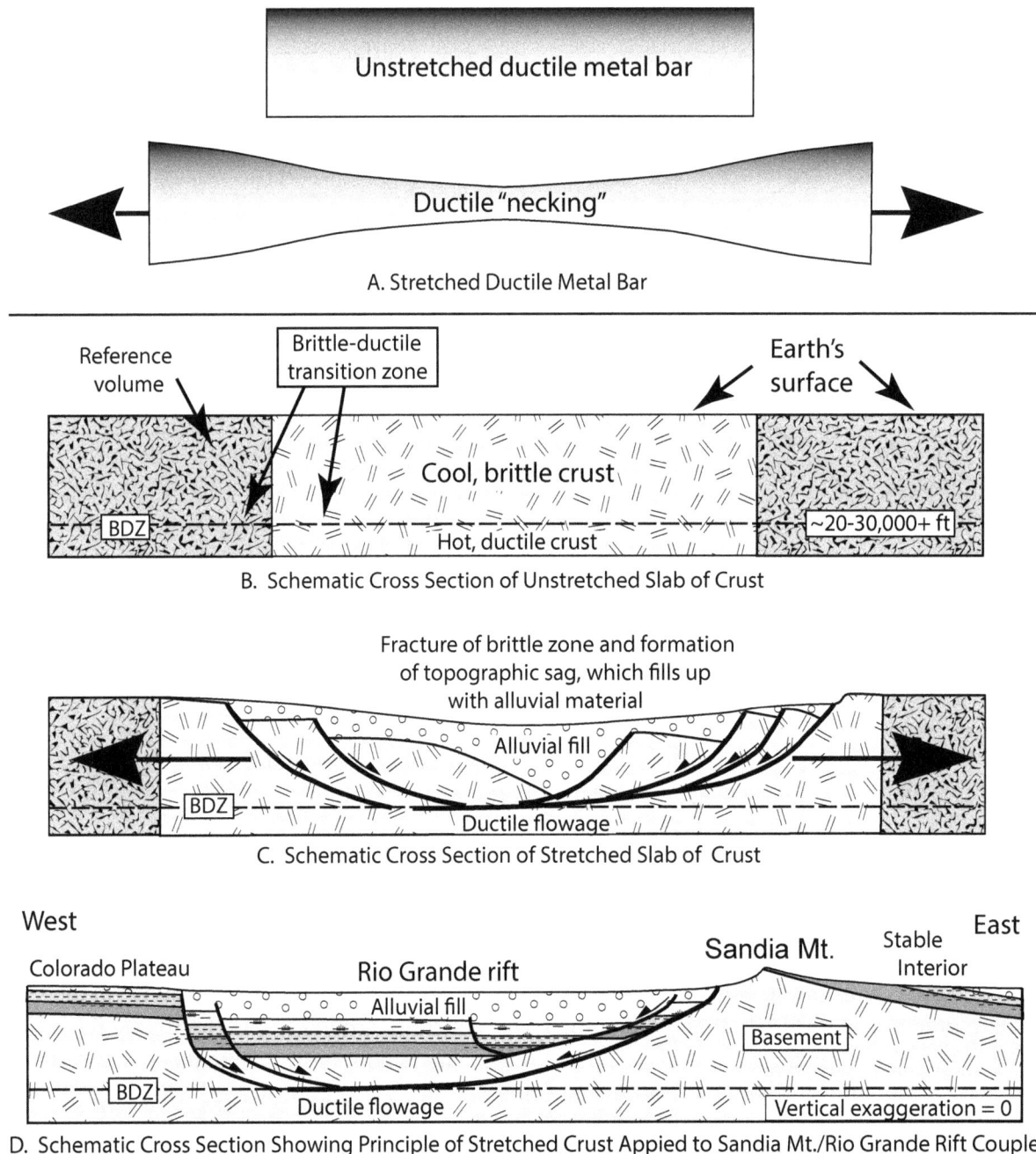

Figure 6-3. Stretching of Earth's Crust

Unloading and Rebound

With the above *Figure 6-2* in mind, we can deal with the second principle: the flexibility of Earth's crust. As a mind experiment, consider a block of ice floating in water. We know that such a block (or an iceberg) floats on water, but just barely. The density of the ice is only slightly less than that of the water, so a small volume of ice (~10%) sticks up above the water (*Figure 6-4A*). The downward force on the ice, i.e., gravity, is equal to, and is *compensated* by, the upward force of the displaced water below the water line. Therefore, the ice does not sink, it floats. The forces balance. If the volume of the exposed ice lessens via melting, the base of the ice in the water rises by buoyancy to a new level where the forces are again in balance.

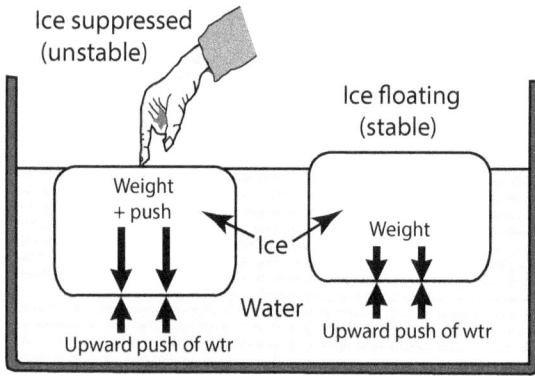

A. Block of Ice in Water

C. Weight of Empty Raft Equal to Upward Force of Water

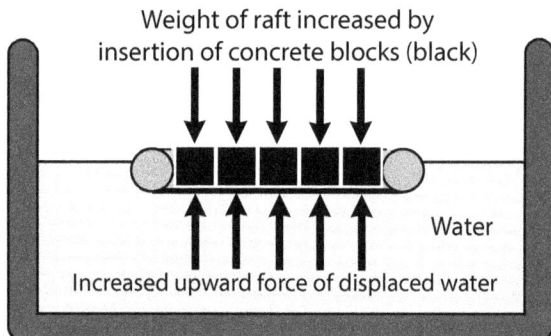

B. Typical Colorado River Raft

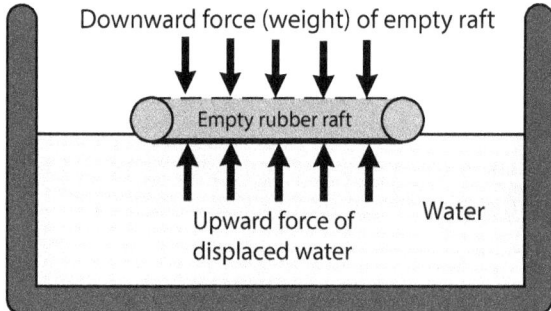

D. Addition of Wall-to-Wall Array of Concrete Blocks

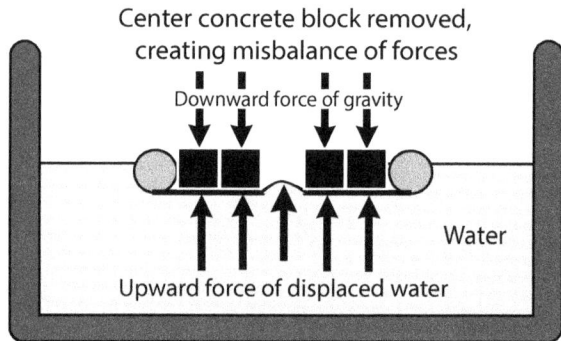

E. Rebound in Center Caused by Unbalanced Forces Above Removed Block

Figure 6-4. Principle of Isostatic Rebound - I

Let's consider another example: a rubber raft floating down the Colorado River (*Figure 6-4B*). In this mind experiment we remove the people and any interior decking so that the raft's floor is rubbery and flexible (*Figure 6-4C*). We now festoon the floor rim to rim with heavy concrete blocks. The raft with its blocks now presses down into the water via gravity, and displaces a volume of water such that the downward force of gravity is again equal to the upward force of the displaced water (*Figure 6-4D*). If we next remove a block from the center of the raft, the flexible rubber raft floor bulges upward because the upward, buoyant force of the water at that spot is no longer being opposed by the gravity from the missing block (*Figure 6-4E*).

We now return to the Earth's crust. Here the downward force (the weight of rigid rock and/or sediment) is perfectly balanced by the equal upward force created by the moveable, ductile (but solid) rock below the BDZ (*Figure 6-5A*).

Earth's surface

Increasing pressure and temperature

Brittle domain

Future fault

Chunk of Earth's crust

~20-30,000 ft

BDZ

Ductile domain

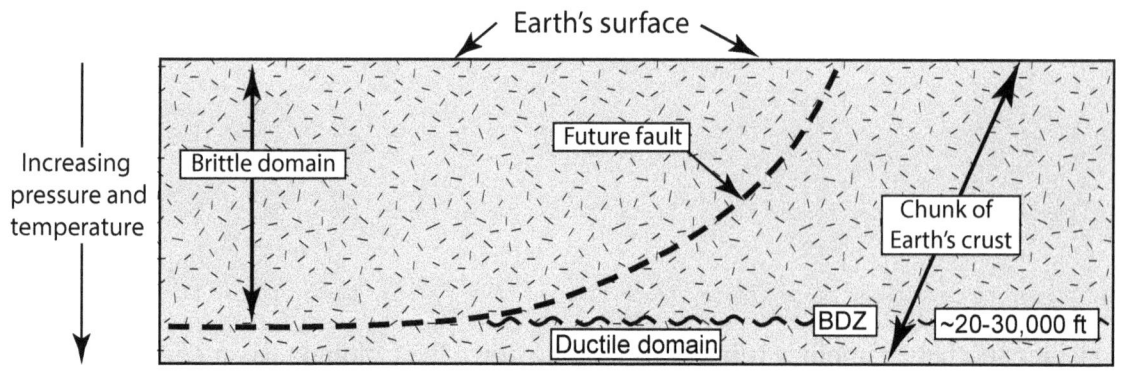

A. Original, Pre-Rift State

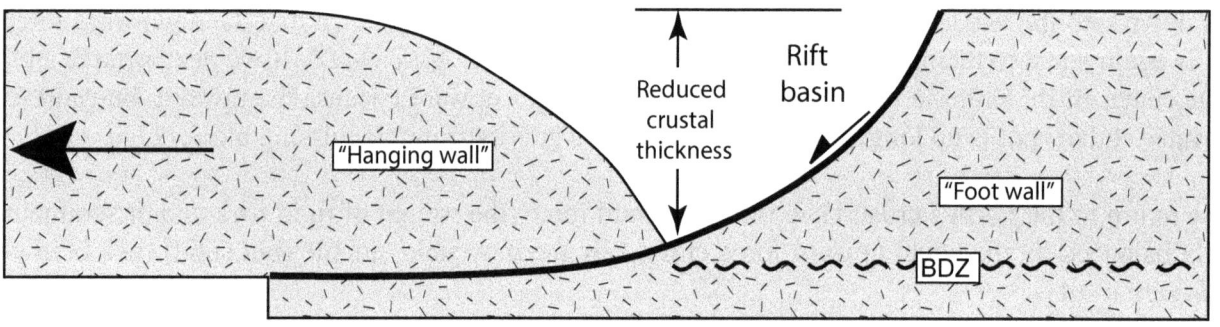

Reduced crustal thickness

Rift basin

"Hanging wall"

"Foot wall"

BDZ

B. Idealized Stretched and Rifted Crust

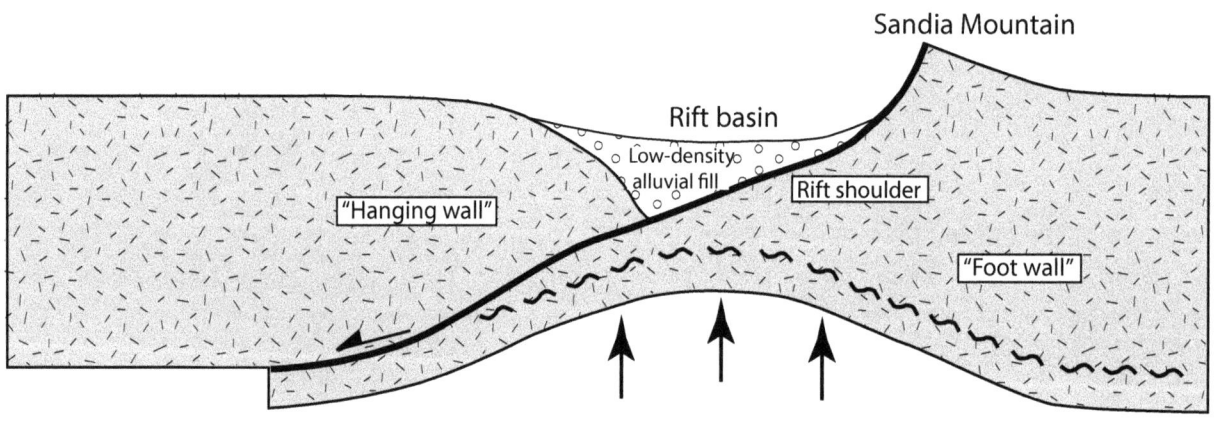

Sandia Mountain

Rift basin

Low-density alluvial fill

"Hanging wall"

Rift shoulder

"Foot wall"

C. Isotatic Rebound of Rift Shoulder Due to Force Imbalance

Figure 6-5. Principle of Isostatic Rebound - II

If the downward gravitational force is locally lessened by stretching and thinning of the upper crust, creating a basin, the forces would be unbalanced (*Figure 6-5B*). Because of the developing imbalance of forces, the ductile rock below the BDZ will bulge upward below the basin, rotating back part of the footwall block (*Figure 6-5C*). This upward bulging is called *isostatic rebound*.

The bounding fault on Sandia Mountain's west side, the Rincón fault plus some associated buried faults farther out in the rift to the west, collectively experienced something between 20,000 ft (Ricketts et al. 2015) to a whopping 27,000 ft (Connell 1995) of down-to-the west displacement. This was accompanied by about 6000 ft (Connell et al. 1999) of down-to-the-north drop on the northern nose along an array of normal faults (the Placitas fault zone). This enormous displacement (see below) created the attendant gravitational imbalance that caused the Sandia Mountain block to *isostatically* rebound and rotate back to the east about 15°, thus forming that east-sloping ramp of the mountain's east side (*Figure 6-5C*).

Huge fault separation: Isleta #2 well. The total amount of Albuquerque basin subsidence vs. uplift of Sandia Mountain is stunning. The maximum depth of the basin ever recorded is by a well drilled by Shell Oil back in 1980. Their Isleta #2 reached a total depth (TD) of 21,266 ft, and never even reached Mesozoic rocks. Using known thickness of the sedimentary section measured in the Hagan basin (Kelley and Northrop 1975), the top of the buried basement at the Isleta #2 location is projected to lie at a depth of almost 32,000 ft, or roughly 27,000 ft below sea level, vs. where basement is roughly at 10,000 ft above sea level at Sandia Crest, for a vertical difference of more than 37,000 ft (*Figure 6-6*).

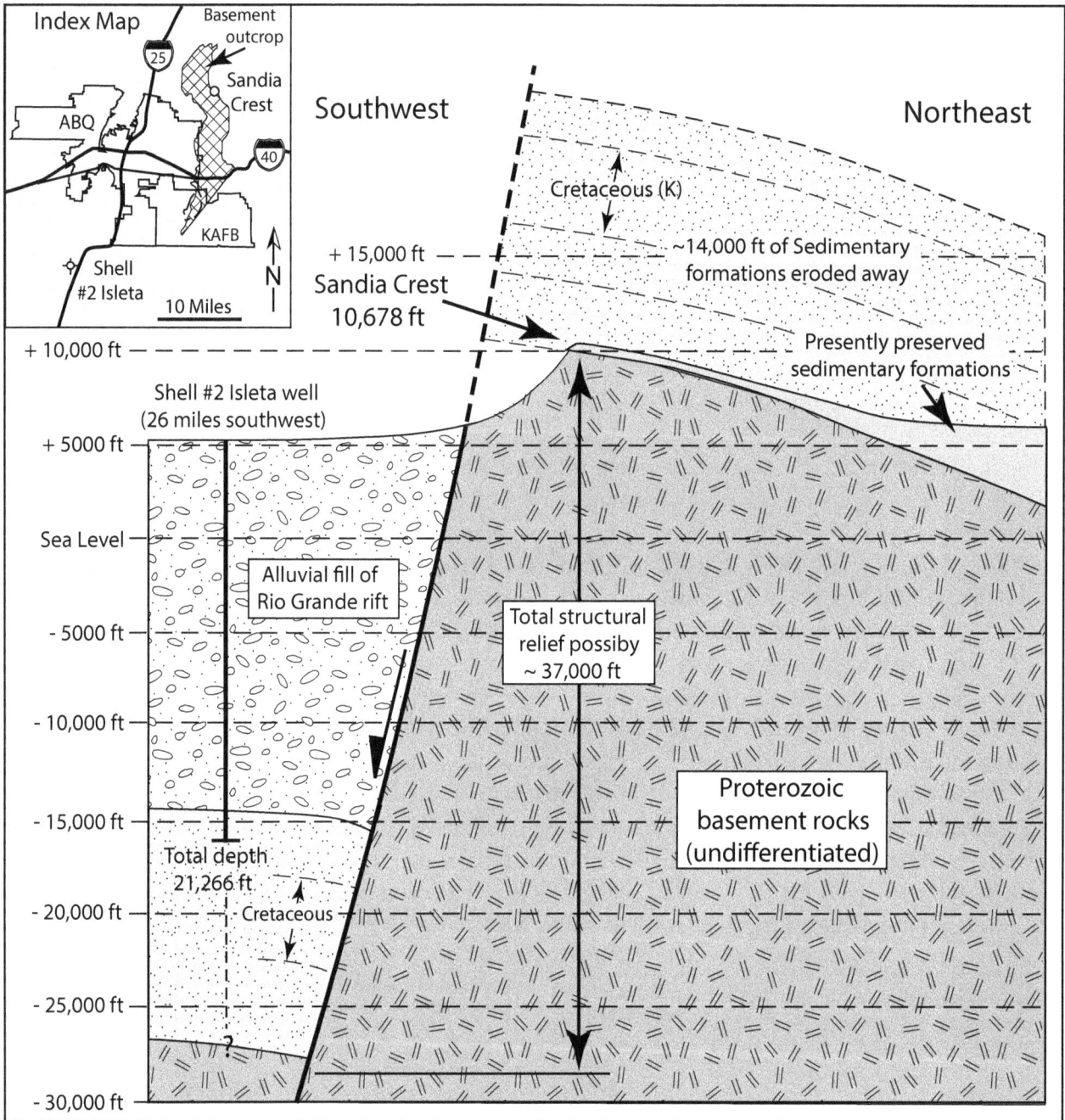

Figure 6-6. Schematic Contrast (Not a Cross Section) Between Highest (Sandia Crest) and Lowest (Isleta #2 Well) Basement Elevations, Sandia Mountain and Albuquerque Basin

To review, after the Laramide compression (55-40 Ma) and the transition period of the volcanic flareup (36-27 Ma), the new environment of regional tension slowly developed (*Figure 4-2*). (We will visit the "big picture" reason for this in *Chapter 15*). The direction of tension was in a SW-NE direction, nearly opposite to the direction of the latest Laramide compression (*Figure 4-3*). Therefore, it is not surprising that many of the pre-existing Laramide faults were "re-tooled" and thus reused. The regional tension pulled the huge crustal block known as the Colorado Plateau southwestward about five miles away from the even larger mass of the Stable Interior, creating the stretched Rio Grande rift in between (Woodward 1977).

History of Uplift

One would think that with more than a century of observation and mapping under our belts we'd have the story of the mountain's uplift down pat. That's not the case. The reason is that this mountain has a complicated, multi-phase origin, complicated by the fact that much of the evidence for especially the earliest (Laramide) uplift has been obscured or eroded away.

We do know that the mountain has had at least a three-phase uplift history. The first was the quasi-uplift back during the Laramide, already discussed in *Chapter 4*. Faults created at that time acted as "scars" that never completely "healed." Think of a leg that has been broken once. If the leg is shocked again and then re-broken, the break will likely be in the same place due to that zone of weakness that never quite went away. Some of the huge "normal faults" bordering the mountain's west side may have had Laramide "reverse fault" parentage and were later reactivated in the opposite direction during formation of the Rio Grande rift.

During this latter event there were two tensional phases: Tensional Phase 1, ca. 20-16 Ma, a pause or "hiccup" 16-14 Ma, and then Tensional Phase 2, 14-3 Ma.

"Timepieces." A clock or "timepiece" of sorts exists in the Placitas area (*Figure 6-7*). One area in particular (circled in *Figure 6-7A*) is a jewel box of timing information. There the Cretaceous Menefee Formation (*Kmf*), ca. 75 Ma, dips to the northeast at about 60°, and the overlying lower Santa Fe Group gravels (*Lsf*), dated at somewhere between 15 and 10 Ma, dip to the northeast at about 30°, creating an angular unconformity (*Figure 3-1*). Within the Menefee is an igneous dike (the "Placitas dike") that has been radiometrically dated at ca. 31 Ma (Lundahl and Geissman 1999).

Legend

Lsf   Tertiary, Lower Santa Fe Grp.

Cretaceous

Kmf   Menefee Formation

Kh   *Harmon* Sandstone

Kpl   Point Lookout Sandstone

Kmu   Upper Mancos Shale

Khd   Hosta-Dalton Sandstone

Kml   Lower Mancos Shale

Formation dip direction and angle

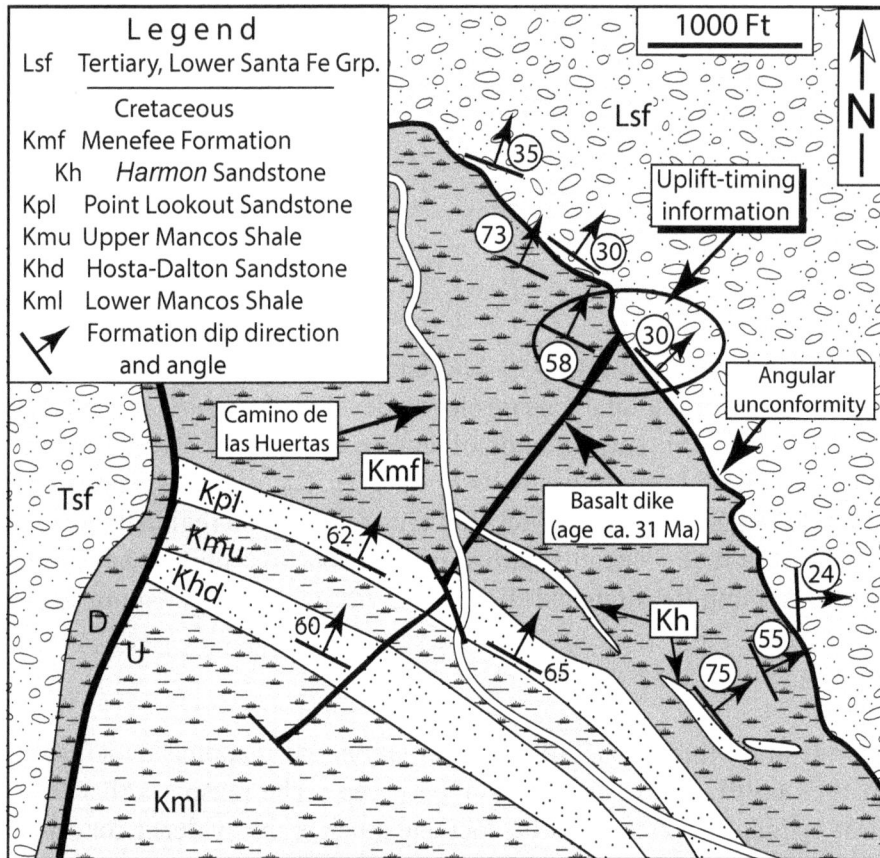

A. Geologic Map of Northern Part of Placitas Area Showing Basalt Dike and Orientation of Formation Bedding (Modified from Connell et al. 1995)

B. East Oblique *Google Earth* Image of Placitas Dike Area

Figure 6-7. Placitas "Timepiece"

Another timepiece is provided by gravels of the Santa Fe Group. At Placitas the lower Santa Fe gravels were deposited as a *fanglomerate* (a bouldery alluvial fan) shed off the growing Sandia uplift to its south. At the base of these gravels are redbed boulders eroded from nearby outcrops of the Permian, Abo Formation. Higher up in the fanglomerate is an increasing abundance of Pennsylvanian, Madera limestone boulders, and finally about 35 ft above the base are chunks, or "clasts" of basement rocks (Woodward and Menne 1995). Out in the Albuquerque basin, basement clasts begin appearing in alluvial-fan deposits eroded from Sandia Mountain that have been dated at ca. 15 Ma to 10 Ma (Lozinsky 1988). Therefore, by 15-10 Ma Sandia Mountain had in places been uplifted, rotated, and eroded down to its basement core.

Some data strongly suggest that the first phase of tension and mountain uplift began slowly, roughly at about the same time as the waning of the Ortiz igneous activity to the east. It started sometime after 30 Ma and at least by ca. 22 or 20 Ma and lasted until roughly 16 Ma.

Tensional Phase 1 (20-16 Ma). Significantly this phase involved *low-angle faulting*. Such faulting is a harbinger of something very important and to describe it we must venture a bit back into the tall weeds.

Low-angle faulting. A brittle block of rock subjected to tension will fracture, gravity will take over, and one piece will drop down along a fault plane dipping roughly 60° from vertical—the "normal" angle (*Figure 0-10B*). A low fault-angle therefore suggests *rotation* back from about 60°. It also suggests that the faulting occurred at depth, i.e., below the BDZ where some plastic flowage could occur, and has been uplifted to the surface (remember *Figure 6-5*). The implication is that the mountain has experienced enormous uplift.

There are four places on the mountain's west side that exhibit low-angle faulting: 1) Piedra Lisa Ridge at the northern rim of the so-called Juan Tabó Cul-de-Sac, 2) an outlier of Mesozoic rocks in that cul-de-sac, 3) along Tramway Road on the way up the hill to the tram, and 4) Caliza Hill (*Figure 6-8*).

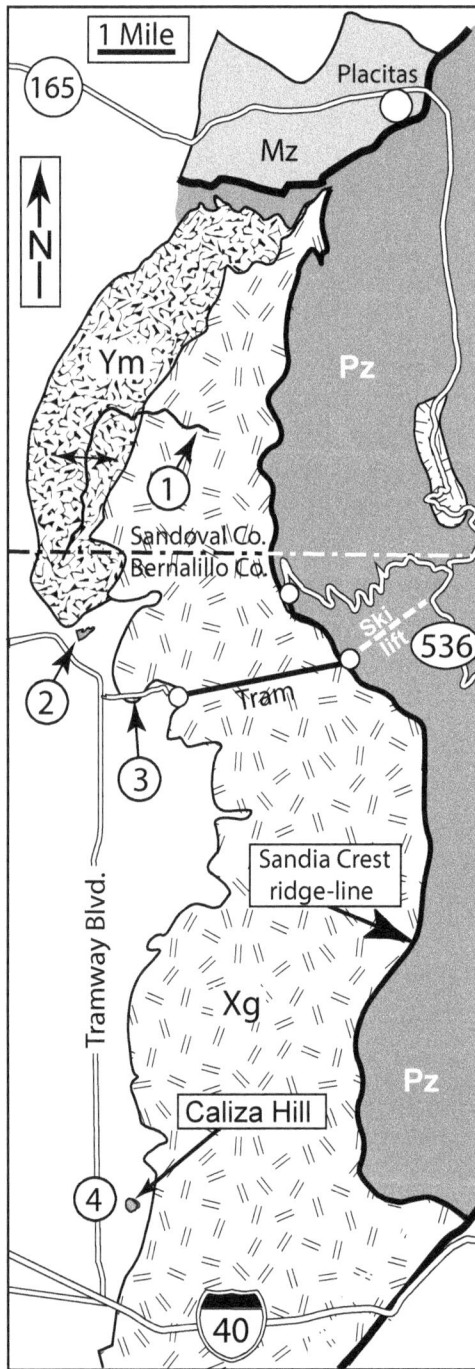

Figure 6-8. Location Map Showing
Sites of Low-Angle-Faulting

Site 1: Piedra Lisa Ridge. This informal name designates the high-up drainage divide that separates the Juan Tabó Cul-de-Sac from the Placitas area to the north (not to be confused with Piedra Lisa Canyon in the foothills, see *Chapter 11*). It is reached by an hour's hike up the South Piedra Lisa Trail, followed by some unpleasant bushwacking eastward along the ridge (*Figures 6-9* and *6-10A*). The ridge sports a number of pinnacles of granite (*Figures 6-10A* and *6-10B*). At the base and south side of the easternmost pinnacle is a nicely-displayed fault that dips 30-40° to the west, called the *Knife Edge Fault* (*Figure 6-9D*; NMBG&MR 2014). Just to the west of it, near the base of another small pinnacle, is another fault, but with dips flattening westward from about 20° to *nearly-horizontal* (*Figures 6-10B* and *6-10C)*. It is frustratingly unclear, but important to know, where the plane of this low-angle fault goes farther to the west: Does it rise and project over the top of the high parts of the ridge or does it extend downward somewhere into the granite, and disappear from sight (*Figure 6-10A*)?

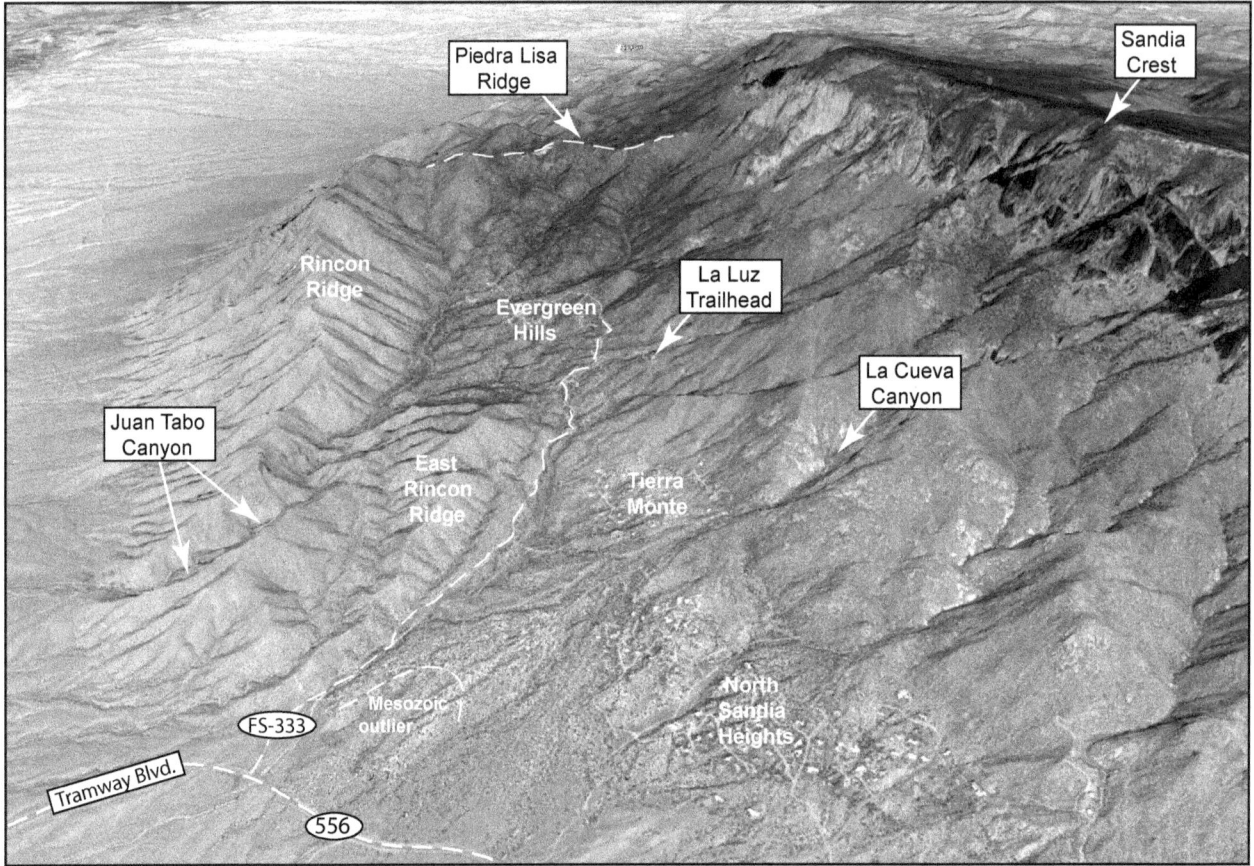

A. North-Northeast Oblique *Google Earth* Image

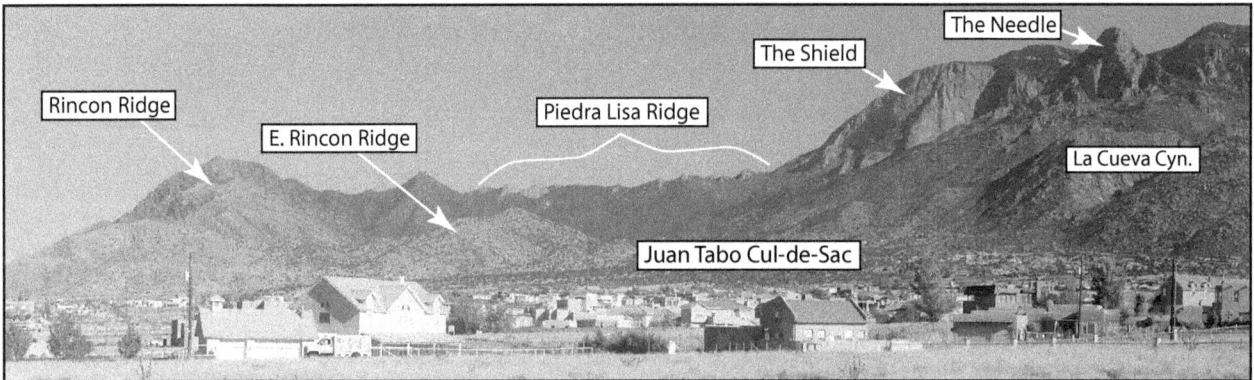

B. North View (Photograph by Author)

Figure 6-9. Juan Tabo Cul-de-Sac

~ 1750 ft

Top of Piedra Lisa Trail just to west

Piedra Lisa Ridge

"Knife-Edge" fault zone

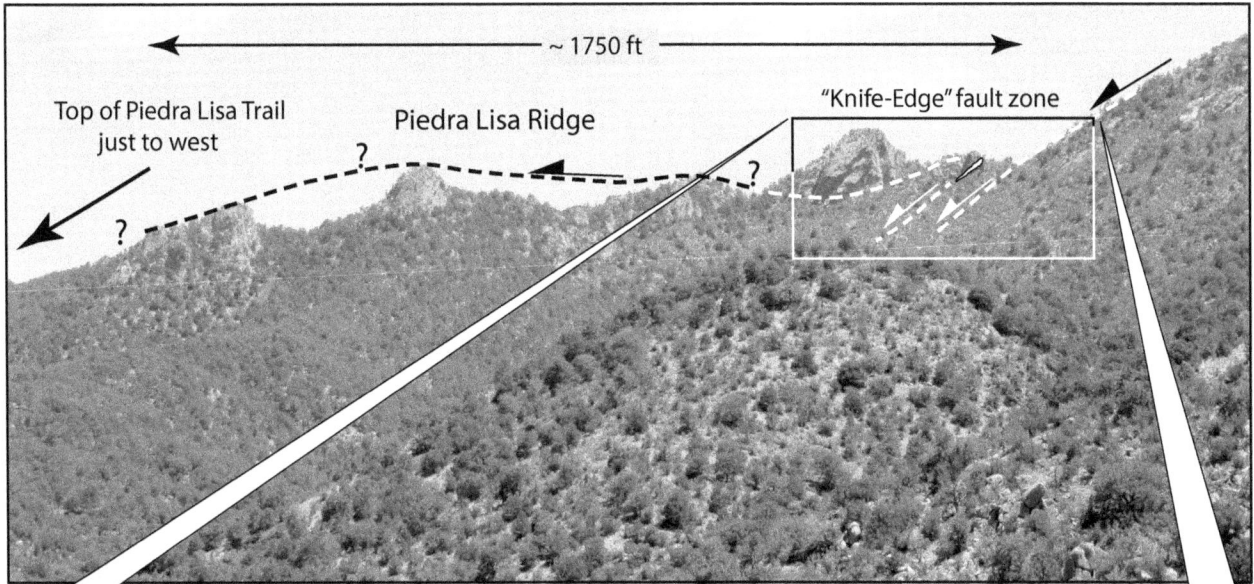

A. North View (Photograph by Author)

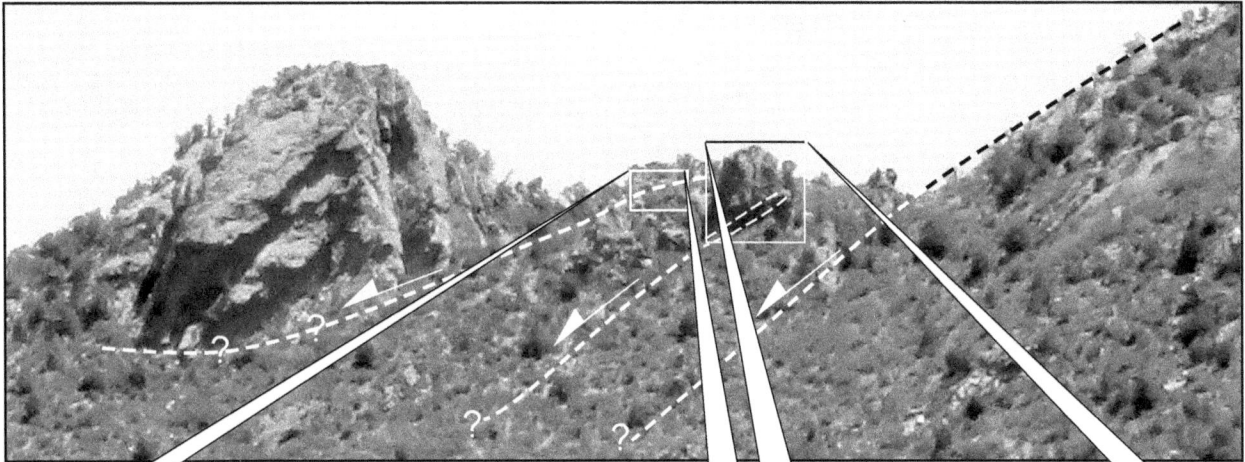

B. Close-Up View (Photograph by Author)

Sandia Granite

Crushed granite

Sandia Granite

C. Detail of Low-Angle Fault (Photograph by Author)

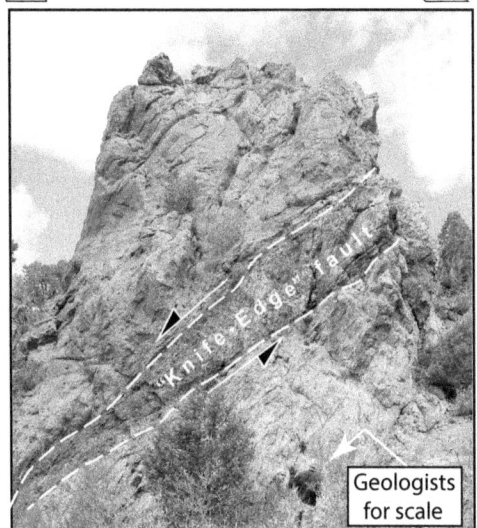

"Knife-Edge" fault

Geologists for scale

D. Rock Pinnacle in "B" Above
(NMBG&MR 2014, and Courtesy of A. Read)

Figure 6-10 Piedra Lisa Ridge and "Knife-Edge" Fault Zone

Site 2: Mesozoic outlier. The second example is located within the cul-de-sac, on the south side of FS-333B leading to the La Cueva Picnic Area (*Figures 6-9A* and *6-11*). Here in an unnamed drainage gulch are easily accessed (and easily overlooked) outcrops of upper Triassic and Jurassic formations. The gulch terminates in two arms of a "Y" pointing north, and the best exposures are within each of the arms. The gulch has eroded through a sheet of young, bouldery alluvium that frustratingly sloughs down from its eroded edges into the gulch. The Mesozoic formations dip mostly about 30 to 40° east to east-northeast—somewhat similar to those strata exposed in the Placitas area to the north, suggesting that at one time both sequences may have been connected.

Legend

Qal Quaternary alluvium

Qt Quaternary terrace

QTf Quaternary/Tertiary
alluvial-fan cover

J Jurassic (Entrada, Summerville,
and Morrison Formations)

Jt Distinctive, Jurassic Todilto Ls.
(shown thicker than actual)

TRc Triassic, Chinle Formation

TRa Triassic, Agua Zarca Sandstone

Xg Sandia Granite

Ym Metamorphic rocks

⊢ Formation dips

"Spoon-shaped"
low-angle fault
(ball on down side)

M Magnetometer lines

500 Ft

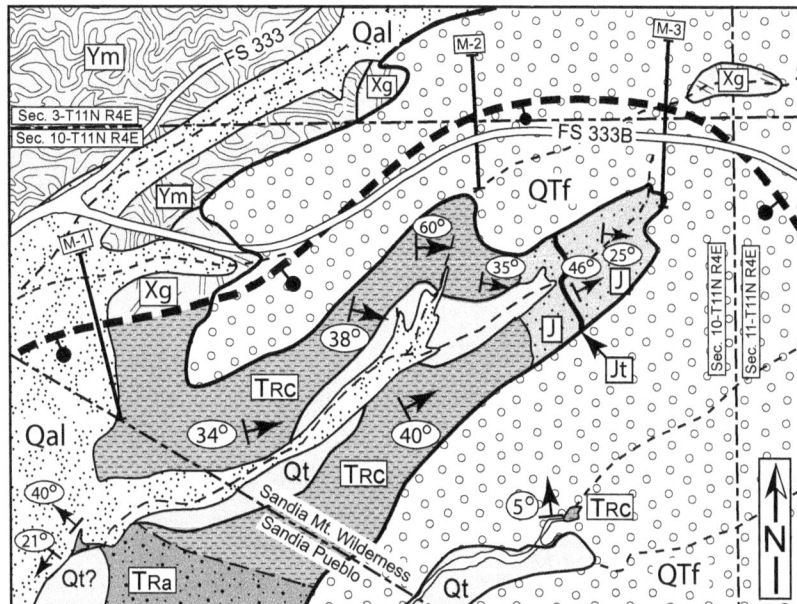

A. Present Outcrops (Much Bedrock Obscured by Slump of Loose Material from Overlying QTf)

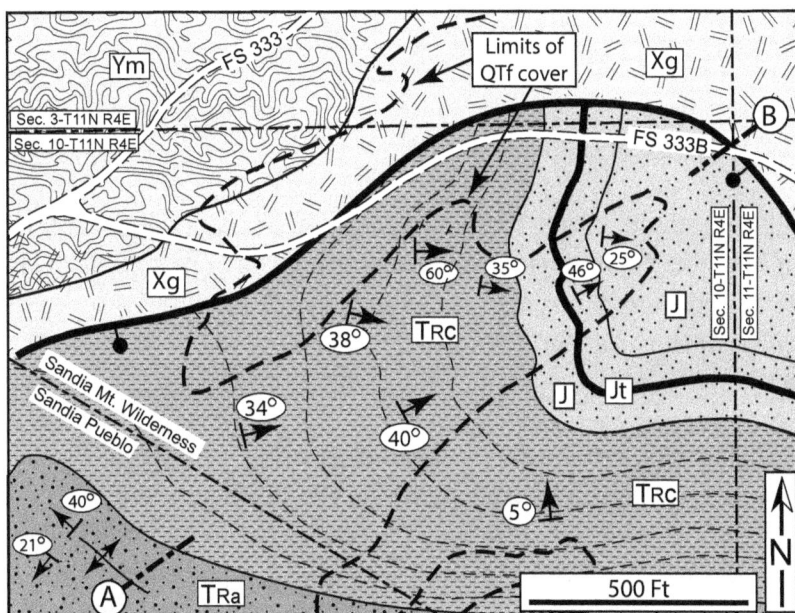

B. Interpreted Distribution of Mesozoic Bedrock

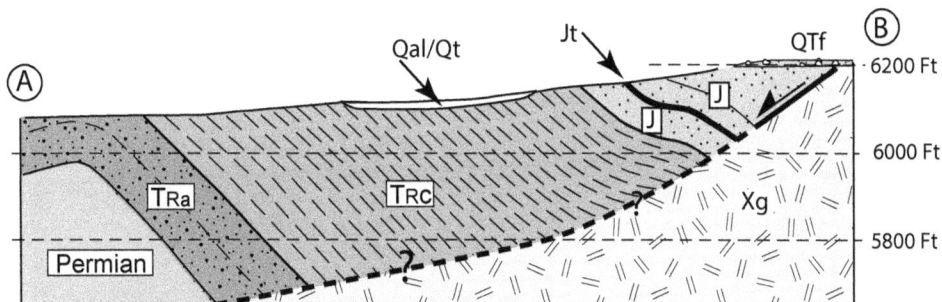

C. Cross Section A-B in "B" Above (Vertical Exaggeration = 0)

Figure 6-11. Outlier of Mesozoic Rocks and Low-Angle Faulting in Juan Tabo Area
(Modified from Van Hart 1999)

On northwest, north, and northeast sides, the Mesozoic exposures are surrounded by Sandia Granite and separated from them by a buried fault (Van Hart 1999). Importantly, the trace of this fault is curved in map view (indicated by three geophysical, magnetometer survey lines), suggesting that it is low-angle, spoon-shaped fault that dips to the southwest under the formations.

Site 3: Tramway Road. This site lies along the south side of Tramway Road (not Tramway Blvd.), down the hill from, and west of, the lower Tram terminal (*Figure 6-8*). The smooth top of the granite here has been sheared by some very large mass moving across it at a low angle (Rhodes and Callender 1983). Unfortunately, these details are only recognizable to a geologist's trained eye. Therefore, the site is indeed important, but not something to visit.

Site 4: Caliza Hill. Because this place lies well (about six miles) to the south (*Figure 6-8*) I postpone discussion of it to *Chapter 11, Sandia Foothills*.

Critical questions are what caused the bizarre low-angle faulting, and what does it all mean? To make sense of the huge uplift this implies I very cautiously introduce a new term: *core complex*. This term arose for the first time in the early 1980s to designate strange, very large, bulge-like uplifts of highly-deformed rock in the American West that have risen to the surface from great depth. I ask the reader to come with me again just a little farther into the tall weeds.

Core complex. This odd type of structure, characterized by low-angle faulting, has developed in provinces that have been severely stretched, such as in the *Basin and Range Province* of western North America (*Figure 6-12A*).Where crust has been severely stretched, the ductile-brittle zone (BDZ, remember *Figure 6-2*) actually reaches the surface. The brittle, faulted blocks of crust that once were on top of the zone slid off the bulge along a master low-angle fault (*Figure 6-12B*). It is important to realize that crustal stretching operates on a continuum—from almost nothing to extreme. (The closest, and very best, example of a core complex is the Catalina Mountains that form the incredible northern backdrop for the city of Tucson, Arizona.)

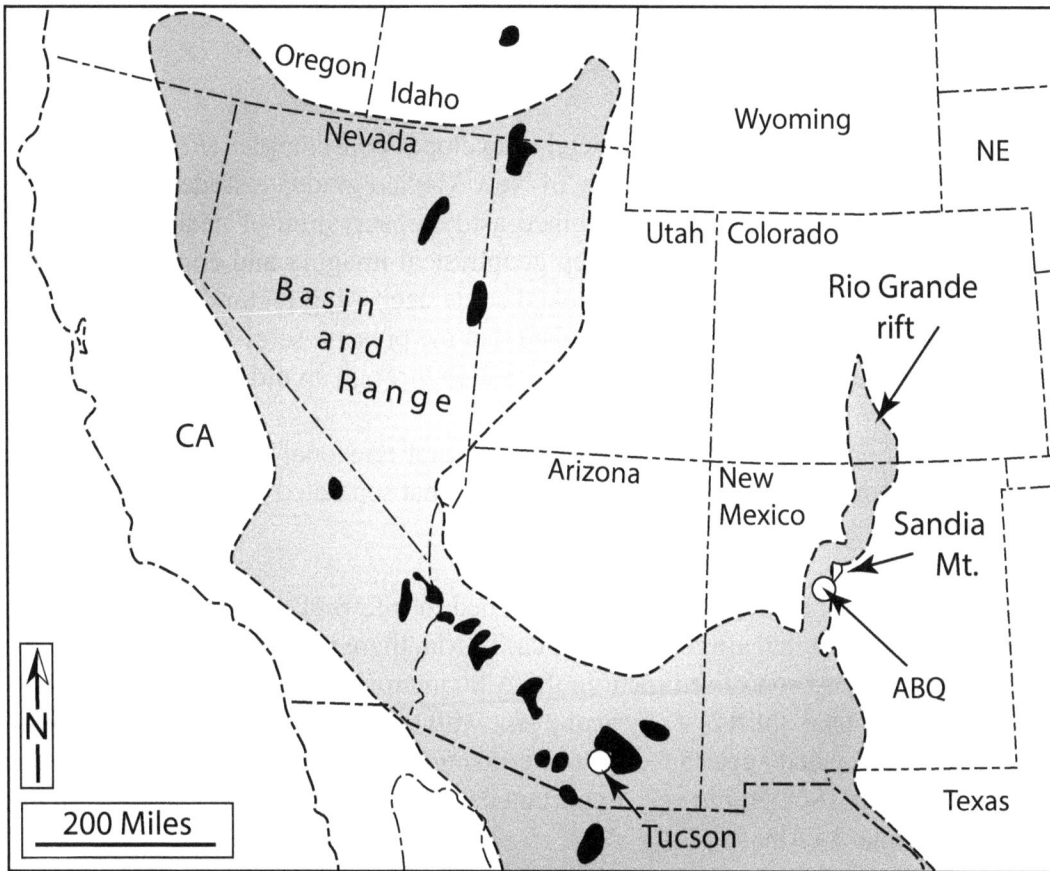

A. Map of **Basin-and-Range** Province of Crustal Stretching (Shaded) and "Core Complexes" (Black), Southwestern North America

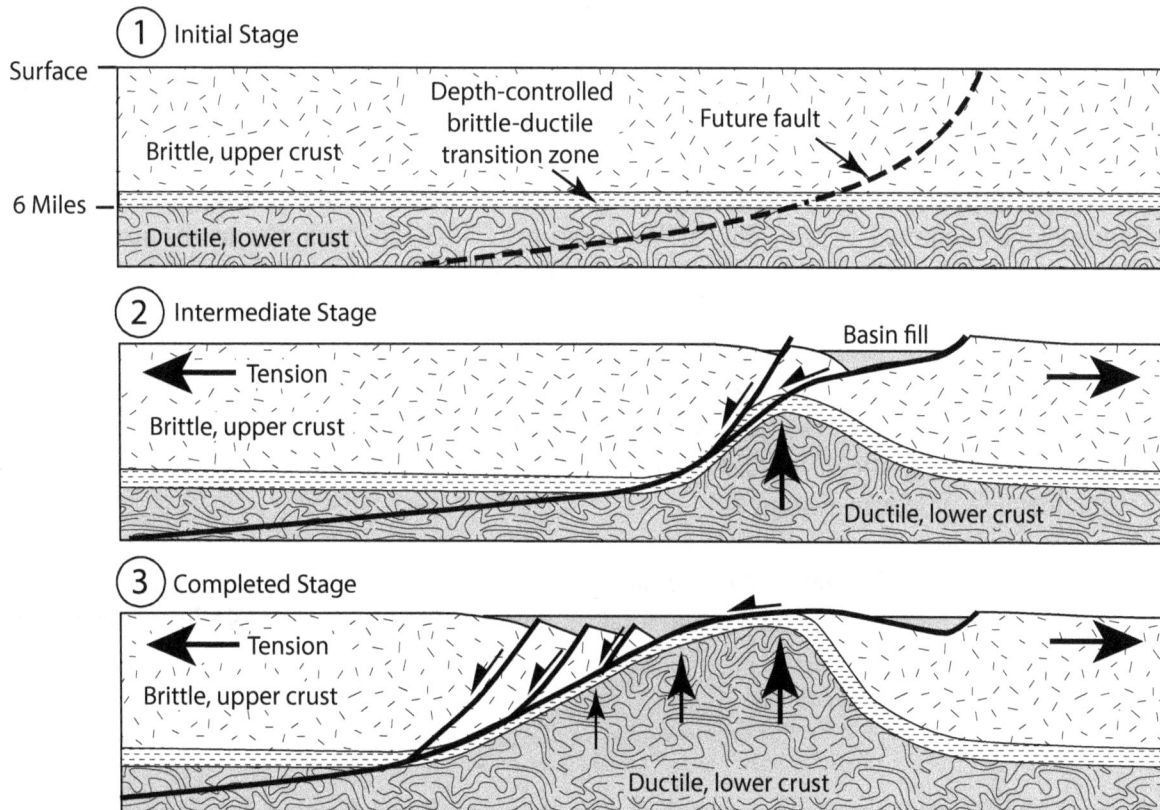

① Initial Stage

Surface

Brittle, upper crust

Depth-controlled brittle-ductile transition zone

Future fault

6 Miles

Ductile, lower crust

② Intermediate Stage

Tension

Basin fill

Brittle, upper crust

Ductile, lower crust

③ Completed Stage

Tension

Brittle, upper crust

Ductile, lower crust

B. Schematic, Sequential Cross Sections Showing Evolution of Core Complex

Figure 6-12. "Core Complexes": Products of Regional Tension

The entire Juan Tabó block may be an incompletely-developed core complex (*Figure 6-13*). This revolutionary view is a very recent idea. In 2013 University of New Mexico graduate student Jason Ricketts (advised by Karl Karlstrom, whom we met earlier) combined astute observation of nearby analogues and Basin and Range structures with recent advances in deep geophysical imaging and computer power to advance this new model (Ricketts et al. 2014). He also invoked some deep, "big picture" processes which we'll visit in *Chapter 15*. I will stick my neck out and speculate that the present, semi-flat-bottomed, south-sloping floor of the cul-de-sac may have something to do with erosion through an old, rolled-over, low-angle fault.

This low-angle faulting seems to have concluded the first tensional phase of uplift (House et al. 2003). After that came a tectonic lull, a "hiccup" ca. 16-14 Ma, that separated the two tensional phases. Its cause is elusive and invokes some major geologic arm-waving.

"Hiccup" (16-14 Ma). Something important happened to interrupt the uplift. Studies of sediments in the Albuquerque basin suggest that after that hiatus, ca. 14 Ma, there was almost a tripling of the rate of erosion of the mountain and deposition of sediment in the Albuquerque basin. What caused such a discontinuity is no trifling question. Crustal uplift is a physical event with enormous momentum that doesn't easily change its course. Think of a loaded supertanker at sea cruising at full throttle. To change its speed or direction requires a massive effort. Something very important caused the uplift rate of Sandia Mountain to slow at 16 Ma and them to resume 14 Ma.

It turns out that a great deal happened globally at about that time. Close to home, the Colorado Plateau of northeastern Arizona and northwestern New Mexico experienced more than 4000 ft of downward erosion starting ca. 27 Ma and ending 16 Ma (Cather et al. 2008). Beginning ca. 17 Ma, the Basin and Range Province in western North America experienced a period of rapid extension (Quigley and Karlstrom 2010). And then there comes the arm-waving.

Far-away impact? Some geologists speculate that the onset of the vast outpouring of Columbia Plateau lavas centered in the state of Washington, 17-16 Ma, was due to the impact of a large meteorite in southeastern Oregon. The *bolide* pierced right down to the Earth's hot mantle, unloaded the confining pressure on it, causing it to melt and to spew out some 42 cubic miles of basalt over a time period of about 1 Ma (Alt and Hyndman 1989). This then might have triggered the hotspot, across which North America migrated, resulting in the volcanic outpourings of the Snake River Plain of Idaho extending to the Yellowstone Plateau of Wyoming. If true, the shock, and the possible effect on global climate, could have been significant. Could such a *possible* impact have jostled a place as far away as New Mexico? Coincidence? Wild arm-waving?

Tensional Phase 2 (14–3 Ma). After 14 Ma the rate of uplift of the mountain almost tripled, adding 11,650 ft of uplift to the existing 4000 ft, giving a grand total of about 15,650 ft of uplift (Kelley and Duncan 1984; Woodward and Menne, 1995).

Pediments

Below the steep upper part of the shear upper cliffs (the "Western Wall" discussed below) are some deeply-dissected, step-like, gently west-sloping surfaces, creating a separate, "lower mountain" (*Figure 6-14*). These are likely remnants of several once-continuous *pediments*.

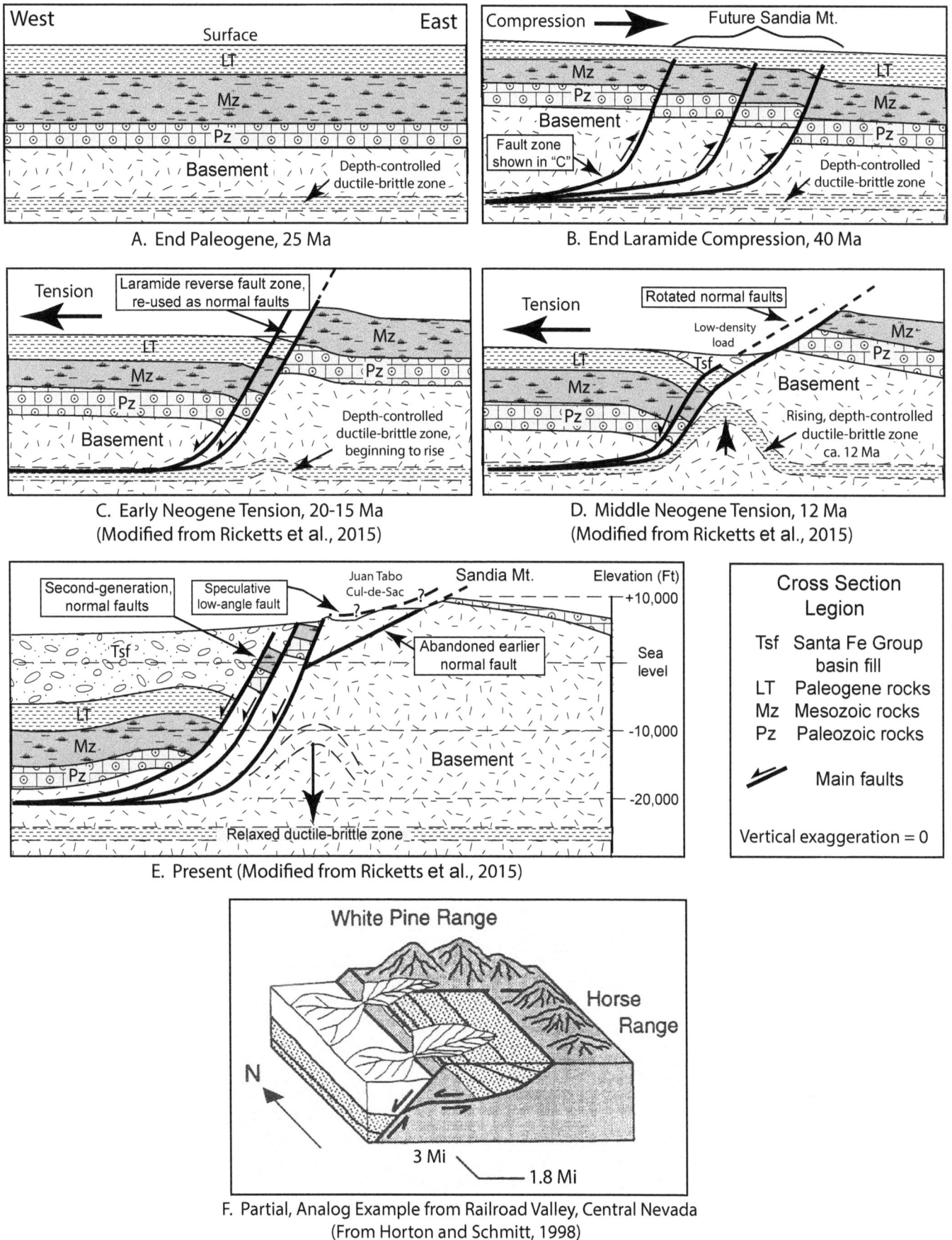

A. End Paleogene, 25 Ma

B. End Laramide Compression, 40 Ma

C. Early Neogene Tension, 20-15 Ma
(Modified from Ricketts et al., 2015)

D. Middle Neogene Tension, 12 Ma
(Modified from Ricketts et al., 2015)

E. Present (Modified from Ricketts et al., 2015)

Cross Section
Legion

Tsf Santa Fe Group
      basin fill
LT   Paleogene rocks
Mz   Mesozoic rocks
Pz   Paleozoic rocks

Main faults

Vertical exaggeration = 0

F. Partial, Analog Example from Railroad Valley, Central Nevada
(From Horton and Schmitt, 1998)

Figure 6-13. Schematic West-to-East Cross Sections Showing Generalized Uplift History of Sandia Mountain

A. South-Southeast, Oblique *Google Earth* Image

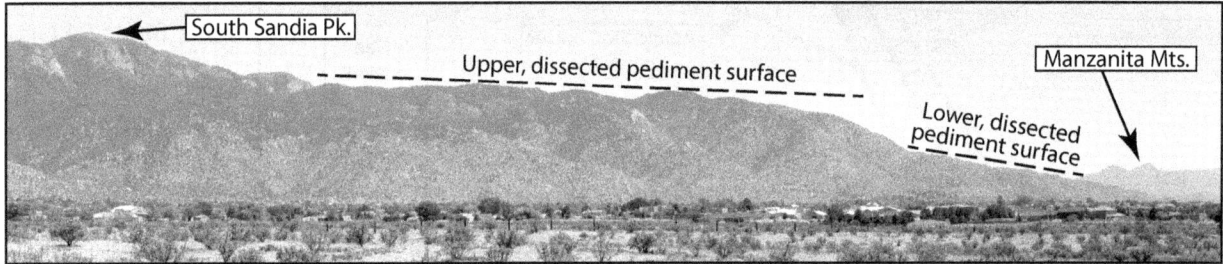

B. South-Southeast View from Tramway Blvd./NM-556 (Photograph by Author)

C. East View from Tramway Blvd./NM-556 (Photograph by Author)

Figure 6-14. Ancient, Dissected Pediment Surfaces

Pediment. The Latin stem *ped*, meaning "foot," is used in such words as "pedal" and "pedestrian." The term denotes a low-relief, gently-sloping erosional surface formed on bedrock, usually at the foot of a receding, abrupt mountain front. Pediments form in arid areas of high relief where the growth of vegetation is limited, and develop over a prolonged period of stability during which streams have had sufficient time to bevel a rock surface to a semi-smooth plain.

To grasp what this "lower mountain" is all about, it is instructive to look to the Four Hills area on Kirtland Air Force Base (KAFB) just to the south of Sandia Mountain (*Figure 6-15*). Four Hills is a little mountain with four bumps, and is the namesake of Albuquerque's Four Hills neighborhood to its north. KAFB has renamed the feature *Manzano Base*, which is restricted territory and completely surrounded by two strands of barbed wire (one of which was once electrified).

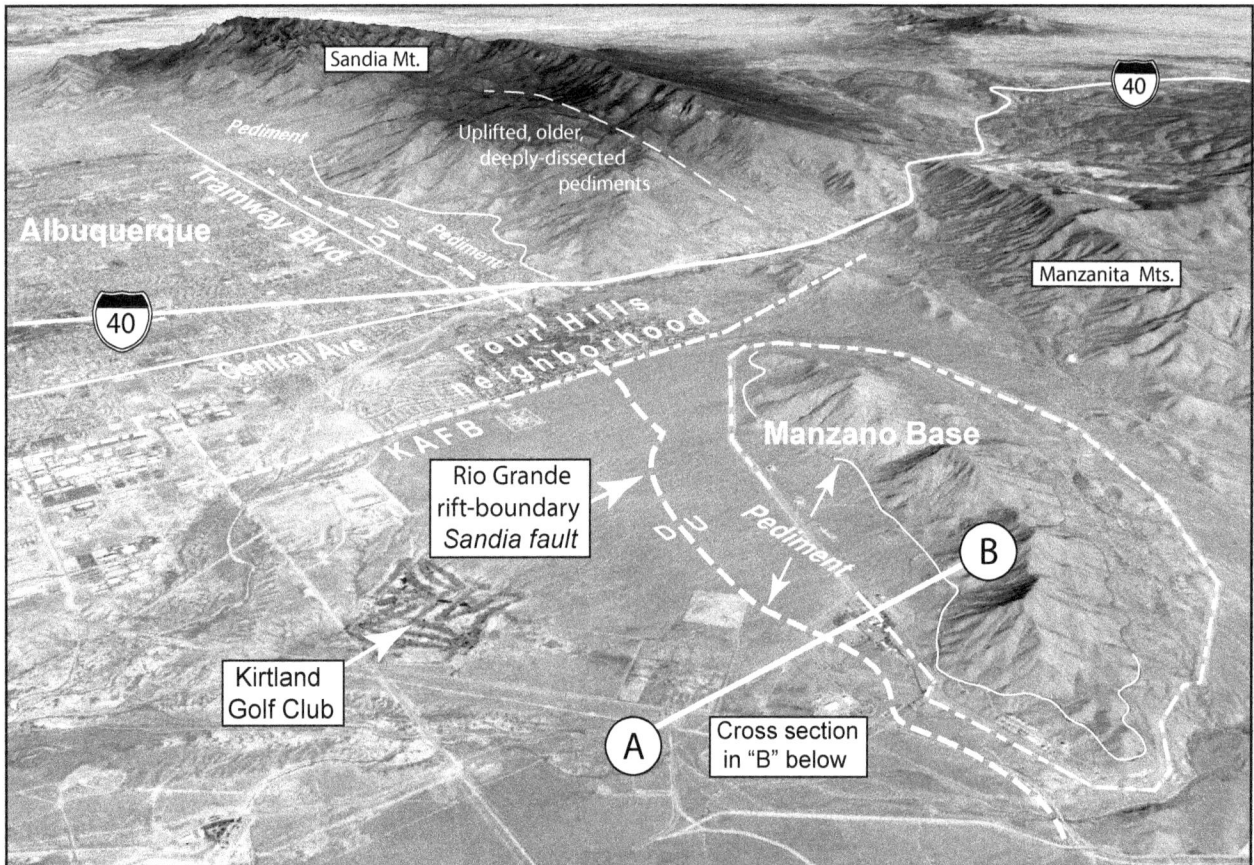

A. North-Northeast Oblique *Google Earth* Image of Sandia Mountain and Manzano Base

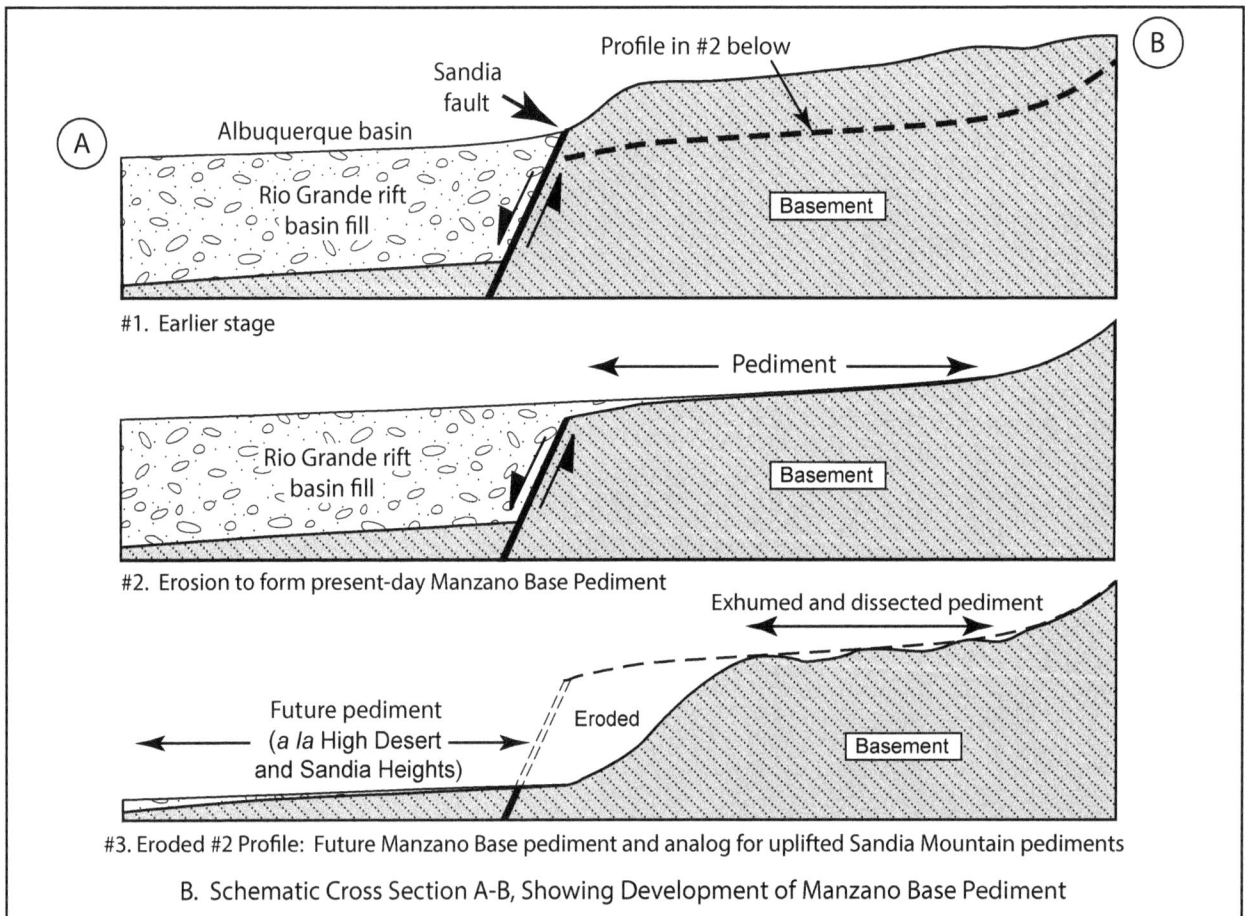

#1. Earlier stage

#2. Erosion to form present-day Manzano Base Pediment

#3. Eroded #2 Profile: Future Manzano Base pediment and analog for uplifted Sandia Mountain pediments

B. Schematic Cross Section A-B, Showing Development of Manzano Base Pediment

Figure 6-15. Pediments

Four Hills/Manzano Base (FH/MB). In 1947 the Air Force renamed the little four-peaked mountain as *Site Able*. In 1952 the site was renamed *Manzano Base* and was still operated by the Air Force. Construction crews hollowed out tunnels and bomb-proof chambers to store nuclear weapons. For a time during the 1950s the site was a "presidential emergency relocation" to shelter President Eisenhower in case of a nuclear attack. Manzano Base was deactivated in 1992, and the nuclear weapons moved westward to an obscure, more secure underground site on the base.

The core of FH/MB is a partially worn-down block of crystalline, basement rock consisting of Proterozoic Sandia Granite and some metamorphics. Its west flank is a smooth, planar pediment surface dipping gently to the west (*Figure 6-15*). Granite is exposed in the bottoms of some of the arroyos cut through the thin veneer of sand and gravel. Because the main Sandia Mountain plunges down to the south and virtually disappears at I-40, north of KAFB, the question arises whether this smooth pediment once was continuous with the lower half of Sandia Mountain before its uplift? Let's look at FH/MB more closely.

Schematic cross sections drawn across FH/MB allow us to speculate how the pediment may have appeared in the past, how it appears in the present, and how it will in the future (#1, #2, and #3 in *Figure 6-15B*, respectively). The latter of the three very much resembles the "lower mountain" mentioned above, and could be an analog for it. The "lower mountain is an array of isolated, semi-level patches of ground that seem to be remnants of ancient pediment surfaces (*Figure 6-16*). This lower mountain thus resembles a flight of giant stairs, with perhaps two main higher surfaces and a number of poorly-developed ones lower down to the west. All of them, when "connected" by mapping their elevations, slope to the south and southwest. Interestingly, the southern projection of the highest surface possibly connects to the top of FH/MB (*Figure 6-16C*).

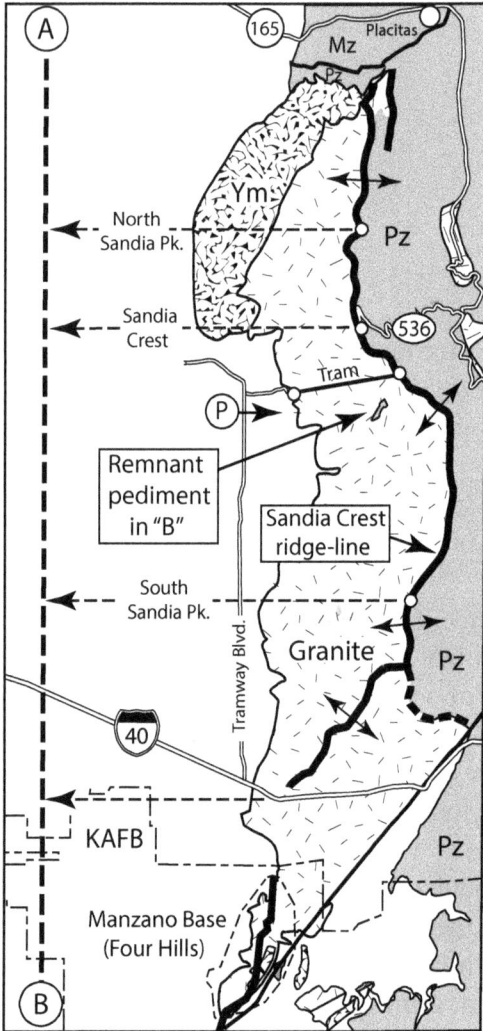

A. Location Map Showing West Projection of Crest-Elevation Points to Plane A-B (P = Photograph in "B")

B. East View (Photograph "P" at Left) of Sandia Crest and Select Remnant Pediment Surface (Photograph by Author)

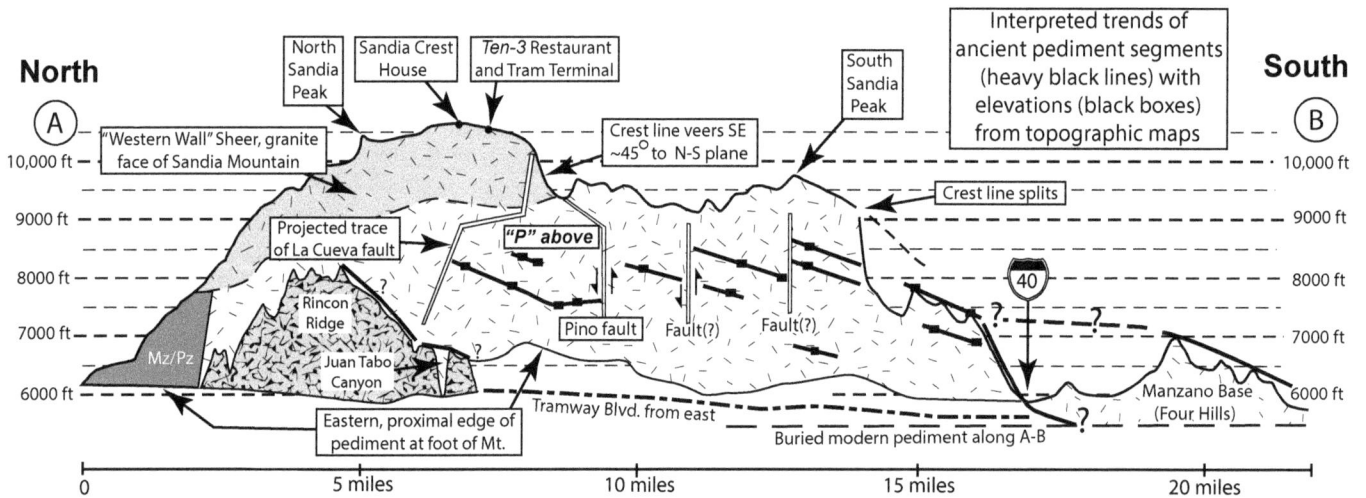

C. Mountain Profile Projected 90° West to North-South Plane A-B (in "A" Above) (Vertical Exaggeration = 5.5)

Figure 6-16. North-South Sandia Mountain Profile and Remnants of Ancient Pediments

Can we somehow date these pediments? Only indirectly. Detailed study of the sequence of sediments in the Albuquerque basin reveals two big interruptions, or time gaps, probably linked to interruptions in mountain uplift. One of those, mentioned earlier in this chapter, occurred at about 16 Ma, but a second was at about 6 Ma. The two main, upper pediments possibly correspond in age to these two interruptions.

Role of Rock-Density

The entire mountain and its pediments rise in elevation to the north and attain a maximum not far from the mountain's northern end (*Figure 6-16C*). It is likely no coincidence that this high point is where the Sandia Granite is near its maximum vertical thickness. A property of the granite, in contrast to that of its enclosing metamorphic rock, probably played a key role in the shaping the mountain, and that property is rock density.

As mentioned in Chapter 2 (*Figure 2-4*), Kirby (1994) interpreted the Sandia Granite pluton as a body dipping about 45° to the northwest. In map view the pluton is interpreted to have a mushroom-like shape. Assuming that this shape is roughly replicated in cross section, a nine-mile (14.5-km) vertical column of granite may be buried under the northern part of the range (*Figure 6-17*).

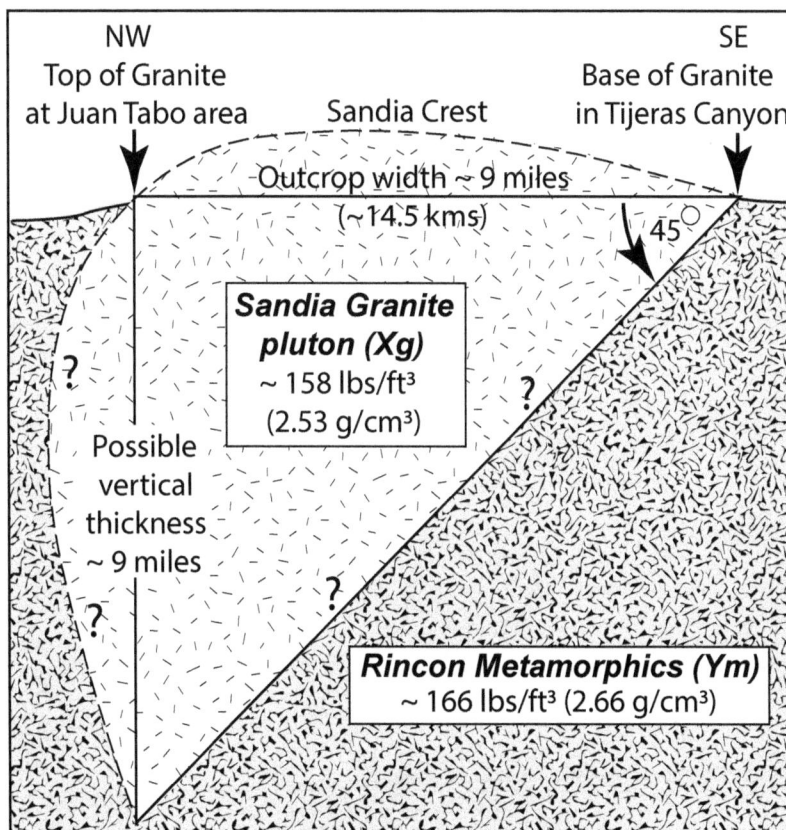

Figure 6-17. Schematic Cross Section Showing
Estimated Densities of Basement Rocks

Densities. Using mineralogy from Kelley and Northrop (1975), published mineral densities, and back-of-the-envelope calculations, the gross average density of the enclosing Rincón Metamorphics is approximately 166 lbs/ft^3 (2.66 g/cm^3) vs. about 158 lbs/ft^3 (2.53 g/cm^3) for the granite (*Figure 6-17*). The approximate 8-lb/ft^3 density difference between the metamorphics and the granite, spread over nine vertical miles, should produce a very significant buoyant force. I believe that this attribute was an important contributing factor to the great differential uplift of the northern part of Sandia Mountain, and that the uplift was facilitated by the enormous normal slip along the Rincón fault on the west, and the Placitas fault system on the north, which "liberated" the Sandia granite block from the enclosing crust.

Geologists love to make maps and to plot data, be it of elevations, thicknesses, colors, etc. to reveal subtle, hidden trends, thus providing "feedback" to the growing body of knowledge. For example, elevation profiles drawn on a number of surfaces in the northern part of the mountain suggest that Rincón Ridge, the cul-de-sac, remnants of the pediments, and even Sandia Crest were bowed up as a cohesive unit on the north and rotated down on the south (*Figure 6-18*). This bending created tensional stresses that led to lateral, transverse faulting that in turn segmented the mountain we see today (*Figure 6-19*).

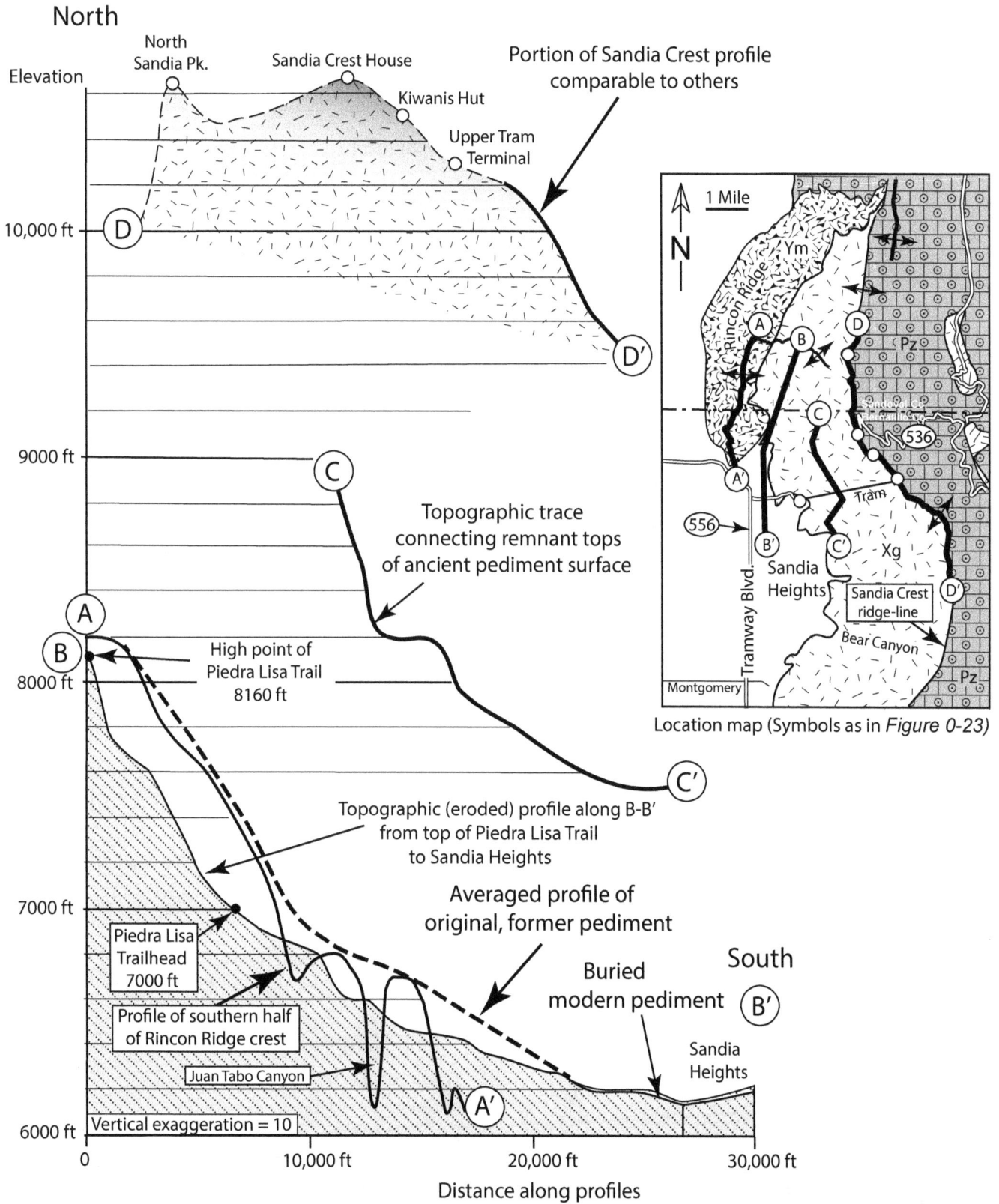

North

Elevation

North Sandia Pk.

Sandia Crest House

Kiwanis Hut

Portion of Sandia Crest profile comparable to others

Upper Tram Terminal

D

10,000 ft

D'

Topographic trace connecting remnant tops of ancient pediment surface

9000 ft

C

A

B

High point of Piedra Lisa Trail 8160 ft

8000 ft

C'

Topographic (eroded) profile along B-B' from top of Piedra Lisa Trail to Sandia Heights

Averaged profile of original, former pediment

South

Buried modern pediment

B'

7000 ft

Piedra Lisa Trailhead 7000 ft

Profile of southern half of Rincon Ridge crest

Sandia Heights

Juan Tabo Canyon

6000 ft   Vertical exaggeration = 10

A'

0   10,000 ft   20,000 ft   30,000 ft

Distance along profiles

Location map (Symbols as in *Figure 0-23*)

1 Mile

N

Ym

Rincon Ridge

A

B

D

Pz

C

Sandoval Bernalillo

536

A'

Tram

B'

556

C'

Xg

D'

Sandia Heights

Tramway Blvd.

Sandia Crest ridge-line

Bear Canyon

Pz

Montgomery

Figure 6-18. Comparison of Topographic Profiles

211

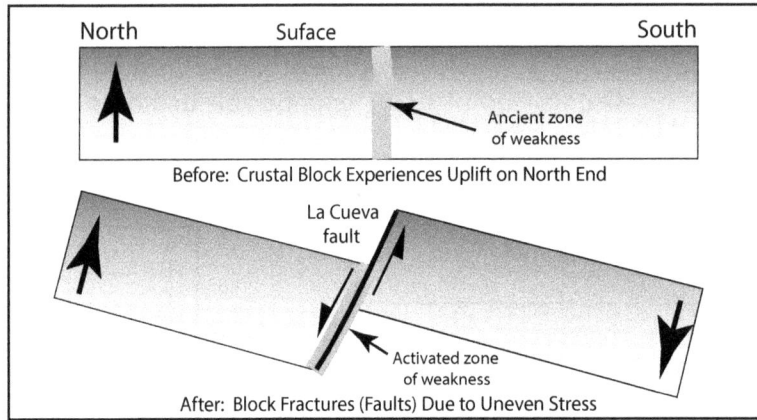

North Suface South

Ancient zone
of weakness

Before: Crustal Block Experiences Uplift on North End

La Cueva
fault

Activated zone
of weakness

After: Block Fractures (Faults) Due to Uneven Stress

A. Block Diagram Showing Normal Faulting in Response to Differential Uplift/Rotation

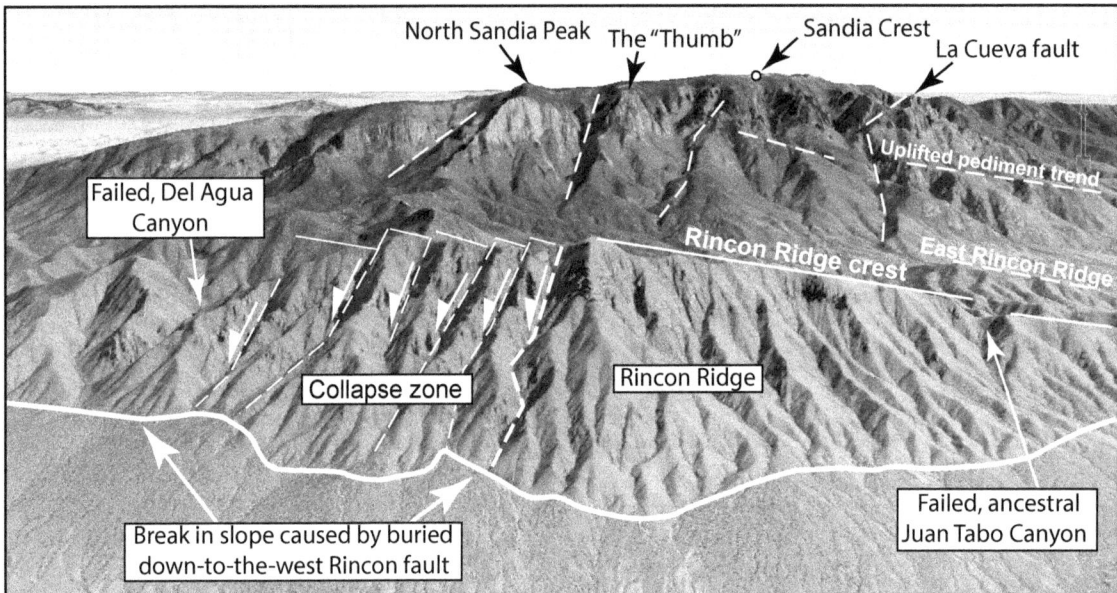

North Sandia Peak The "Thumb" Sandia Crest La Cueva fault

Uplifted pediment trend

Failed, Del Agua
Canyon

Rincon Ridge crest East Rincon Ridge

Collapse zone Rincon Ridge

Break in slope caused by buried
down-to-the-west Rincon fault

Failed, ancestral
Juan Tabo Canyon

B. East-Southeast Oblique *Google Earth* Image of Northern Part of Rincon Ridge

Crest of
Rincon Ridge

North Sandia Peak The "Thumb"

The "Shield"

C. East Ground-Level View of Collapsed Portion of Rincon Ridge (Photograph by Author)

Figure 6-19. Faulting of Rincon Ridge

The "Western Wall" – I

Viewed from the west the mountain seems to consist of a shear, west-facing wall (*Figure 6-20)*. The greatest, single piece of the "wall" is the 1000-ft rampart just below the crest at North Sandia Peak dubbed the *Shield*, which for decades has been a magnet for serious rock climbers. The Shield seems fresh and unweathered and thus suggests a geologically young feature. There are two schools of thought about its origin. About a century ago the entire wall was thought to be the remnant surface of a single, segmented, master fault plane, running the length of the mountain (Ellis 1922). On the other hand, Kelly and Northrop (1975) suspected that the base of the cliffs was not the base of a fault at all, but rather the base of a type of erosion typical of high altitude that produces cliffs rather than the rounded forms of lower altitudes." With caution, I lean toward the former. We will elaborate on this feature (Western Wall – II) in *Chapter 10*.

A. East-Southeast View of Sandia Mountain and Rincon Ridge

B. Interpretion of "A" Above

Figure 6-20. Sandia Mountain's "Western Wall" and Possible Normal Fault (Patterned)
(Photograph by Author)

The Remaining Mystery About that "Lower Mountain"

The entire lower mountain certainly seems to be a complex, down-dropped block that is separated from the crest by that shear granite escarpment mentioned above. Down-dropped blocks usually "protect" their original upper surfaces by safely dropping them down away from the ravages of erosive forces that strive to destroy them. So why are rocks of the Sandia Formation and Madera Group present near the crest but not seen anywhere on the lower mountain as well (except at Caliza Hill)? They certainly existed there at one time.

To answer this, we must allude back to the earlier, Laramide phase of compression. In *Chapter 4* we saw that what would later to become Sandia Mountain got shoved around and "damaged" during the Laramide, but not so much as to be raised up and formed into a mountain. Even so, many workers have noted that the Laramide effects on Sandia Mountain are probably much more influential than is generally suspected (for example Karlstrom et al. 1999), and I wholeheartedly agree. I suggest that the Western Wall might be a reactivated Laramide fault, one converted from a reverse fault to a normal fault (*Figures 6-21A* and *6-21B*). To boot, what is today's "lower mountain" (with its dissected pediments) may have once been an *uplifted* block such that the once-intact sedimentary cover had been eroded off down to the Sandia Granite. Prolonged exposure and erosion then resulted in development of a pediment atop the granite (*Figure 6-21C*). Later, when regional Rio Grande rift tension took over < 26 Ma, the first and easiest thing to snap and accommodate the tension was that old damage zone, the reverse fault plane, which obliged and thus became the big normal fault seen today as the Western Wall (*Figure 6-21D*).

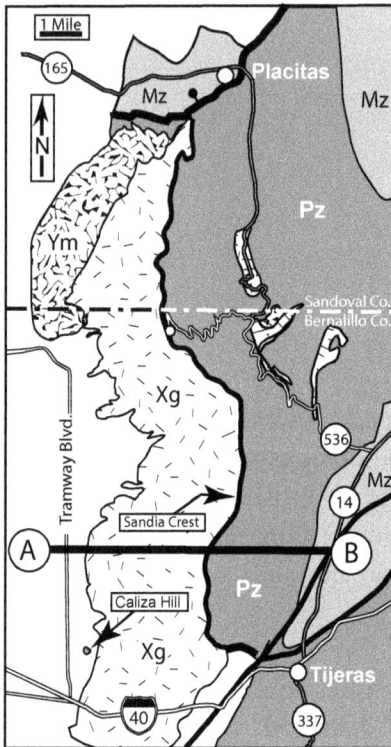

Location Map

Legend

SF   Santa Fe Group
       (basin-fill sand
       and gravel)
Mz   Mesozoic Units
    Kmv   Cretaceous, Mesa
         Verde Formation
    Km   Cretaceous, Mancos
         Shale
    J/Tr   Jurassic and Jurassic
         formations
Pz   Paleozoic Units
   P   Permian formations
   IP   Pennsylvanian,
        Madera Group and
        Sandia Formation
Xg   Sandia Granite
       basement
Ym   Rincon Metamorphics

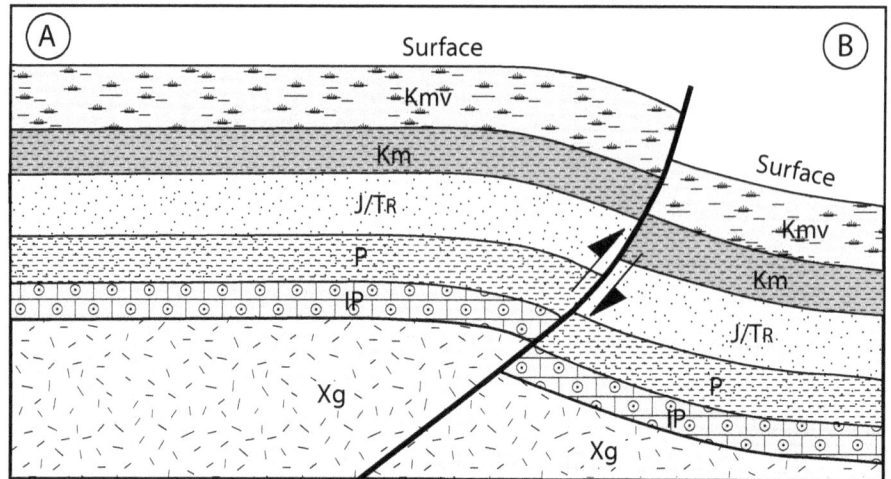

A. After Laramide Reverse Faulting (Restored to Non-Eroded State), 40 Ma

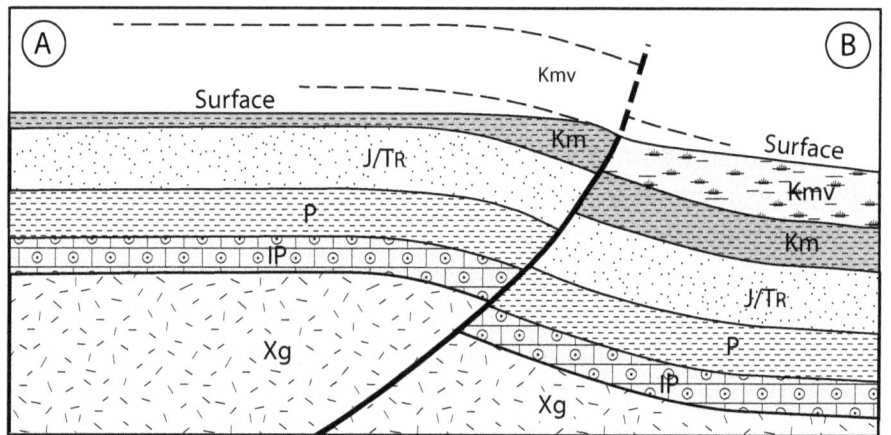

B. After Post-Laramide Erosion, < 40 Ma

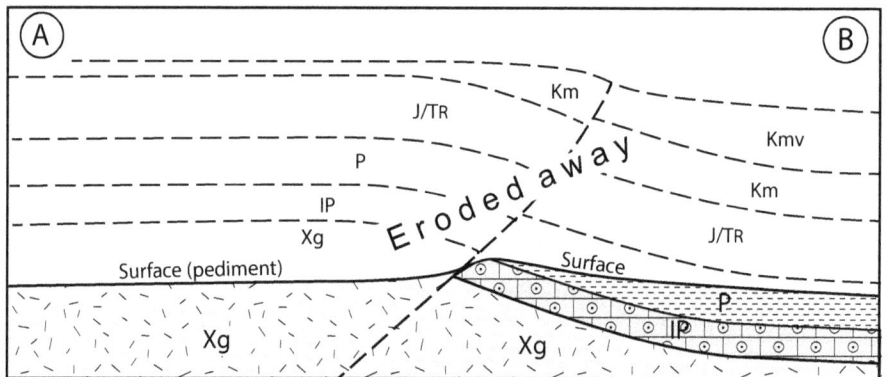

C. After Uplift and Deep Erosion, 15-10 Ma

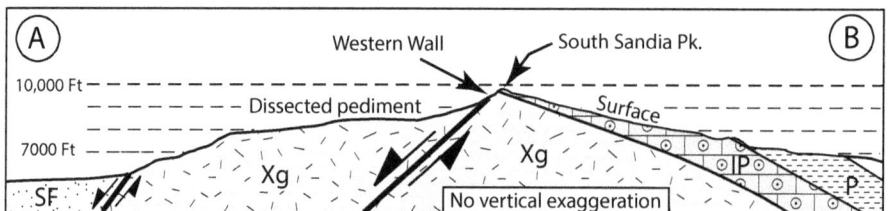

D. Present Landscape After Final Uplift, Down-to-East Rotation, and Reversal of Old Laramide Fault, <10 Ma

Figure 6-21. West-to-East Cross Sections A-B Showing Speculated Development of Sandia Mountain

7:
THE RIO GRANDE

Synopsis

Opening of the Rio Grande rift, somewhere after around 26 Ma, created a topographic catchment depression called the Albuquerque basin located on the western flank of the rising Sandia Mountain. The basin, lacking an outlet, received all its drainage and sediment load from its flanking uplands. At this time the basin perhaps resembled present-day Death Valley, with its limited drainage ending in shallow, playa lakes.

This changed ca. 5.3 Ma when other catchment basins along the rift to the north filled up and overflowed, ushering a newly-charged river, the *Ancestral Rio Grande*, down into the southern part of the Albuquerque basin where it ended in a playa lake, called *Lake Albuquerque*. Eventually even the Albuquerque basin filled up and the ancestral river spilled over to the next basin to the south, and by 4.5 Ma the river had reached the region of today's southern New Mexico near Las Cruces and El Paso. There it terminated in an enormous playa lake called *Lake Cabeza de Vaca*.

By ca. 1.5 Ma the southern basin had also filled up, the Ancestral Rio Grande spilled over, draining Lake Cabeza de Vaca, and advanced into West Texas. Only ca. 800 ka or perhaps a little later did the river reach the Gulf of Mexico. Once it had reached its new, "ultimate base level" there the river adjusted its upstream gradient and began to incise down into its old deposits. At about 440 ka Lake Alamosa, far to the north in southern Colorado, overflowed its southern barrier, spilled over and reinforced the Ancestral Rio Grande to its south. In the Albuquerque basin the lowered base level and newly increased stream power resulted in downcutting some 400 to 450 ft, forming the modern Rio Grande. The entrenchment of the river caused entrenchment of flanking drainages, particularly the formation of the modern Tijeras Arroyo.

$\sim\sim\sim$

Introduction

As we've seen from the previous chapter, Sandia Mountain and the Albuquerque basin together comprise a couplet. They are linked together to form one system: a source of sediment and a place to dump it. Subcomponents of the couplet are the beveled-off modern pediment of the "lower mountain," a lower apron of alluvial fans, and the Rio Grande itself (*Figure 7-1*). It's important to note that the Albuquerque basin is but one segment of the much more extensive Rio Grande rift, a linear zone of stretched crust extending from near Leadville in southern Colorado to southern New Mexico (*Figure 7-2*). The Rio Grande drains the rift and eventually reaches the Gulf of Mexico, 1860 river miles away. The behavior of the Rio Grande has greatly affected the shaping of Sandia Mountain's western flank, i.e., the *piedmont* (Latin *pedimontium,* for "at foot of mountains").

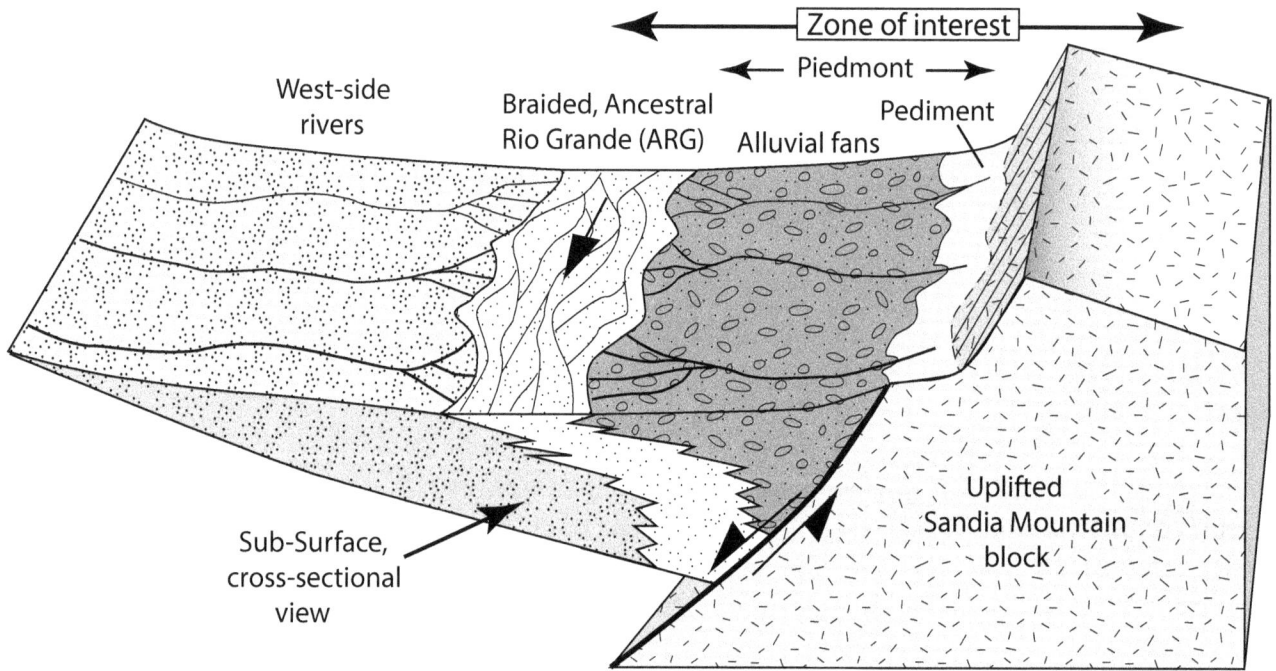

Figure 7-1. Schematic Block Diagram of Mountain/Alluvial-Fan/Braided-River System
(Modified from Mack and Seager 1990)

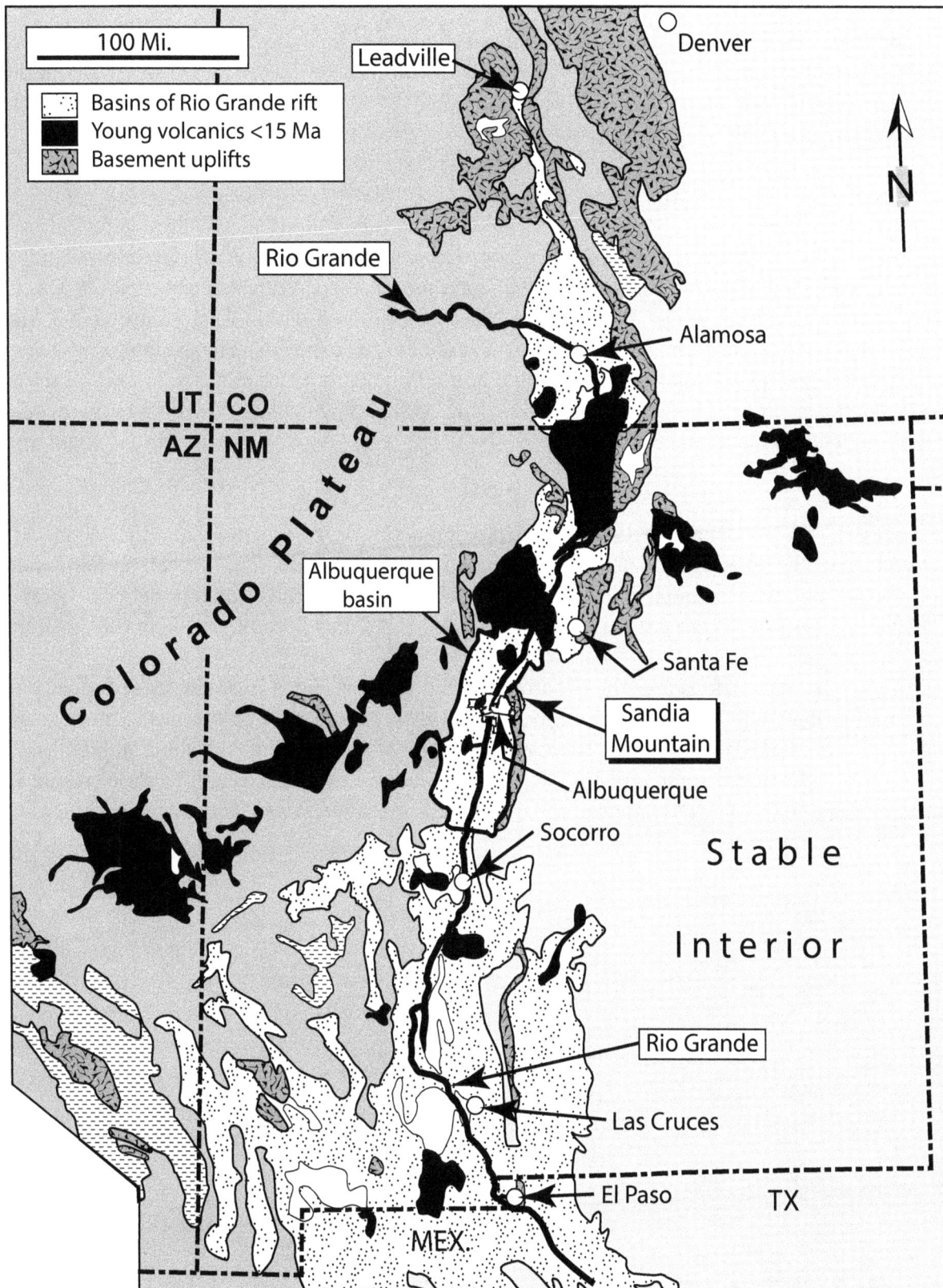

Figure 7-2. Rio Grande Rift

It is time for some context: How has the Rio Grande evolved and how did that evolution shape the Sandia Mountain's piedmont? The Rio Grande is an *exotic stream* in that the source of the water lies in the well-watered mountains far to the north, and its lower 1600 miles (86% of a total of 1860 miles) receives a minor contribution. (The exemplar of such a stream is the Nile River of north Africa.)

March to the Gulf

The river's history is a fascinating one. It may come as a surprise that the Rio Grande did not always reach the Gulf of Mexico. As the Rio Grande rift began to develop, somewhere after ca. 26 Ma, isolated topographic depressions such as the Albuquerque basin were formed within it. Up until ca. 5.3 Ma, these depressions received their sparse stream flowage and sediment from the adjacent mountains (*Figure 7-3A*). By ca. 5.3 Ma, though, significant flowage from the northern uplands as far afield as today's Alamosa, CO, spilled over and reached the Albuquerque basin, bringing its sediment load along with it (*Figure 7-3A*). The river then terminated in a shallow, ephemeral *playa* lake (dubbed *Lake Albuquerque*) in the southern part of the basin.

? **Ancestral vs. Modern Rio Grande**. These competing terms create some confusion. All early versions of the river, from before ca. 800 ka, are typically called the "Ancestral Rio Grande." It flowed at a higher elevation than today's river. After ca. 800 ka the river entrenched to a new, lower elevation due to a lowering of base level (see below) and thus became our modern Rio Grande.

Think of the basins of the rift as a chain of bathtubs lined up from north to south: as sediment filled and topped off the northern tub, the river overtopped it and then spilled over to the next, lower basin. This overtopping of basins gave the Ancestral Rio Grande an episodic advance to the south, tub by tub. By ca. 4.5 Ma the river had reached southern New Mexico near today's Las Cruces and El Paso, where it again terminated, or dead-ended, this time in an enormous playa lake called *Lake Cabeza de Vaca* (*Figure 7-3B*).

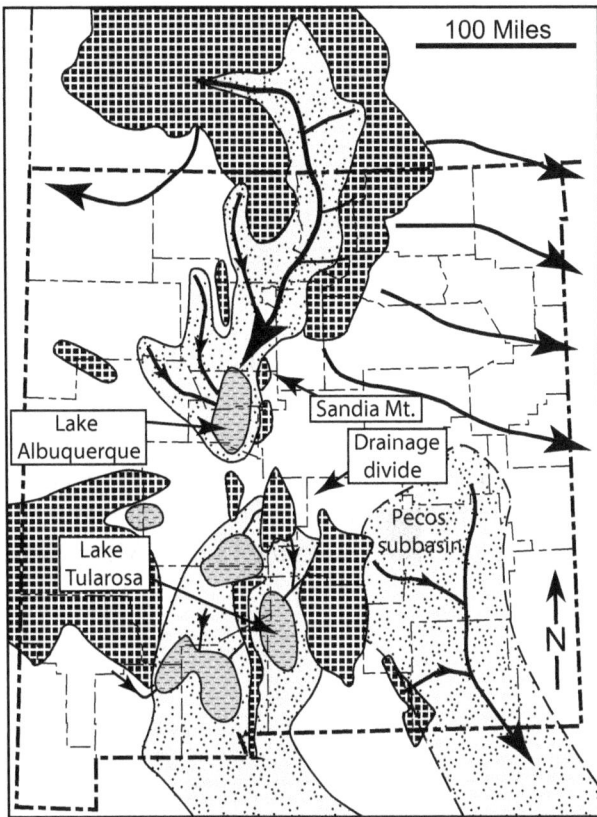

A. 5.3 Ma

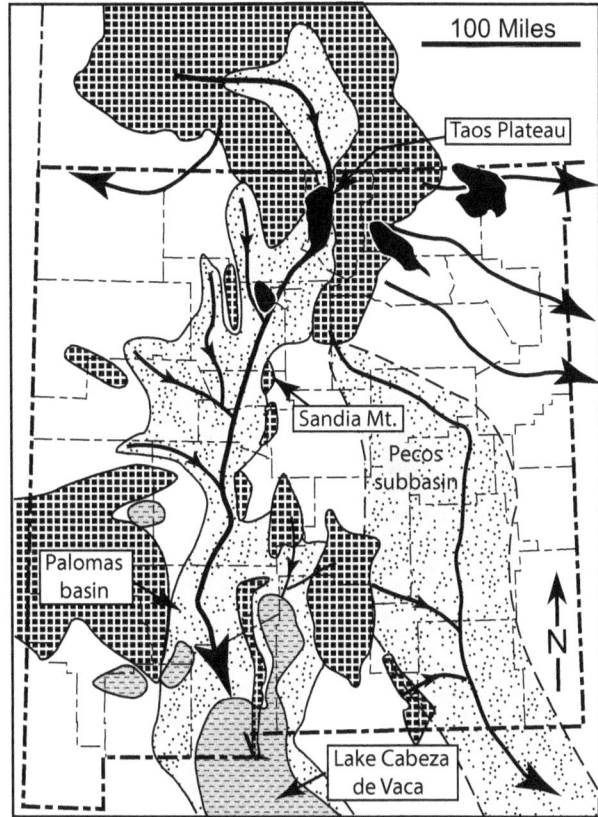

B. 4.5 Ma: Taos Plateau Forming, Rio Grande Extends South to Palomas Basin

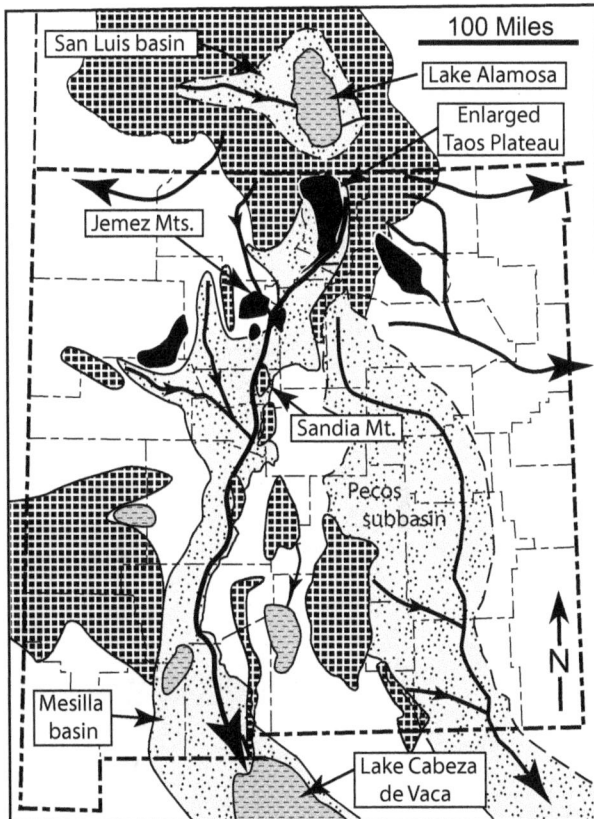

C. 3.1-2.5 Ma: Taos Plateau Volcanics Isolate San Luis Basin

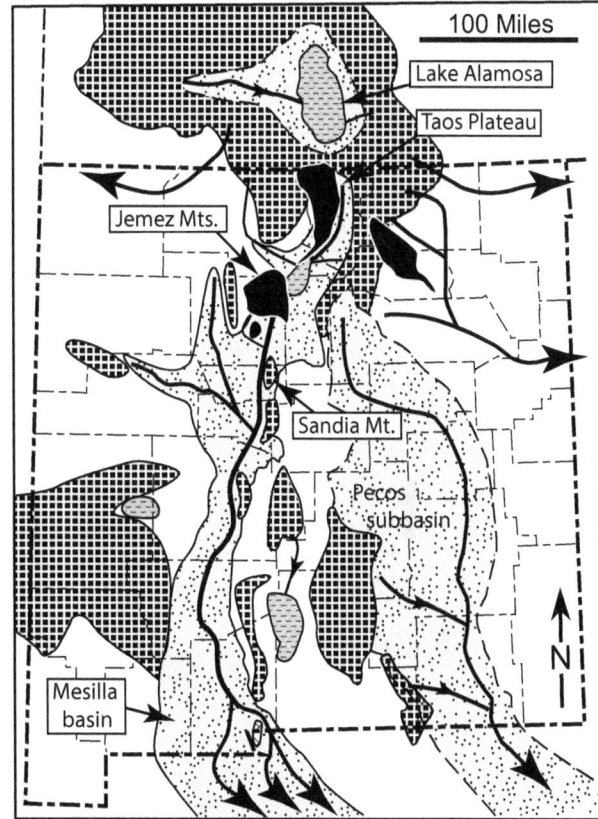

D. 1.5 Ma: Through-Flowing Rio Grande to West Texas

Figure 7-3. Evolution of Rio Grande Drainage System in New Mexico
(Modified from Connell et al. 2005; Machette et al. 2007; Repasch et al. 2017;
Rivers = Heavy Black lines, Basins = Dotted, Volcanics = Black, Highlands = Checkered, Lakes = Dashed)

Dead-end rivers. Rivers normally don't come to an end on dry land, but examples do exist. Consider a famous one, the *Okavango River/Delta* system in Botswana, southern Africa, a favorite subject of TV nature shows (*Figure 7-4A*). This vast delta becomes a seasonal lake, watering enormous herds of large animals. Closer to home, consider the Animas River/Delta system of southwestern New Mexico (*Figure 7-4B*). The Animas River surges south out of the Gila Mountains and then completely fizzles out as the terrain flattens, creating a marshy delta around the city of Deming and the Florida Mountains.

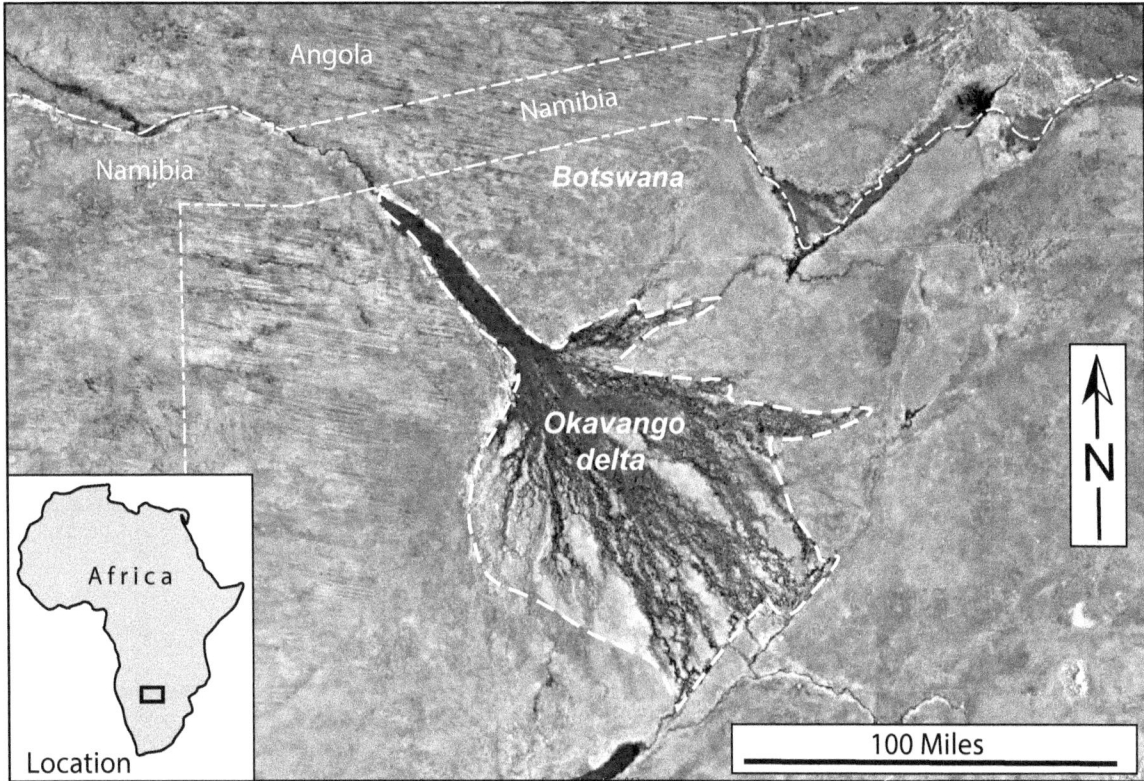

A. Image of Okavango River and Delta, Botswana, Africa

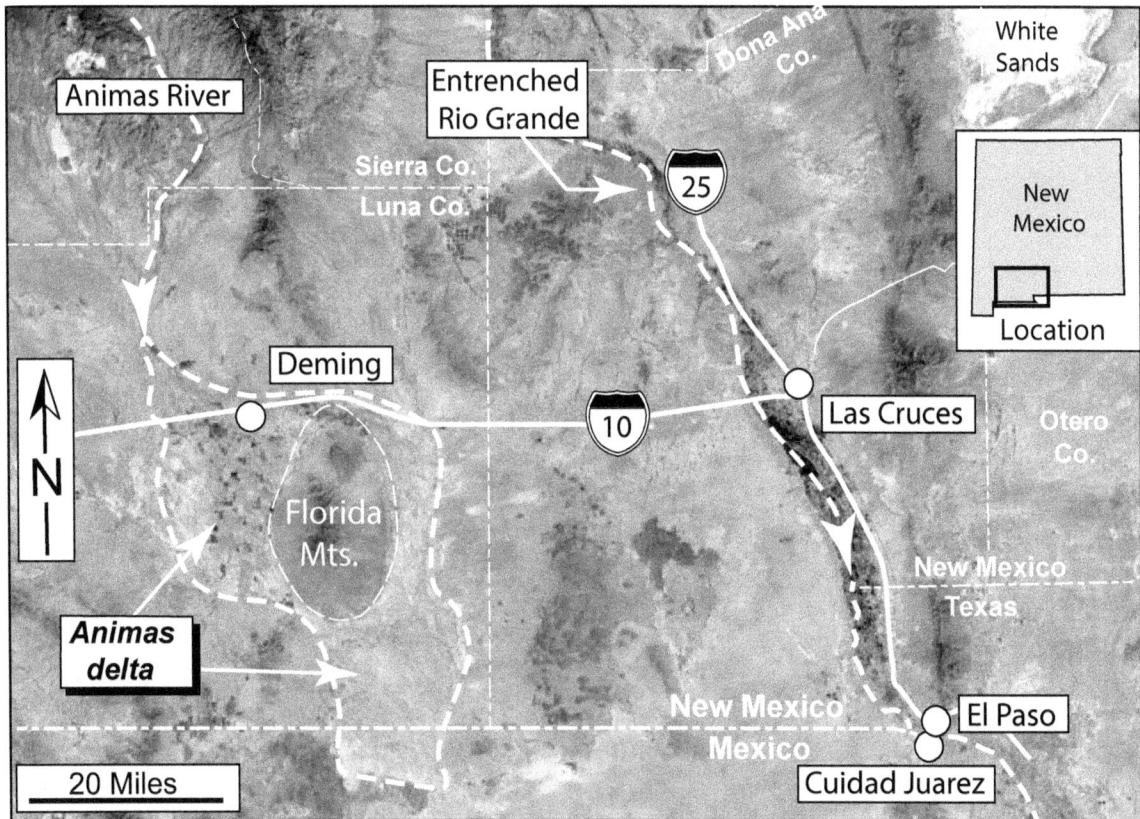

B. Image of Animas River and Delta, Southwestern New Mexico

Figure 7-4. *Google Earth* Images of "Dead-End" Rivers

By ca. 2.5 Ma outpourings of voluminous volcanics to the north near Taos completely blocked the drainage pouring down from the San Luis basin, backing up the drainage there and creating Lake Alamosa (*Figure 7-3C*). By ca. 1.5 Ma the ancestral river had advanced south into West Texas and had drained Lake Cabeza de Vaca (*Figure 7-3-D*). From today's Big Bend National Park to the Gulf of Mexico things get more complicated, but the Ancestral Rio Grande finally made its way to the Gulf, reaching it by ca. 800 ka or perhaps a little later (*Figure 7-5*).

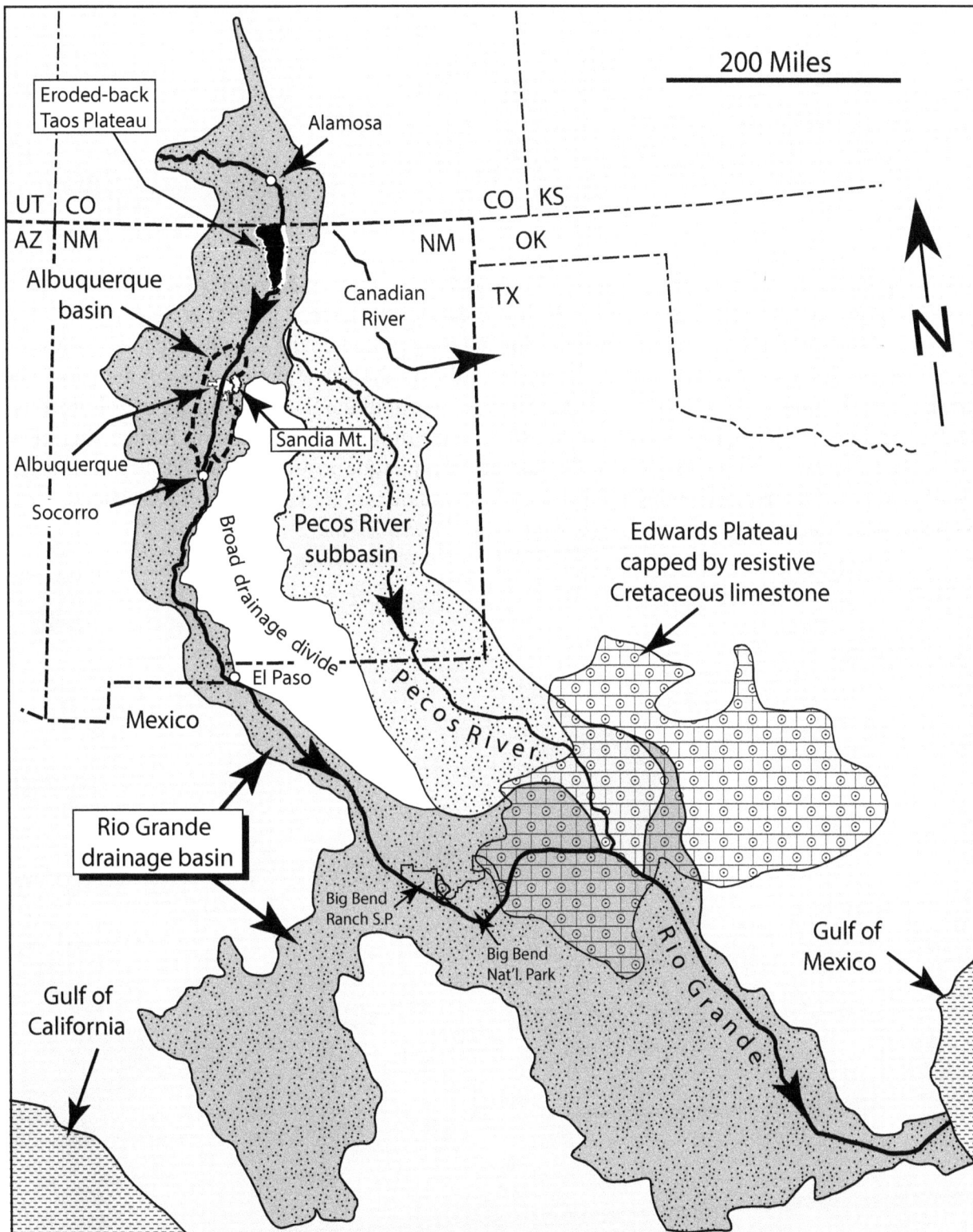

Figure 7-5. Rio Grande Drainage Basin Fully Integrated with Gulf of Mexico, ca. 800 Ka

Upon reaching the Gulf, the river experienced a very significant drop in *base level* (see below). The river therefore had a greater vertical distance to fall, thus creating in increase in stream velocity and erosive power. Starting from the Gulf, a wave of erosional downcutting pulsed up the Rio Grande to adjust the entire river to the new base-level (*Figure 7-6*). On its northward path the wave passed through the Albuquerque basin area. Complicating things, ca. 440 ka Lake Alamosa, north in the San Luis Basin, overflowed its low southern sill and once more spilled over into the Rio Grande drainage basin, adding a huge slug of water to its flow. The increased power of the Rio Grande at Albuquerque resulted in about 400 to 450 feet of downward erosion of the river, creating the modern Rio Grande and its valley. The upstream edge of this moving pulse is now at the Rio Grande Gorge, near Taos (*Figure 7-7*).

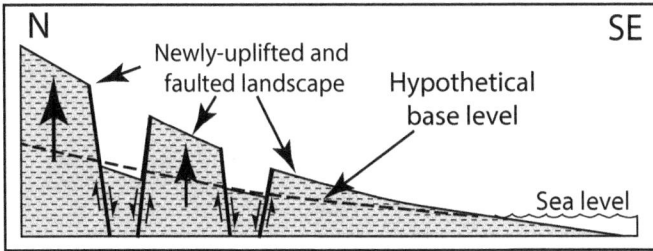

A. Initial Conditions: Uplifted and Block-Faulted Uplands

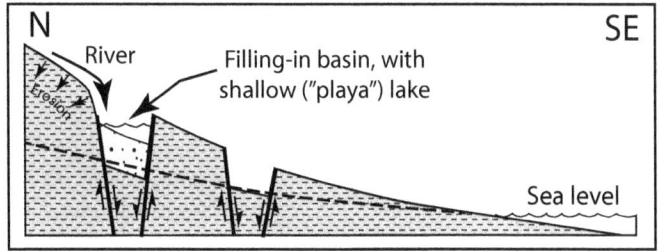

N — SE
Newly-uplifted and faulted landscape
Hypothetical base level
Sea level

B. River from North Begins to Fill First Basin

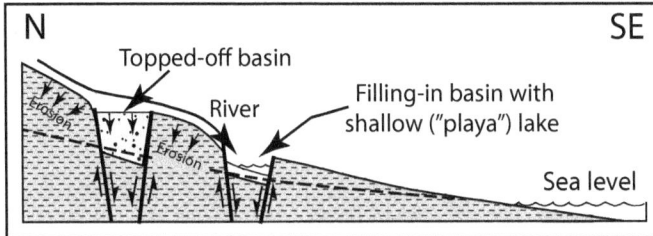

N — SE
River
Erosion
Filling-in basin, with shallow ("playa") lake
Sea level

C. River Tops Off and By-passes First Basin

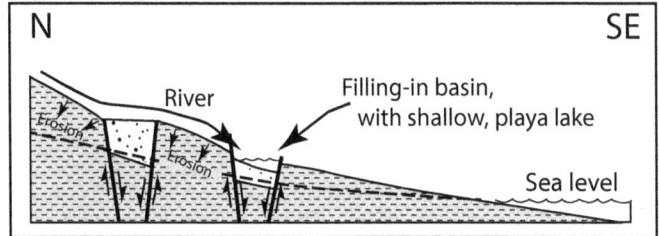

N — SE
Topped-off basin
River
Erosion
Filling-in basin with shallow ("playa") lake
Sea level

D. River Begins to Fill Up Next Basin

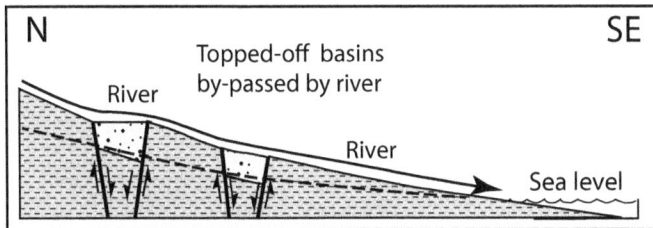

N — SE
River
Erosion
Filling-in basin, with shallow, playa lake
Sea level

E. River Fills Up and By-Passes Basins, and Reaches Gulf of Mexico, ca. 800 Ka

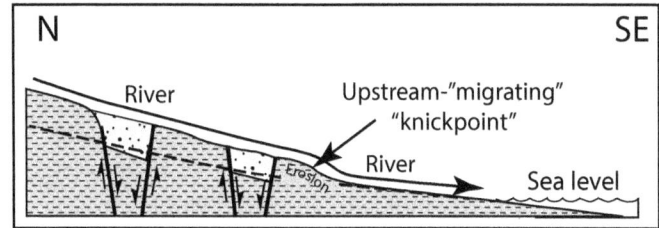

N — SE
River
Topped-off basins by-passed by river
River
Sea level

F. River Erodes Headward to Establish New Base Level

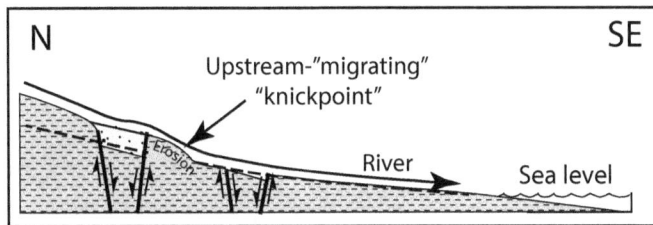

N — SE
River
Upstream-"migrating" "knickpoint"
River
Erosion
Sea level

G. Continued Headward Erosion to Establish New Base Level

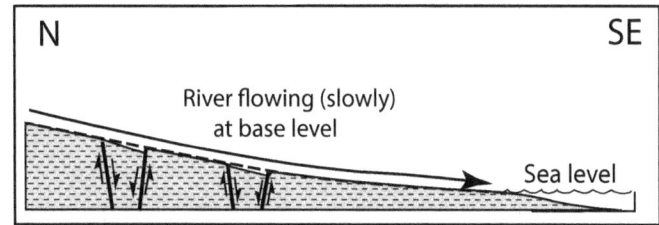

N — SE
River
Upstream-"migrating" "knickpoint"
Erosion
River
Sea level

H. New Base Level Established

N — SE
River flowing (slowly) at base level
Sea level

Figure 7-6. Schematic Development of Rio Grande and Its Topped-Off Basins

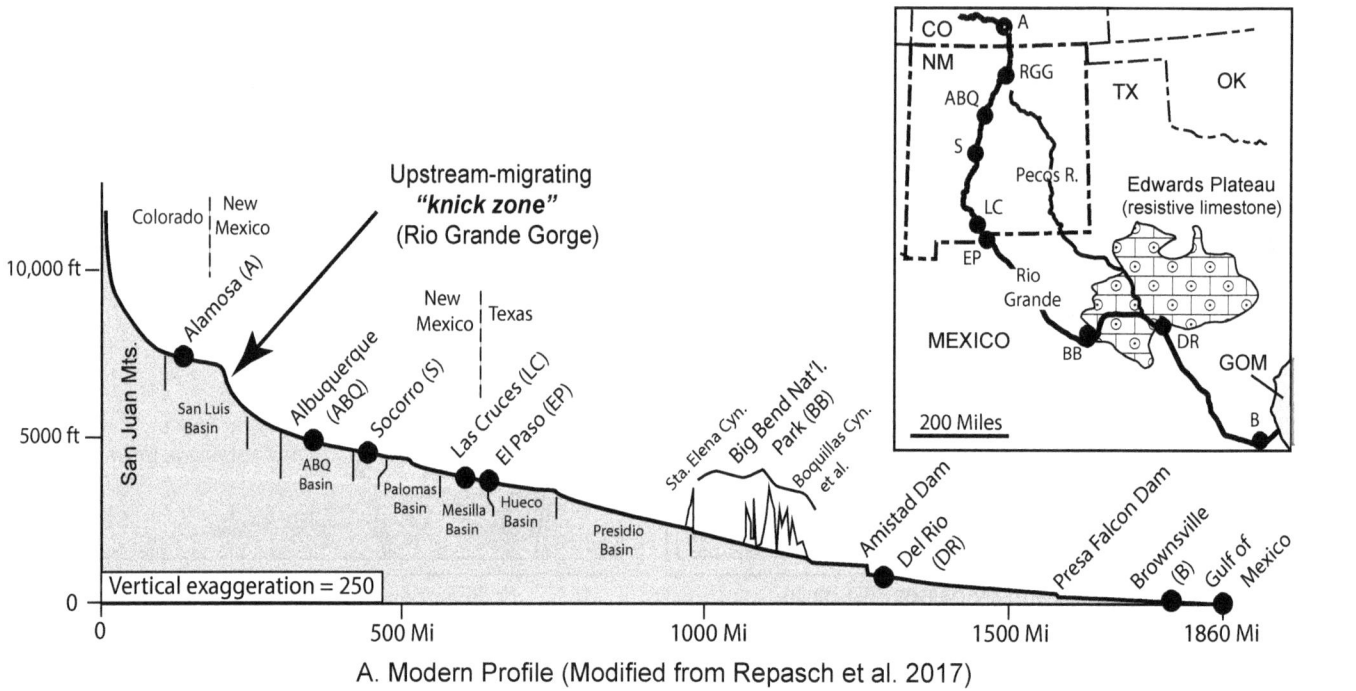

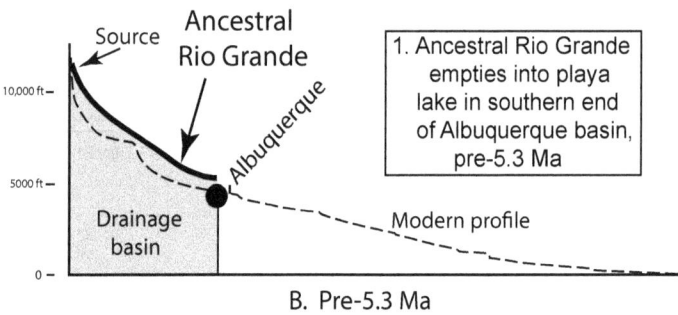

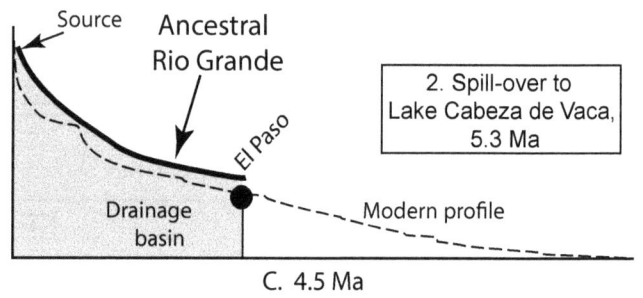

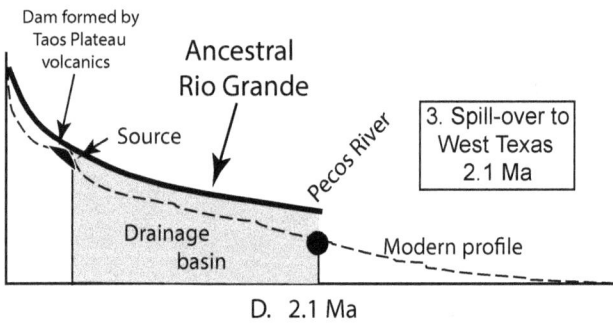

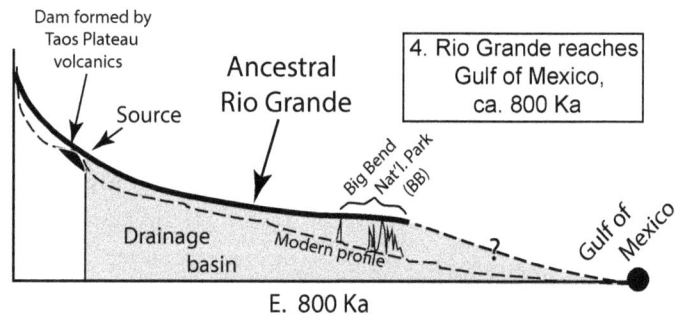

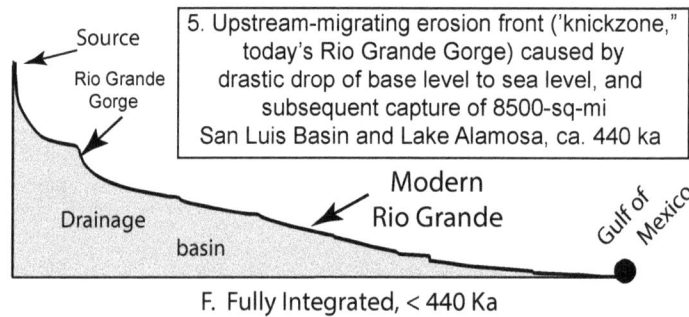

Figure 7-7. Profile of the Rio Grande from Its Headwaters to Gulf of Mexico

Base level: What is it? Let's back up. To understand the behavior of the Rio Grande, as well as that of rivers and streams in general, the concept of base level is central. Base level is defined as the lowest elevation to which a stream can erode its bed.

Base level. At base level the stream lacks the power to dig down any farther and so lazily sloshes around. It is important to realize that river water does not erode rock; it's the stuff that rivers drag along (the sediment) that does the cutting, something akin to the teeth of a bandsaw. The rate of cutting is dependent on both stream-velocity and hardness of the underlying rock. Stream velocity, in turn, is dependent on stream-gradient—the rate of elevation-drop per unit of distance, i.e., the steepness. If stream velocity increases, either due to uplift or to a drop of base level, a stream will erode and deepen its bed until it reaches the new base level (*Figure 7-8*).

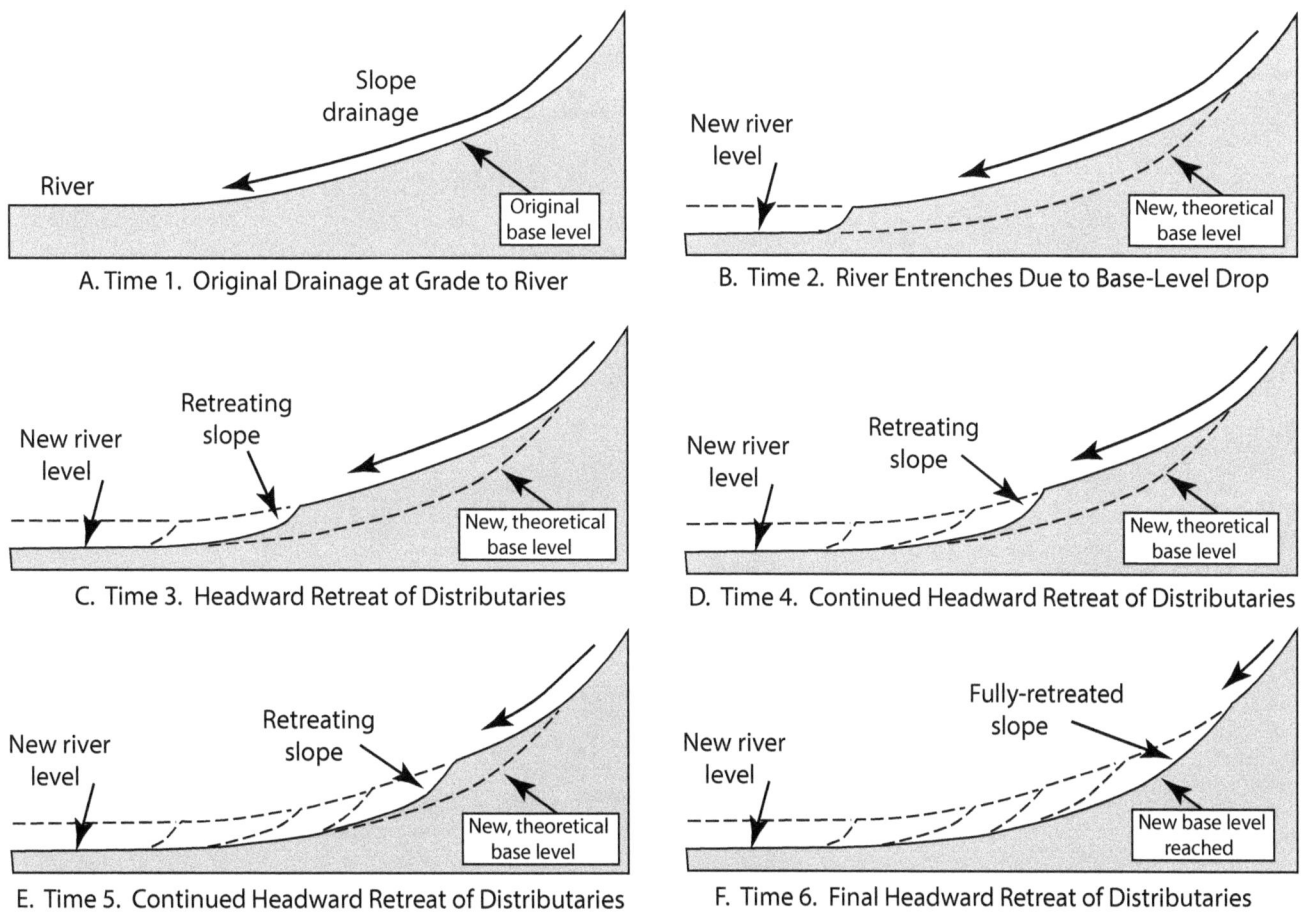

Figure 7-8. Schematic Cross Sections Showing Base-Level Fall, Causing Retreat of Piedmont Slopes

If the downcutting stream encounters hard, resistive beds in its channel, waterfalls can form (*Figure 7-9A*). The lip of the waterfall will then slowly retreat upstream until the gradient is smoothed. Once the stream or river slows down because of a loss of gradient (a flattening out), it loses its ability to transport sediment, dumps it in place, backfills the channel, and then slowly meanders around, "choking" on its own sediment (*Figure 7-9B*). At that point depositional base level has been reached. *Stream velocity is the key ingredient.*

A. River Well Above Base Level: (Spokane River, Spokane, Washington; Photograph by Author)

B. *Google Earth* Image of River at Base Level (Dale Bumpers White River N.W.R., Arkansas)

Figure 7-9. Rivers and Base Level

In the arid Southwest it is common to see *arroyos*—deep-floored, rapidly-forming gullies in fields. They stubbornly defy the best efforts of ranchers to slow their development by tossing obstructions (tires, refrigerators, etc.) into them to stem the flow. The greatest point of erosion is at the arroyo's upstream lip (a mini-waterfall) where the water rapidly drops into the arroyo floor. That point of abrupt steepening of the drainage profile, a.k.a. the "knickpoint," migrates upstream as a traveling pulse (despite the rancher's best efforts!), relentlessly extending the new local base level upstream.

It is notable that when flowing above base level, rivers keep reasonably well-defined channels because the rivers are eroding. When base level is approached, stream channels stop their downward cutting, meander sideways and frequently change course (*Figure 7-9B*). This is the reason why rivers at base-level, such those as on southern North America's coastal plain, make poor political boundaries. Of course, ultimate base level is sea level, but there are also local, temporary base levels. Lakes also produce these, slowing down upstream flows, until the lakes overflow.

The Pivotal Time, 800 Ka: The Great Denudation

At ca. 800 ka the pulse of base-level adjustment of the Rio Grande system had reached north to the area of Albuquerque. Filling of the basin stopped, and incision by the river and its tributaries began. The phase called the *Great Denudation* had begun, sculpting the modern landscape. Estimates of the amount of river incision in the Albuquerque basin vary, ranging from about 250 ft (Kelley 1982) to as much as 700 ft (Connell 2004). For its final act the river has backfilled its incised channel some 75 ft to produce the modern floodplain (*Figure 7-10*).

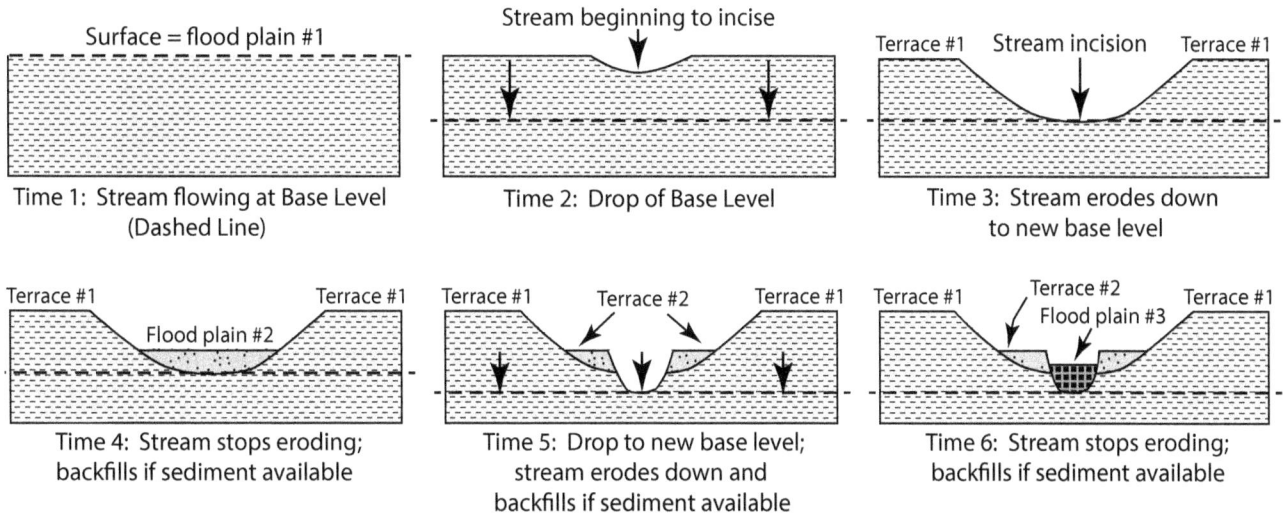

Time 1: Stream flowing at Base Level (Dashed Line)

Time 2: Drop of Base Level

Time 3: Stream erodes down to new base level

Time 4: Stream stops eroding; backfills if sediment available

Time 5: Drop to new base level; stream erodes down and backfills if sediment available

Time 6: Stream stops eroding; backfills if sediment available

A. Schematic Cross Sections Showing Drainage's Step-by-Step Response to Drop in Base Level

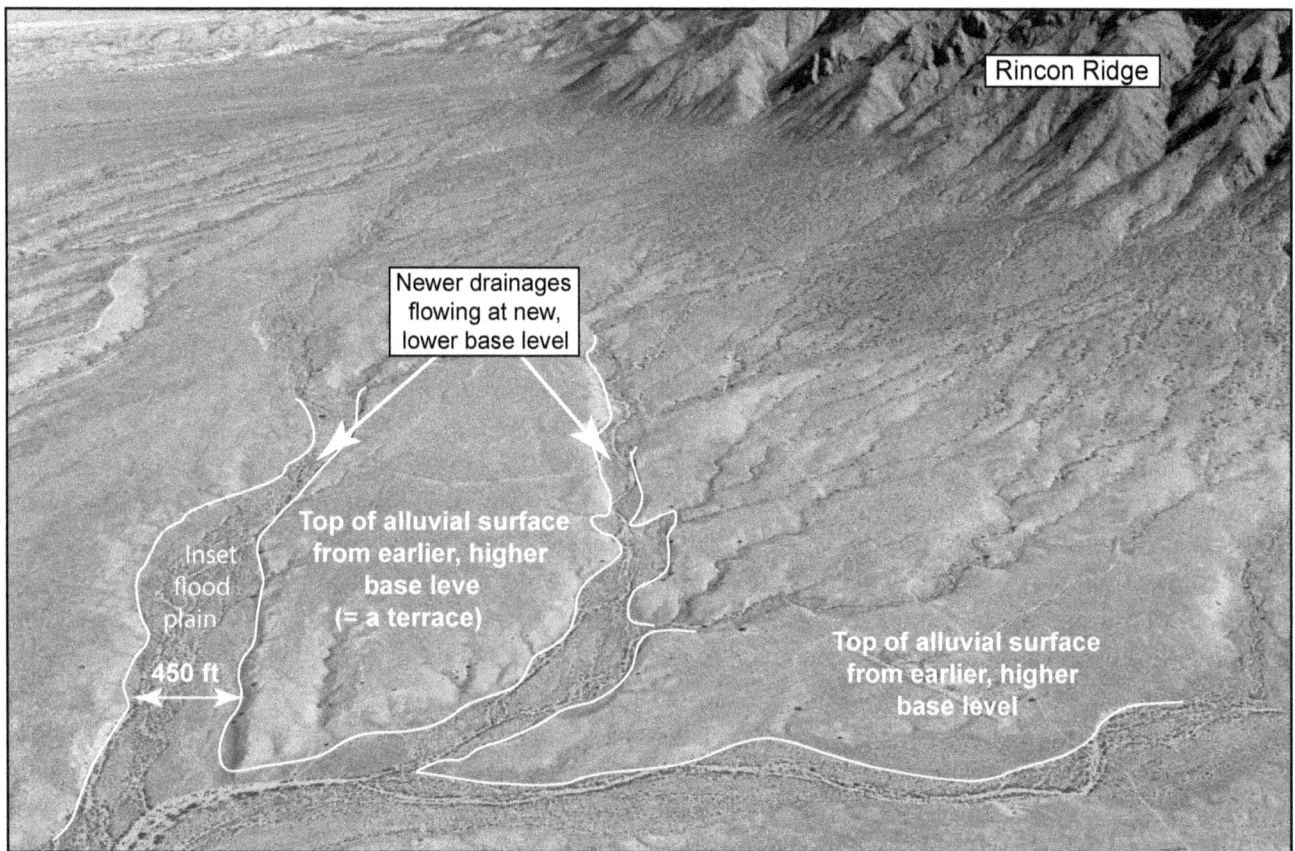

B. East-Northeast Oblique *Google Earth* Image of Sandia Mountain Piedmont West of Rincon Ridge

Figure 7-10. Drainages Adjusting to Lowered Base Level

233

The Great Denudation was not a continuous process, but rather one of sporadic swings between downward erosive episodes and partial back-fillings, which suggest a guiding hand that couldn't quite make up its mind. The reason for this peculiar behavior of rivers deserves a look at climate, which perhaps was the primary cause.

Climate cycles: the global cause. What can these swings of river behavior possibly mean? In recent decades the reason for this *denudo-interruptus* has finally been revealed via a magnificent achievement of cooperative science: the establishment of a link between cores of deep-sea sediments to river *terraces* (see Glossary) in New Mexico. More precisely, this geologic information links to the climatic fluctuations of the Pleistocene Period's ice ages. Details of this relationship is far beyond the scope of this book.

The gist is that fine-grained deep-sea sediments recovered from thousands of deep-sea cores acquired by the very successful *Deep-Sea Drilling Project* (DSDP) during 1968 to 1983 tell a wonderful story of global climate swings during the past 2.6 Ma. Microscopic fossils made of calcium carbonate (calcite) found within the cores contain information about the ancient atmosphere. The chemical nature of the calcite serves as a proxy for the volume of glacial ice that accumulated on land during each ice age. Linked to the global glacial ice volume was a corresponding rise and fall of global sea level (large glacial ice volume = low sea level; small glacial ice volume = high sea level). These swings affected base levels almost everywhere.

The core data can be plotted on a time scale. They show that there was a major shift in the duration of the glacial-to-interglacial cycles (each cycle includes a glacial and interglacial stage), from 41-ka to 100-ka, occurring at roughly 800 ka ago. This is the hugely important *Mid-Pleistocene Transition*. The cycles are controlled by slight shifts in the shape of Earth's orbit around the sun and to a lesser extent in Earth's axis of rotation. These shifts control the amount of solar radiation Earth receives, and therefore climate. Affected are not just in the areas actually being glaciated but also those areas out in front of the glaciers, the *pluvial* world—like much of New Mexico.

To gain some perspective, since the beginnings of the ice ages ca. 2.6 Ma there have been 52 glacial-interglacial cycles: 43 41-ka cycles from 2.6 Ma to 800 ka, and nine 100-ka cycles from ca. 800 ka to the present (Lisiecki and Raymo 2005). The glacial stages of the periods generally last almost 90 ka whereas the interglacial stages last about 10 ka. We presently live near (or at) the end of the last interglacial period that started about 15 ka ago.

The most important climate event in these cycles is the swing from cool/wet (glacial) to warm/dry (interglacial) conditions. During the former the western hillslopes of Sandia Mountain were heavily forested and carpeted with a thick soil, anchored in place by vegetation (*Figure 7-11A*). When things dried out, during the transition to interglacial conditions, the vegetation withered and the soil—with nothing to hold it in place—was eroded away and washed down to the backfill the river valley (*Figure 7-11B*). With the soil gone, the "unroofing" and exposure of the underlying "corestones" began (for more, see *Chapter 11*), characterizing today's foothills.

234

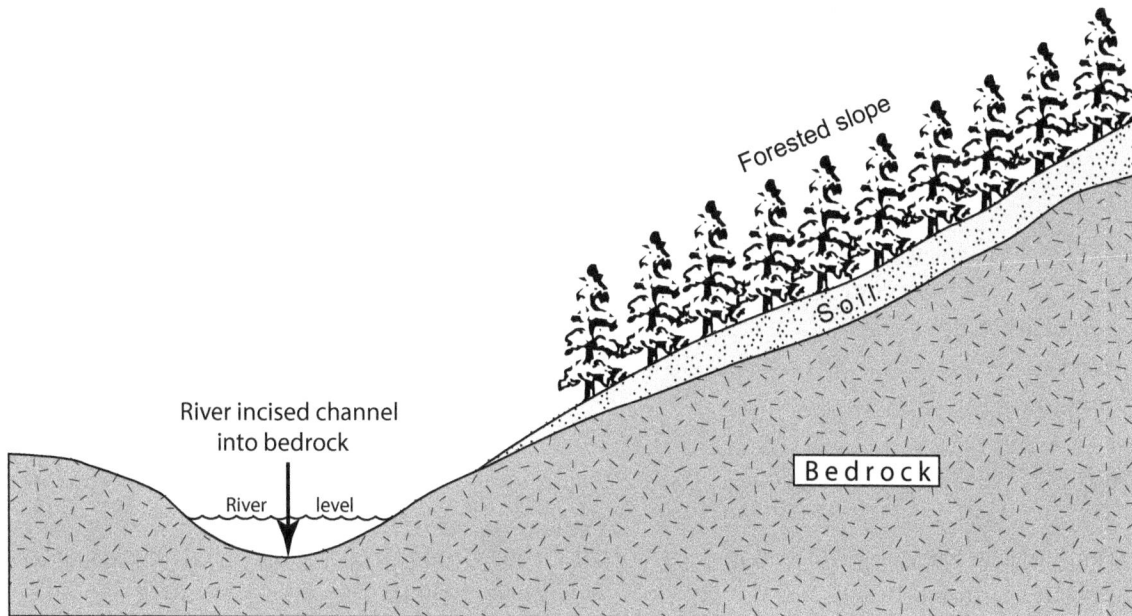

A. Pluvial Period: Cold and Wet
(Soil Held on Hillslopes by Vegetation as River Incises Channel Down into Bedrock)

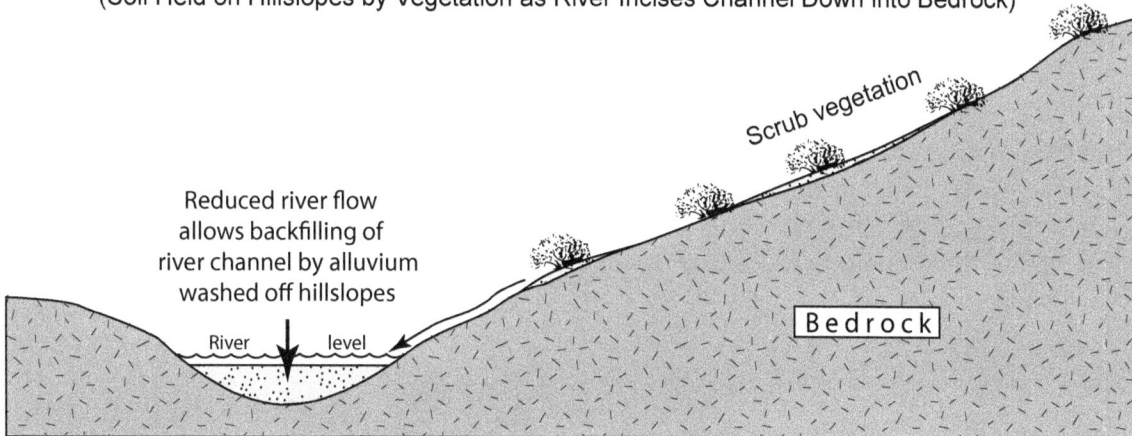

B. Interpluvial Period: Warm and Dry
(Vegetation on Hillslopes is Reduced; Soil is Stripped to Become Alluvium that Backfills River Channel)

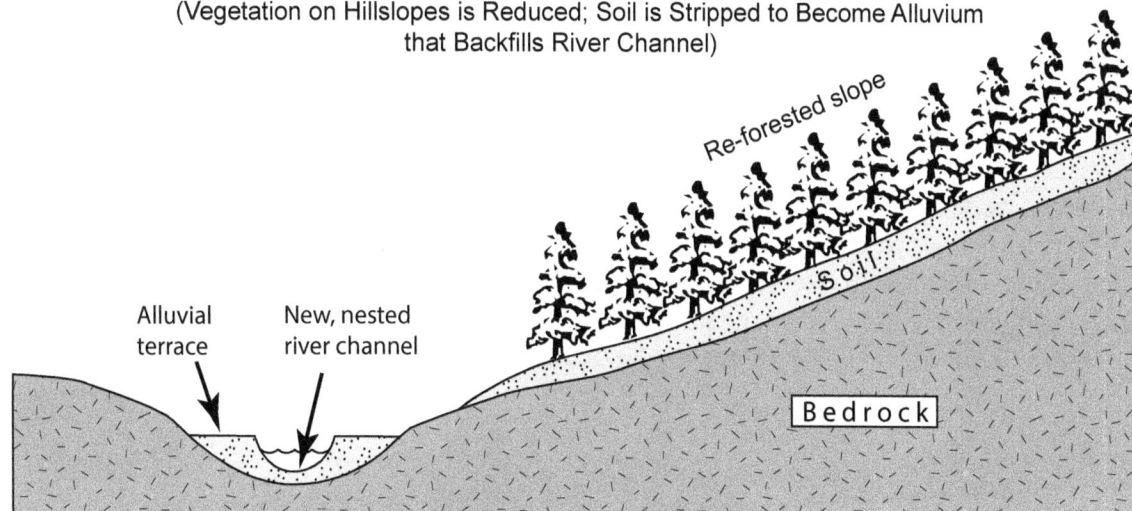

C. Pluvial Period: Cold and Wet
(Vegetation Is Re-Established; Soil Forms, Thickens, and is Retained on Hill-Slopes; River Incises into Old Alluvium, Abandoning its Previous Floodplain, Forming a Terrace)

Figure 7-11. Cyclic Development of Stream Channels

Later, when cool and wet glacial conditions returned, the vegetation on the slopes was restored and the soil was once again built up and held in place. The river down below, now vigorous and relatively free of sediment, incised a new channel, nesting into and abandoning the previous floodplain and leaving it behind as a terrace (*Figure 7-11C*). This nesting of stream-channel deposits is typical of many rivers in the Southwest, including the Rio Grande.

Debris Flows

The "unroofing" of the Sandia granite slopes, from their sandy tops to the corestone bottoms, is mirrored in reverse order by certain alluvial aprons on the slopes down below. The sandy bases of the latter grade upward from sand and cobbles to enormous car-size granite corestones, all transported by sudden, violent *debris flows* (Pazzaglia et al. 1999). They differ from alluvial fans (see below) in that the former are sudden, often singular events, rather than sustained over longer periods.

It is truly difficult to envision how enormous boulders can somehow be transported significant distances, sometimes miles, downslope. The process is enabled by a combination of great potential energy (steep slope plus high elevation), an abundance of loose material, and plenty of lubricant (water and air). Once this combination is in place, all it takes is a little shake like a jolt of an earthquake, or a little too much water. Once things are set loose, an underlying cushion of entrapped water and air lowers the friction and off the stuff goes.

Alluvial Fans

These depositional features are formed in tandem with the uplift of the adjacent mountains. Alluvial fans are characteristic of arid and semi-arid regions where what little drainage exists comes in sudden, torrential surges. Alluvial fans are land forms consisting of unconsolidated sand, gravel, cobbles and boulders, sloping gently down from an outlet at the foot of a mountain where stream velocity abruptly slows.

Modern alluvial fans have smooth, uneroded upper surfaces (*Figure 7-12A*). When a fan becomes inactive it begins to slowly erode away. Fans merge, forming an apron (a *bajada*) of sloping composite fans. Only general ages can be assigned to them and dating is based largely on the degree of fan destruction by erosion. At the Sandia Mountain piedmont, the fans become smoother and younger on the north, such as on the west flank of Rincón Ridge where they are younger than ca. 50 ka. To the south the fans are not nearly as smooth, are all inactive and deeply eroded, and are thus older, in the range of ca. 300-120 ka (*Figure 7-12B*; Connell 1995).

A. Northeast Oblique *Google Earth* Image of Modern Alluvial Fans, Death Valley, CA

B. East Oblique *Google Earth* Image of Mature Alluvial Fans, Northern Portion of Sandia Mountain Piedmont

Figure 7-12. Modern vs. Mature Alluvial Fans

The alluvial-fan bajada has been sourced by a number of small catchment basins on the mountain's west side. The irregular bajada can be best appreciated along Tramway Blvd., the main north-south drag along the foothills, located about one to ½ mile west of the mountain front. Vestiges of the individual fans make Tramway Blvd. a gentle up-and-down rollercoaster as one drives north, from an elevation of about 5700 ft just north of I-40, up to 5850 just north of Indian School, a little dip down to 5800 ft at Candelaria, and a gradual rise to 6080 ft north of Paseo del Norte (*Figure 7-13*).

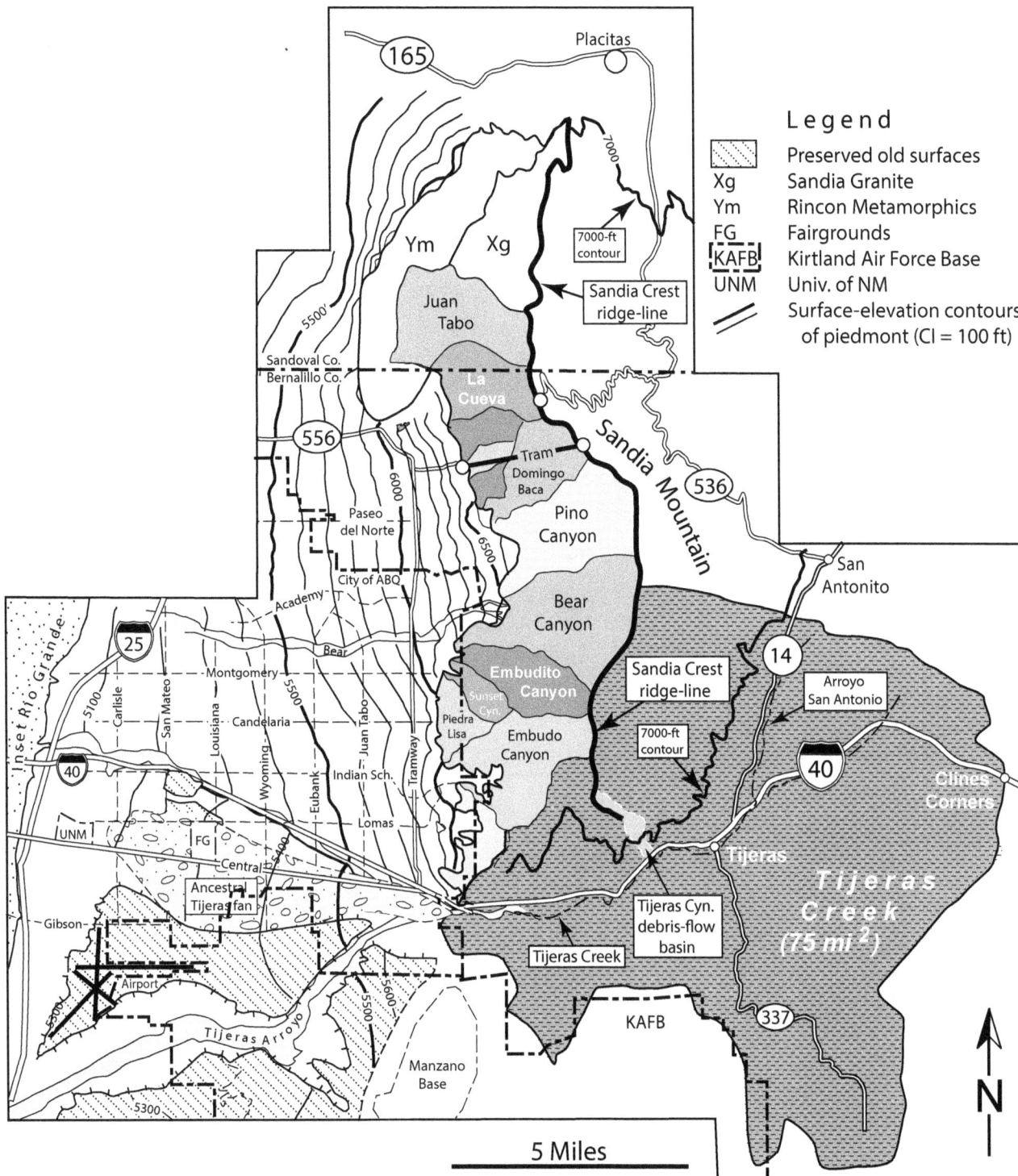

Figure 7-13. Drainage Sub-Basins (Gray Shades) of Tijeras Creek and Sandia Mountain
(Modified from Connell 1995)

A Tale of Two Fans: Ancestral and Modern Tijeras

On the south side of I-40 in the Albuquerque basin is found a subtle, inactive fan derived from Tijeras Canyon and known as the Ancestral Tijeras fan. The canyon's drainage is sourced by a large (~75 mi2) catchment basin on the east side of Sandia Mountain *(Figure 7-13)*. Significant flow from that side has supplied a large volume of coarse sediment for the Albuquerque basin *(Figure 7-14)*. Boulders more than a foot in diameter have been recovered from excavations at and near UNM, some seven miles out into the basin (Kelley 1982).

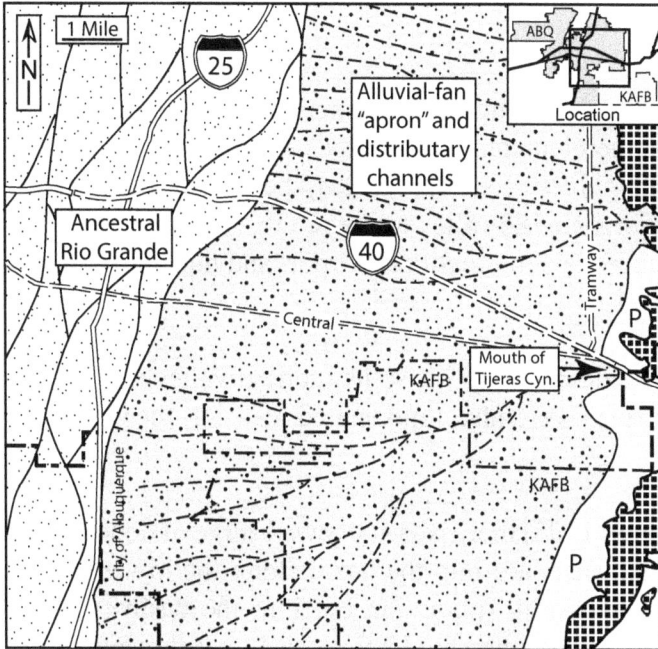

A. Basin Filled to Spill Point, Early Pleistocene, 800 Ka

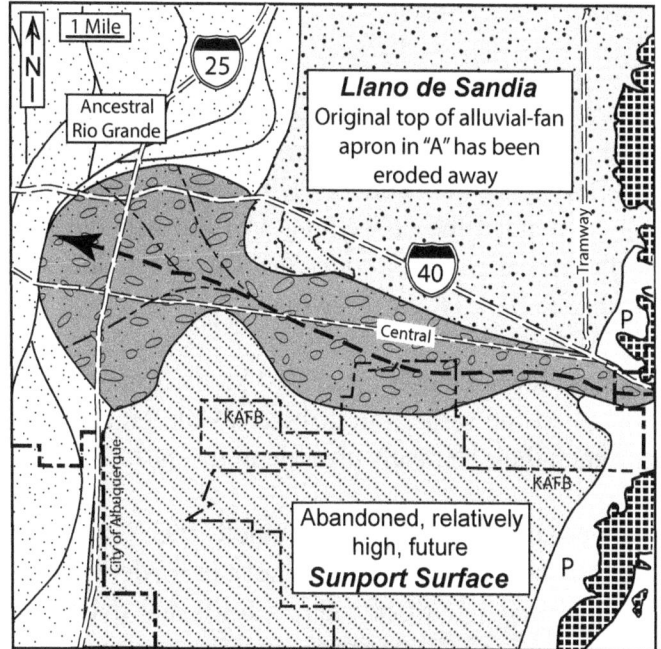

B. Ancestral Tijeras Creek and Its Fan (Dark Gray)
Middle Pleistocene, 28-21 Ka
(Modified from Lambert et al. 1982)

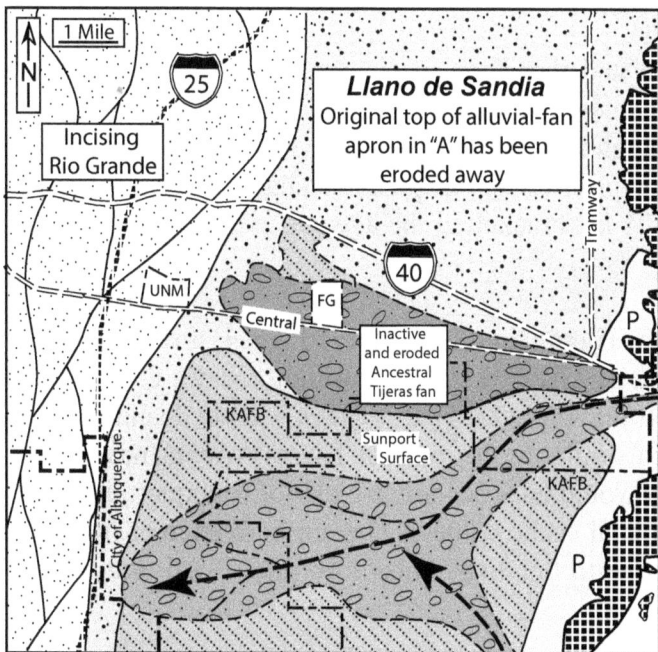

C. Tijeras Creek Capture and Drainage Shift to South,
Late Pleistocene, 21 Ka

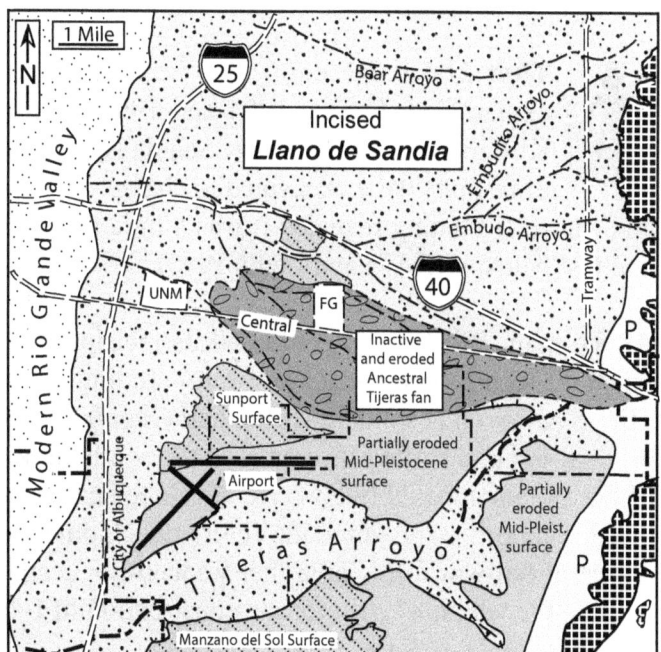

D. Headward-Erosion of Tijeras Creek, and Incision
of Modern Tijeras Arroyo, Late Pleistocene, 20-10 Ka

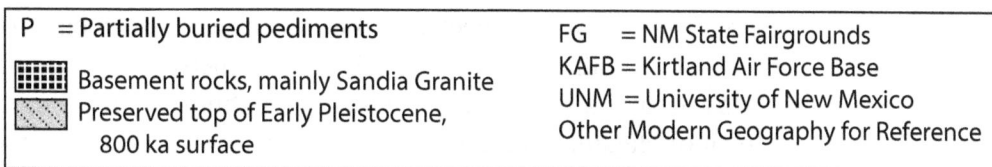

P = Partially buried pediments

▦ Basement rocks, mainly Sandia Granite

▨ Preserved top of Early Pleistocene,
800 ka surface

FG = NM State Fairgrounds
KAFB = Kirtland Air Force Base
UNM = University of New Mexico
Other Modern Geography for Reference

Figure 7-14. Ancestral Tijeras Alluvial Fan
(Select Modern Geography Shown for Reference)

241

By ca. 800 ka the Albuquerque basin had been filled to its brim, and sediment accumulation ceased. The land surface sat there, free from erosion and exposed to the elements for hundreds of thousands of years, developing a thick, *caliche*-rich soil (more on caliche in *Chapter 10*). This surface, the ancient basin floor, once extended the entire length of the Albuquerque basin some 80 miles to the south. A remnant of this surface is called the *Sunport Surface*, which floors Albuquerque's International Airport and most of Kirtland Air Force Base (*Figure 7-14A*).

As base level dropped after 800 ka the Ancestral Rio Grande began to incise its modern valley more than 400 ft. At the same time tributaries on the sides of the river were also rejuvenated, causing them to also entrench their channels upslope to the east.

Between ca. 128 and 21 ka (Connell 1995; GRAM and Lettis 1995) there was a renewed surge of deposition. The Ancestral Tijeras Creek (not the modern one) flowed westward into the basin along the lowest route it could find (south of today's I-40) and deposited the Ancestral Tijeras fan out into the basin (*Figure 7-14B*). This ancestral fan then slowly built up its top surface to a threshold height such that flowage in its distributary channels began to slow down and fill up, the entire edifice therefore becoming unstable. Around 21 ka, the Tijeras drainage, now seeking a new, lower path, abruptly broke loose from its old channel, shifted south to a lower area and deposited a new Tijeras fan in the area of today's Tijeras Arroyo (*Figure 7-14C*).

That 21-ka age date. This is a rather precise and therefore suspicious date, and warrants an explanation. During 1994 and 1995 a comprehensive geologic investigation was carried out on Kirtland Air Force Base by Sandia National Laboratories. During the course of field work a number of test pits were excavated to characterize the soil profiles. At one pit some charcoal was recovered and carbon-14 dating revealed it was a little older than 21 ka, bracketing the ages of sediments just above and just below (GRAM and Lettis 1995).

As the main Rio Grande continued its incision, the new Tijeras fan distributary channel, now becoming inactive, became entrenched eastward, eroding away much of its new fan and forming today's 250-ft deep Tijeras Arroyo that splits KAFB in half (*Figure 7-14D*). If the Ancestral Tijeras Creek had not abandoned its original fan and shifted its course to the south, an arroyo similar to the modern Tijeras Arroyo (now south of the Sunport) would likely have developed across what would later become the city of Albuquerque north of the Sunport, vastly affecting future development (and real estate values). The northern edge of this old, ancestral fan was a low area that frequently became prone to flooding. This "low value" real estate later became the "default" path for the construction of I-40.

Denudation of East Mountain Area

Drainages on the east side of Sandia Mountain that connected to the Rio Grande also keenly felt the drop in base level. Part of the alluvial cover that had built up and buried the Ortiz Igneous Belt "up to its chin" was partially stripped away (*Figure 7-15*). Bedrock in the Hagan basin was exhumed, including the coal beds at what later became the mining town of Hagan. It's arguable whether that exposure was a good thing or a curse. Its suddenly-exposed coal led to construction of a town that had an unhappy end. (More about this in *Chapter 14*.)

242

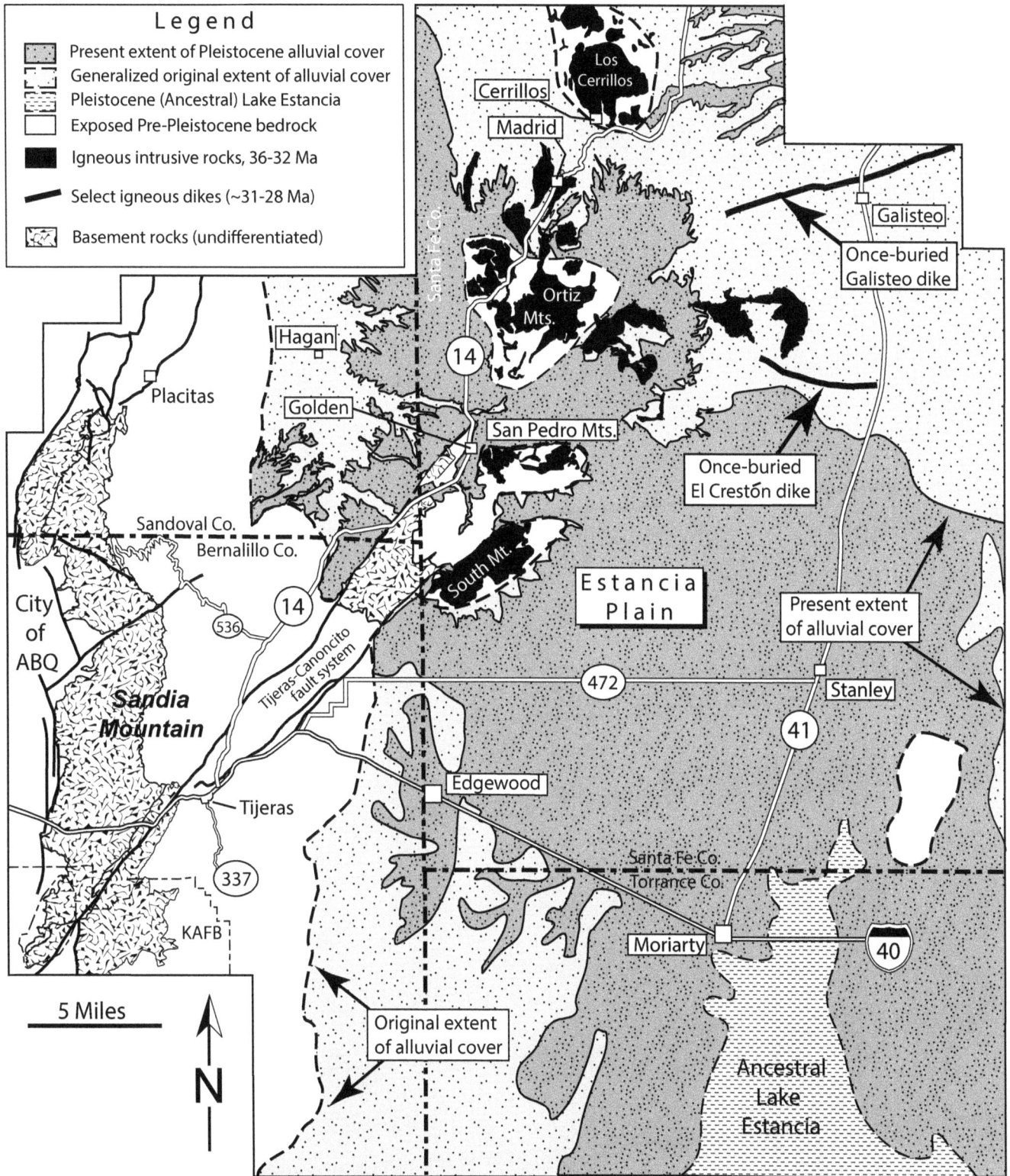

Figure 7-15. Alluvial Cover of Ortiz Igneous Belt and Ancestral Entancia Valley, East Side of Sandia Mountain (In Part Modified from Allen 2007)

243

PART II:
PLACES OF SPECIAL INTEREST

8:
NORTHERN NOSE/PLACITAS AREA

Introduction

Sandia Mountain's northern nose is geologically a very strange place. Part of the area is occupied by the village of Placitas and much is private property, with only a few places of interest being accessible. The southern, higher parts are very rugged and for practical purposes rather inaccessible.

? **Village of Placitas**. The northern portion of this area contains the little community of Placitas, with about 5000 people in it and the surrounding area (2020 census). Once the sleepy, traditional village of *San José de las Huertas* (St. Joseph of the Orchards) dating from the late 1700s, it experienced a surge of "counter-culture" settlers ("hippies") during the 1960s and early 1970s. These folks set up settlements, especially up Las Huertas Creek to the south, given such whimsical names as "Little Farm," "Tawapa," and "Manera Nueva," and somehow eked by. Age and drugs took their toll, and a few of the hippies still remain. Later came an influx of the more affluent, including day-commuters to the cities of Santa Fe and Albuquerque, and, of course, retirees. Within the accessible areas are 1) an array of strange, flat-topped hills, each with a story, 2) what I call the Alcove, 3) the Strip Mine Trail, 4) the Ranchos fault, and 5) the Placitas Dike (*Figure 8-1*)

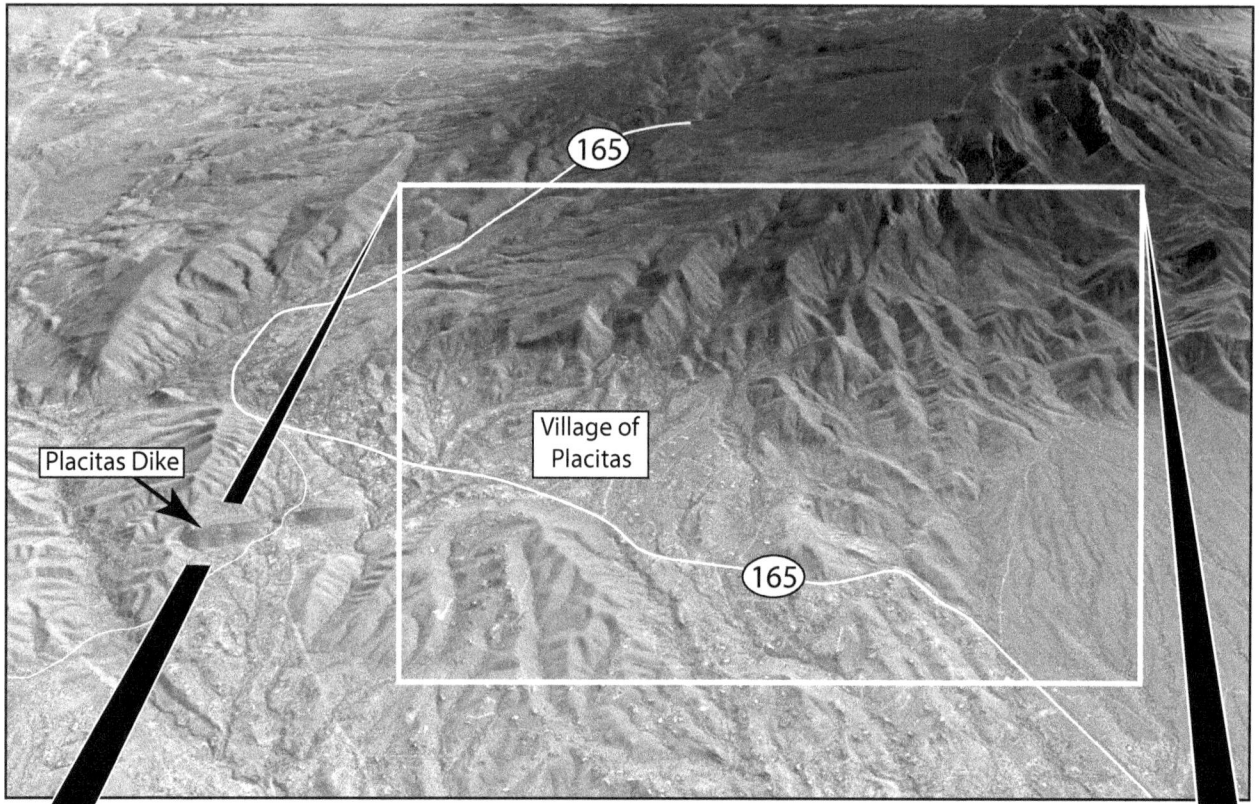

A. South-Southeast Oblique *Google Earth* Image

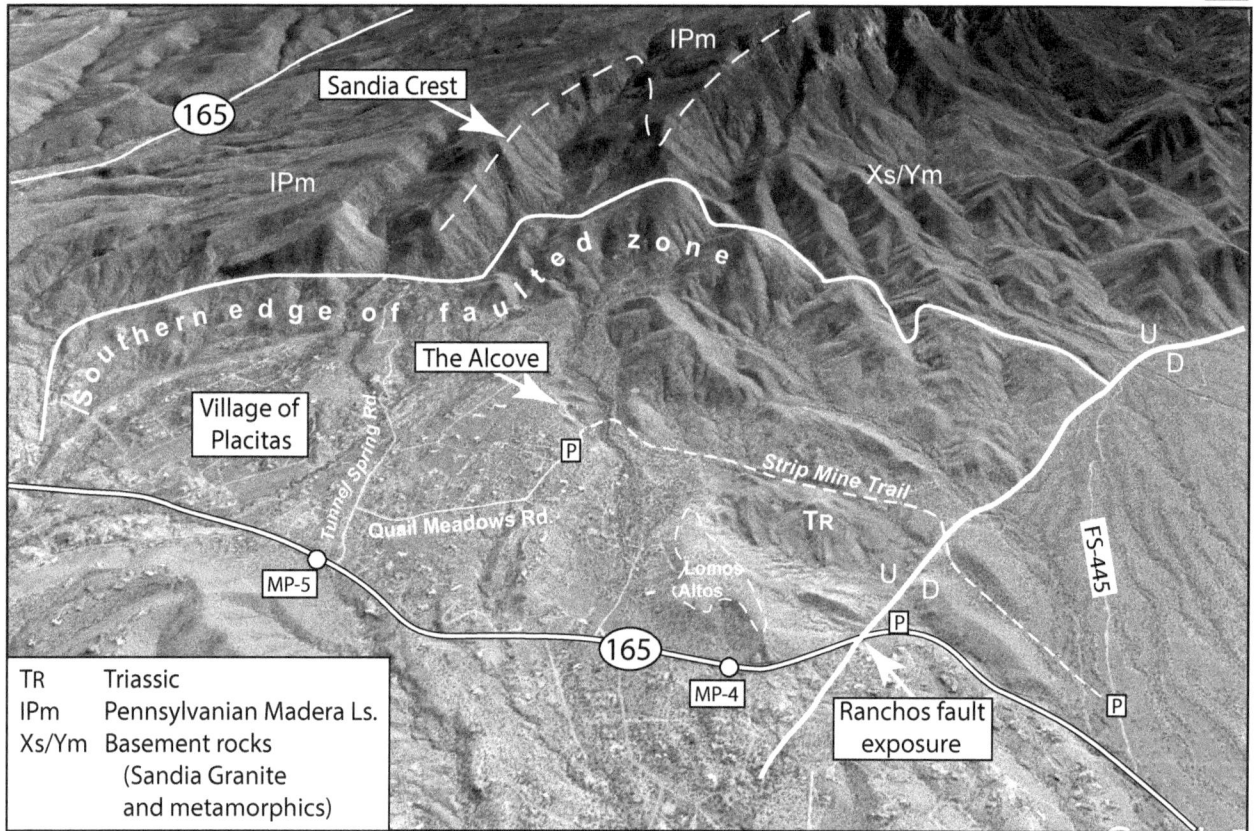

Placitas Dike

Village of Placitas

165

165

Sandia Crest

IPm

IPm

Xs/Ym

Southern edge of faulted zone

The Alcove

Village of Placitas

Tunnel Spring Rd.

Strip Mine Trail

Quail Meadows Rd.

TR

U
D

FS-445

U
D

P

MP-5

165

Lomos Altos

P

MP-4

Ranchos fault exposure

P

TR   Triassic
IPm   Pennsylvanian Madera Ls.
Xs/Ym Basement rocks
        (Sandia Granite
        and metamorphics)

B. Close-Up of "A" Above (P = Parking Areas)

Figure 8-1. Overview of Placitas Area, Northern End of Sandia Mountain

Flat-Topped Hills, or Mesas

Strikingly obvious here are some flat-topped hills, *mesas*, that seem very much out of place. These features are the remnants of former depositional surfaces that have since been abandoned and left high and dry by the incision of the Rio Grande after 800 ka. By "depositional surfaces" I mean ancient, former drainage-system floors that now form the tops of mesas perched 100 to 200 feet above the village of Placitas. These mesa-top heights provide choice sites for some of the more expensive homes in Placitas, offering expansive views off to the west.

The mesas are capped by bouldery gravel, in sharp contrast to the sediment in the drainages at the lower elevations, signaling a different geology. The gravels atop the mesas are all younger than about 800 ka, and collectively form what geologists call "surficial" geologic deposits (*Figure 8-2*; see note in the INTRODUCTION).

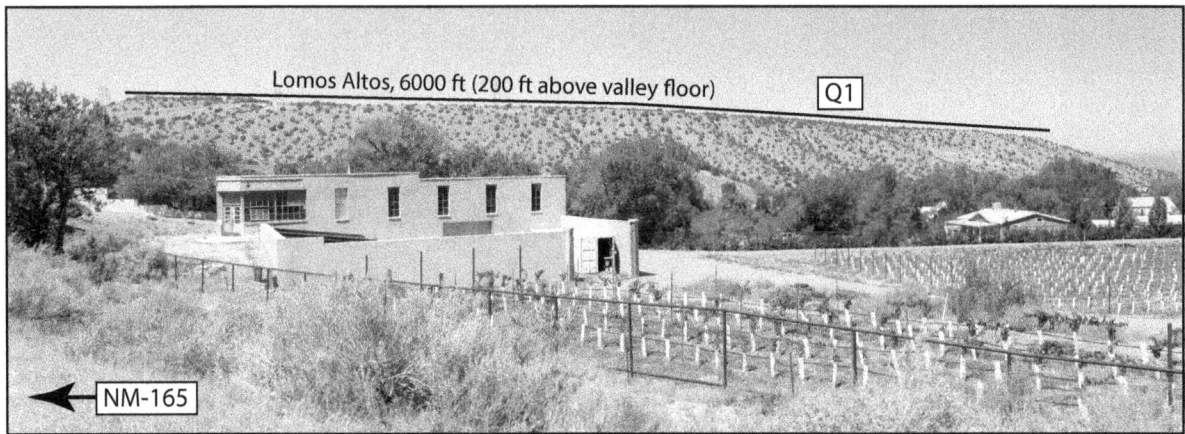

A. West-Northwest View of Lomos Altos (Q1; Photograph by Author)

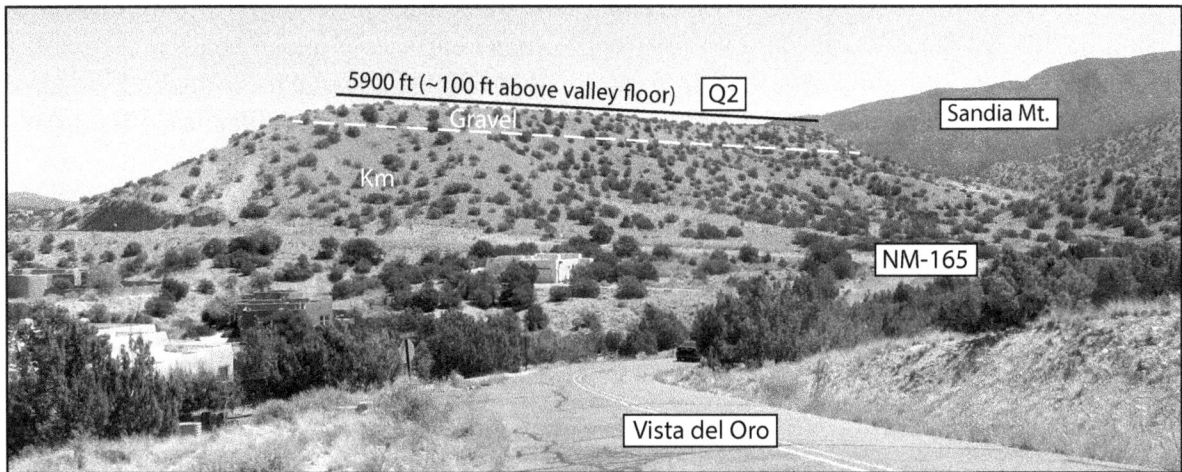

B. East View of Q2 Mesa Located South of NM-165 (Km = Cretaceous Mancos Shale; Photograph by Author)

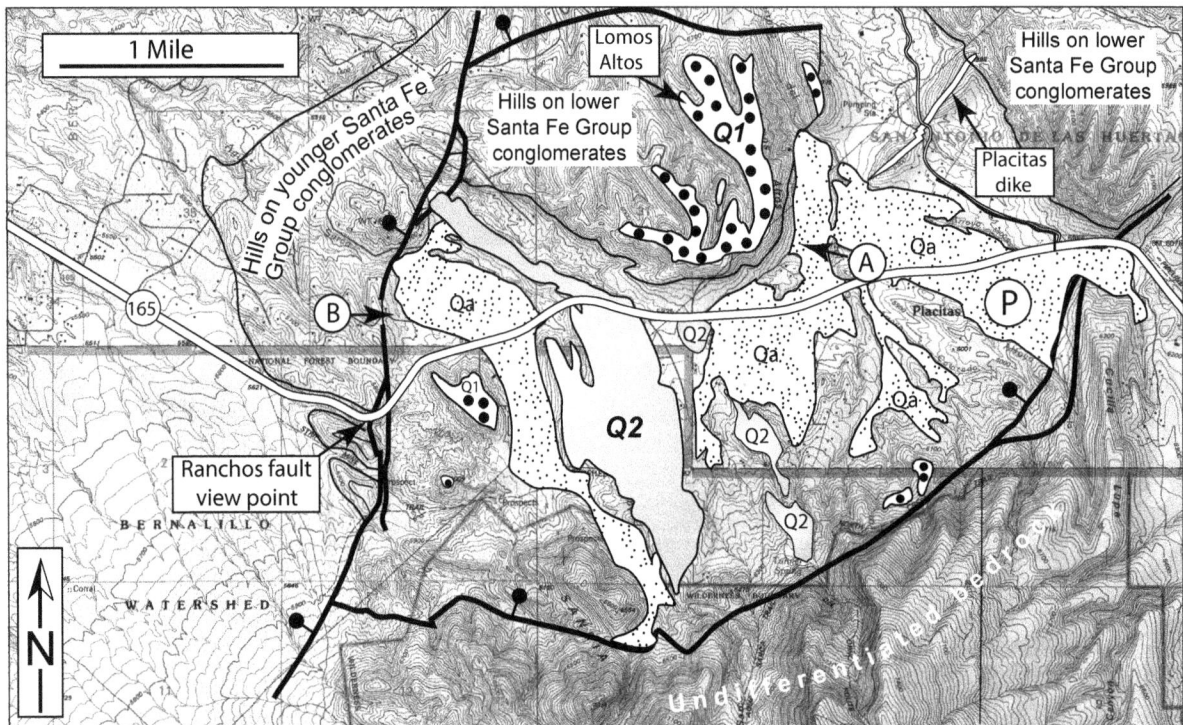

C. General Topography of Placitas Area ("P" = Village) and Locations of Photographs in "A" and "B" Above
(Q1 = Highest Mesas; Q2 = Intermediate-Height Mesas; Qa = Alluvial Lowlands; Modified from Connell et al. 1999)

Figure 8-2. Mesas of Placitas Area

Surficial units and their "mesas." The four surficial units at Placitas are from highest elevation (oldest) to lowest elevation (youngest): 1) *Q1 = Lomos Altos*, 2) *Q2 = Las Huertas*, 3) *Q3* = an unnamed level, and 4) *Qal* = the modern stream alluvium, respectively (*Figure 8-3*).

For the past 800 ka this area has been in the process of denudation, i.e., erosional lowering of the landscape, and the individual mesas document the stop-and-go nature of the denudation. These levels are in turn genetically linked to the cyclic glacial-interglacial phases mentioned in the previous chapter. There are four of these levels at Placitas, but only the first two (Q1 and Q2) are of interest.

Q1, Lomos Altos. This bouldery to cobbly gravel-capped mesa hosts an array of up-scale houses in a subdivision called *The Overlook*, named for its excellent views (*Figure 8-2A*). The mesa tops off at about 6040 ft, 200 ft above the surrounding low areas, and slopes gently to the northwest. The mesa cap is a remnant of one of the earliest (~ 2 Ma; Connell and Wells 1999) surges of coarse sediment transported off the north flank of Sandia Mountain. Its cobbles are almost entirely of Madera limestone. Rare chunks of granite indicate that the mountain's granite core had just about been breached by that time.

Q2, Las Huertas. This is a mesa of intermediate height above the valley bottom. It is quite prominent to one approaching Placitas from the west (*Figure 8-2B*).

Q3. This surface is unnamed and does not form an obvious land form.

Qal. The old village of Placitas is nestled down in the lowest region on top of modern stream alluvium, Qal, and therefore is close to the accessible water.

The "Alcove"

This is my informal term for an interesting, accessible little cliffy recess (*Figure 8-4*). It is a little "jewel box" containing the same Jurassic rocks that form the spectacular scenery of the Ghost Ranch area of Rio Arriba County, sometimes known as "Georgia O'Keeffe country," but at a more modest scale. The Alcove provides an interesting diversion on a day hike along the Strip Mine Trail (see below).

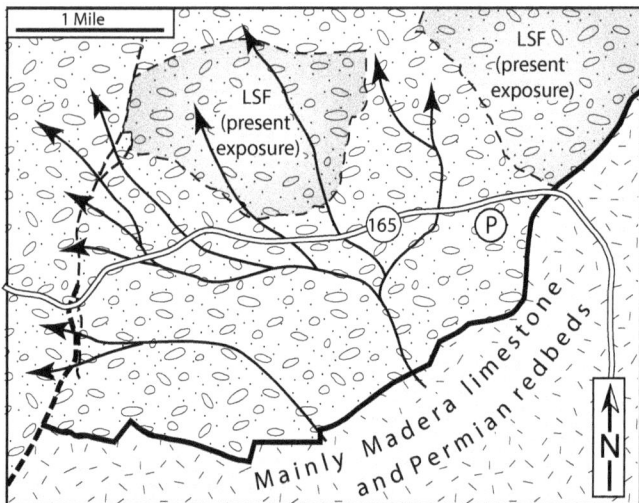

A. Lower Santa Fe (LSF) Fanglomerate, 15(?) - 10 Ma
(Present Exposures of LSF in Dashed Outlines)

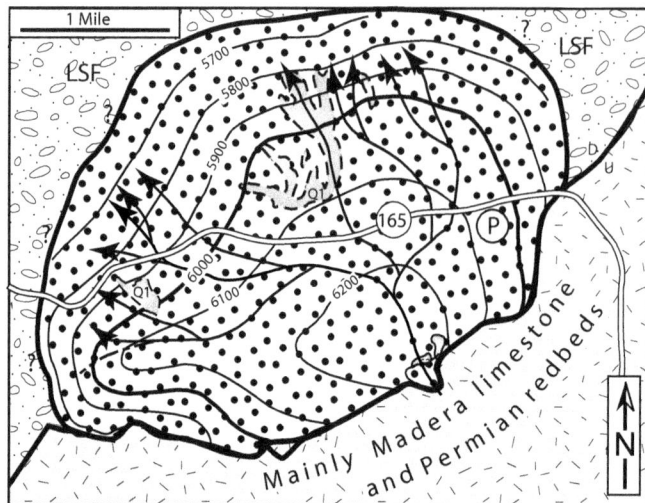

B. Lomos Altos Fanglomerate (Q1), >800 Ka
(Present Exposures of Q1 Shaded; Modern and
Restored Elevation Contours for Q1, CI = 100 Ft)

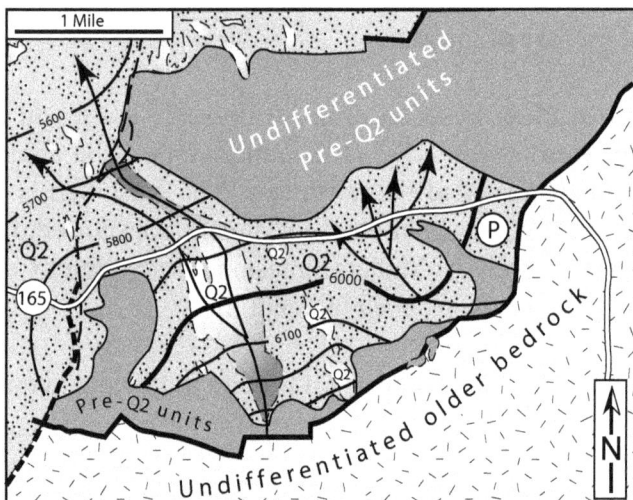

C. First Incision of Rio Grande (to West): Las Huertas Fanglomerate
(Q2, ca. 800 Ka; Present Exposures of Q2 in Dashed Outlines;
Modern and Restored Elevation Contours for Q2, CI = 100 Ft)

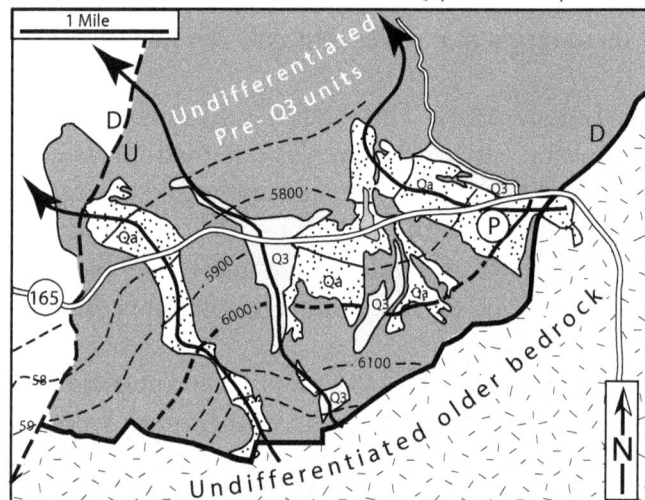

D. Q3 (ca. 800-400 Ka) and Modern (Lowlands) Alluvium (Qa)
(Modern and Restored Elevation Contours for Qa)

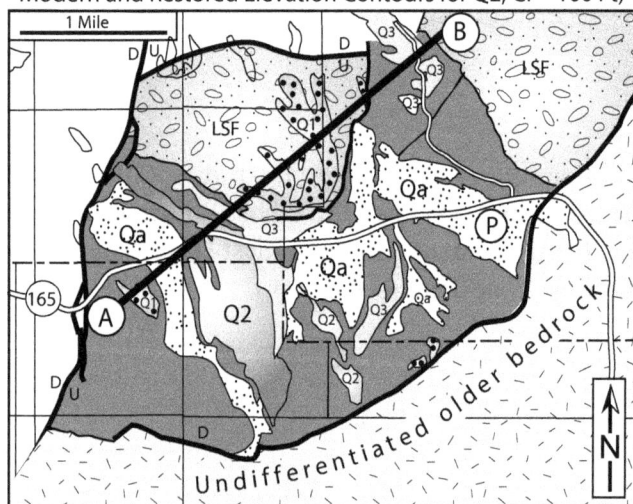

E. Present Surficial Geology Obscuring Underlying Bedrock

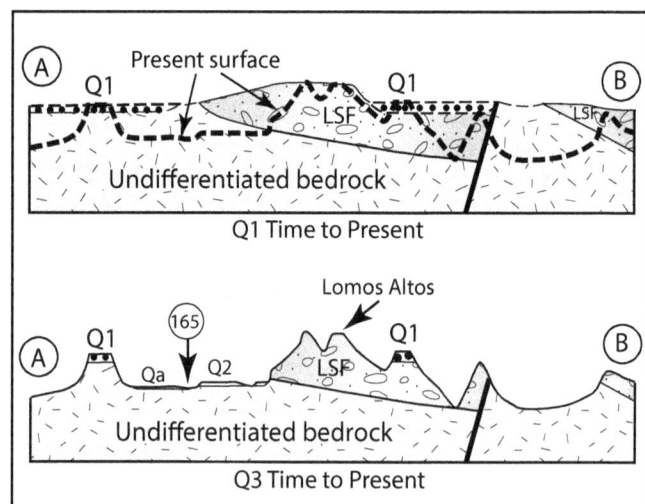

F. Schematic Time-Sequential Cross Section A-B (Location in "E")

Figure 8-3. Surficial Geology of Sandia Mountain's Northern Nose: Placitas Area
(Select Modern Geography and Select Modern Surface Elevation Contours Shown; CI =100 ft;
P = Old Village of Placitas; Arrowed Lines = Paleo-Drainages)

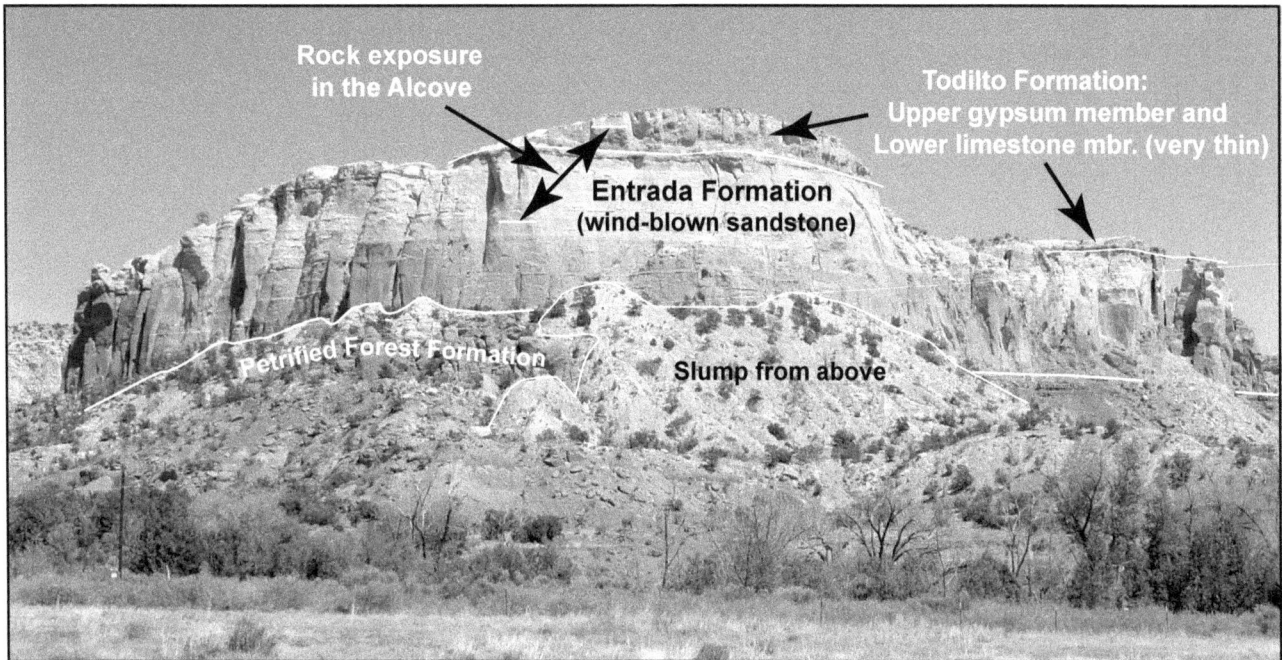

Rock exposure in the Alcove

Todilto Formation: Upper gypsum member and Lower limestone mbr. (very thin)

Entrada Formation (wind-blown sandstone)

Petrified Forest Formation

Slump from above

A. East-Southeast View of *Kitchen Mesa*, Ghost Ranch's Section of O'Keeffe Country

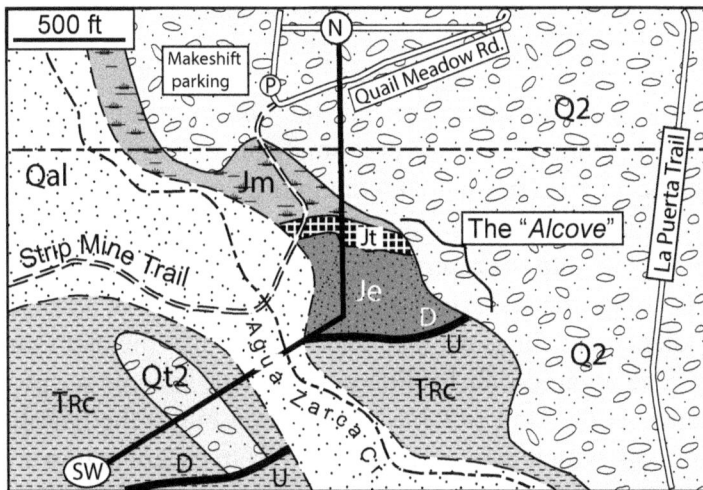

B. Geologic Map of the Alcove Area (from Connell et al. 1999)

Legend

Qal	Quaternary (modern) alluvium
Q2	Early Quaternary, high gravel platform
Jm	Jurassic Morrison Formation
Jt	Jurassic Todilto Formation
Je	Jurassic Entrada Ss. Formation
TRc	Triassic Chinle Formation
P	Parking area

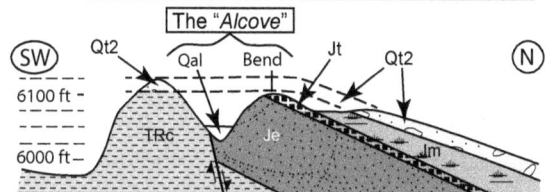

C. "Elbow" Cross Section in "B" (V/H = 3.25)

D. East View of Todilto Fm. (Jt) and Entrada Sandstone (Je)

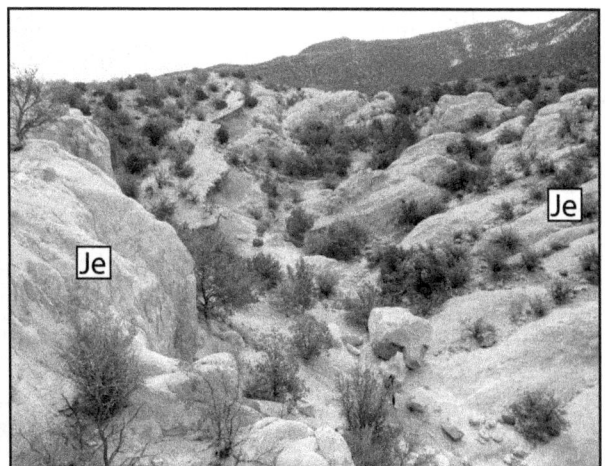

E. East-Southeast View of Entrada Sandstone (Je)

Figure 8-4. The "Alcove," Placitas Area (Photographs by Author)

Parking. At one time there was a little parking area just north of the trailhead at the National Forest boundary. That has been eliminated, probably due to complaints from residents who have built their homes right up to the line. Parking now is anywhere that can be found near the southwest end of Quail Meadows Road (*Figure 8-4B*). From there it is a mere 0.1 mile walk south along the old jeep road leading to the west end of an alcove containing the Jurassic rocks.

The floor and walls of the Alcove consists of the whitish to pinkish, poorly-cemented, wind-blown sand of the *Entrada* sandstone. The walls are capped by the distinctive *Todilto Formation*. This latter unit consists of a lower limestone and an overlying gypsum unit. The latter two parts are quite thin here, but nicely displayed. This limestone is truly unique and is made up of a thinly-laminated rock that doesn't look at all like limestone.

Entrada and Todilto Formations. This pair is significant in the San Juan basin of northwestern New Mexico where it is deeply buried and responsible for several small oil fields. Oil and gas accumulations all have three key ingredients: 1) a reservoir rock containing pore space in which fluids can accumulate, 2) a source rock containing organic matter that will "crack" to mobile oil and gas with increased heat and that is then able to migrate into the reservoir rock, and 3) a sealing rock that prevents the oil from escaping to the surface. The Entrada (the reservoir) is another ancient *erg* or sand sea. In places in the San Juan basin some of the erg's ancient sand dunes were perched higher than the main sand body, forming a localized "paleotopographic high." The organic-rich Todildo member (the source rock) was deposited on top of, and draped, over the high dunes as the land subsided. With deep burial the heat has "cracked" the organic matter in the limestone to form oil, which then was prevented from escaping upwards by the impermeable Todilto gypsum member (the seal) on top. The oil was therefore forced downward into the porous sandstone (the only direction the oil could go), forming small oil fields. Here at the Alcove, though, the limestone was never buried deeply enough, and its organic matter has not been broken down to oil, so the original organic matter is still present in the rock, giving it a dank, fetid smell when crushed in the hand.

Strip Mine Trail (#51)

This foot trail runs west for an easy 1.6-mile loop from the Alcove to a parking area on Forest Loop Road, FS-445 (*Figure 8-1B*). (Multiple vehicles help for this little hike.) The trail traverses the Triassic redbeds of the Petrified Forest Member of the Chinle Formation, plus several abandoned pits that had been bulldozed by prospectors. This is the same redbed unit as at Petrified Forest National Park in eastern Arizona with its fabulous display of fossilized logs. The prospectors here were looking for "redbed copper."

Redbed copper. The Petrified Forest Member is known for its petrified wood, but not here. Shortly after deposition of the river sands and muds, woody debris carried along and buried with the sediment simply became decaying organic matter. Copper carried in solution by later migrating groundwater through the sands encountered the buried organic matter, reacted with it (became reduced), precipitated, and replaced the wood with a blackish copper oxide mineral. Where these mineralized logs were raised to near the surface they became oxidized to the beautiful green mineral *malachite* and less commonly to the equally beautiful blue mineral *azurite*. The ores therefore are simply copper-replaced wood. Near Placitas the prospectors knew the association and scraped open some pits with bulldozers where they saw the redbeds. They didn't find any copper, possibly because of the lack of wood.

Ranchos Fault

This is a favorite stop for geology students (*Figure 8-5*). Exposures of normal faults are not common, and this excellent one is a "gift" from the highway department. Seen here are coarse, unconsolidated Santa Fe Group sediments in fault contact with the underlying Cretaceous-age Mancos Shale. The fault angle is about an apparent 15° dip to the west, but the road cuts across it at an angle such that the true west dip-angle is a tad more, about 20°. As mentioned earlier, most normal faults bracket an angle of about 60°, so this is much less than a "normal" normal fault should be, suggesting that this area has been rotated back to the east about 40°.

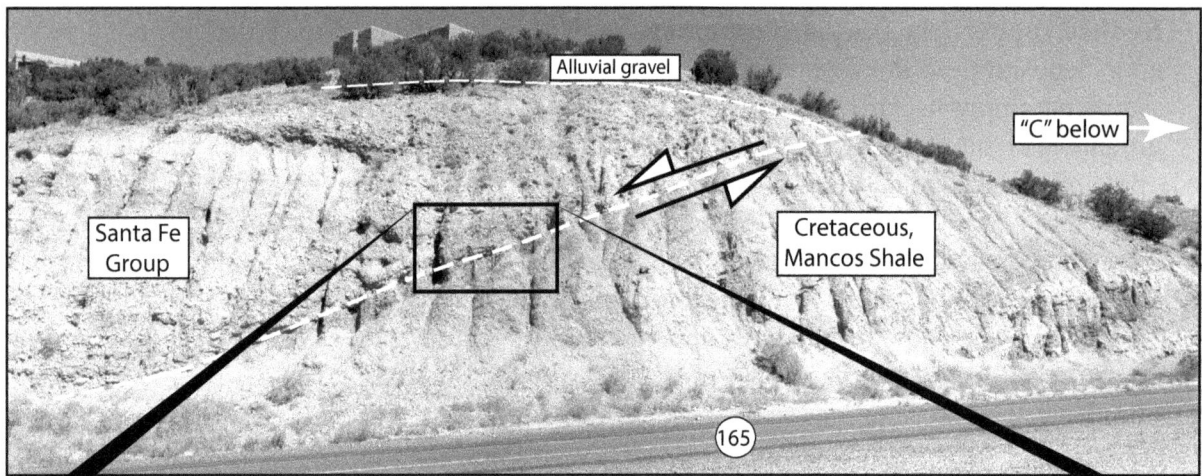

A. North View (Left of "C" Below)

B. Detail of "A" Above

C. North-Northwest View (Right of "A" Above; Based in Part on Interpretation of Bauer et al. 2003)

Figure 8-5. Ranchos Fault Zone, NM-165 West of Placitas (Photographs by Author)

Placitas Dike

Although this feature is geologically important, providing the important date of 31 Ma, it is not a tourist stop (*Figures 6-7* and *8-1A*) and one barely notices it while driving north on Camino de las Huertas through the gap in the dike. It can be skipped.

9:
RINCÓN RIDGE
AND JUAN TABÓ CANYON

Introduction

Now we begin a north-to-south sweep down the west side of Sandia Mountain. We start off with the prominent spine called Rincón Ridge (*rincon* is Spanish for "corner"). The ridge forms the western rim of the Juan Tabó Cul-de-Sac. Most notably it also hosts an extremely bizarre feature called Juan Tabó Canyon.

Rincón Ridge

This sharp, topographic ridge is made up of the relatively hard, Rincón metamorphic rock. Its southern end plunges gently, 5 to 10° to the south and disappears beneath Tramway Blvd. (NM-556; *Figure 0-21*).

Juan Tabó Canyon/Water Gap

When approaching Sandia Mountain from the west along NM-556 (Tramway Blvd.) one is struck by the sharp, knife-like gash in Rincón Ridge (*Figure 9-1A*) known as Juan Tabó Canyon. This is a textbook example of what geologists call a *superimposed stream.* A second, subdued and apparently "unfinished" partial gap lies along the ridge just to the north (*Figure 9-1B*). The pathetic little trickle of modern Juan Tabó Creek clearly did not excavate this canyon, a *water gap*, and so there's clearly more to the story.

A. East View of Juan Tabo Canyon (Center Foreground), Southern Rincon Ridge (Photograph by Author)

B. East Oblique *Google Earth* Image of "Unfinished" and Present Juan Tabo Canyons

Figure 9-1. Rincon Ridge and Juan Tabo Canyon

Superimposed streams and their canyons are common in areas that have experienced geologically-recent regional uplift with subsequent downcutting by streams (*Figure 9-2*). What they all have in common is that 1) an original stream flowed at a higher elevation on a gently-sloping surface of soft material, 2) upon uplift the stream, with its fixed channel, cut down through the soft material until it encountered the harder rocks buried below, and 3) with its course already fixed, the stream superimposed itself onto and through the harder material.

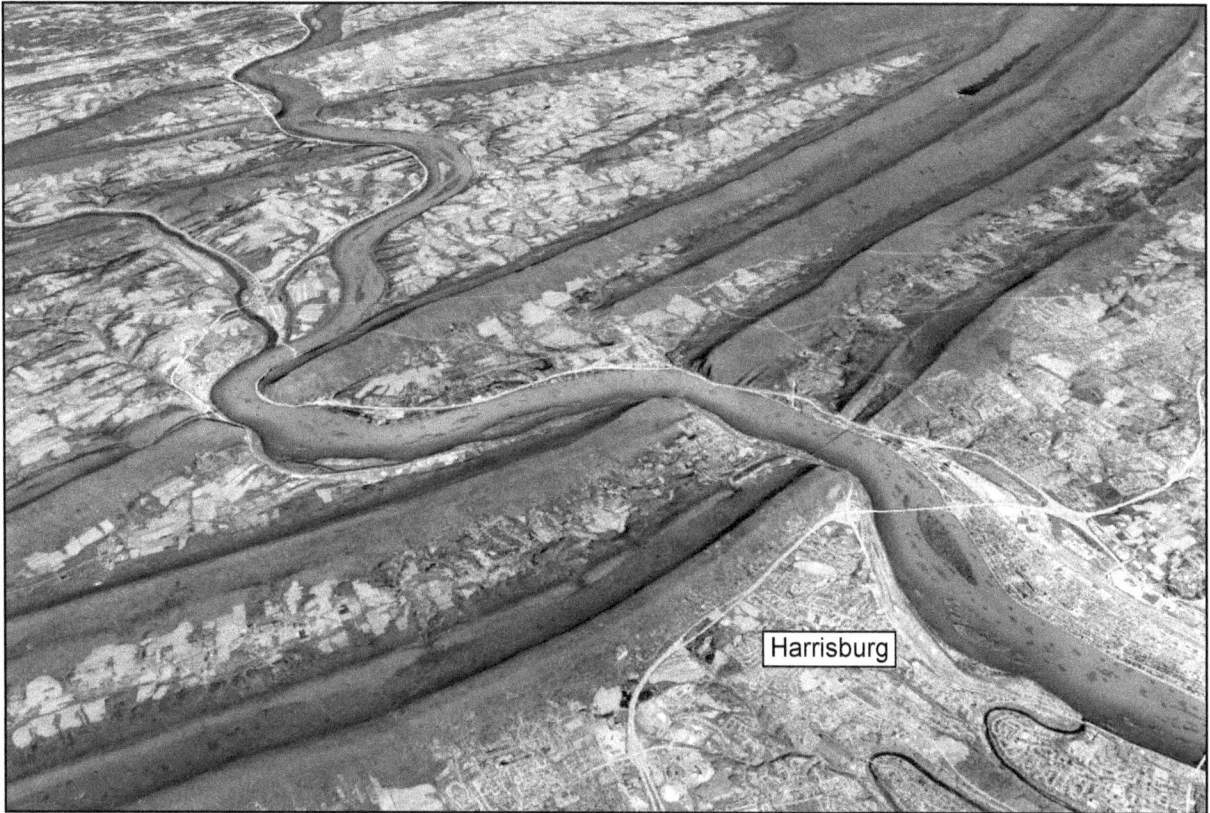

A. North-Northeast Oblique *Google Earth* Image of Harrisburg, PA

B. Northeast View of *Devil's Gate*, Oregon Trail, WY (Photograph by Author 1991)

Figure 9-2. Water Gaps

However simple that explanation sounds, the origin of Juan Tabó Canyon is more complicated and presents a genuine mystery that requires some serious geologic sleuthing. Let's approach this mystery like detectives by first listing what we know:

1. Water flows downhill!
2. The rock of Rincón Ridge is of metamorphic rock that is much harder and therefore more resistant to erosion and weathering than the Sandia Granite lying to its east.
3. The evidence at hand suggests that the entire block of the Juan Tabó Cul-de-Sac has rotated up to the north and down to the south, providing a natural southerly gradient for the streams (*Figure 6-19*).
4. Something had to *force* the original Juan Tabó drainage to flow in a westerly direction instead of taking the natural, "easy" way south.
5. Juan Tabó Creek had to have been once flowing in a fixed channel on soft alluvium, high above the top of the then-buried metamorphic rocks, allowing unimpeded westward flow.
6. From west to east, the unfinished partial gap, the abrupt northern edge of East Rincón Ridge and the so-called *La Cueva lineament* in the main mountain all line up nicely.

The upshot of all the above is that *something* forced the ancestral Juan Tabó Creek to flow west instead of south before the creek cut its canyon. But what was the culprit? Before delving into this mystery let's consider this famous chestnut:

"Once you eliminate the impossible, whatever remains,no matter how improbable, must be the truth."(Arthur Conan Doyle, British author/creator of Sherlock Holmes, 1859–1930)

With Doyle's wisdom and the six "givens" above in mind, the La Cueva lineament cited in #6 above emerges as the possible key to the puzzle. The lineament, including the rim of East Rincón Ridge, divides the Juan Tabó Cul-de-Sac into two sub-realms: a topographic bowl on the north that today drains west through Juan Tabó Canyon, and a catchment basin on the south that drains south.

Back in *Chapter 2* (Emplacement of the Sandia Granite) I suggested that the modern La Cueva lineament is an ancient flow-discontinuity formed within the granite magma as it was being intruded 1.4 Ga (*Figure 2-12*). Along the lineament, high up the mountain to the southeast, any sense of up or down fault displacement becomes indistinct, if not zero, but the lineament does deeply erode to form a gorge. The lineament runs southeast to cross a drainage divide, then past the western base of a huge granite buttress called the "Thumb," and terminates at the base of the Pennsylvanian rocks capping Sandia Crest where it disappears (*Figure 9-3A*). Hence that high southeastern part of the La Cueva lineament is clearly pre-Pennsylvanian in age.

A. East Oblique *Google Earth* Image of Full Length of La Cueva Lineament

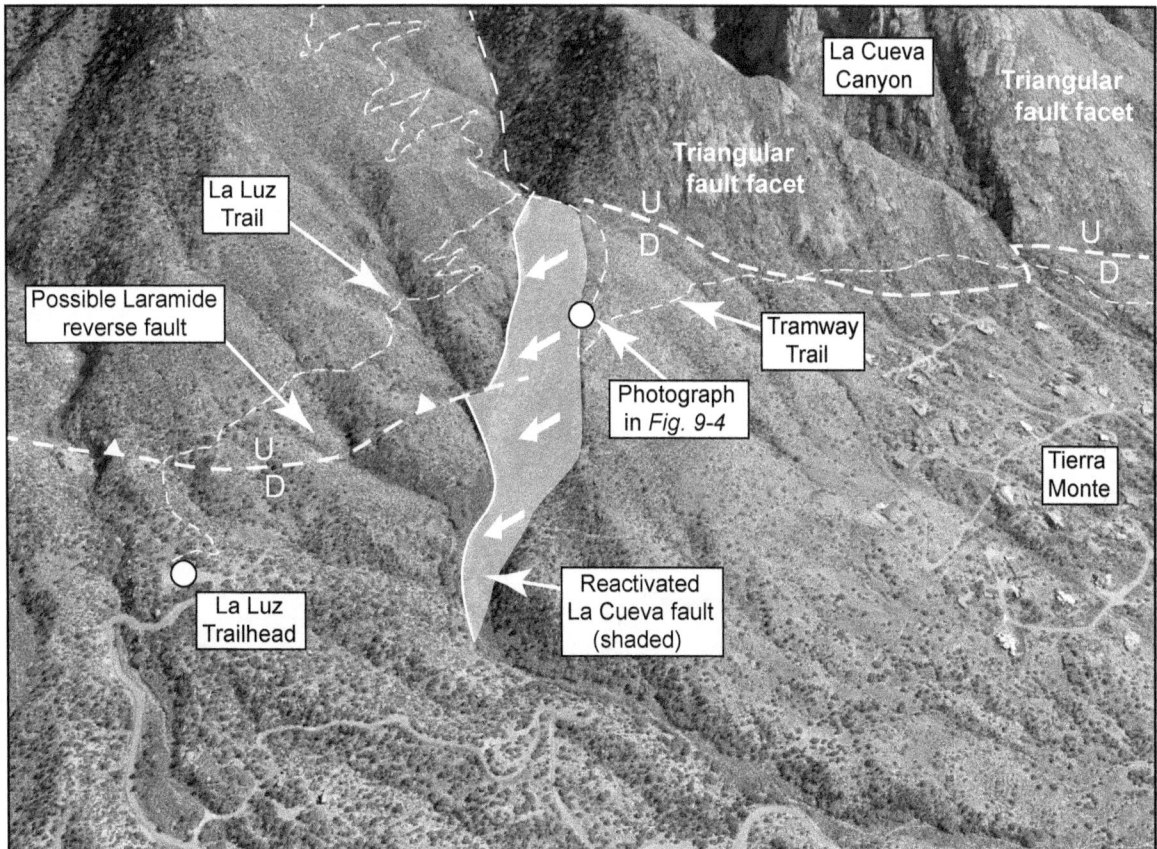

B. East-Southeast Oblique *Google Earth* Image of Lower, Reactivated Part of La Cueva Lineament

Figure 9-3. La Cueva Fault and Lineament

I suggest that the lower part of the lineament (south of the La Luz Trailhead) has been selectively and more recently reactivated to become a down-to-the-north normal fault as the Juan Tabó block was rotating down to the south (*Figure 9-3B*). Supporting the notion of more recent, localized fault movement are the north-facing topographic escarpment (*Figure 9-3*) and polished planer surfaces ("slickensides") seen near the top of the fault (*Figure 9-4*). I further suggest that this fault escarpment possibly created a north-facing barrier or dam that, before being partly eroded away, once extended farther west across the cul-de-sac. This blocked the drainage in the northern bowl and forced it to escape westward. The ancestral Juan Tabó Creek then 1) flowed across a land surface over and above the buried mass of metamorphic rock, 2) deepened its channel until it encountered the metamorphics, 3) began to cut a canyon through the metamorphics (a "false attempt"), and 4) then shifted slightly to the south and slowly dug the present Juan Tabó Canyon (*Figure 9-5*).

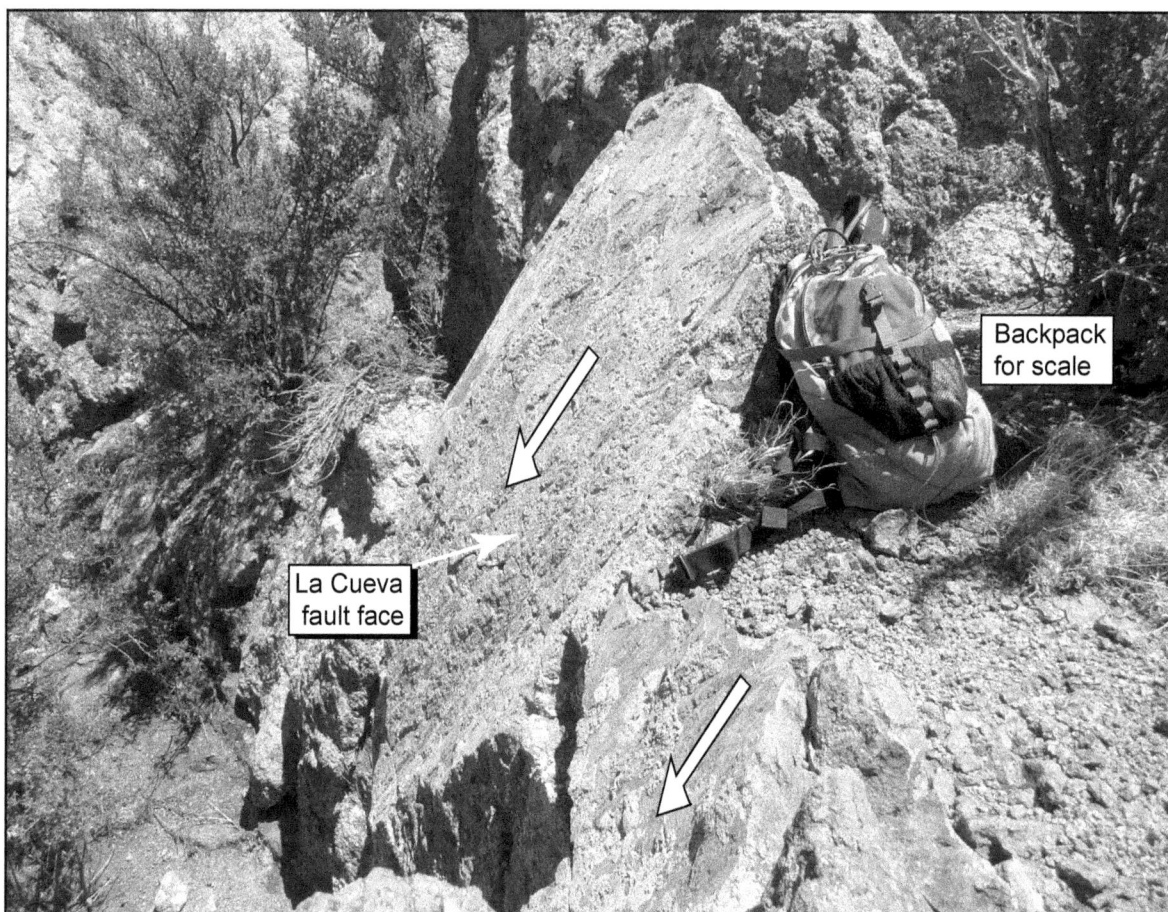

Figure 9-4. Southeast View of Slickenslided La Cueva Fault Face
(Location in *Figure 9-3*; Photograph by Author)

West

East

Future Rincon Ridge

Ancestral Juan Tabo Creek

Santa Fe Group alluvium

Rincon Metamorphics

Sandia Granite

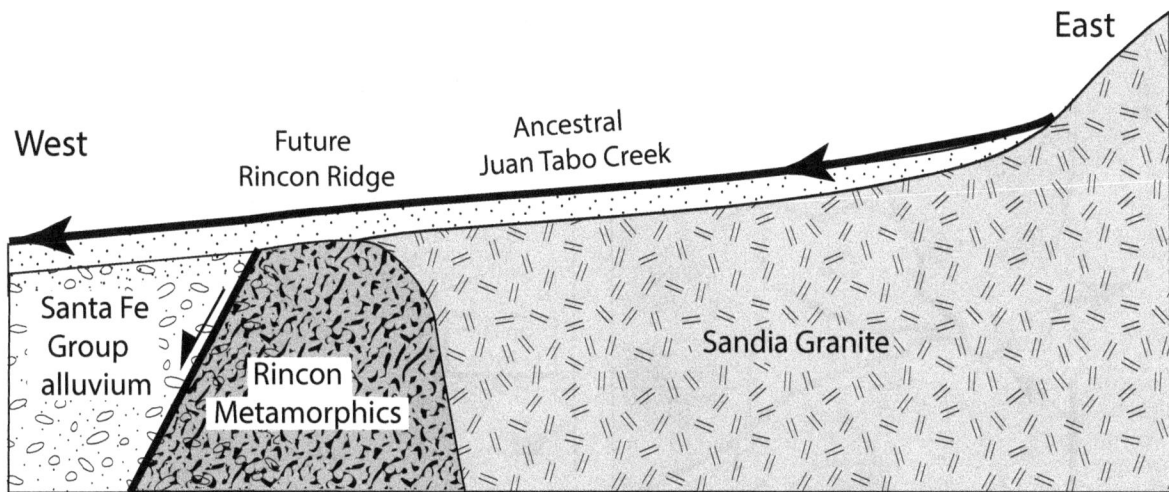

A. Pre-Canyon Drainage

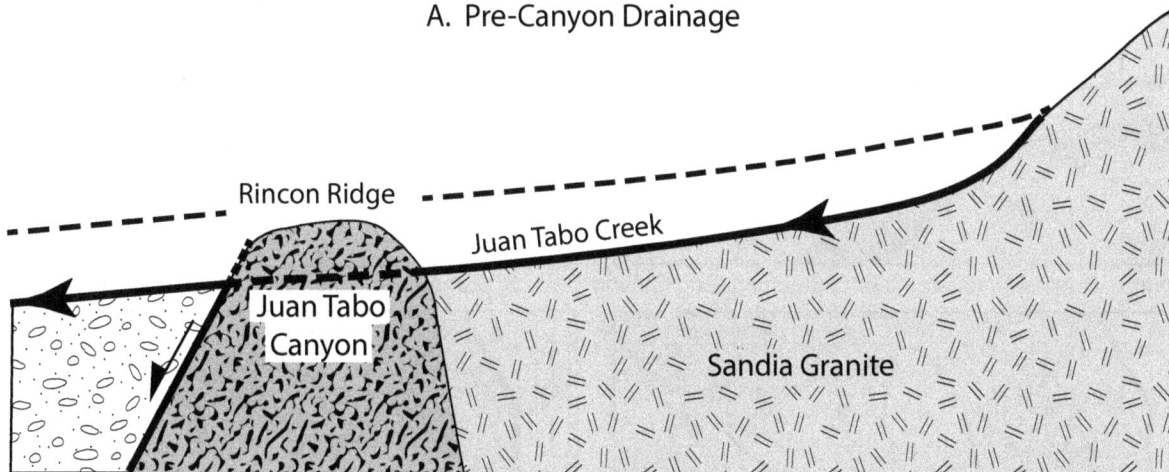

Rincon Ridge

Juan Tabo Creek

Juan Tabo Canyon

Sandia Granite

B. Through-Canyon Drainage

Figure 9-5. Schematic Cross-Sections Showing Origin of Juan Tabo Canyon

We can now sketch a speculative scenario of the Juan Tabó Canyon's development (*Figure 9-6*). In this busy figure I affirm my interpretation that the controlling influence was the "dam" caused by the activated lower part of the La Cueva lineament (the modern fault itself), and the westward extension of that dam (since eroded away) to link with the north-facing bulwark of East Rincón Ridge.

Legend (Panel A):
Dam formed by partially-activated down-to-the-north La Cueva fault
Ym = Rincon Metamorphics; Xg = Sandia Granite;
Mz = Mesozoic outlier; Qa = Quaternary alluvium

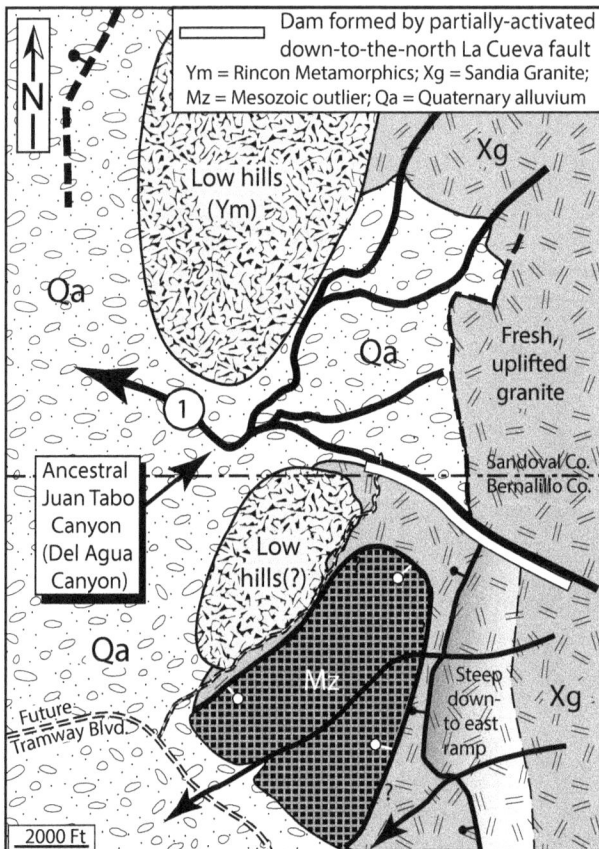

A. Stage 1: Original Drainage via Ancestral Juan Tabo Canyon

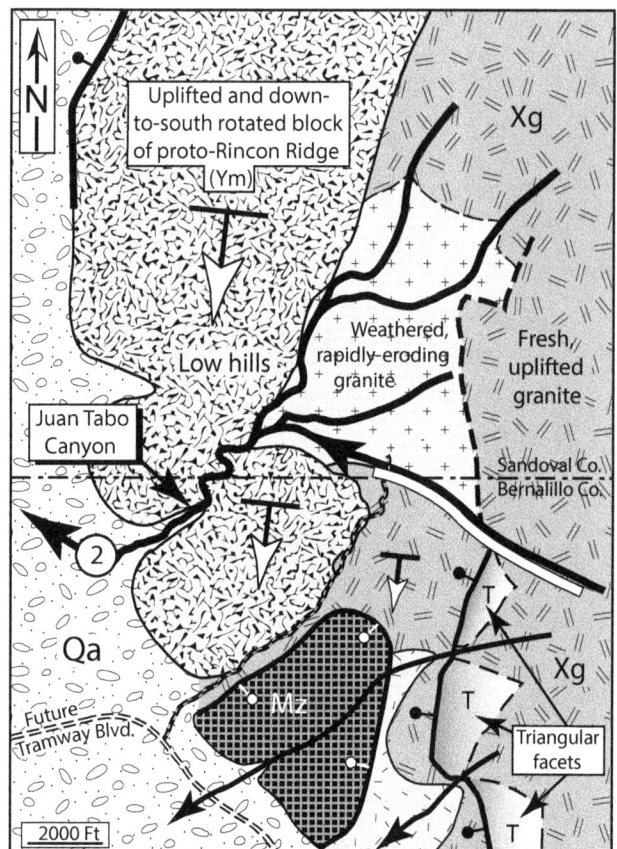

B. Stage 2: Drainage via New Juan Tabo Canyon

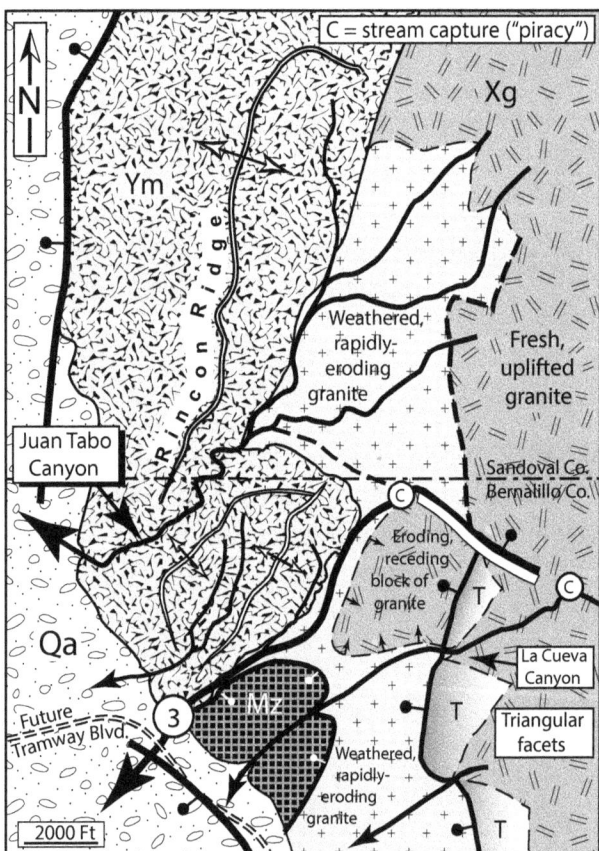

Legend (Panel C): C = stream capture ("piracy")

C. Stage 3: La Cueva Dam Breached, Juan Tabo Canyon Semi-Abandoned

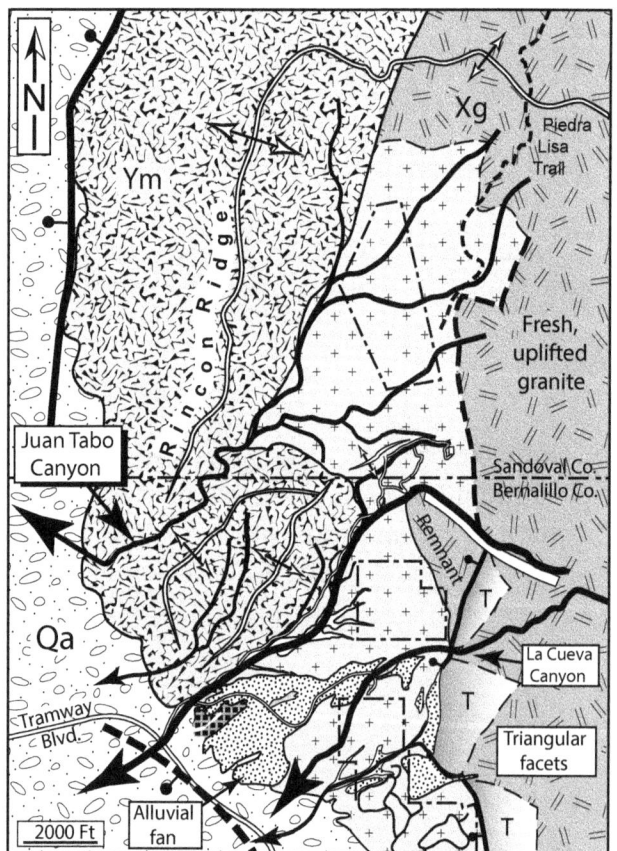

D. Stage 4: Modern Drainage

Figure 9-6. Schematic Origin of Juan Tabo Canyon
(Drainage = Black-Arrowed Lines; Select Modern Geography for Reference)

The terms *La Cueva fault* and *La Cueva Canyon*. There is an unfortunate duplication of names here. The La Cueva fault is the locally reactivated portion of the La Cueva lineament discussed above. La Cueva Canyon is the spectacular, wide-mouthed opening in the main mountain front located about 2000 ft to the south (*Figure 9-3A*).

10:
JUAN TABÓ CUL-DE-SAC

Introduction

At one time this place was on the remote outskirts of Albuquerque. Today it is a popular recreation area (*Figure 10-1*). It is a bowl-like topographic cul-de-sac excavated into the Sandia Granite, bordered on the west by the Rincón Ridge and East Rincón Ridge, on the north by Piedra Lisa Ridge, and on the east by the main mass of Sandia Mountain. The area includes two popular picnic areas, La Cueva and Juan Tabó, and I've applied the latter name to the entire cul-de-sac.

A. 1945 (Postcard from Collection of Richard Melzer)

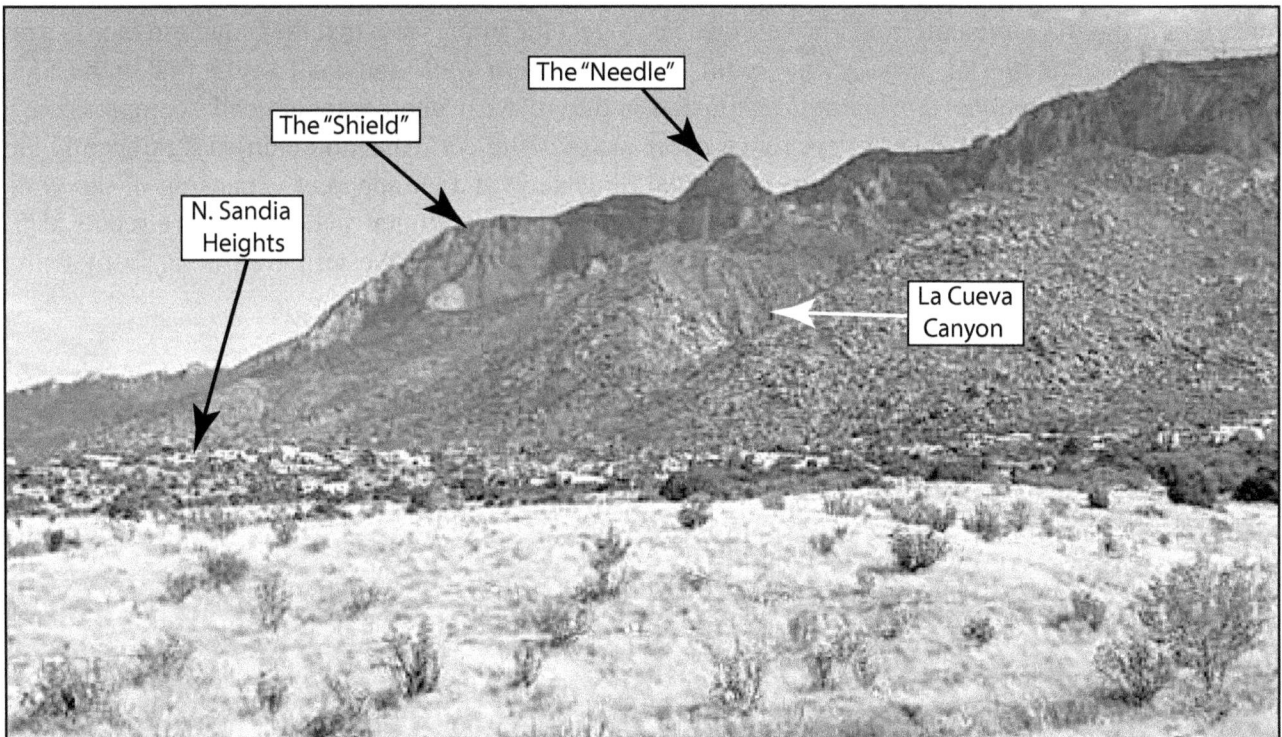

B. Modern View of "A" Above (Photograph by Author)

Figure 10-1. Northeast View of Sandia Mountain and Juan Tabo Cul-de-Sac
from Tramway Road

As mentioned in the previous chapter, the cul-de-sac is a basin subdivided into two parts by the La Cueva lineament: 1) a northern basin drained by Juan Tabó Creek through Juan Tabó Canyon to the west, and 2) a southern, less-dissected area that serves as a catchment basin for drainage exiting to the south. The southern segment has an assortment of interesting features, covered below.

Mesozoic Outlier

This anomalous patch of Mesozoic rock is surrounded on three sides by basement rocks. Because of its relevance to the mountain's uplift history, I covered it earlier in *Chapters 6* and *9*.

The "Western Wall" -- II

Approaching Sandia Mountain from the west one easily gets the impression that the mountain face is a near-vertical wall (*Figure 10-2*). I casually refer to this as the "Western Wall" and dealt with it first in the context of faulting back in *Chapter 6*. This optical illusion is due to what photographers call "compression." In short, when the distance between the eye (or camera) and the "subject" (the mountain) is significantly large, it creates an apparent compression of the horizontal dimension and an apparent expansion of the vertical dimension. A topographic profile drawn with equal vertical and horizontal scales therefore seems at odds with a photograph (*Figures 10-2B* and *10-2*C). As we'll see below, the Western Wall is far from a simple monolith!

A. "Compressed" East-Northeast View (Photograph by Author)

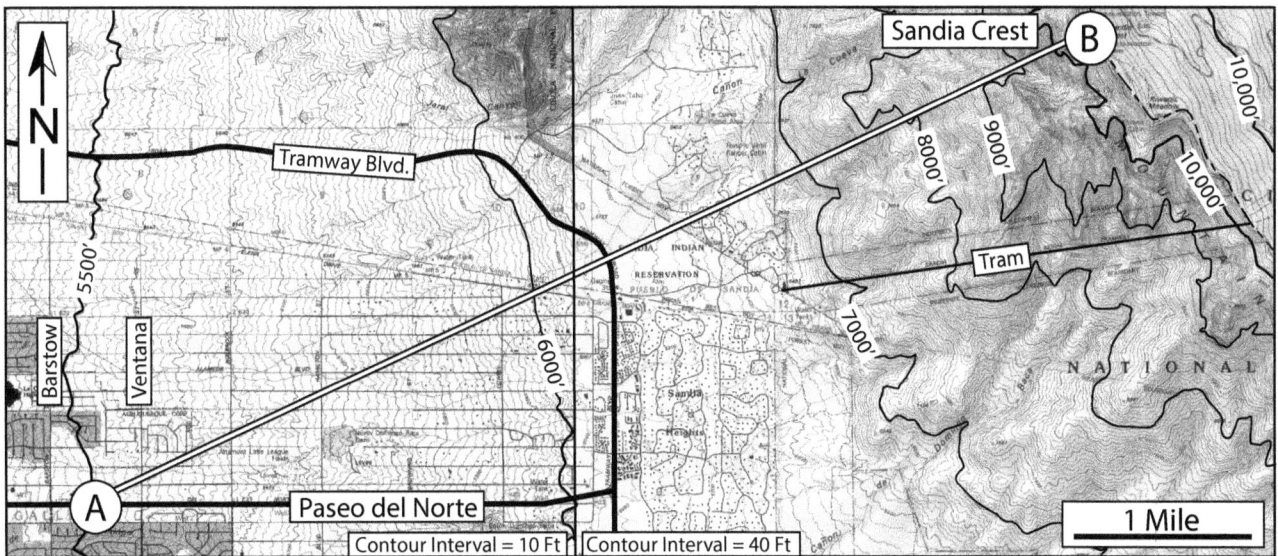

B. Topographic Map and Line of Cross Section A-B

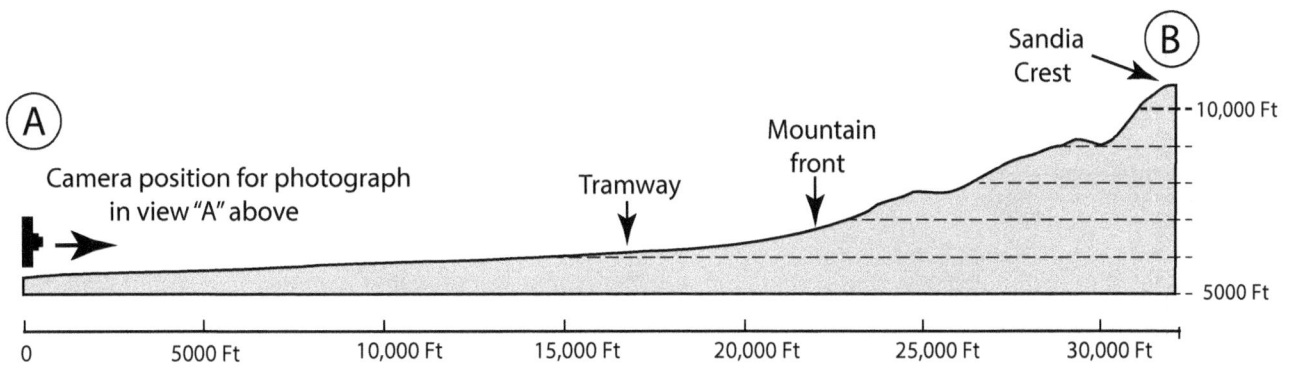

C. Topographic Profile Along Line of Cross Section A-B Above (No Vertical Exaggeration)

Figure 10-2. Optical "Compression" of Northern Sandia Mountain's "Western Wall"

Triangular Facets

When the face of a normal-fault has been exposed to the elements and partially eroded, it often becomes carved up into a series of triangular spurs or *facets* (*Figure 10-3*). Two of these are beautifully exposed on either side of La Cueva Canyon (*Figure 10-3A*). The fact that the facets are preserved implies geologic youth.

A. East Oblique *Google Earth* Image of Mountain's West Front Showing Triangular Fault Facets

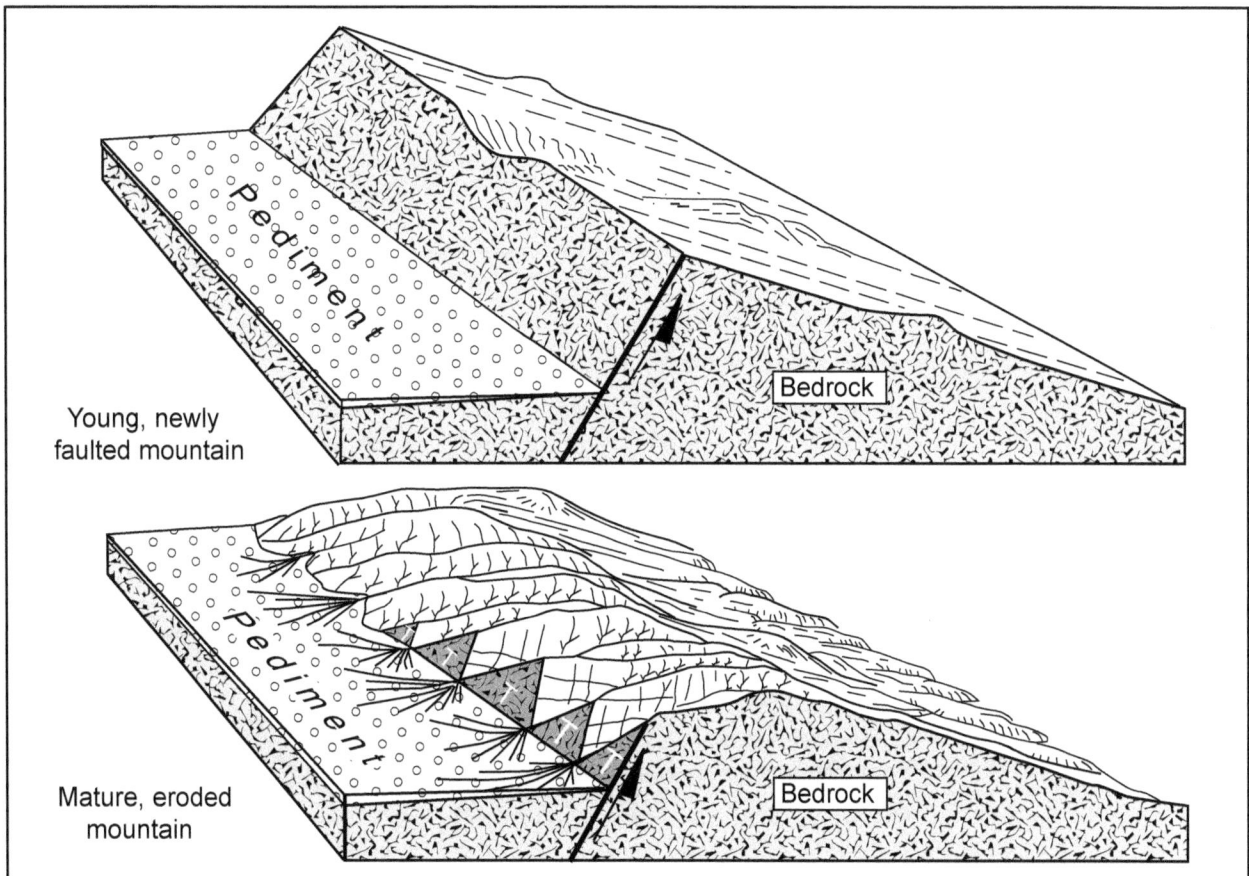

B. Idealized Sequential Block Diagrams Showing Development (Modified from Lobek 1939)

Figure 10-3. Triangular Fault Facets (T)

Juan Tabó Boulder Debris-Flows and Granite "Orphans"

Back in *Chapter 7* we talked about debris flows in general, and how and where they're formed. Two fine examples occur in the cul-de-sac. The first is at the Juan Tabó Picnic Area (*Figure 10-4*). The facility is built atop a chaotic pile of granite boulders, giving the area its unique, hummocky character (*Figure 10-5A*). Some of the boulders are truly enormous. Back-of-the-envelope calculation, using the boulders' approximate size times Sandia Granite density, results in masses of as much as 100 tons and probably more (*Figure 10-5B*).

A. East Oblique *Google Earth* Image of Northern Sandia Mountain Front

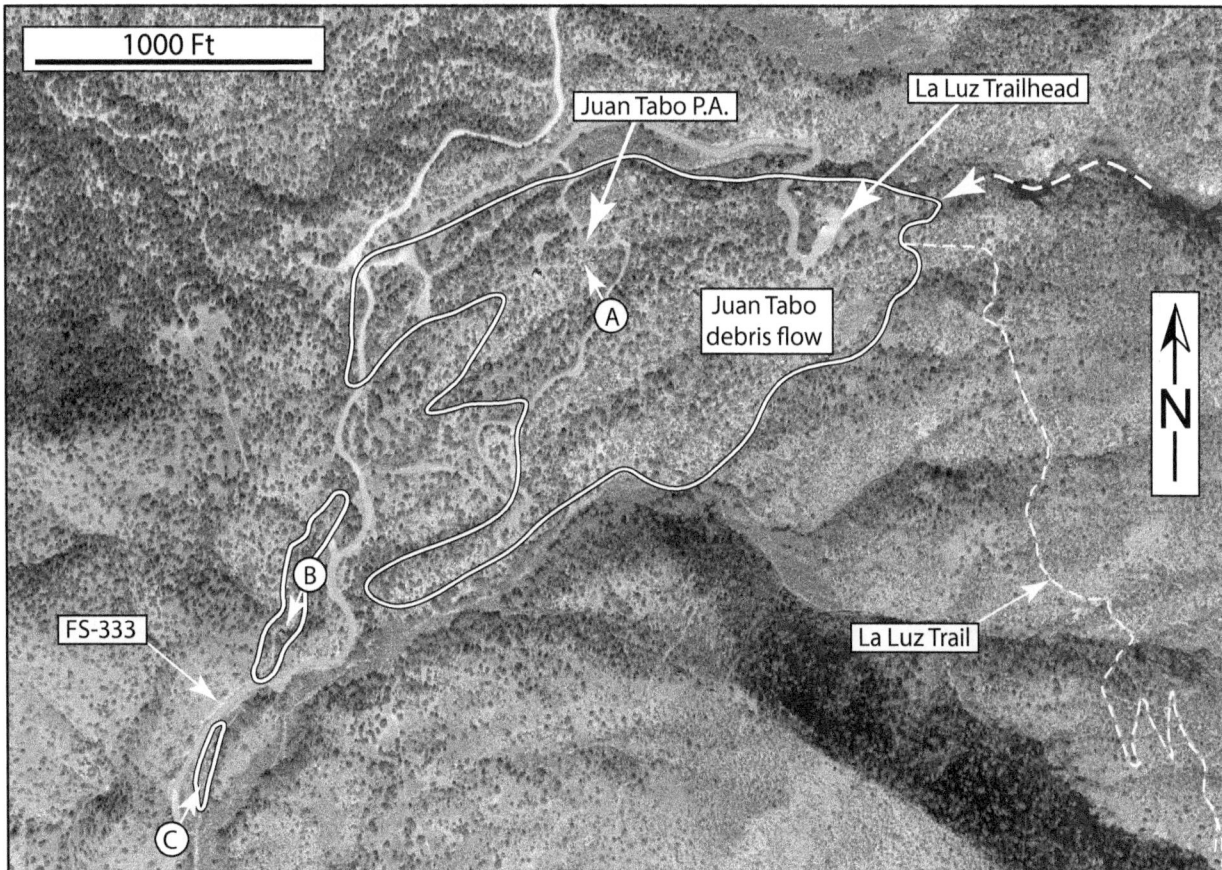

B. North *Google Earth* Image (Letters = Photographs in *Figure 10-5*)

Figure 10-4. Juan Tabo Granite-Boulder Debris Flow - I: Juan Tabo Picnic Area

A. North-Northwest View of Juan Tabo Picnic Area
(Vegetation for Scale)

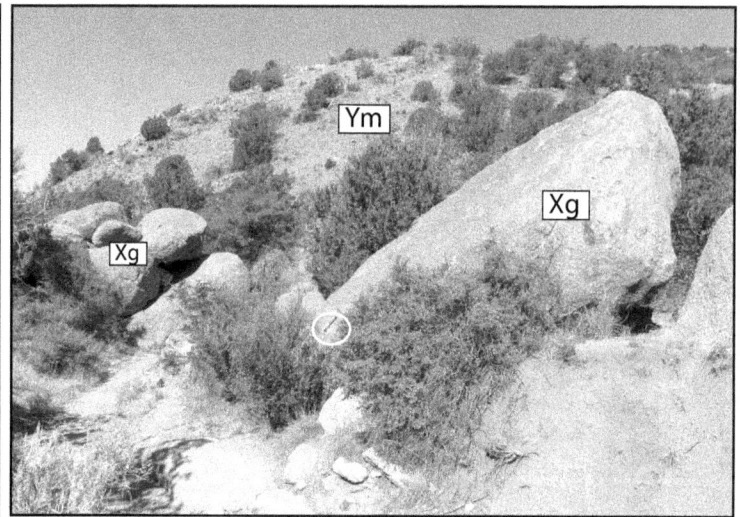

B. Northeast View of Boulders along FS-333
(1-Ft Hammer in Circle for Scale)

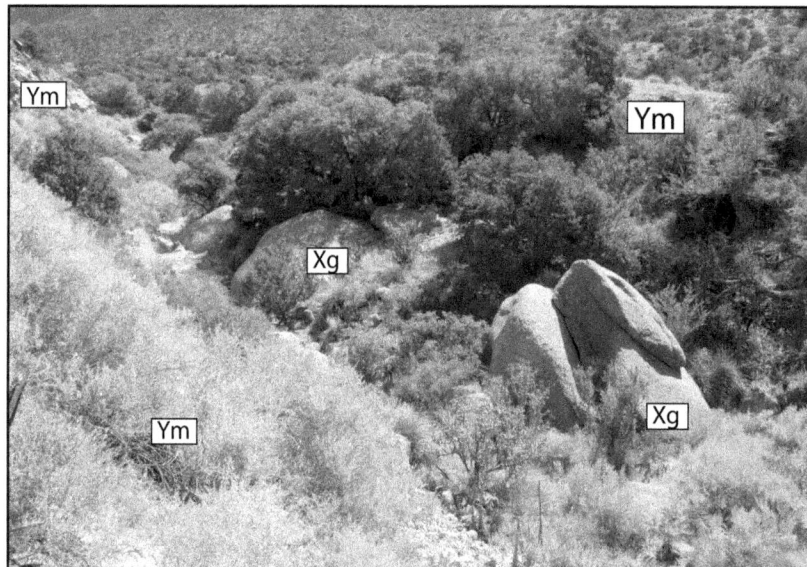

C. West View of Undermined Boulders in Ravine (Vegetation for Scale)

Figure 10-5. Juan Tabo Granite-Boulder Debris Flow - II
(Photographs by Author; Locations in *Figure 10-4*)

The boulders were transported via a sudden, catastrophic surge from high enough up on the mountain to provide the necessary potential energy. Perhaps the granite mass was dislodged by an earthquake, or simply by normal mass-slippage, well-lubricated by a cushion of abundant water and air. Some of the boulders rushed far down the cul-de-sac and now rest in some unlikely places. One example is a cluster of huge granite boulders that occupies a ravine southwest of the picnic area along FS-333 (*Figure 10-5C*). Both sides of the ravine, flanking the boulders, are of metamorphic rock. The boulders don't belong here. The small but active little stream has evidently undermined the finer material from around and under the granite boulders, easing them down (like lowering a jack) into the ravine inch by inch.

A second example is to the south, the La Cueva debris flow (*Figure 10-6*). Unlike the first flow that lies directly atop granite, the second has a thin alluvial fan sandwiched between a granite pediment below and the boulder flow above. This fan was either never deposited as far north as the Juan Tabó Picnic Area or was eroded away before the northern flow as the cul-de-sac rotated down to the south. The boulders are clearly not nearly as plentiful here and seem not as fresh (therefore older), but they are still every bit as huge, indicating tremendous kinetic energy.

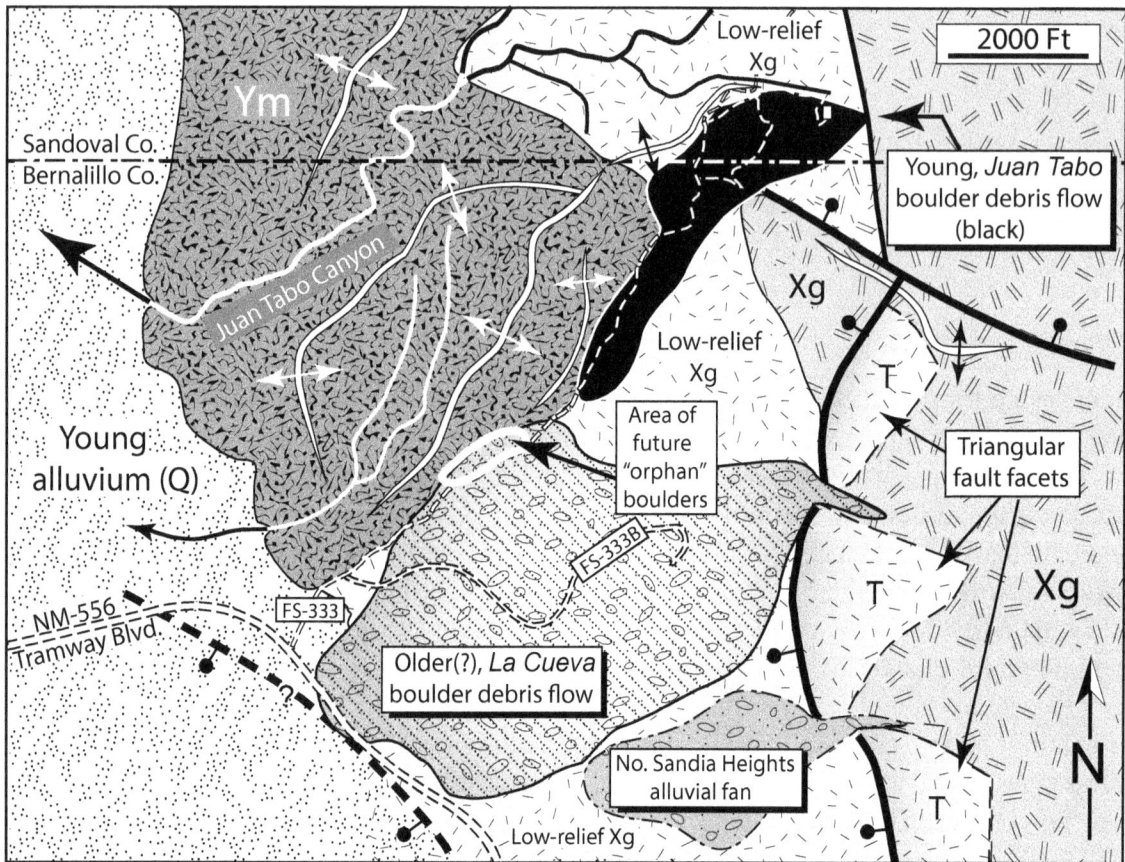

A. Geology After Boulder Debris Flows (Select Geography Shown for Reference)

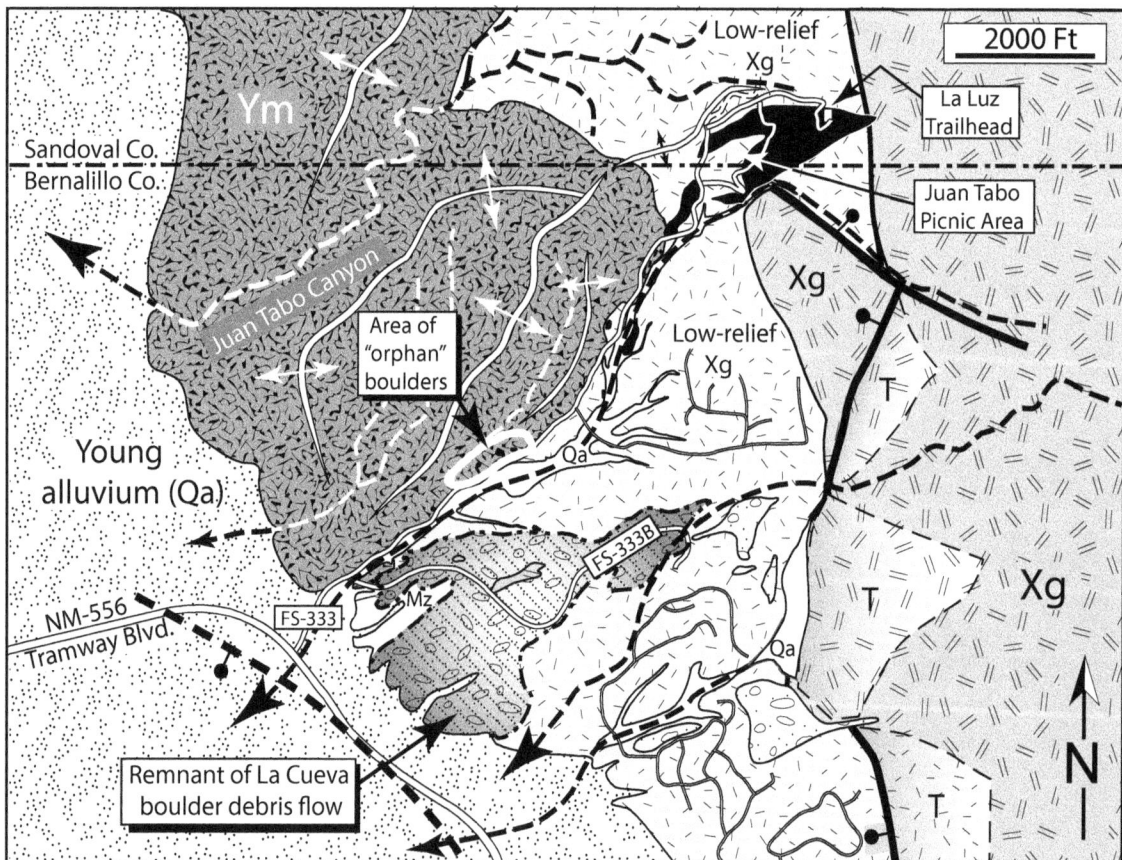

Labels in image A:
- Low-relief Xg
- 2000 Ft
- Ym
- Sandoval Co. / Bernalillo Co.
- Young, *Juan Tabo* boulder debris flow (black)
- Juan Tabo Canyon
- Xg
- T
- Low-relief Xg
- Young alluvium (Q)
- Area of future "orphan" boulders
- Triangular fault facets
- Xg
- FS-333B
- NM-556 / Tramway Blvd
- FS-333
- Older(?), *La Cueva* boulder debris flow
- No. Sandia Heights alluvial fan
- T
- Low-relief Xg
- N

B. Generalized Present Geology ("Orphan" Boulder Area in *Figure 10-7* Outlined; Mz = Mesozoic Rocks)

Labels in image B:
- Low-relief Xg
- 2000 Ft
- Ym
- Sandoval Co. / Bernalillo Co.
- La Luz Trailhead
- Juan Tabo Picnic Area
- Juan Tabo Canyon
- Xg
- Area of "orphan" boulders
- Low-relief Xg
- T
- Young alluvium (Qa)
- Qa
- Xg
- FS-333B
- NM-556 / Tramway Blvd
- FS-333
- Mz
- Qa
- T
- Remnant of La Cueva boulder debris flow
- T
- N

Figure 10-6. Granite-Boulder Debris Flows and Granite "Orphans," Juan Tabo Cul-de-Sac
(Ym = Rincon Metamorphics; Xg = High and Low-Relief Sandia Granite; Qa = Quaternary Alluvium)

This flow, more than the first, has been incised by modern arroyos, leaving some isolated boulders in unlikely places. The best examples lie atop the crest of East Rincón Ridge, where we find several isolated "orphan" granite boulders perched atop the ridge of metamorphic rock (*Figures 10-6* and *10-7*). Just how did these granite "orphans get up here? They had to flow downhill over a surface of some sort, but that surface is difficult to find. This is a fascinating conundrum and the geologist's creative juices are called into play. Only by mentally "restoring" the missing surface, i.e., filling in the unnamed arroyo east of the ridge and FS-333 to a pre-flow stage, is it possible to connect these exotics to their source.

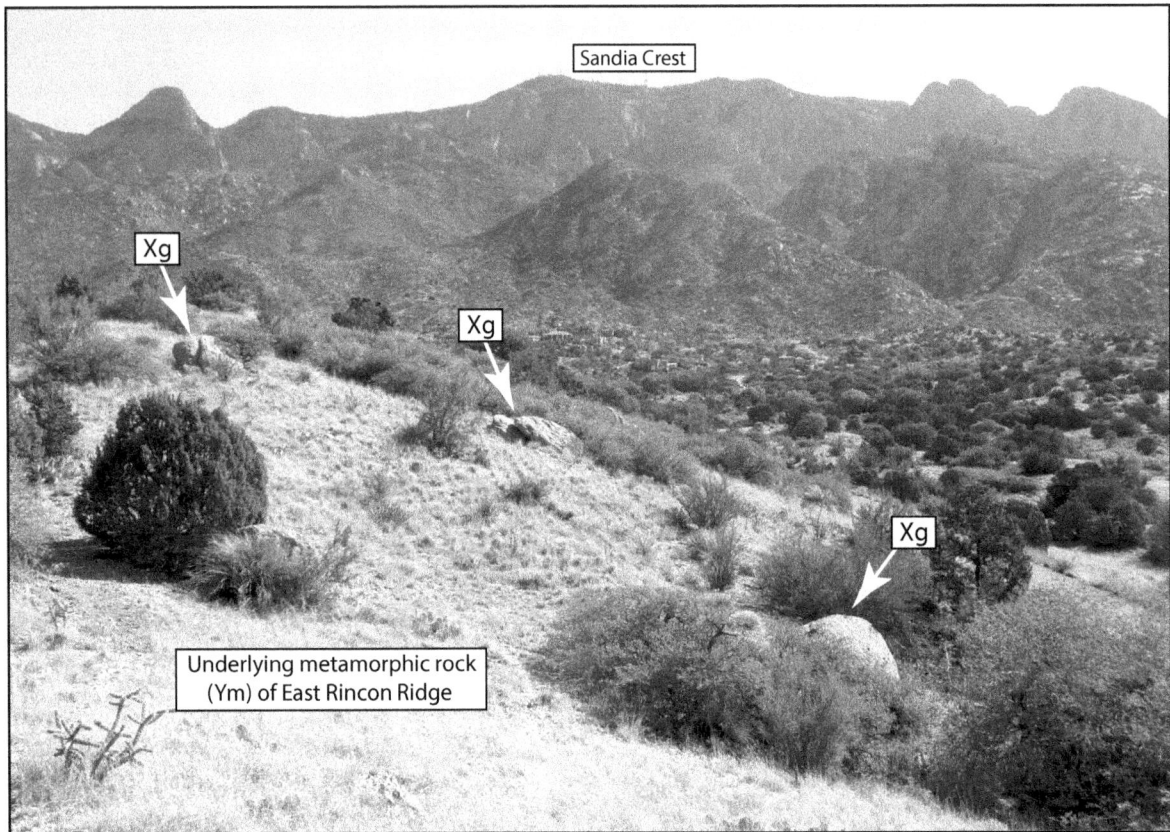

A. East View of Orphans (Xg)

B. South View of Orphan (Xg; Rock Hammer for Scale)

Figure 10-7. Orphaned Granite Boulders on East Rincon Ridge
(Photographs by Author)

I speculate that before the debris flow, the top of the eastern part of the Rincón metamorphic rocks, what has been since eroded into today's East Rincón Ridge, was level with the La Cueva alluvial fan (*Figure 10-8A*). The boulder debris flow surged west across the underlying fan out from the mouth of La Cueva Canyon to the east, and piled up against the higher slopes of what is now East Rincon Ridge (*Figure 10-8B*). When the new stream later cut its channel along today's FS-333 remnants of the boulder flow were left high and dry atop the ridge (*Figure 10-8C*).

West East

Allluvial fan from La Cueva Canyon to east

Rincon metamorphics (Ym)

Sandia Granite (Xg)

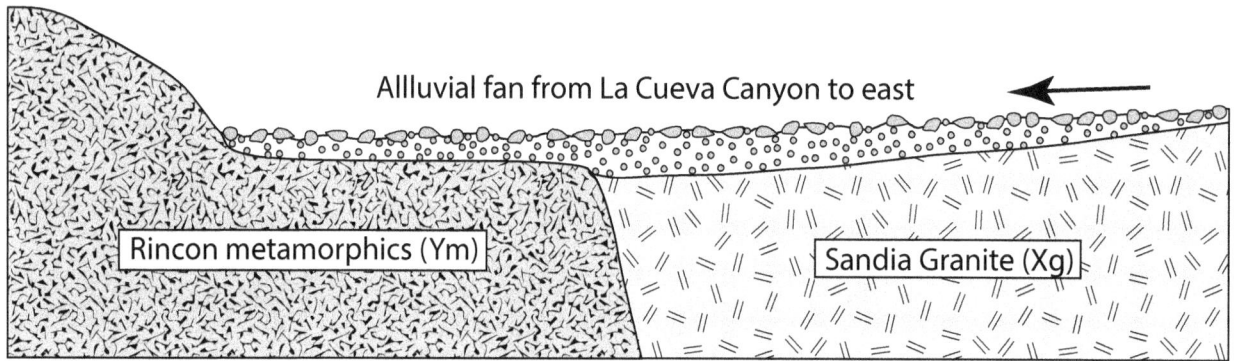

A. Deposition of La Cueva Alluvial Fan

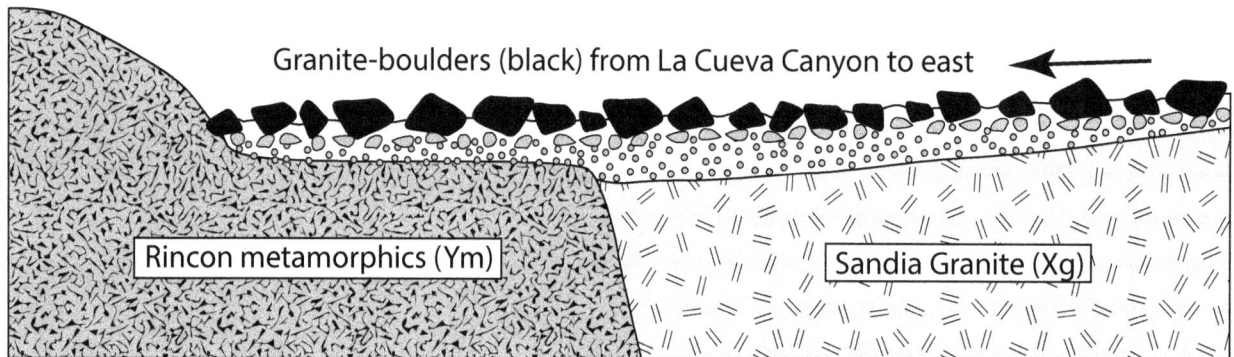

Granite-boulders (black) from La Cueva Canyon to east

Rincon metamorphics (Ym)

Sandia Granite (Xg)

B. Granite-Boulder Debris Flow onto Alluvial Fan

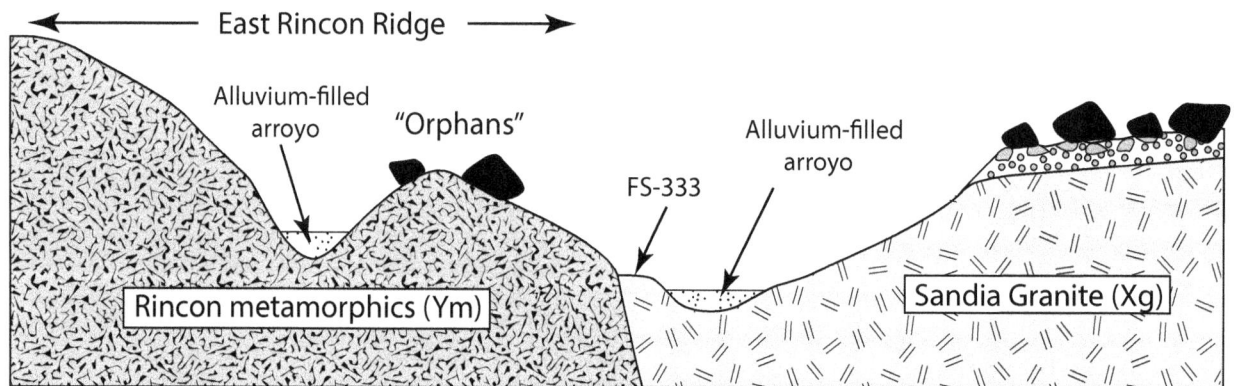

East Rincon Ridge

Alluvium-filled arroyo

"Orphans"

Alluvium-filled arroyo

FS-333

Rincon metamorphics (Ym)

Sandia Granite (Xg)

C. Inset of Arroyos and Marooning of "Orphaned" Boulders

Figure 10-8. Cartoon Cross Sections Showing Origin of "Orphaned" Granite Boulders

"Split" Boulders

A strange sight occasionally seen on the west side are some large, rounded granite boulders split down the middle in roughly a north-south direction (*Figure 10-9A*). Why the north-south trend? A professor of geology at UNM, Les McFadden, who specializes in the field of geology called *geomorphology* (origin of landscapes) with three colleagues, tackled the problem (McFadden et al. 2005). By observing split boulders located in arid southern California, southern Arizona, and New Mexico (admittedly not as common here) they concluded that the sun has everything to do with it via differential *diurnal* heating and cooling from morning to evening (*Figure 10-9B*). Strong heat gradients that shift in direction from east to west from mornings to evenings develop strong tensile stresses at the center of the boulder. This results in a fracture that cleaves the boulder in two at a point midway between the stresses, i.e., in a north-south direction.

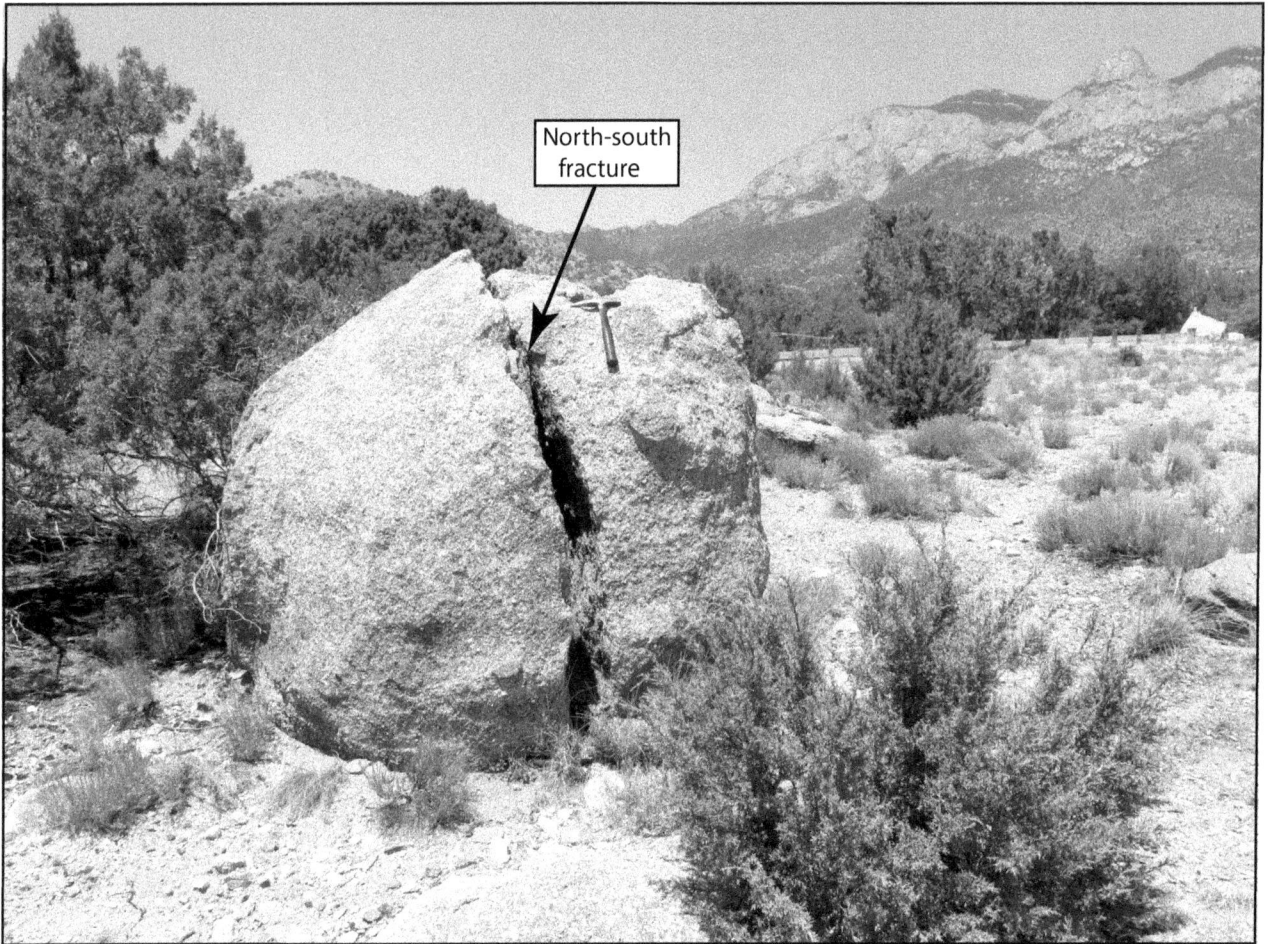

A. North-Northeast View of Split Boulder, Juan Tabo *Cul-de-Sac*

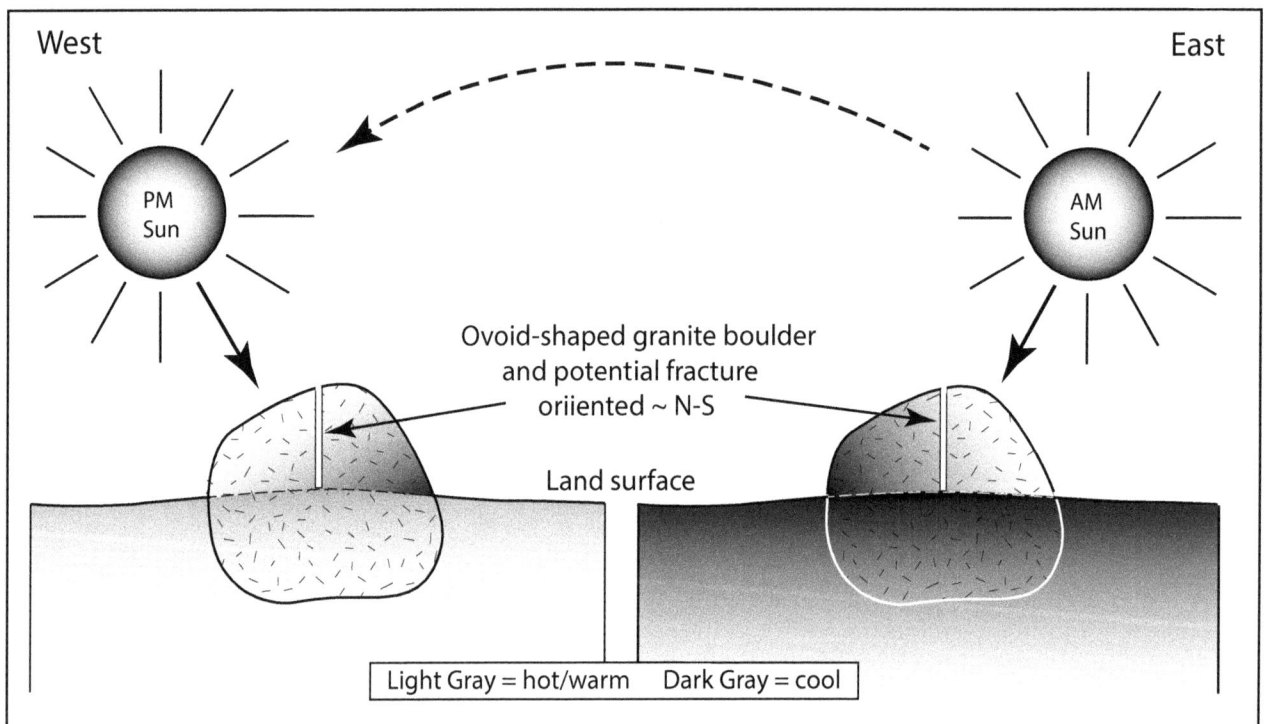

West

East

PM Sun

AM Sun

Ovoid-shaped granite boulder
and potential fracture
oriiented ~ N-S

Land surface

Light Gray = hot/warm   Dark Gray = cool

B. Schematic View of Differential,Diurnal Solar Heating of Boulder (Modified from McFadden et al. 2005)

Figure 10-9. Diurnal Solar Splitting of Granite Boulders

Colluvium

This term contrasts with "alluvium," which in turn refers to unconsolidated sediment deposited by running water. "Colluvium" though refers to heterogeneous, unconsolidated, surficial material creeping downslope over the top of a bedrock surface by the pull of gravity. It is difficult to appreciate just how important this process is in moving weathered rock material downslope because it usually only becomes visible via man-made excavations such as road cuts. In an example *par excellence*, the Grand Canyon, slow downslope movement of colluvium by gravity creep is more than 40 times that of transport by stream erosion (Shelton 1966). This is actually the primary process of erosion.

? **Origin of terms "colluvium" vs. "alluvium."** I'm fascinated about the origin of words. In geology, as in everyday language, many of the words we use are composites made up of roots with various appended prefixes and suffixes. Most go back to Latin and some to Greek. Our two terms here both rely on the Latin *leure*—to wash. There is *alluere*—to wash against, and *colleure*—to wash out thoroughly. The all refer to stuff (in our case sediment) that is somehow washed.

An excellent display of colluvium is seen at about mile 1.4 along the slope on the west side of FS-333 (*Figure 10-10*). We can thank the Forest Service for this totally different artificial exposure that appears suddenly amidst the long bedrock outcrop. Most bedrock surfaces of moderate to low relief (elsewhere as well) are blanketed by colluvium that is in a state of slow-motion gravity creep, downward to stream bottoms where it can be transported away.

Flying Buttresses, Fractures, and Crash of TWA Flight 260

As mentioned earlier, Sandia Mountain's Western Wall is far from being a simple cliff. Rather it exhibits numerous deep subsidiary valleys and crannies that are unrecognizable from a distance. This is especially true of the highest, northern sector of the mountain, east of the Juan Tabó Cul-de-Sac. Here the "wall" is characterized by semi-isolated spires, towers, buttresses, and fins of granite (*Figures 10-11A* and *10-11B*). These are pulling loose from the main mass of the mountain along deeply-weathered, north-trending fractures because of the lack of lateral structural support on the west side (*Figures 10-11C* and *10-11D*). A few of these features have names, e.g., (from north to south) the "Needle," which looks like a thumb visible from just about everywhere on the west side, and the "Thumb," which looks more like a needle and is a prominent feature of the upper part of the La Luz Trail (*Figure 10-12*).

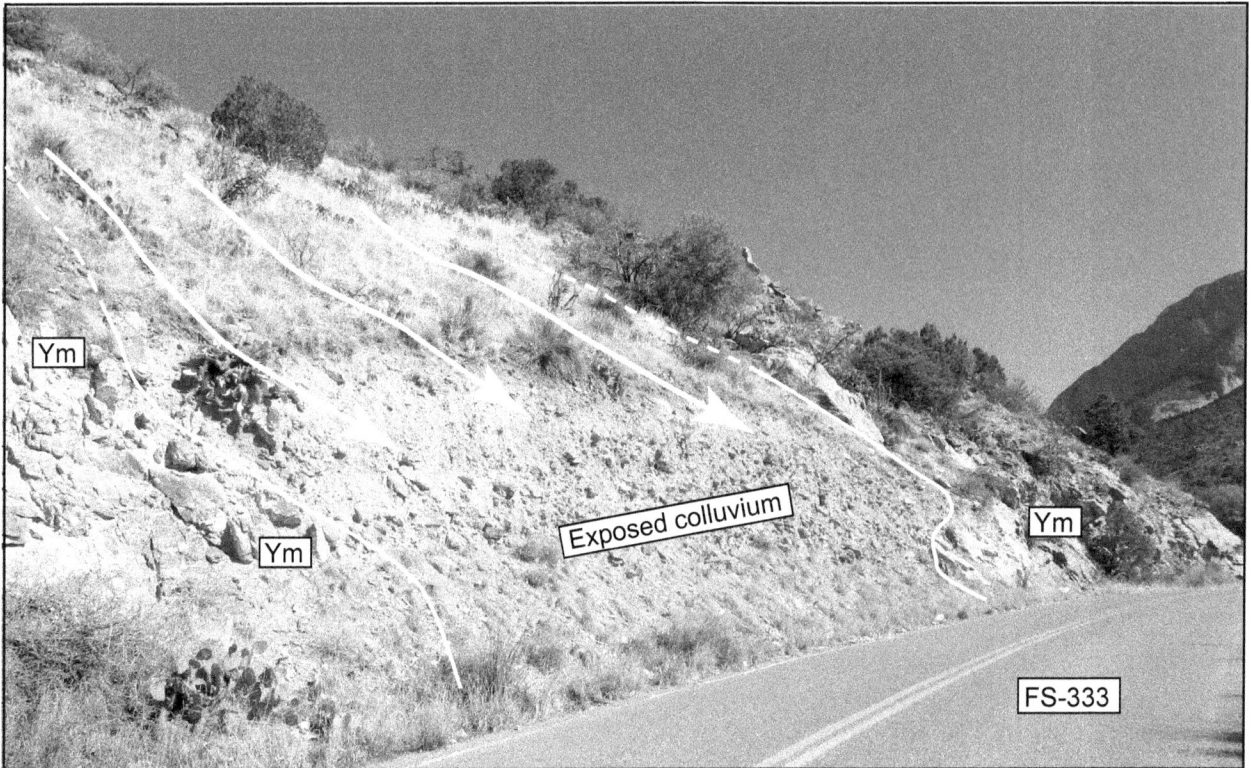

Figure 10-10 North View of Colluvial Slump, FS-333, Juan Tabo Cul-de-Sac
(Ym = Rincon Metamorphics; Photograph by Author)

A. North View of Buttresses from Tram
(Photograph by Author)

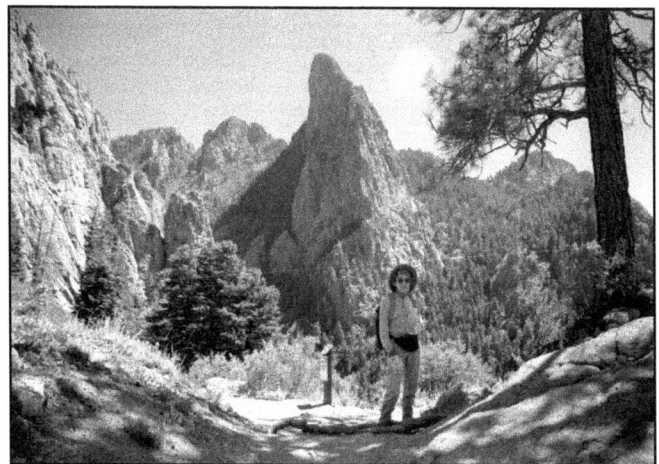

B. The Most Famous Buttress: The *Thumb* from La Luz Trail
(Photograph by Author)

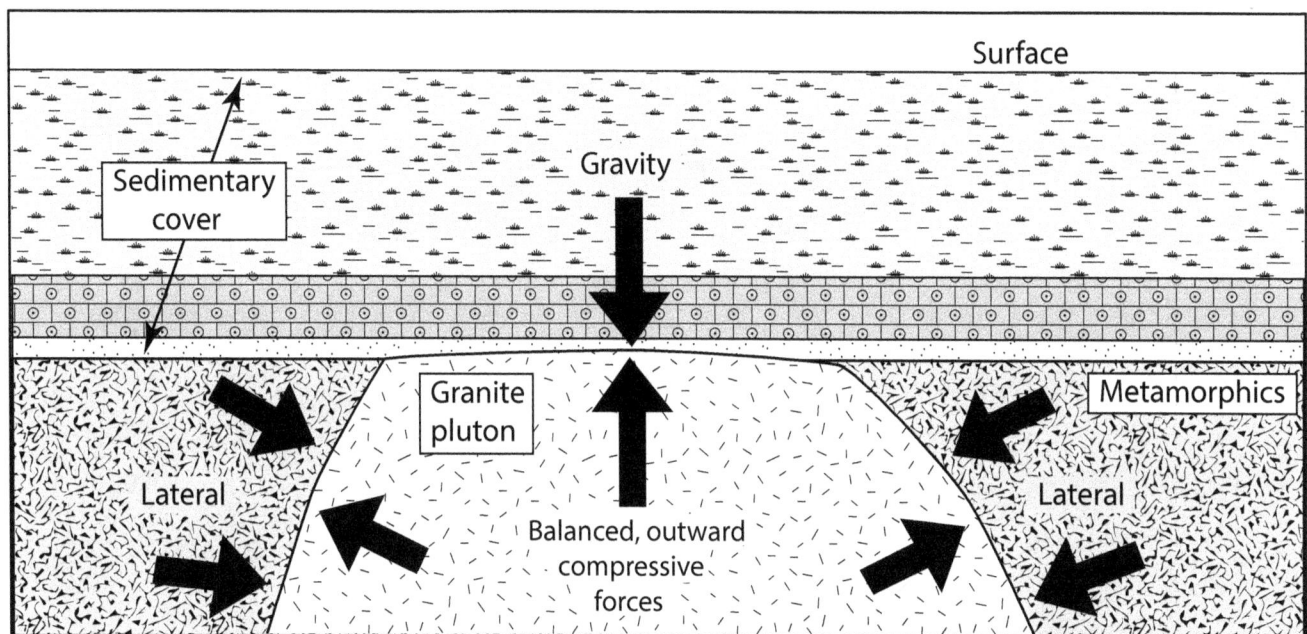

C. Schematic Cross Section of Buried Granite Pluton, Subject to Balanced Forces

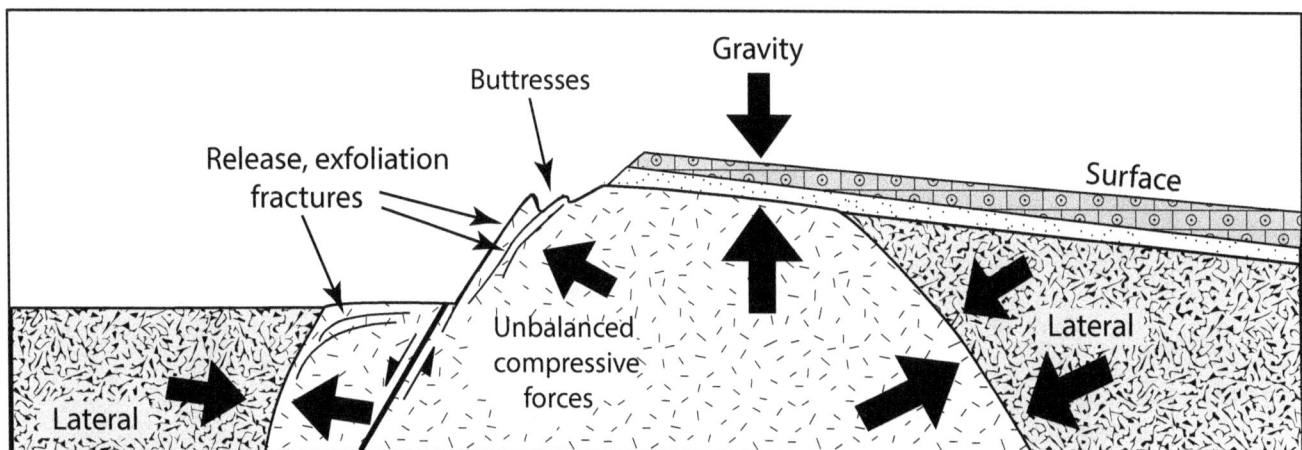

D. Schematic Cross Section of Exhumed Granite Pluton, Subject to Unbalanced Forces

Figure 10-11. The "Buttresses"

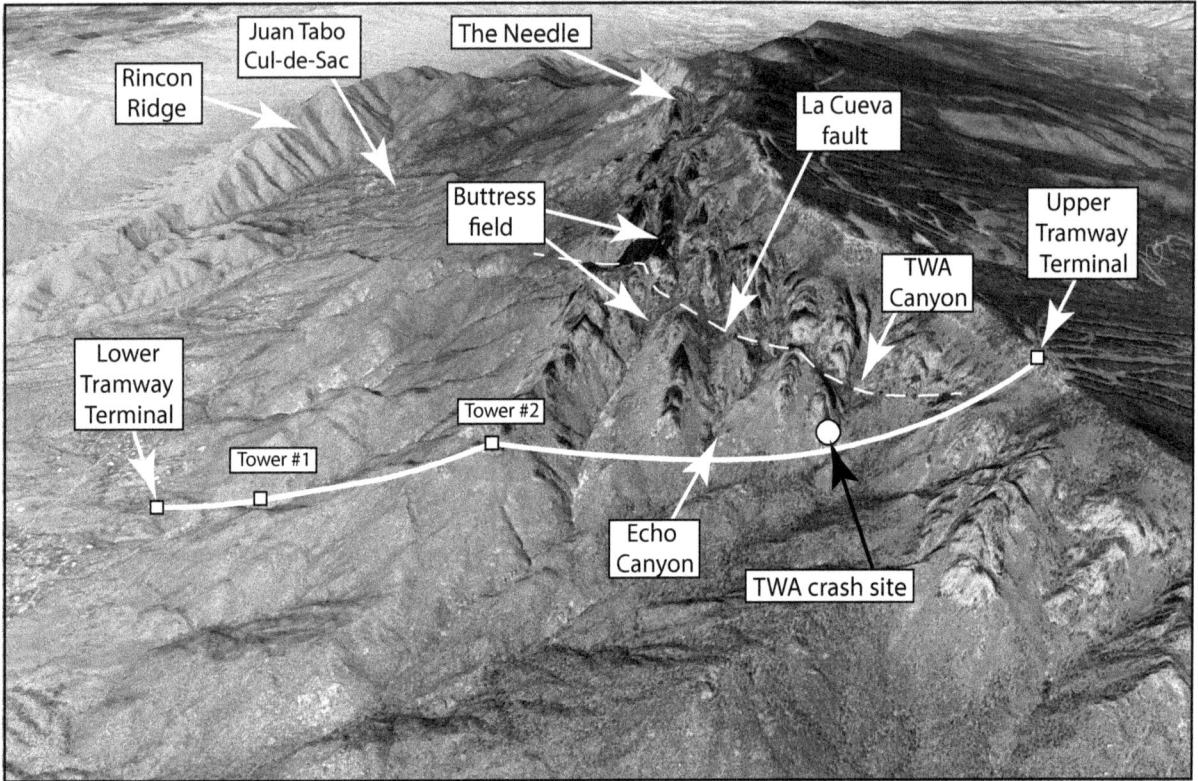

A. North-Northwest Oblique *Google Earth* Image of Sandia Peak Tram and the Buttresses

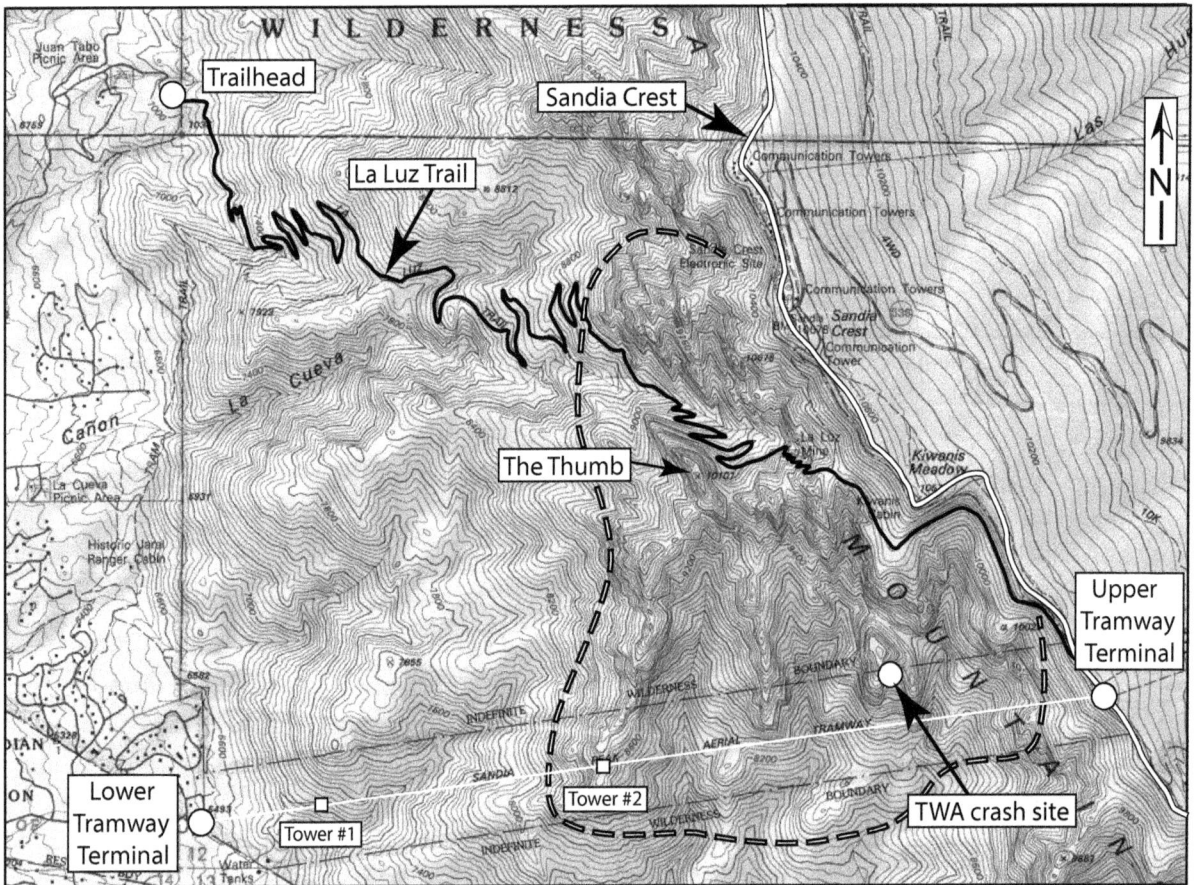

B. Topographic Map of the Buttress Area (Outlined by Dashed White Line) and the La Luz Trail

Figure 10-12. The "Buttresss" Area and TWA Flight 260 Crash Site

This labyrinth of deadly obstacles provided a venue for a tragic event (*Figures 10-12* and *10-13*). Shortly after sunrise on the morning of February 19, 1955, an unpleasant overcast day, TWA Flight #260 took off from Kirtland Field in Albuquerque on a half-hour flight bound for Santa Fe. The plane was a *Martin 404*, a short- to medium-range propeller craft that was replacing the aging workhorse, the *DC-3*, with three crew members and 13 passengers (*Figure 10-14A*). This particular plane was the *Skyliner Binghamton*, #40416. The flight plan was to head north-northwest to clear Sandia Mountain, then dogleg northeast to Santa Fe. For an unknown reason at the time, soon after takeoff the plane (flying on instruments) headed instead northeast *into* the mountain. When the pilot suddenly realized where he was, he frantically tried to double back west but instead plowed into the southern end of one of the buttresses (*Figure 10-13B*). All 16 people on board were killed.

A. North *Google Earth* Image

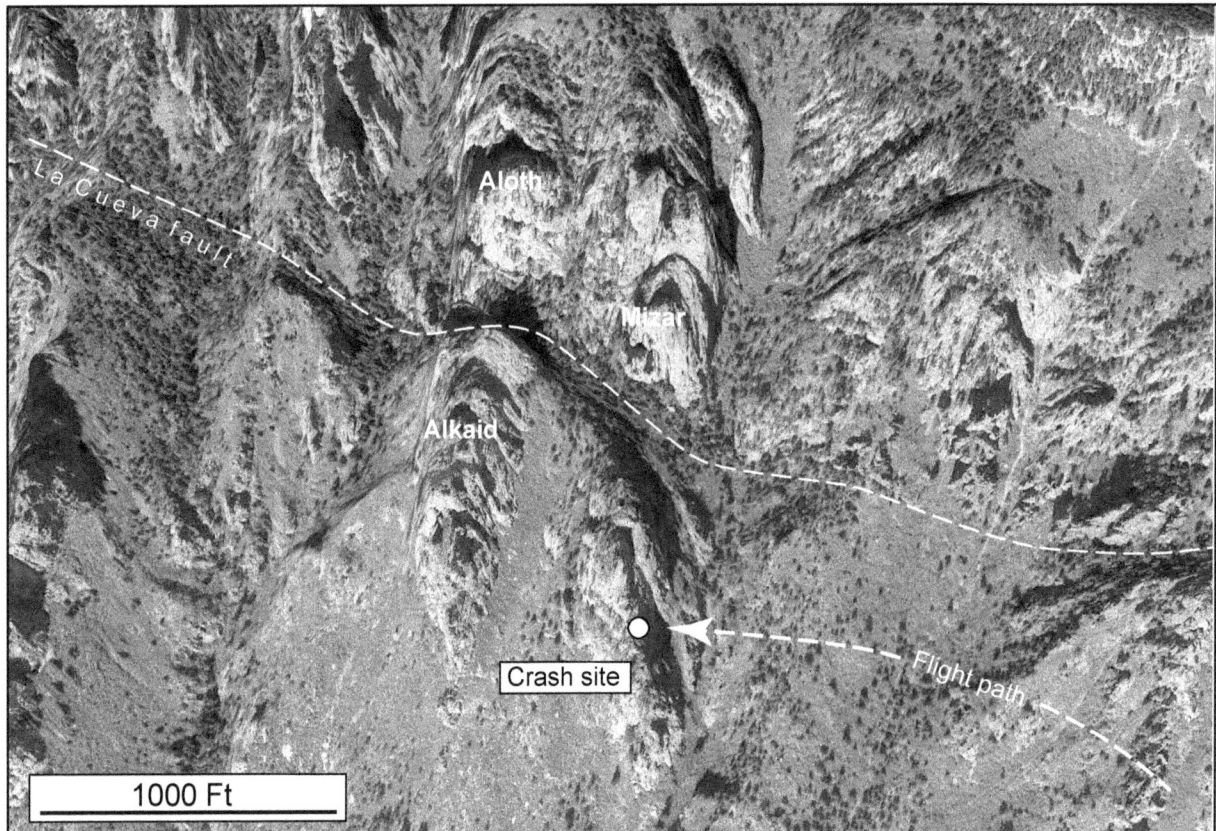

B. North Oblique *Google Earth* Detail Image

Figure 10-13. 1955 Crash of TWA Flight 260 - I

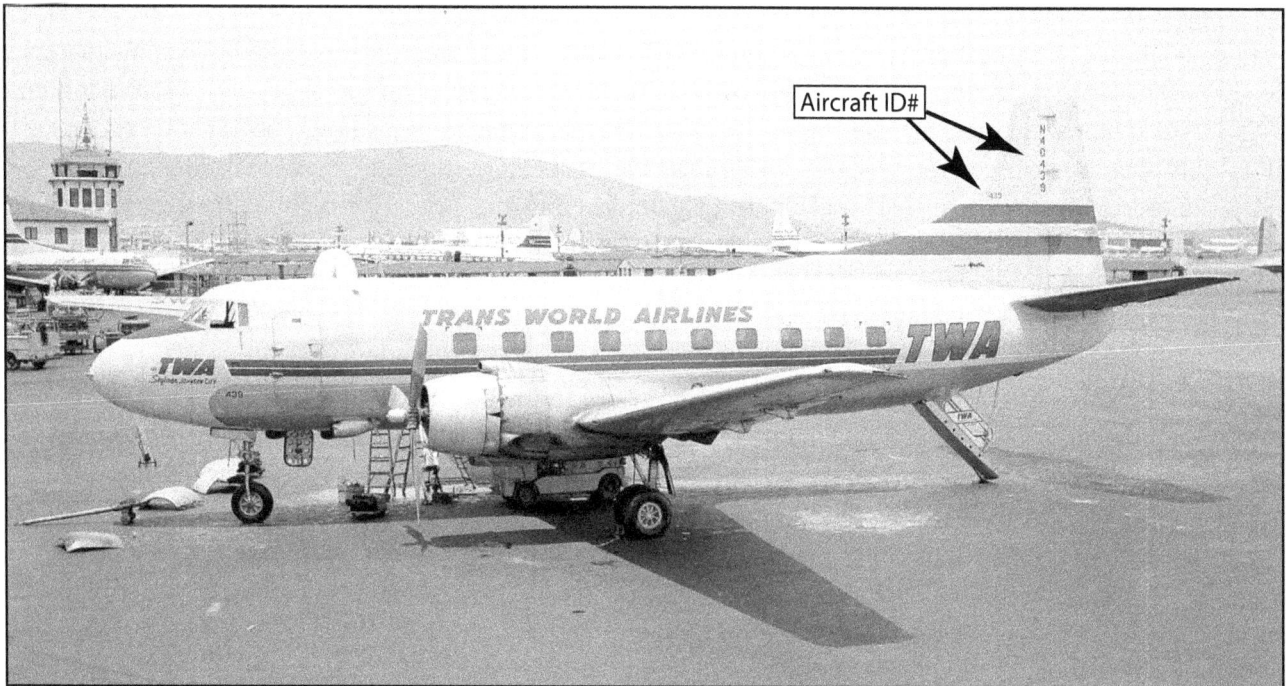

A. Example of a *Martin 4-0-4* (Aircraft #40439 in San Francisco; Photograph from Internet)

B. Crash Site in 1990 (Photographs by Author)

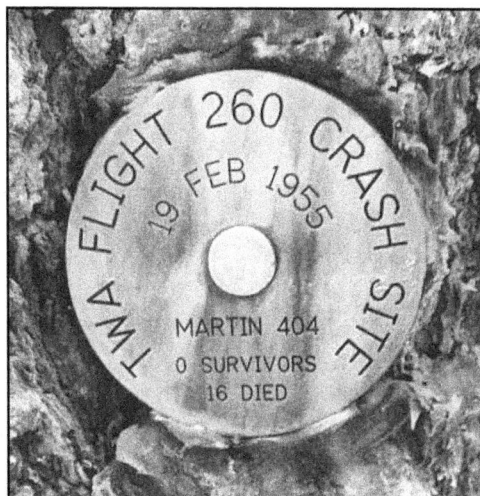

C. Commemorative Medallion

Figure 10-14 1955 Crash of TWA Flight 260 - II

At first, bureaucrats from the Civilian Aeronautics Board (CAB) placed the cause of the crash squarely on the shoulders of the pilot and copilot, whom in turn were accused of trying to short-cut over the mountain to Santa Fe. They were even accused by some of having participated in a suicide pact, to the agony of the pilots' families. Only much later was it determined that a special "fluxgate" compass, inherent to the 404 and mounted on the wingtip, had malfunctioned, misleading the pilots (remember they were flying on instruments), thus clearing their names. In 2010 a comprehensive book was published about the accident (Williams 2010). Williams was a member of the original recovery team that trudged up the mountain to the site a mere two days after the crash.

? **Crash-Site Buttress names**. Some accounts refer to the buttress of the crash as "Dragon's Tooth." Another account assigns that name to another buttress to the west. Due to this confusion (Hill 1983) I have avoided "Dragon's Tooth" in the figures. Referring to the original field notes for his book, Williams (2010) decided to dub the rock-spires near the sites with names preferred by rock climbers. These were taken from the first stars (from left to right) in the handle of the constellation *Big Dipper* (*Ursa Major*): Alkaid, Mizar, and Alioth (*Figure 10-13B*). These bizarre names derive from ancient Mesopotamia and the tale of a coffin (the dipper) attended to by three mourners (the handle stars). But that's enough of this or I'll have to change my "?" icon to an "in-the-weeds" one.

The site is today accessible via Domingo Baca Canyon for the intrepid hiker who is inured to bushwacking. I visited the site with friends in 1990 and found a considerable amount of wreckage, fortunately too massive for souvenir-scavengers to haul out. Prominent is a part of the landing gear, and a large metal fragment from the tail section showing the number "416" for aircraft #40416 (*Figure 10-14B*). This plane was the 16th "404" (103 were built in all). In 2006, a memorial plaque listing the names of the deceased was installed on part of the "416" panel, and a medallion was placed on a nearby tree to commemorate the crash (*Figure 10-14C*). Parts of the wreckage are just barely visible from the Sandia Peak Tram, looking almost straight down and to the north from the long span between Tower #2 and the crest (*Figure 10-13A*).

Juan Tabó CCC/WPA Camp F-26-N, and Juan Tabó Cabin

This is the site of one of the state's earliest *Civilian Conservation Corps* (*CCC*) camps, F-26-N, established in 1933 (F = Forest Service, 26 = sequential number of Forest Service camps, N = New Mexico). The CCC, a massive national relief-work program during the Great Depression, operated from 1933 through June 1942 and did extensive conservation work on the state's public lands (*Figure 10-15*). A 200-man "company" of CCC "boys" (young men aged 18-25 plus support staff) lived and worked here through the winter of 1933/1934. The CCC company's main chore was erosion-control work, constructing the Loop Drive (now derelict), initial development work on the Juan Tabó Picnic Area, construction of the Jaral Ranger Station (now a ruin), and fencing off tracts to keep the goats out because previously a number of goat ranches occupied this area. In April 1934 the CCC moved on and the camp was partially decommissioned, but some of the camp's numerous buildings were reused by the *Works Progress Administration* (*WPA*) for several years from 1935 (Van Hart 2020). These men, typically older than the "boys," often married and usually trucked to the area, completed construction of both the Juan Tabó and La Cueva Picnic Areas.

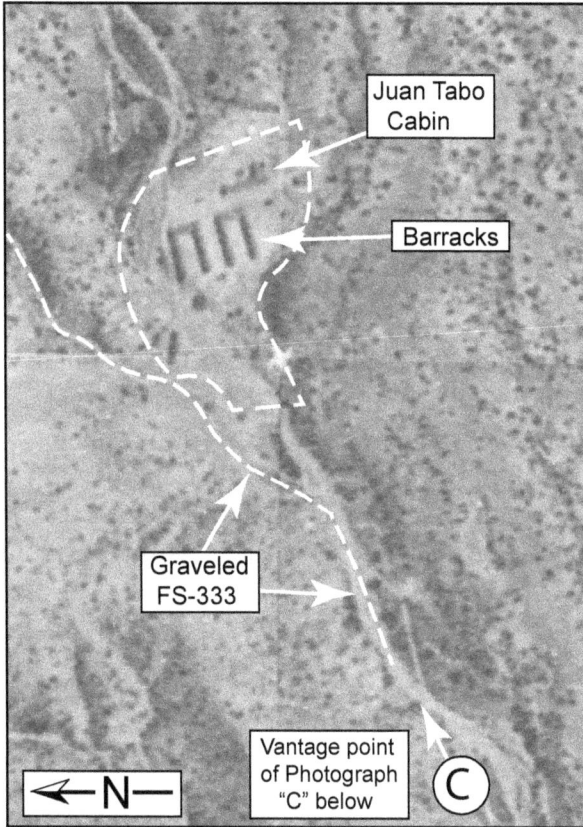

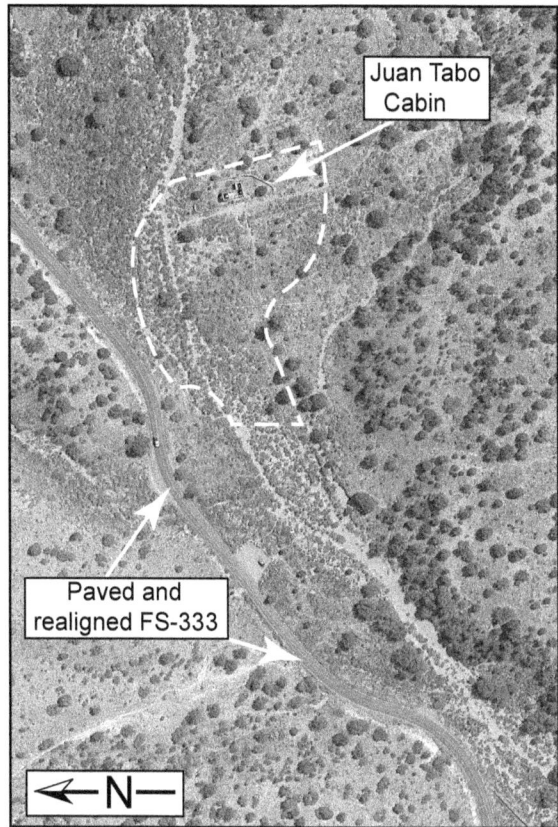

A. Aerial Image of Partly-Decommissioned Campsite, 1935 (SCS Aerial Photograph #179-045)

B. *Google Earth* Image of Campsite Today

C. East-Northeast View, 1930s (Photograph Courtesy of Cibola National Forest

Figure 10-15. Civilian Conservation Corps (CCC)/Works Progress Administration (WPA) Juan Tabo Camp F-26-N

The most visible trace today of the camp is the lonely stone cabin visible east from FS-333, informally called Juan Tabó Cabin (*Figure 10-16A*). This building is an enigmatic, solidly-built, three-room structure (two offices and a bathroom). The structure clearly was built in at least two phases. An early 1930s photograph of the camp from the southwest (*Figure 10-15C)* shows the cabin, but only with the door and window for its main (west) room (a bathroom is hidden behind on the east side. A 1944 photograph shows the cabin with two front rooms, so the extra, north room was added sometime in the late 1930s/early 1940s (*Figure 10-16*).

A. South-Southeast View

Labels in photo A:
- Cabin built of Rincon Metamorphics
- Water fountain
- Perimeter wall of Sandia Granite

B. Northeast View, 1944
(Photograph Courtesy of Cibola National Forest)

C. Northeast View Today

D. West View of Hastily-Built East Wall
Made of Metamorphic Blocks

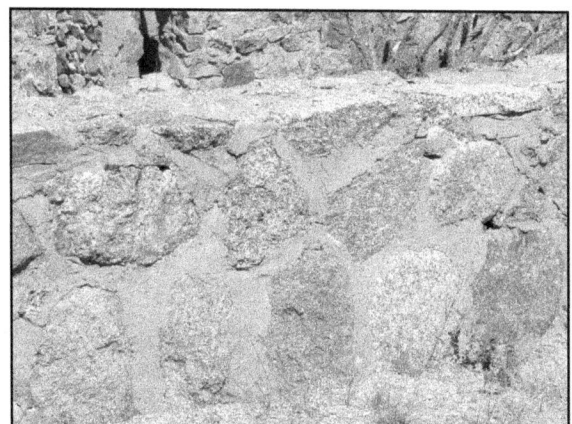

E. West View of East Perimeter Wall
Made of Granite Blocks

Figure 10-16. Juan Tabo Cabin (Photographs Except "B" by Author)

Juan Tabó Cabin construction. The structure seems to have been built in a hurried manner. In fact, the entire cabin construction seems to be inferior to that of the surrounding stone wall, the front porch, the two benches on the west side, and the water fountain. The cabin is made of irregular, undressed blocks of Rincón metamorphic rocks (*Figure 10-16D*). There being no obvious quarry sites, I suspect that these blocks were taken from rock displaced by the Forest Service during construction of FS-333. The latter features mentioned were constructed from blocks of Sandia Granite, obviously acquired from a different source, with craftsmanship typical of the CCC (*Figure 10-16E*).

The Forest Service believes that the cabin possibly predates the CCC camp that was established in 1933 (Benedict 1994a). There is a glaring contrast between the stone cabin and the other more normal CCC camp structures made of wood that were intended to be removed when the camp was fully decommissioned (*Figure 10-15C*). Perhaps the original cabin was a part of a homestead or ranch, built sometime prior to 1933 and having fallen into disrepair before the CCC arrived and then converted it into an administrative office. With all this history and intriguing mystery, it's disappointing that there is no explanatory sign posted here.

11:
SANDIA FOOTHILLS

Introduction

The Sandia foothills generally are that area at the lower west side of the mountain front, reaching from the southern end of the Juan Tabó Cul-de-Sac at Tramway Road leading up to the Sandia Tram, south to about I-40. It includes the eastern fringe of the city of Albuquerque. The west side of Sandia Mountain is utterly different from the forested east side due to its high relief and extensive exposures of "naked" granite bedrock. It sports some odd granitic features and never seems to lose its lure to both the amateur and professional geologist. But first some words about how granite weathers to form these features.

Weathering of Sandia Granite

To recap, granite is a coarse-grained rock formed by crystallization of a hot liquid (*magma*) miles below the Earth's surface under conditions of great heat and pressure. Its essential minerals are mainly two kinds of feldspar, with decreasing amounts of quartz, biotite, and a trace of the iron oxide magnetite thrown in. Fresh granite is a crisp, light gray color that the human eye averages from the whitish feldspar, gray quartz, and black specks of biotite mica.

When uplifted and exposed at the surface the minerals of granite enter a wholly new chemical environment consisting of low pressure, mild temperature, and abundant water. There they become unstable and strive to reach equilibrium with the new conditions, i.e., they *"chemically weather"* (*Figure 11-1*).

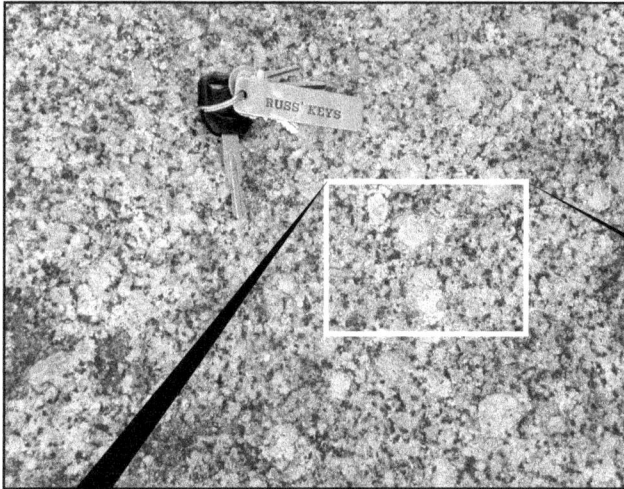

A. Fresh, Unweathered Sandia Granite

B. Granite Boulder Weathered to *Grus*
(Grus = K-Feldspar, Quartz, and Clay)

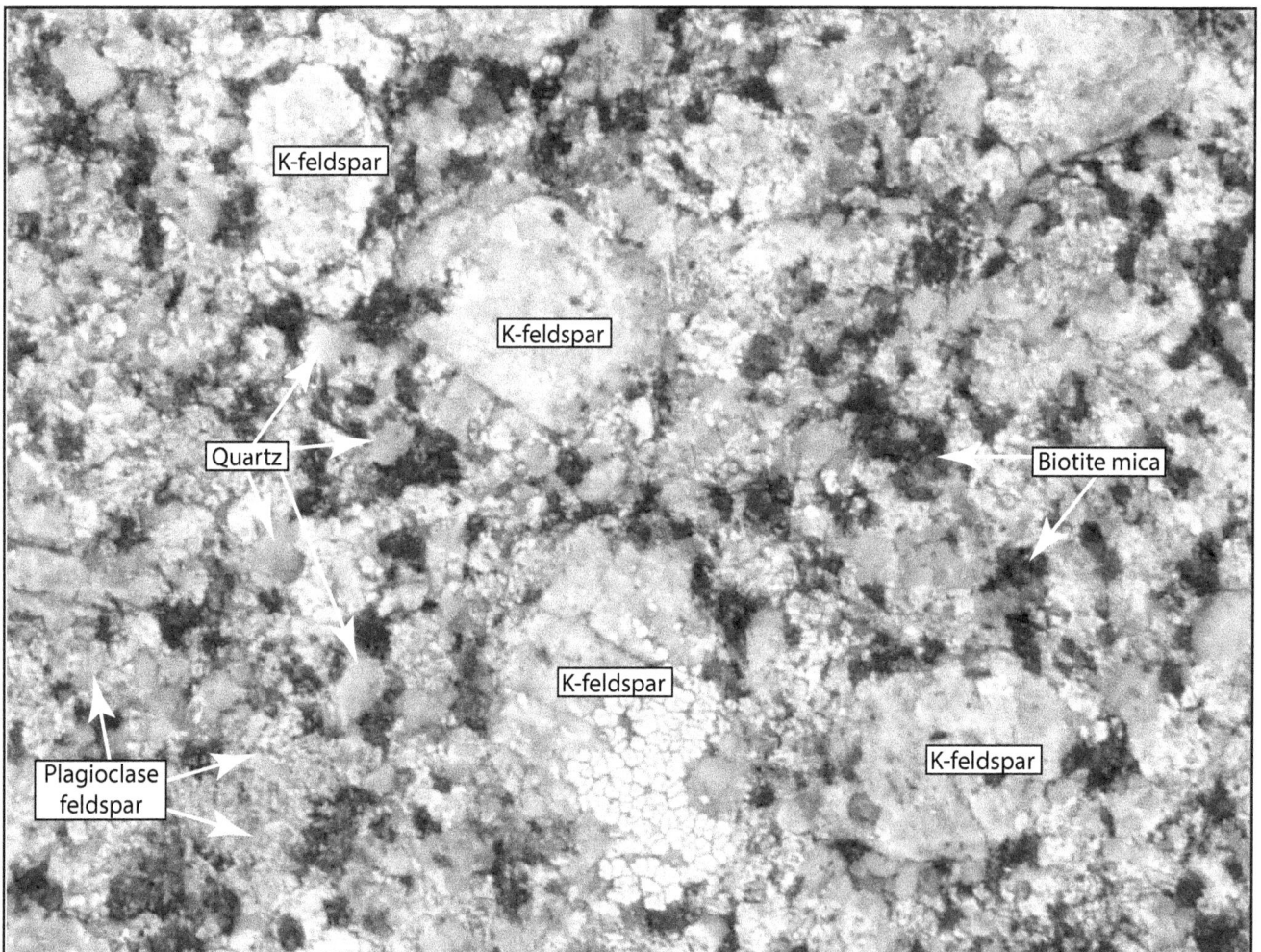

C. Enlargement of "A" Above, with Four Constituent Minerals Selectively Indicated
(K-Feldspar + Quartz = Main Ingredient of Grus; Biotite and Plagioclase Weather to Clay)

Figure 11-1. Sandia Granite and *Grus*

Carbonic acid. The real beast of chemical weathering is acidic water. Rainwater readily picks up and dissolves carbon dioxide in the air, becoming the very weak *carbonic acid* (H_2CO_3). This acid is so weak that it creates no sensation whatever on human skin. However, over as little as a few centuries its effect on rock is noticeable (as anyone visiting tombstones in old graveyards can attest), not to mention over hundreds or thousands of years.

During chemical weathering the whitish plagioclase feldspars and the dark, flaky biotite grains are the first to achieve equilibrium by weathering to clays that easily wash away. The coarser, whitish to pinkish grains of potassium feldspar (K-spar) weather more slowly and stay behind (*Figures 11-1A* and *11-1C*). These grains thus become isolated by the weathering and removal of their neighbors and tend stick out and pop loose as coarse sand grains (*Figure 11-1B*). The glassy, gray quartz grains, being most stable, remain unweathered and also pop loose to join the sand. The coarse quartz and K-spar sand produced by granite weathering is called *grus*, from the German meaning grit, or debris. The magnetite is also relatively stable and is also released intact, forming the black placer streaks in arroyo bottoms (see magnetite placers below).

Grus. This substance develops on the surfaces of the outcropping granite as the minerals progressively react with the acidic rainwaters. The unweathered inner rock grades outward to the surface stained a dull brownish from the iron of oxidized biotite and from a clayey matrix from the weathering of plagioclase feldspars (Smith et al. 1982). The biotite content drives and regulates grussification. This platy mineral splits along cleavage planes, reacts with water and oxygen, and then expands. Microfractures within the biotite are propagated outward to other mineral grains and eventually merge, creating a less-dense rock that sloughs off to create the grus (Isherwood and Street 1976).

Corestones. Most granite is fractured, both from tectonic movement at depth and by later expansion due to the unloading of overlying material (*Figure 11-2A*) The fractures provide pathways for water and air to penetrate the rock. Weathering to grus occurs at the fracture surfaces and works its way inward toward the unfractured granite (*Figure 11-2B*). The process tends to form spheroidal shapes of fresh granite called "corestones."

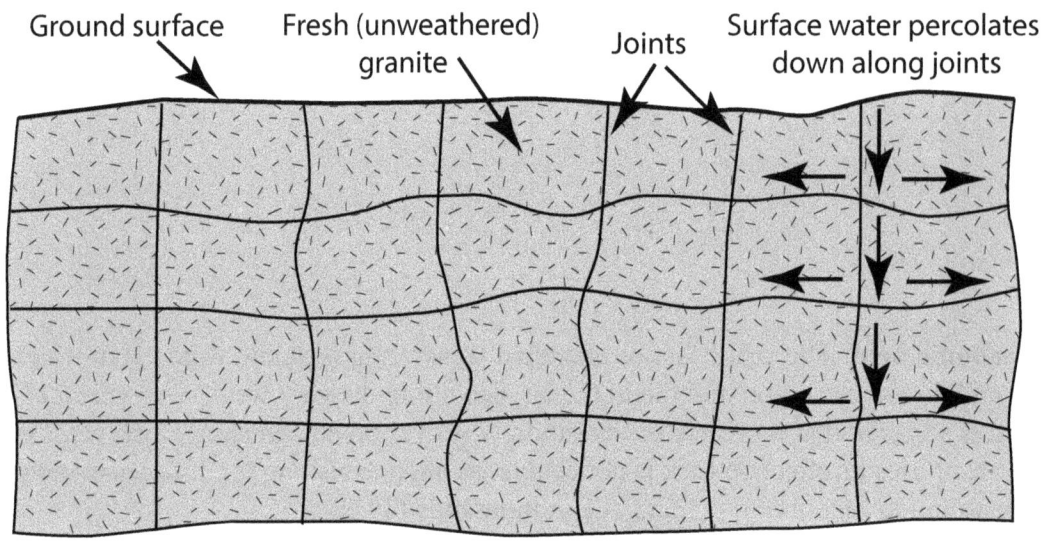

A. Fresh Granite Terrain with Network of Joints (Fractures Without Displacement)

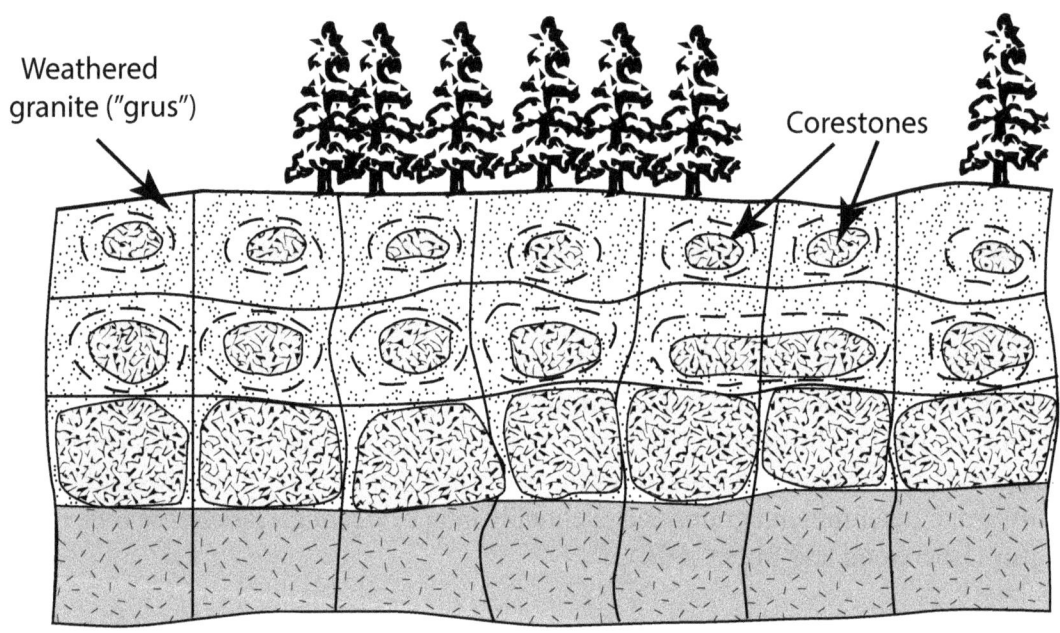

B. Deeply Weathered Granite Terrain with Forested Soil and Subsurface Corestones

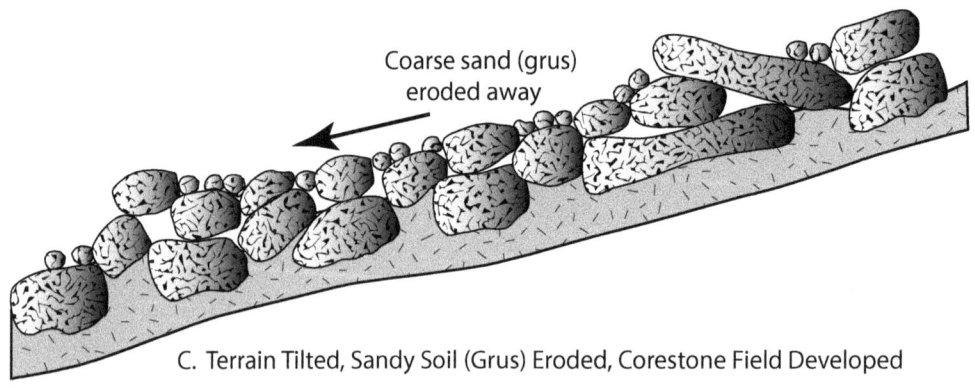

C. Terrain Tilted, Sandy Soil (Grus) Eroded, Corestone Field Developed

Figure 11-2. Corestone Development

If the grus is carried away by running water, the corestones remain behind, forming fields such as those that blanket much of the Sandia Mountain foothills (*Figures 11-2* and *11-3*). The corestone fields are the typical weathering product of the arid to semiarid conditions that prevailed in the American Southwest during the past 12 ka of the final interglacial stage. Most of the weathering, though, occurred during the wetter, glacial period before ca. 12 ka, and the granitic terrain back then was probably covered by a thick forest, long since gone. The present corestone fields are testament that the original land surface was once located across the top of the highest corestones and that all the loose material has been removed.

A. Typical Corestone Field, West-Side Foothills (Vegetation for Scale)

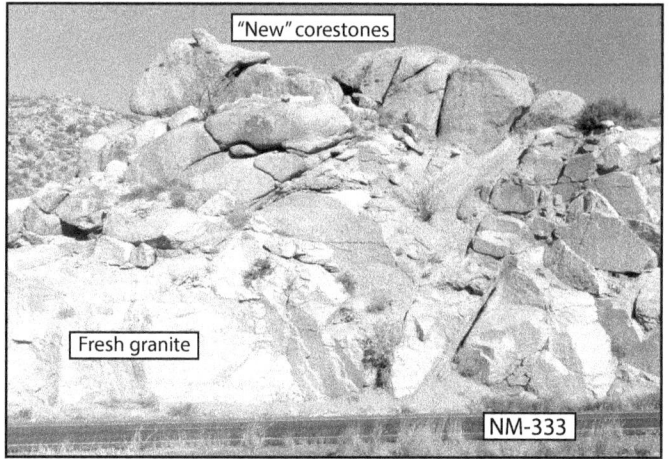

B. North View of Corestone-Development Profile, NM-333, Mile 2.7

C. Corestones, West-Side Foothills (Vegetation for Scale)

D. Marooned Corestone

E. "Old Man of the Mountain," West Side Foothills (Vegetation for Scale)

Figure 11-3. Corestones (Photographs by Author)

Magnetite Placers. Anyone who has walked the mountain's many sandy arroyos, especially after a rain, has noticed black streaks of the mineral magnetite *(Figure 11-4A)*. Magnetite is an iron oxide (Fe_3O_4) that occurs in trace amounts (< 1%) and as small grains in the Sandia Granite. Despite the low concentration, its density exceeds that of the other minerals, and as the granite disintegrates and the grus washes away, the magnetite lags behind, just as gold was concentrated in the 1800s as placer deposits in California streams. If you refuse to grow up, like me, you can amuse yourself for long periods by collecting the magnetite with a magnet held inside a plastic bag *(Figure 11-4B)*.

A. Typical Dark Magnetite Streak in Arroyo After Heavy Rain, Juan Tabo Area

B. Magnetite Granules Strongly Adhering to Magnet

Figure 11-4. Placer Magnetite from Weathered Granite in West-Side Arroyos

Magnetite. Being strongly magnetic, the grains of magnetite that are still in place within the parent granite retain a trace of the Earth's magnetic field that existed at the time the granite crystallized. The linear direction of this ancient magnetic field (Earth is like a bar magnet with north and south poles) and its intensity can be detected and measured by both hand-held instruments as well as by low-flying aircraft with the proper equipment. The magnetic intensity can be mapped. Aeromagnetic maps are an enormously important reconnaissance exploratory tool, thanks to magnetite.

Caliche. This is a light-colored to white, calcareous substance usually found two to three feet below the surface within soils of arid or semi-arid regions. The term is from the Latin *calx* for lime, and *caliche* is Spanish for porous materials that have been cemented by the mineral $CaCO_3$.

Caliche, sometimes called *calcrete*, is ultimately derived from finely-divided $CaCO_3$ dust in the atmosphere that has been captured by rain and delivered in solution down to the surface (Machette 1985). There the rain percolates into the soil about two to four feet, evaporates, and precipitates its $CaCO_3$ on the bottoms of the large particles. If these large particles are later dislodged (such as boulders used for xeriscaping) the tell-tale coating of white caliche indicates at which end the original bottom was (*Figure 11-5*).

Figure 11-5. Caliche on Granite Boulder (Boulder 26" High; Photograph by Author)

If the soil remains stable and undisturbed, i.e., is neither being eroded away nor buried by new material, the $CaCO_3$ accumulates, and from tens of thousands to many hundreds of thousands of years gradually merges to become like a rock. Being a time-dependent property, caliche concentration becomes a mapping tool used to assign rough dates to surfaces in the Sandia foothills and the Albuquerque basin. A notable example of highly-developed caliche is the vast, elevated platform of the "caprock" on the *Llano Estacado* of the High Plains of eastern New Mexico and the Texas Panhandle.

Some Features of Interest

The foothills are largely devoid of vegetation except for a gaggle of struggling junipers and shrubs. This area of mostly "naked rock" nicely reveals a number of interesting features, including the six mentioned below.

Bear Canyon Re-Entrant and its Pediment. (*Figure 11-6*). This re-entrant of the mountain front is of particular interest because we know more about it due to a number of small outcrops and from four shallow boreholes. The area also provides a broad, even expanse of valuable real estate for most of the South Sandia Heights and the High Desert subdivisions, with great views of the mountain and access to city amenities. The pediment itself is a smooth, eroded granite surface thinly buried under a veneer of coarse debris, and occasionally pops through as small outcrops near the foot of the mountain. A pair of down-to-the-west normal faults creates a stair-step arrangement to the top-granite surface, over which most of the housing is sited (*Figure 11-7*). The first fault on the east is constrained by a borehole that topped granite (*Xg*) at 42 ft, within sight of outcropping granite on the east, other side of the fault. To the west, running north-to-south though the subdivisions is the second normal fault constrained in part by three boreholes (only one shown in the figure) that each had total depths (TDs) of about 500 ft within the basin-fill gravels of the Quaternary (*Qal*) and Santa Fe Group, indicating that granite bedrock is deeper still. This amount of depth does not seem significant until it is realized it equals 50 stories.

The southern end of the re-entrant is bordered by the transverse *Pino fault*, a strong lineation that extends northeast to Sandia Crest and into the Paleozoic rocks of the east mountain, indicating that at least a part of the fault is relatively young (*Figure 11-6B*). West of Sandia Crest "damage" to the Sandia Granite along the fault seems to be responsible for the excavation of Pino Canyon. East of the crest the fault leads to an interesting story, to be visited later (*Chapter 12*).

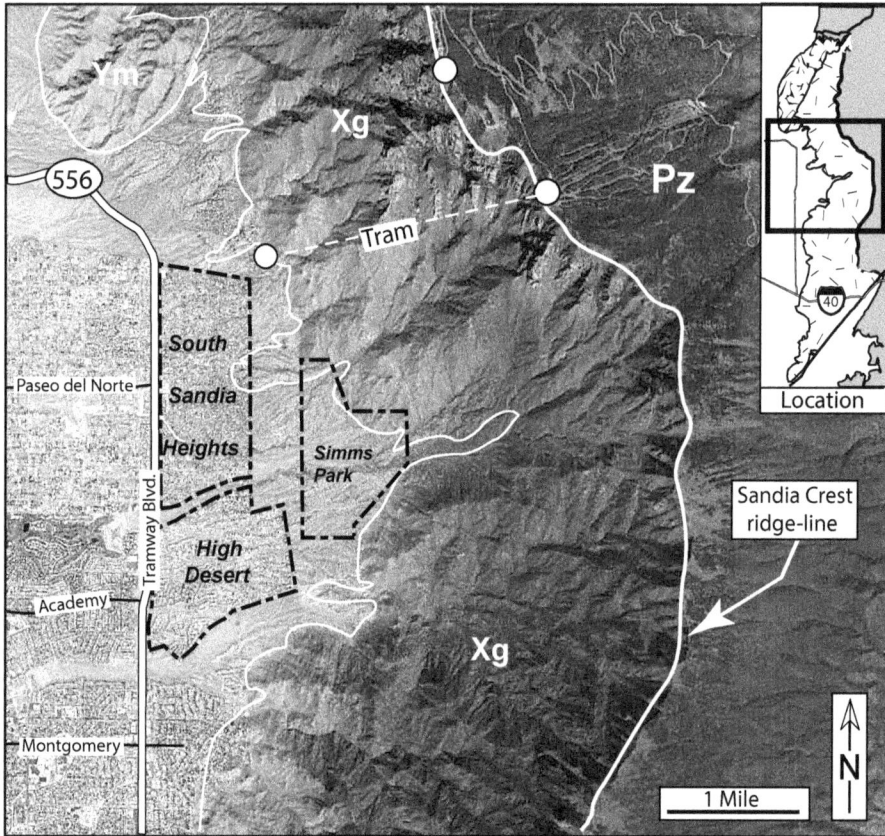

A. *Google Earth* Image

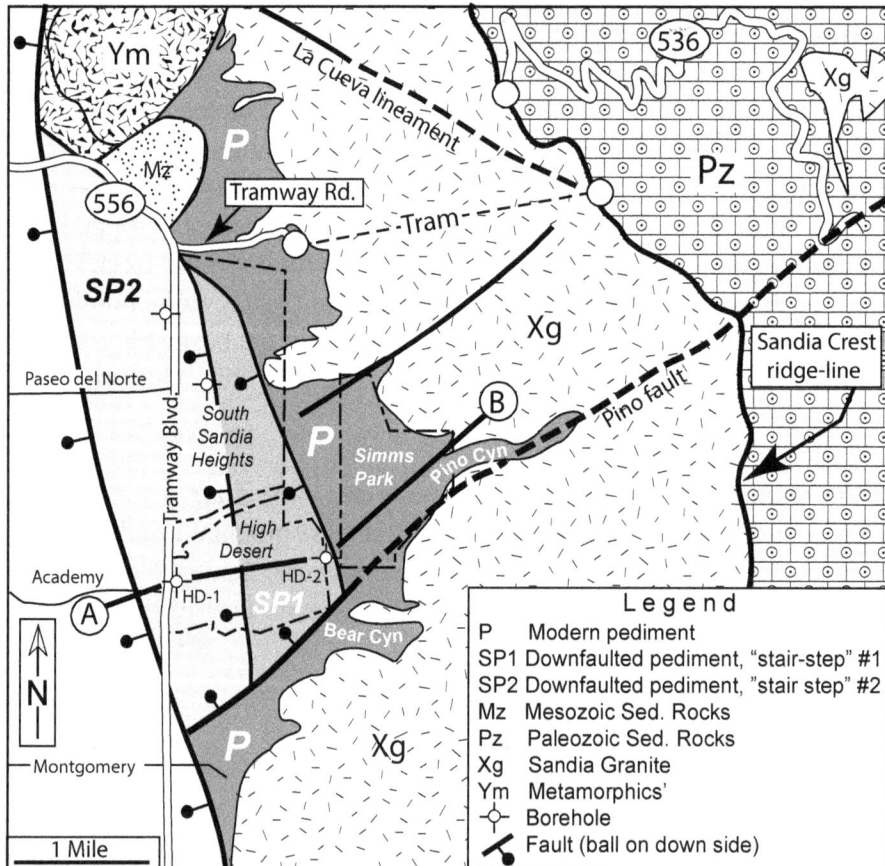

Legend

P	Modern pediment
SP1	Downfaulted pediment, "stair-step" #1
SP2	Downfaulted pediment, "stair step" #2
Mz	Mesozoic Sed. Rocks
Pz	Paleozoic Sed. Rocks
Xg	Sandia Granite
Ym	Metamorphics'
⊕	Borehole
⬤▬	Fault (ball on down side)

B. Bedrock Geologic Map and Select Faults (Cross Section A-B in *Figure 11-7*)

Figure 11-6. Bear Canyon Re-Entrant (Select Modern Geography for Reference; Modified from Read et al. 1999)

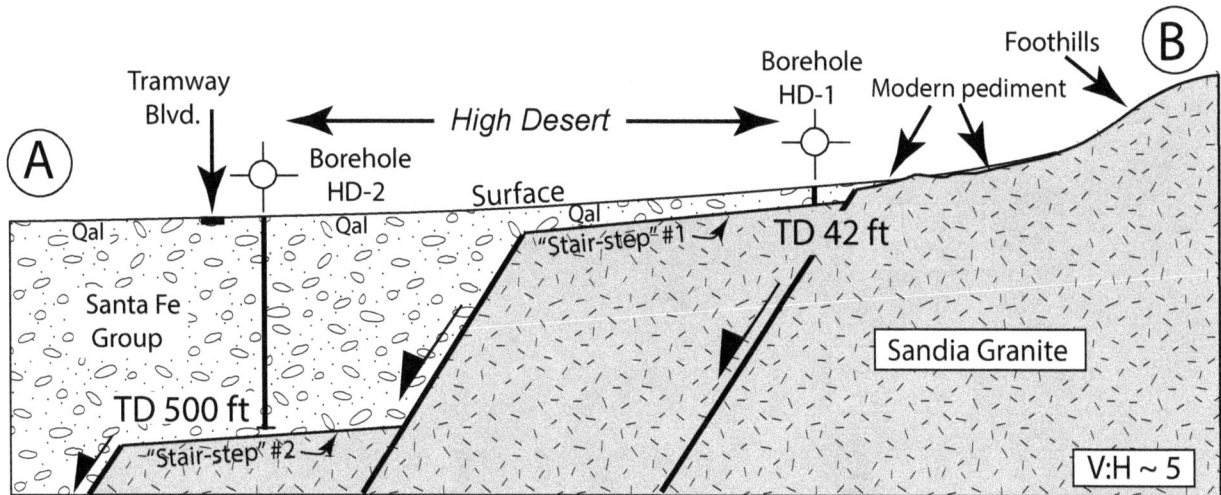

Figure 11-7. Schematic West-to-East Cross Section A-B (Location in Figure 11-6)

Switchback Road. This bulldozed series of switchbacks up the lower mountain is not geological. It is very prominent and occurs between the features noted above and below, and so I have included it. Some 30 years ago its trace was extremely visible from just about anywhere in the city as an ugly scar. Today it has been so obscured by desert vegetation that it is best seen via *Google Earth* (*Figure 11-8A*).

A. East Oblique *Google Earth* Image (Xg = Sandia Granite)

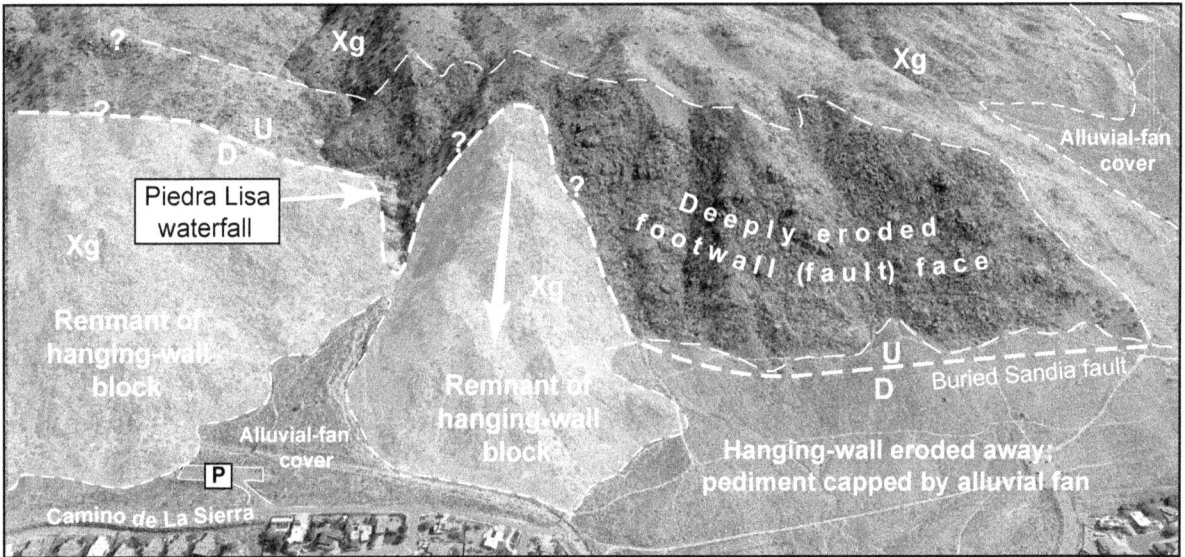

B. Slightly Realigned Inset Image in "A" Above

C. East View of Waterfall

D. Close-Up View of Waterfall (Vegetation for Scale)

Figure 11-8. Piedra Lisa Canyon and Its Waterfall (Photographs "C" and "D" by Author)

The feature's story involves the Glenwood Hills/Casa Grande Estates subdivision between Montogomery Blvd. to the north and Candelaria Road to the south, and between Tramway Blvd to the west and Cíbola National Forest in the foothills to the east, The land had been acquired by the city of Albuquerque in 1969, and the subdivision was developed from 1969 to 1971. Proposals at the time were to extend the subdivision up the slope east of the residential street Camino de la Sierra (CS), and a rough switchbacked path was bladed up the hill in anticipation. Unfortunately, or perhaps fortunately, it was learned that the cost to extend utilities up there would have been prohibitively expensive, so the road was abandoned. In 1973 the city of Albuquerque purchased the 520-acre strip between CS on the west and the national forest land on the east, which included the road, as Open Space and administratively added it to Cibola National Forest (Holben 2021). The switchbacks, in New Mexico's arid climate, will remain partly visible for many years.

Piedra Lisa Canyon and Its Waterfall. This feature (name is Spanish for "smooth rock") is unique along the western foothills. The feature displays a stream-polished granite escarpment over which water surges prolifically after a rare heavy rain. But why a waterfall exists at all in this place and not in in others is an intriguing question.

Here is one of those cases where *Google Earth* imagery stands out because it reveals details not at all evident on the ground. Importantly, the waterfall is clearly on trend with, but north of, the north-trending range-front Sandia fault (*Figures 11-8A* and *11-8B*). At the latter area, the fault's hanging wall (the "down" block) has been eroded away, exposing the footwall. To the north, but before the waterfall, the hanging wall is again preserved, and the trace of the Sandia fault has been scalloped in map view due to erosion, greatly complicating the picture. The waterfall appears to have exposed, and developed across, the Sandia fault's composite face. A visit to the site, however, detects no traces of fault movement such as slickensides (compare to *Figure 9-4*), perhaps because of erosion and polishing by the running water.

Caliza Hill. This is an informal name that I've applied to an anomalous, 120-ft-high hill in the southern Sandia foothills, near the east end of Lomas Blvd. (*Figure 11-9*). The hill is notable because it sports two large, very-up-scale homes that haughtily look down onto the surrounding neighborhood (*Figure 11-10A*). The developer who built these homes ca. 2000, having some knowledge of Spanish, named the road to the top *Loma de Caliza* (*caliza* is Spanish for limestone), and I have co-opted the name for the hill. Caliza Hill is formed by an 11-acre, flat-lying slab of Madera Limestone that rests directly atop Sandia Granite (*Figure 11-10B*). The rock unit that normally occurs between the two formations, the shaley Sandia Formation, is ominously missing (*Figure 0-12*). Caliza hill, with its missing Sandia Formation, has perplexed geologists for decades. Ellis took note of the feature but didn't give it a name (Ellis 1922), and others have struggled to explain it (e.g., Kelley and Reed 1961). A compelling hypothesis involves low-angle faulting, where a chunk of Madera limestone glided across a weak bed of Sandia Formation to rest directly atop the underlying Sandia Granite. If true, this feature may be part of the larger low-angle faulting regime that is documented to the north in the Juan Tabó area (*Figure 6-7*).

A. *Soil Conservation Service* Aerial Photograph, 1935 (Photograph SCS #179-049)

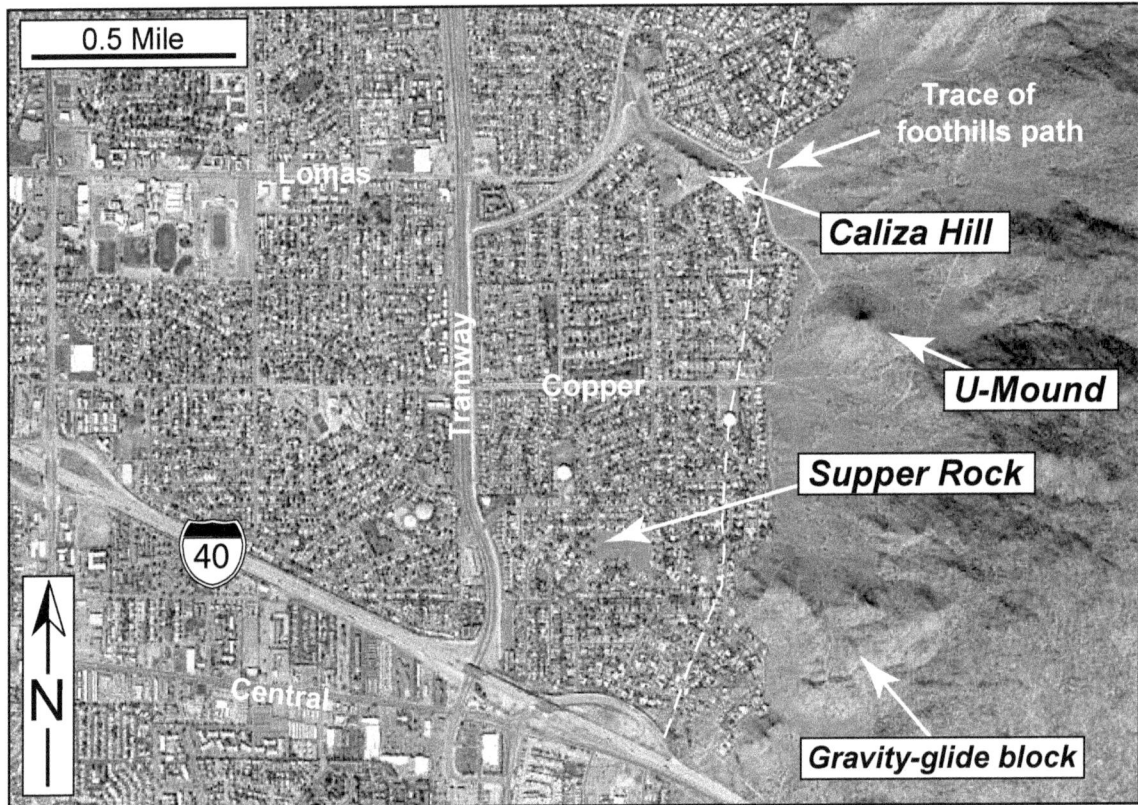

B. Contemporary *Google Earth* Image of Area in "A" Above

Figure 11-9. Southern Sandia Foothills

A. Southwest View of Caliza Hill (Center; Photograph by Author)

B. *Google Earth* Image of Caliza Hill and its Southern Satellite

Figure 11-10. Caliza Hill - I

The possible culprit: Sandia Formation. The Sandia Formation, with its beds of relatively soft mudstone (*Figure 3-2*), is a mechanically weak unit that may have provided a necessary lubricant for the overlying, rigid Madera Limestone. Any unit of rock, even one consisting of alternating rigid and soft layers, is extremely stable when horizontal, but quite unstable when inclined and unsupported in the lateral direction. Weak horizons can provide a sliding, "escape" surface for lateral movement. The adage, "a chain is only as strong as its weakest link," applies nicely here. It is possible that such breaking loose and lateral sliding played a key role here and allowed that slab of Madera Limestone to glide down and across its "greasy skid" to land directly atop the Sandia Granite (*Figure 11-11*). (Recall that I also called on the weak Sandia Formation for help to explain faulting behavior in the Placitas area, *Figure 4-9*.)

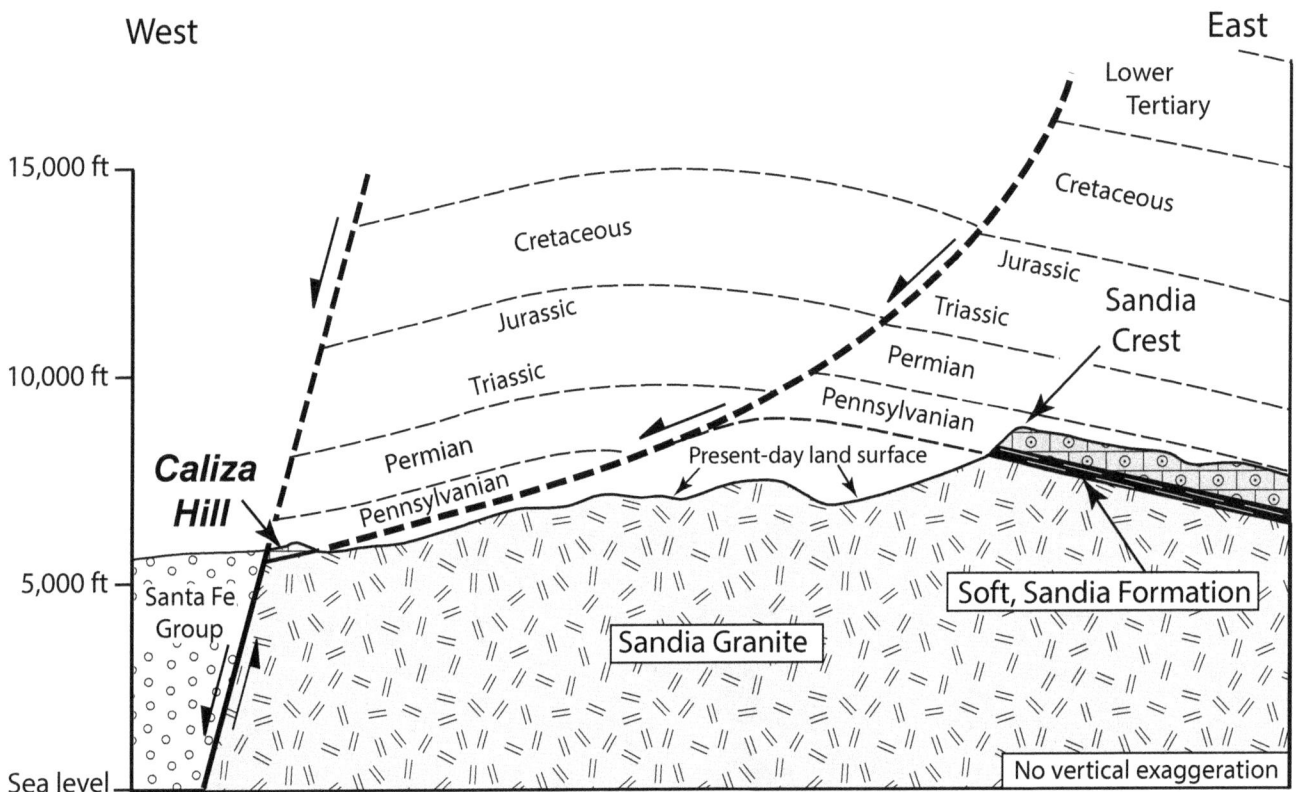

Figure 11-11. Caliza Hill - II: Schematic Cross Section Showing One Possible Interpretation

U-Mound. At an elevation of 6307 ft and a relief of only about 160 ft, this roughly conical little hill is located just north of the eastern end of Copper Ave. (*Figure 11-9A*). The hill's name comes from a huge letter "U" on its west side formed from rocks painted white in the 1920s by University of New Mexico engineering students (*Figures 11-12A* and *11-12B*). The "U" beamed west until the 1960s when a new generation of students, in a more environmentally conscious mood, turned the rocks over and restored the hill to its original appearance (*Figure 11-12C*; Smith 2006; Bannerman 2008).

A. East view, 1930 (Postcard from Bannerman 2008)

B. Close-Up of U-Mound's West Face, 1940s (From Smith 2006)

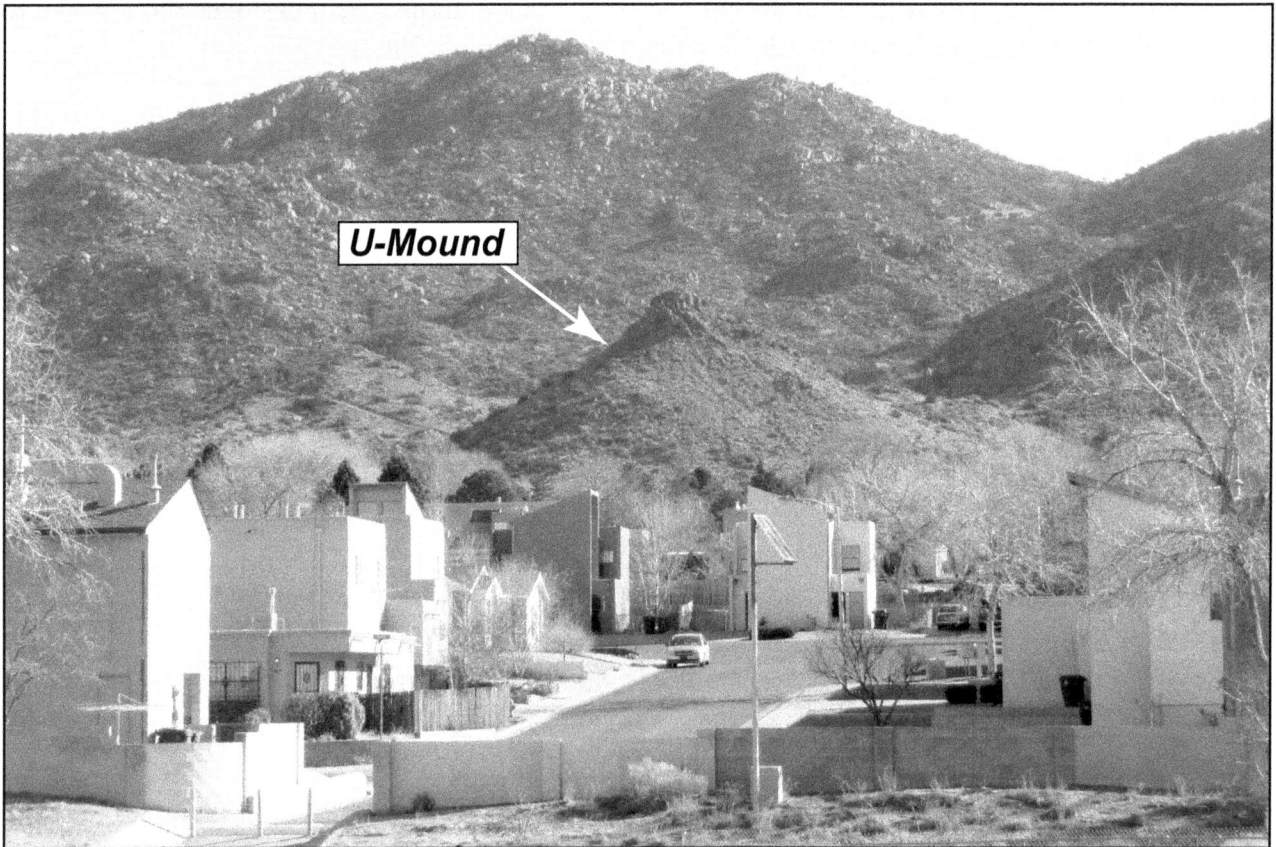

C. East View up Grady Court from Tramway Blvd. (Photograph by Author)

Figure 11-12. U-Mound - I: Southern Sandia Foothills

U-Mound is one of a scattering of anomalous granite hills along the southern foothills. These are residual rock masses still attached to the granite pediment below and stand out in bold relief because they are more resistant to erosion than the surrounding granite. Here and elsewhere, the lesson is that if something is sticking up above its surroundings, there is a geologic reason, and here the reason is *hematite*.

The dark-colored iron oxide mineral hematite (Fe_3O_4) has chemically replaced the granite's plagioclase feldspar and biotite mica (*Figure 11-13A*). Within the zones of alteration only the large pinkish crystals of potassium feldspar and some of the glassy quartz grains are unaffected and prominently stick out, utterly unlike normal Sandia Granite (*Figure 11-13B*; Shomaker 1965; Kelley and Northrop 1975). Hematite is usually dark red in color, whereas here it is black, making it difficult to identify. Fortunately, the clincher is that hematite is slightly magnetic, and a super-powerful magnet (neodymium type) will tellingly cling to its surface (*Figure 11-13C*). Hematite is the world's most important iron-ore mineral and supports the massive mining activity in the Mesabi range of upper Minnesota and Michigan's Upper Peninsula.

A. North Summit Area Showing Hematite-Rich Zones (Hammer for Scale;)

B. K-Feldspar Crystals in Dark Hematite-Rich Matrix

C. Rock Sample Attracting Magnet (M)

Figure 11-13. U-Mound - II: Hematite-Rich Granite
(Photographs by Author)

U-Mound is located at the western terminus of what appears to be a west-to-east fault (*Figure 11-14*). The age of this fault is not known but the apparent association of the fault with the mound is suspicious. The fault may have served as a conduit for mineralizing fluids much later than the intrusion of the granite. One possibility is that mineralization occurred back during the volcanic flareup of the late Paleogene (*Chapter 5*) when all the rocks were much hotter than now and igneous material gushed to the surface and near-surface. Perhaps the emplacement of hot hematite-bearing fluids along fractures was associated with that episode.

A. East Oblique *Google Earth* Image

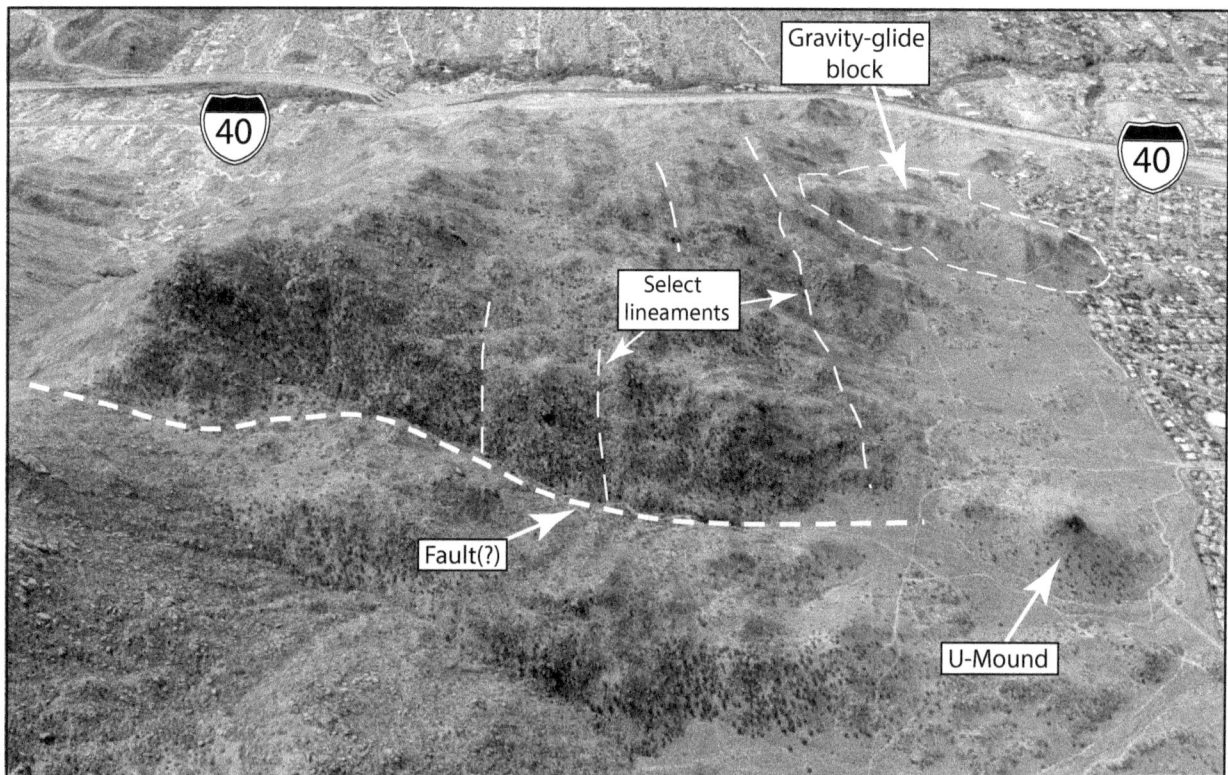

B. South Oblique *Google Earth* Image of Select Lineations in Sandia Granite

Figure 11-14. Images of Southern Sandia Foothills

U-Mound acquired a dubious notoriety back on March 16, 2014 when a troubled, homeless man named James Boyd, was found illegally camping on the mound's west flank. He was confronted by police and tragically shot and killed.

Supper Rock. Supper Rock is a little knob of granite about 0.8 miles southwest of U-Mound, made slightly more resistant by emplaced hematite, much like U-Mound. It lies within a little park amidst a subdivision also named Supper Rock (*Figure 11-14*). Long before this area was developed it was a favorite picnic spot for weekenders, thus the name.

Gravity-Glide. Not far north of I-40 is what appears to be a large slab of granite that has slid off the main mountain (*Figure 11-15A*). This large feature is not at all obvious on the ground and is only appreciated by that wonderful tool, *Google Earth*. A visit to the glide plane at the base of the slab reveals that an upper mass of granite, highly fractured and hence more deeply weathered, has slid down to the west across much-less-fractured granite below (*Figure 11-15B*). Upon viewing the *Google Earth* image, the reader can be excused for suspecting that the gravity mass has shoved the adjacent housing subdivision to the west (*Figure 11-15A*). The houses, of course, simply avoided the slab and were built around it.

A. East Oblique *Google Earth* Image Showing Extent of Gravity-Glide Block

B. North View of Gravity-Glide Fault (Photograph by Author)

Figure 11-15 Gravity-Glide Block, Southern Sandia Foothills

12:
TIJERAS CANYON
AND THE EAST MOUNTAIN

Introduction

Tijeras Canyon provides the link between the western foothills and the so-called East Mountain. Most motorists blast through it along I-40 without a thought. I instead prefer passage along old Route 66, now NM-333, and the comments below are for that route.

Unusual Corestones

Elephant Rock. Elephant Rock (a.k.a. Balanced Rock) was a huge corestone located on the south side of old US-66 (NM-333) at mile 1.2, serving for many years as a landmark at the western mouth of Tijeras Canyon, (*Figure 12-1A*). It was a popular tourist attraction in the early 1900s when the route was merely a wagon road. By about 1920 the road was straightened and improved, and passed right by the north edge of Elephant Rock. By 1921 the rock was used as a billboard, but the signs disappeared when the Forest Service objected. In 1926 the route was given the designation US-66, and the rock became even more popular with the increased tourist traffic. In May 1947, the University of New Mexico was in the process of moving some buildings from Santa Fe to the UNM campus in Albuquerque via Lamy and Moriarty, and from there through the canyon. When the first building being hauled came to Elephant Rock there did not seem to be enough space to pass. On their own, the crew toppled the rock with hydraulic jacks. Many locals were shocked and raised money to reposition the rock (probably not possible because it weighed something on the order of 50 tons!), but the highway department had no interest in helping and the plan was scrapped (Holben 2020). Today the rock remains in its toppled spot (*Figure 12-1B*)

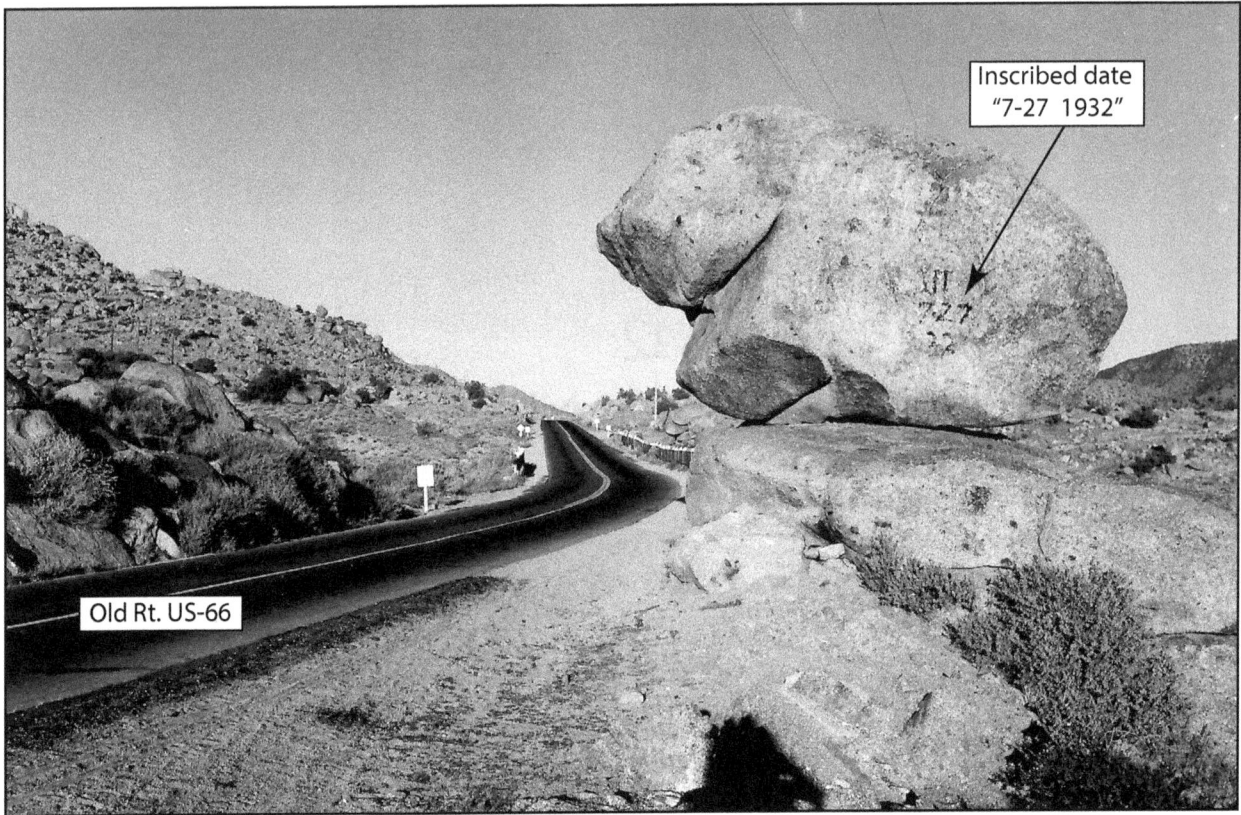

A. East View of the Most Famous Corestone: "Elephant Rock" (Tijeras Canyon, NM-333 Mile 1.1,; ca. 1930s (Postcard Courtesy of Rick Holben)

B. Modern View of "A" Above (Photograph by Author)   C. "Dinosaur Rock" (From Oates 1979; Vegetation for Scale)

Figure 12-1. Notable Corestones in Tijeras Canyon

Dinosaur Rock. One can search high and low in Tijeras Canyon for this weird corestone, which at one time was prominent and quirky enough in cross section to acquire a name (*Figure 12-1C*). A nearby archeological site named for it, *Dinosaur Rock Site LA 14857*, was surveyed in 1976 in advance of a new right-of-way that was to be routed for I-40 (Oates 1979). Not a trace of the rock nor the site remains today and the inevitable conclusion is that they were both bulldozed away.

Igneous Dikes

Immediately opposite Elephant Rock, on the north side of NM-333, is a nice example of an igneous dike, filled with an unusual rock type called *lamprophyre* (*Figure 12-2A*).

A. North View of Dike Filling Vertical Fracture in Sandia Granite (Xg), Tijeras Canyon, NM-333, Mile 1.1
(Photograph by Author; Vegetation for Scale)

B. North Oblique *Google Earth* Image Showing Select Linear Trends

Figure 12-2. Igneous Dike and Fabric of Southern Part of Sandia Mountain

Lamprophyre. The majority of the dikes within Sandia Mountain are filled with this unusual igneous rock type with the strange name (Greek *lampros* = bright). The rock is probably of a similar age as that of the Ortiz igneous belt, but it is compositionally different. In outcrop the rock has a distinctive sparkle, produced by very abundant grains of the black, sheet-like mica mineral, *biotite*. The mineral contains the hydroxyl ion OH^-, indicating that the magma was water-rich, held under great pressure, and this in turn implies that a large magma body existed at depth here at the time.

The dike is part of a swarm of dikes in the mountain spanning an area of some 2 x 12 miles (*Figure 5-2B*; Kelley and Northrop, 1975). The southern portion of Sandia Mountain has a strong, north-to-south, linear fabric that may in part be due either to fracturing, to some original emplacement-flow pattern developed during the long-ago intrusion of the Sandia Granite, or perhaps to some combination of the two (*Figure 12-2B*). Remember from *Chapter 5* that during the late Paleogene this rock was buried below about by 12,000 ft of overburden and that large volumes of igneous material were intruding and spewing out to the surface nearby to the east. Whatever the cause of the lineation, the fabric seems to have at least guided the emplacement of the dikes. There are few natural fresh exposures because the dikes weather easily, but the few that exist, like this one, typically can be attributed to the highway engineers and their bulldozers.

Expansion (Exfoliation) Joints

Many outcrops of Sandia Granite exhibit a confusing array of cracks, or partings, sub-parallel to the present land surface (*Figure 12-3*), and therein lies the clue. The granite had been buried for millions of years, and only during the past million or so has it been exposed at the surface. Rock is very slightly elastic, and when great thicknesses of overburden have been removed, the granite "exhales with a sigh" and slightly expands or *exfoliates* along partings or fractures roughly parallel to the surface. The expansion is greatest at the very surface and it decreases downward. The pattern of these "expansion joints," can be regular or chaotic, but always interesting.

A. Confused Pattern of Expansion Joints
(Tijeras Canyon, Frontage Road Mile 1.9; Bushes for Scale; Photograph by Author)

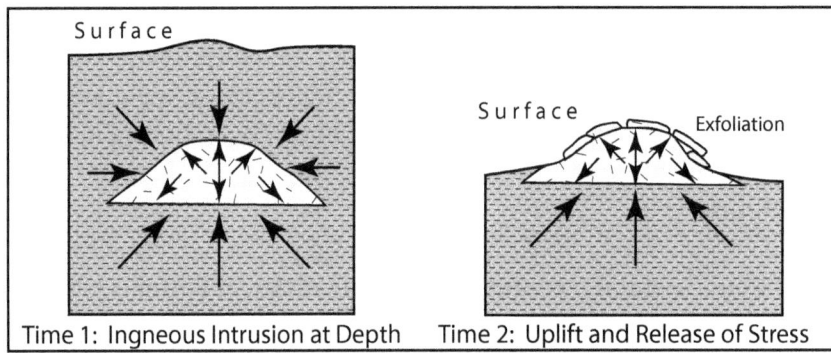

Time 1: Ingneous Intrusion at Depth   Time 2: Uplift and Release of Stress

B. Cartoon Showing Direction of Forces

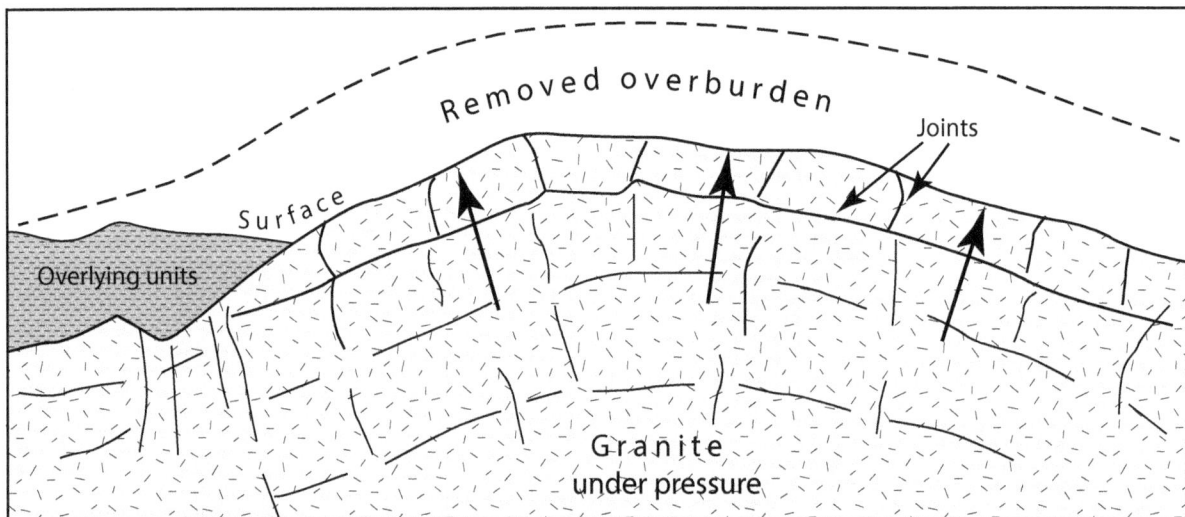

C. Schematic Cross Section Showing Expansion Joints Being Formed by Release of Downward Stress as Overburden is Removed

Figure 12-3. Expansion (Exfoliation) Joints in Granite

Orbicular Granite

An outcrop of orbicular granite (see INTRODUCTION) has been reported in Tijeras Canyon north of I-40, northeast of the Montecello subdivision, and within the Sandia Mountain Wilderness (*Figure 12-4*). Discovered only in the late 1970s, the site has been written up in the technical literature (Affholter 1979; Affholter and Lambert 1982). An apparently different site is located on an unimpressive little knoll, some 40-50 ft or so high at an elevation of about 6720 ft, that I here informally call "Orbicular Granite Hill" (*Figures 12-5A*, *12-5B*, and *12-5C*)). Its geographic coordinates put it about 1100 ft to the west-southwest of the published site, but I believe that this second exposure is the same as the first.

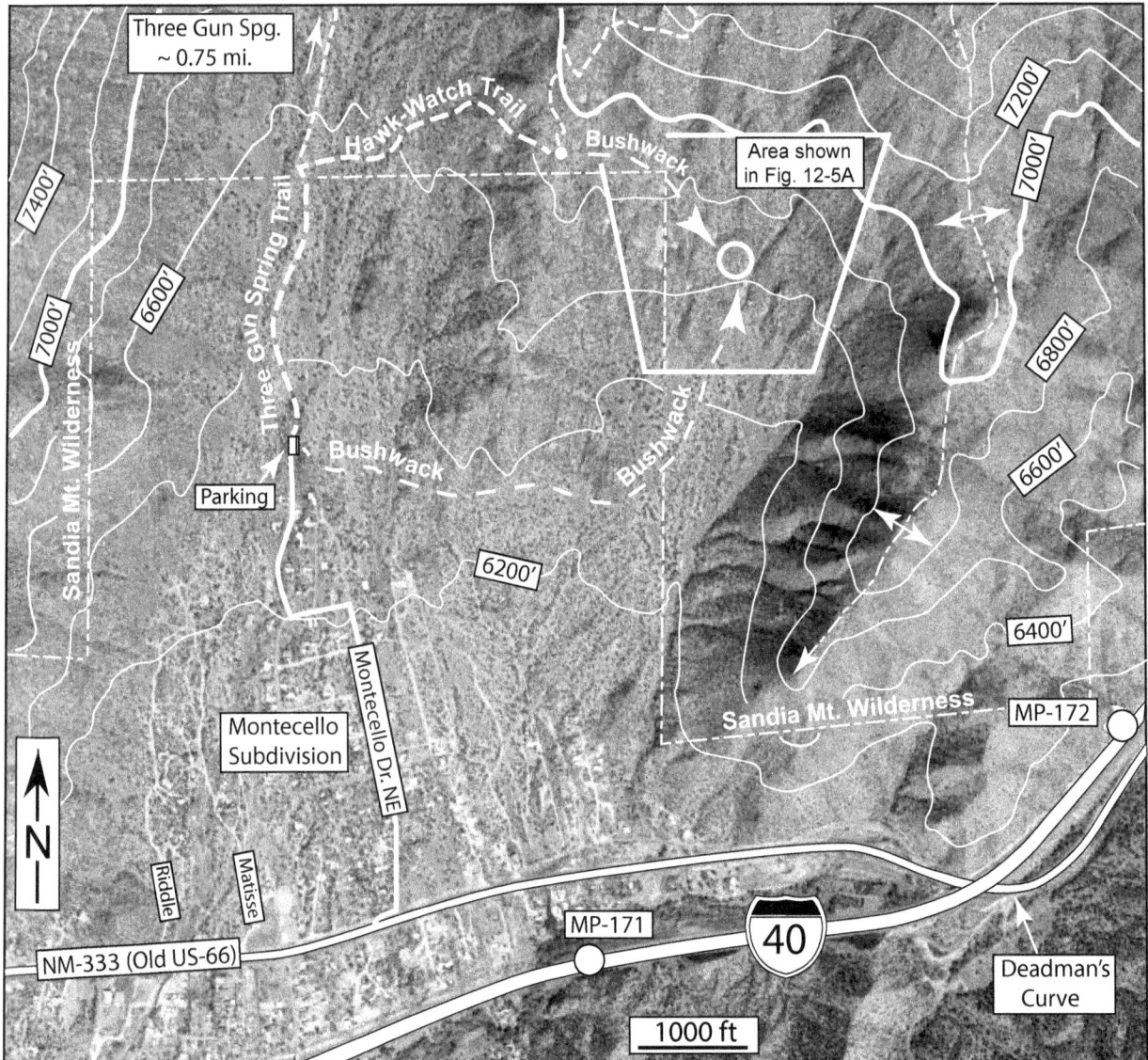

Figure 12-4. *Google Earth* Image Showing Location and Topography (CI = 200 and 500 Ft) of "Orbicular Granite Hill" (Circled), Tijeras Canyon Area

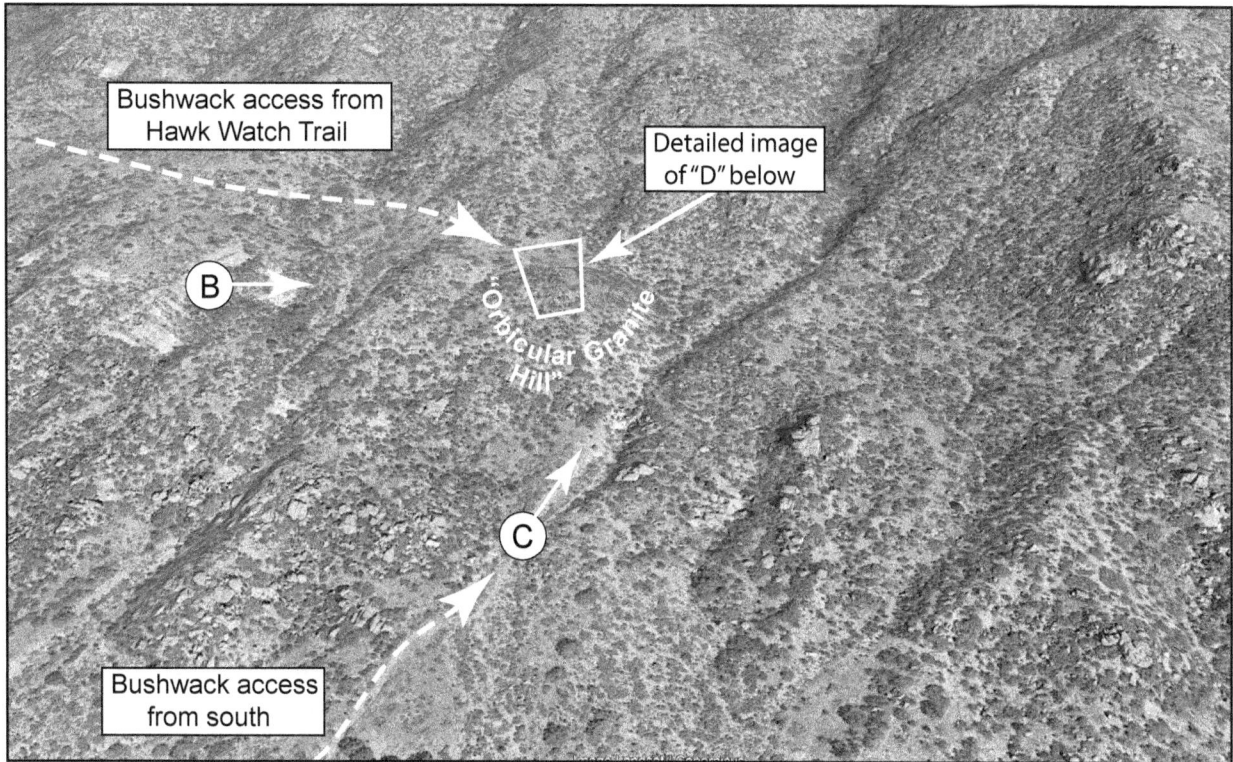

A. North Oblique *Google Earth* Image (Locations of Photographs "B" and "C" Shown)

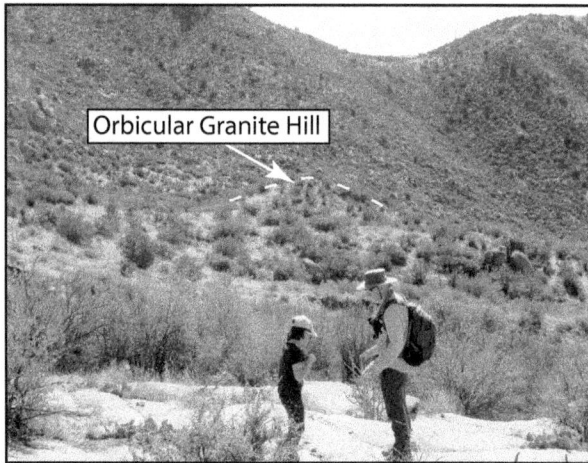

B. East View (Location in "A" Above)

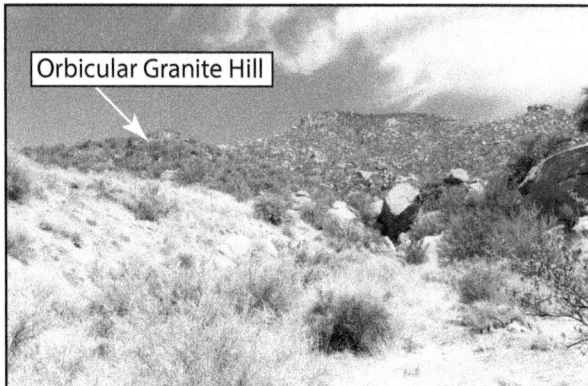

C. North-Northeast View (Location in "A" Above)

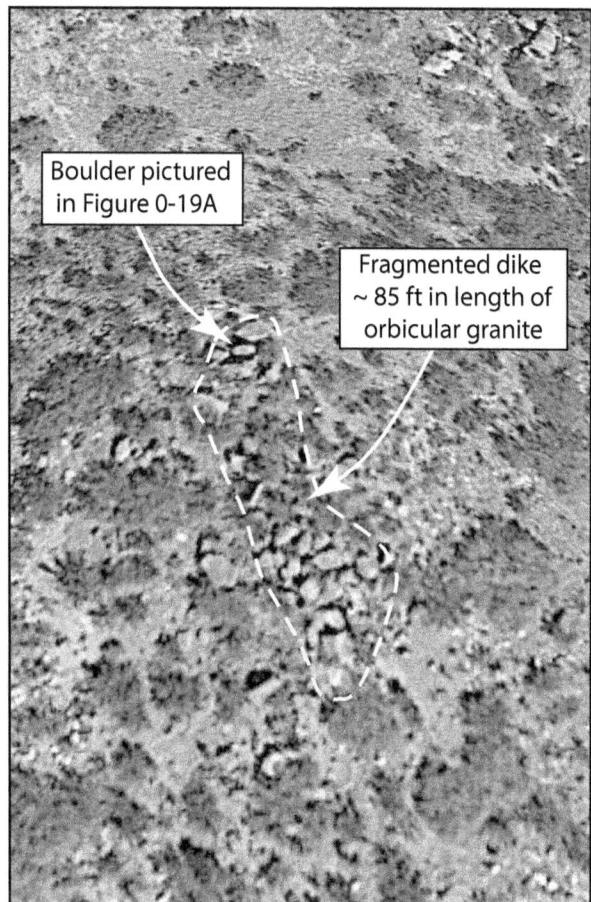

D. Detailed North Oblique *Google Earth* Image

Figure 12-5 Images of "Orbicular Granite Hill"
(Photographs in "B" and "C" by Author)

Access requires some effort, via two routes. One is via a hike north from the parking area at the Three Gun Spring Trailhead to the junction with the Hawk Watch Trail, then east on the latter trail to where it begins to switchback up the slope, and finally by bushwack southeast to the little hill with its outcrop (*Figure 12-4A*). The total distance one-way is about 1.2 miles. A second access is via an up-and-down bushwack due east from the parking area, along the south side of a low ridge of highly fractured granite, and then around the ridge and north along part of an abandoned 4WD road to the hill, for a total one-way distance of about a mile. The abandoned road was likely once used to access the orbicular rocks, but has since been blocked off by private development.

The outcrop is a dike-like, north-northwest trending body, about 85 feet in length. The dike is fractured into blocks encased by abundant cactus and brush (inset in *Figure 12-5D*). It does take some careful scrambling to examine.

Finally, here are two recommendations for those considering a visit to this site. First, by all means avoid hiking during the warm months. The entire Three Gun Spring area lies in a large, flat-bottomed, topographic basin that tilts to the south. It therefore receives the full brunt of the warm-weather sun, providing little shade and thus becoming a very uncomfortable furnace.

Second, take a GPS device (for GPS coordinates see APPENDIX V). Orbicular Granite Hill is quite easy to locate on *Google* Earth but it is quite another issue on the ground. The hill is a small feature, and without well-established access trails it is very easy to miss. Because the feature lies within the Sandia Mountain Wilderness, rock collecting is prohibited.

Ancient Debris Flow

An unexpected and intriguing feature is located at mile 5.5 along NM-333 on the south side, about a mile west of downtown Tijeras. Here is an outcrop showing the toe of a tremendous, ancient flow of bouldery gravel that once roared across here from the north (on the other side of I-40) and gouged a valley into old basement rock, before the final incision of the modern Tijeras Creek (*Figure 12-6A*). Near the top of the flow are some enormous boulders (including one sandstone block measuring 14 ft across), testifying to the great energy this drainage system had, in sharp contrast to the pathetic little trickle of the modern Tijeras Creek. The upper part of this deposit (some has been quarried away) is about 50 ft higher the modern stream channel, requiring engineers of US Route 66 and the later I-40 to excavate a passage through its neck in order to establish a suitable grade.

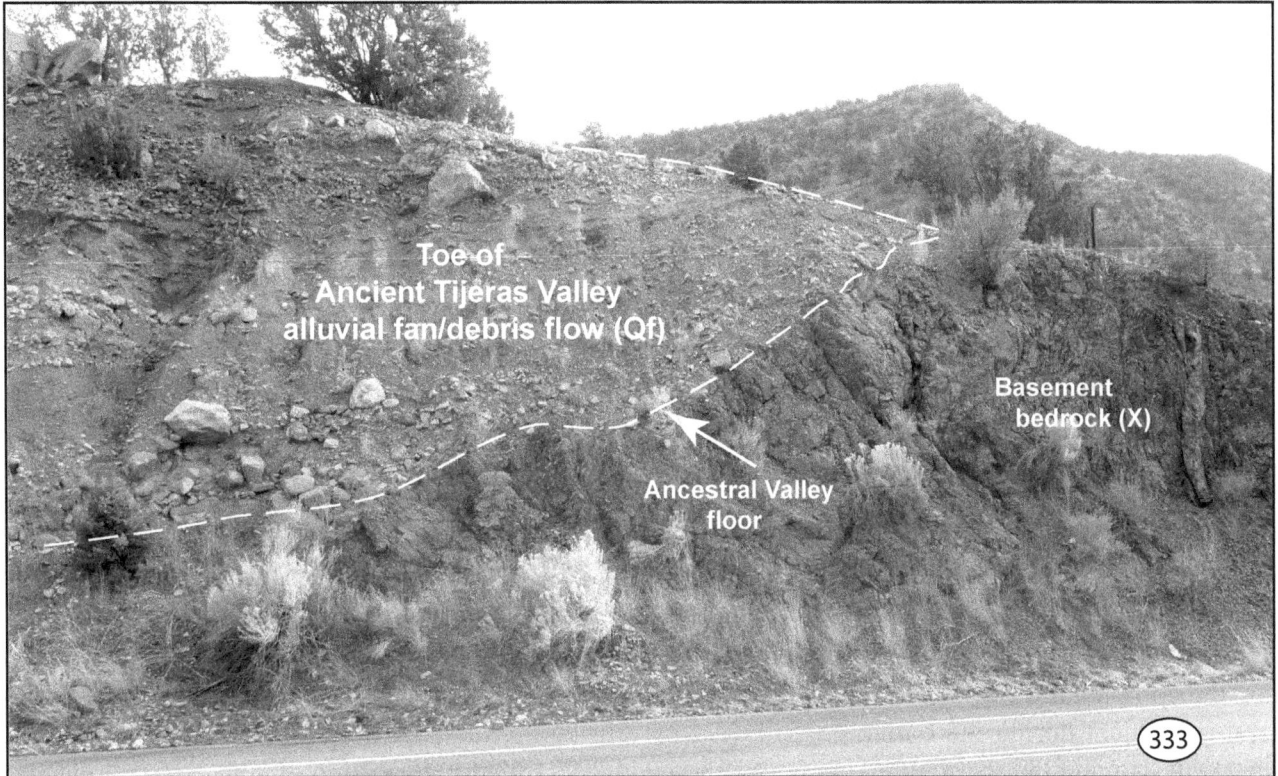

A. South View of Outcrop (Containing Huge Boulders of Sandia Formation (IPs)

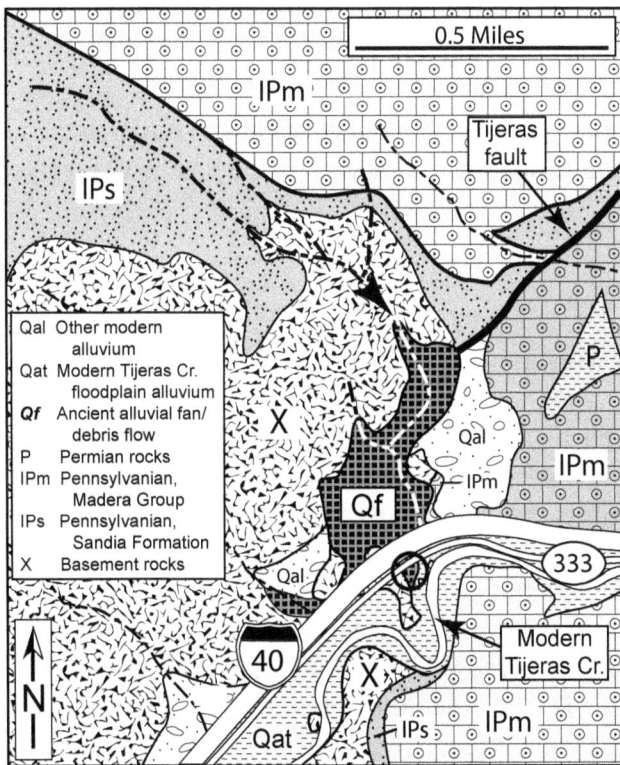

Qal Other modern
     alluvium
Qat Modern Tijeras Cr.
     floodplain alluvium
Qf   Ancient alluvial fan/
     debris flow
P   Permian rocks
IPm Pennsylvanian,
     Madera Group
IPs Pennsylvanian,
     Sandia Formation
X   Basement rocks

B. Geologic Map (Outcrop in "A" Circled)

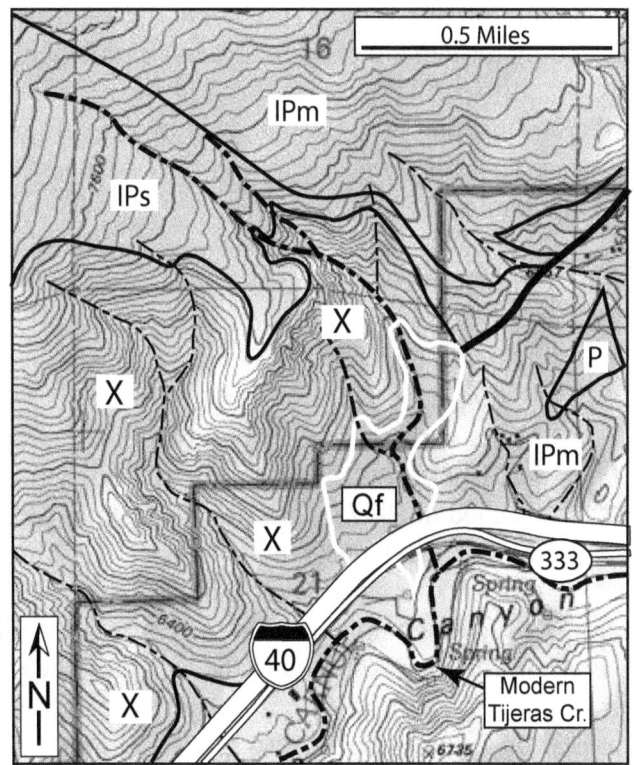

C. Topographic Map (CI = 40 Ft) with Select Geology from "B"
North of I-40

Figure 12-6. Ancient Alluvial Fan/Debris Flow (Qf), Tijeras Canyon (NM-333 Mile 5.5)

333

A geologic map of the debris-flow's source area that was about 1300 ft higher up on the mountain reveals that it was made up of a large swath of gently-dipping Sandia Formation exposed where the Madera limestone has been stripped back (*Figures 12-6B* and *12-6C*). The Sandia Formation is much more susceptible to erosion than the Madera limestone and provided the principal source of the flow. This interesting outcrop demonstrates that conditions in the ancient canyon were more energetic than those of today.

NM-14, *Turquoise Trail*

The *Turquoise Trail* today is considerably built up by human activity. The southern part of the highway passes along the complex trace of the Tijeras fault system. Remember, this master fault system had suffered right-lateral movement during the Laramide, followed by left-lateral movement during the opening of the Rio Grande rift, leaving the rocks along the fault churned and scrambled. A location of some interest is located behind the Mountain Christian Church at mile 3.6 on the east side of NM-14. Here is a very chopped-up, whitish outcrop of Todilto gypsum that had once been quarried for domestic whitewash ca. 1900, and later ca. 1960 by the cement plant in Tijeras (East Mountain Historical Society 2020).

Farther north along NM-14 some outcrops are of great interest to the geologist, but not so much to the non-geologist. Therefore, I will skip over most of this stretch until we arrive at NM-536, the Sandia Crest Byway at San Antonito.

Sandia Crest Byway (NM-536)

This road, *Sandia Crest Byway*, is a New Mexico jewel with an interesting history.

In the early 1920s a "loop road" was proposed to link up Albuquerque, Tijeras, San Antonito, Las Huertas, Placitas, and Bernalillo, and it was finally completed in 1925. Sandia Crest was first reached in 1927 by a road called Rim Road, built by private groups and the Kiwanis Club. In 1936 the Civilian Conservation Corps (CCC) improved the road from their camp F-8-N at Sandia Park (see *CCC camps* below). In 1950s and 1960s the Sandia Crest parking area was expanded, and the array of steel communication towers was installed. In 1959 the state highway department improved the Byway from San Antonito to the ski area. The project involved paving the entire length and realigning some of it to eliminate some curves and steep grades. This is when the little community of Sandia Park was bypassed on its north and a new grade from Doc Long Picnic Area to Tejano Canyon was put in. Before 1988 the entire northern section of the loop road, from Bernalillo to Placitas, across the mountain and down to San Antonito at NM-14, was designated part of state highway NM-44. In 1988 part of the road became NM-536, and in 2000 it also took on the name Sandia Crest Byway—a National Scenic Byway. From the road's zero-mile marker at San Antonito, the Byway ascends 3814 ft along a reach of 13.3 miles to the Sandia Crest at 10,678 ft.

Permian Red Stuff, Mile 0.9 to 1.0. On the north side of the Byway, from mile 0.9 to 1.0, are two fine exposures of the Permian-age Abo Formation, consisting of stacks of reddish river channel sandstones and their associated flanking floodplain muds (*Figure 12-7*. These *fluvial* sediments were formed by rivers as they meandered across their muddy floodplains, all building upward in tandem with the land subsiding to create the necessary accommodation space (inset in *Figure 12-7B*). Farther west up the mountain these beds have been stripped off.

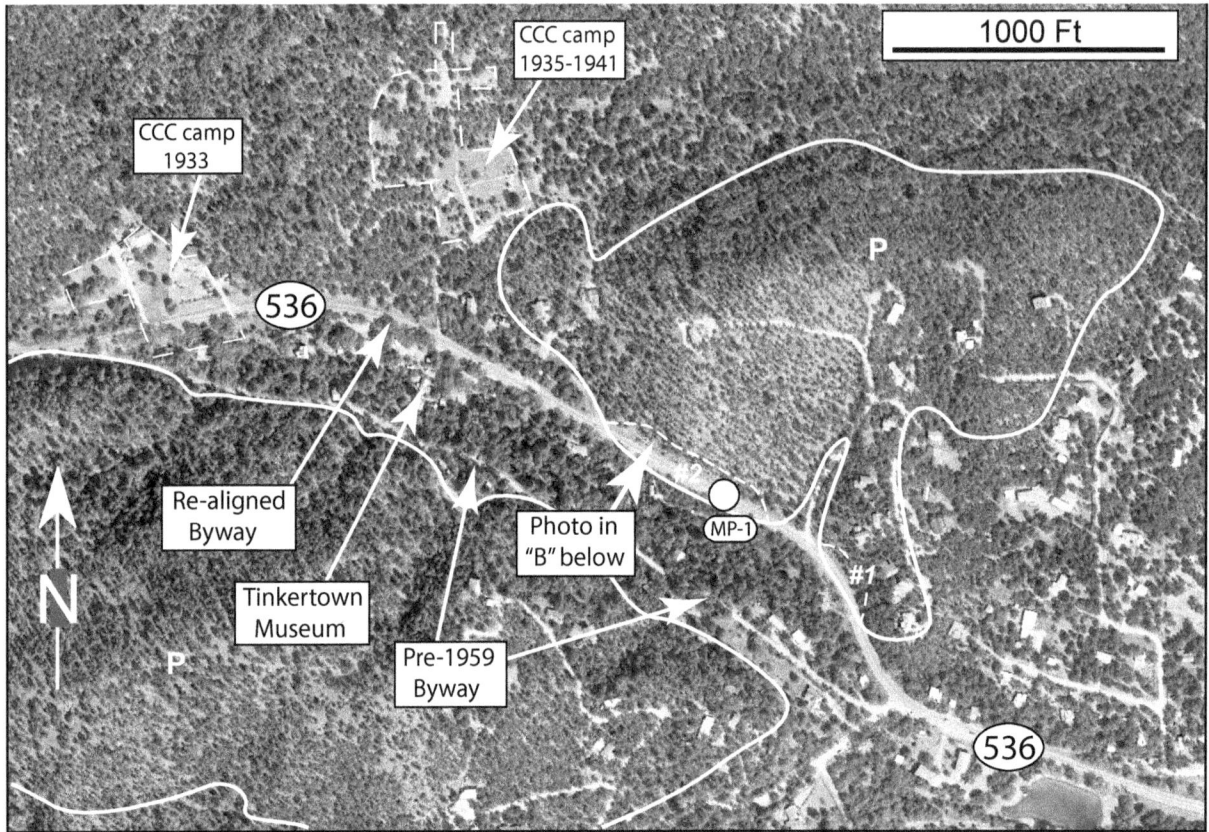

A. *Google Earth* Image Showing Outcrops (#1 and #2) of Permian Redbeds (P)

B. Excellent Outcrop (#2) at MP-1 of "Stacked" Channel Sandstones (C) Encased in Floodplain Muds (F) (Vegetation for Scale; Photograph by Author)

Figure 12-7. Permian Redbeds along Lower End of Sandia Crest Byway (NM-536), Sandia Park

CCC Camps. I mention these two features in passing because they receive little to no attention elsewhere (*Figure 12-7A*). On the south side of the Byway at mile 1.25 is the very visible Tinkertown Museum. Not so obvious though is a turnoff opposite, to the north, that leads about 60 yards past a residence to a locked gate. Beyond the gate is the site of the second of two CCC camps, named Sandia Park F-8-N. (Unfortunately, this site is now on private land and is best seen on *Google Earth*). This was a tent camp for about 200 young men that was active during the summers from 1935 through 1941. About 500 ft farther along the Byway is a cluster of three houses on the north side. This is the center of an earlier CCC camp, named Sulphur Springs F-8-N, that was active only in 1933. One can detect some terracing that had been constructed for an array of tents (Van Hart 2020).

The *Great Unconformity* (GU). This subtle yet profoundly important feature is displayed for miles along the entire west face of the mountain just below the crest, but it is covered there, not to mention difficult to get to (*Figure 3-2*). On the east side, though, it can be seen in two, and only two, convenient sites (*Figure 12-8*).

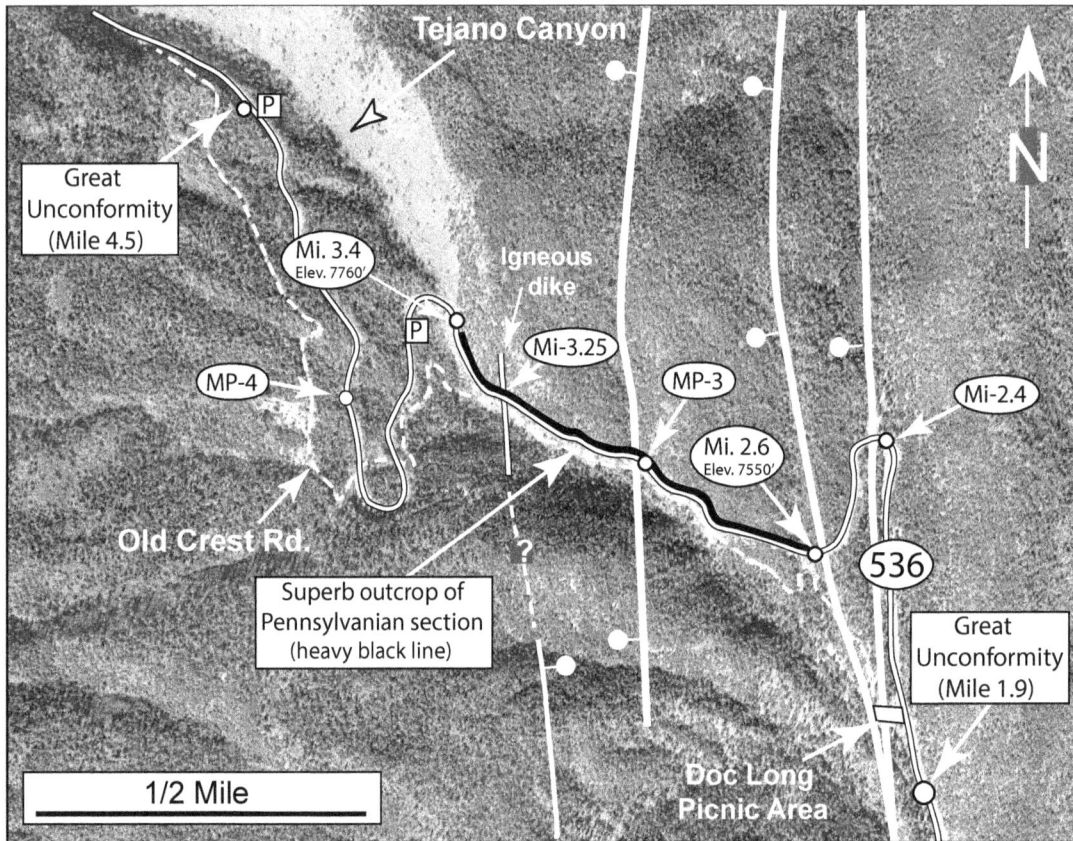

A. *Google Earth* Image (P = Parking Spot)

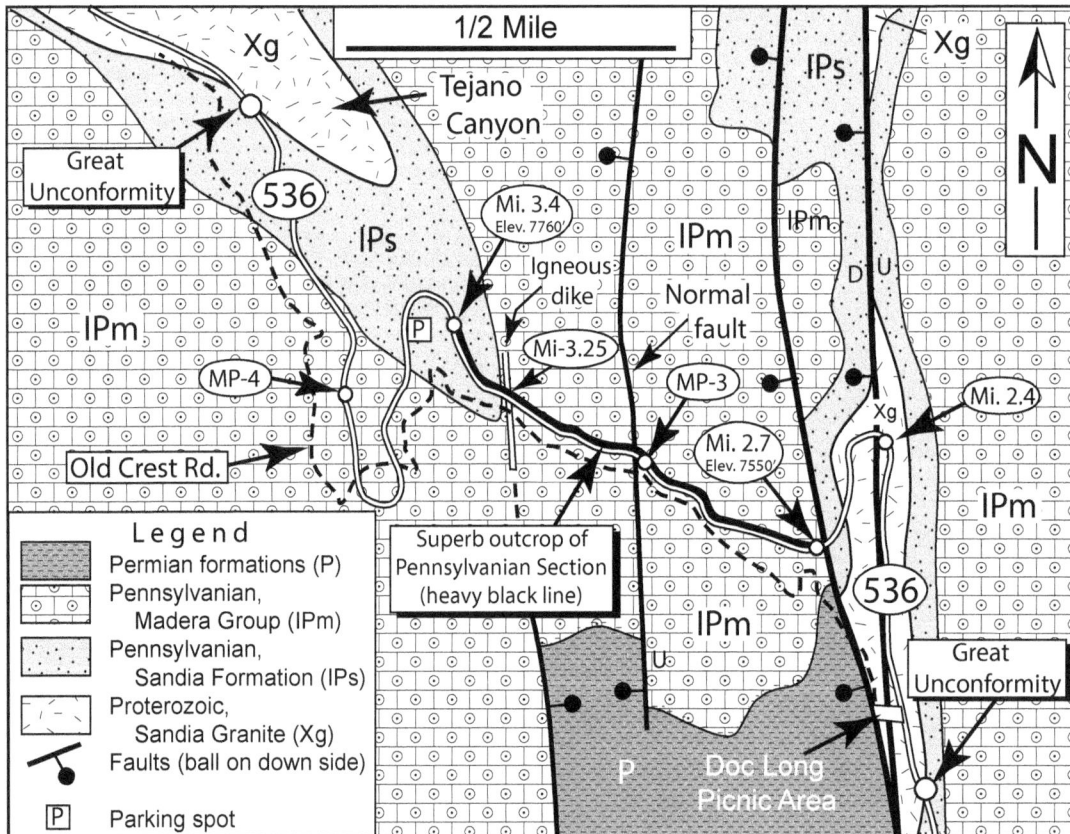

A. Geologic Map of "A" Above (Modified from Read et al. 1999)

Figure 12-8 Pennsylvanian Section Along North Side of Sandia Crest Byway (NM-536)
(Mileages and Mile Posts Noted)

GU Site 1 (Mile 1.9, Near Doc Long Picnic Area). The first of these important exposures is located just south of the Doc Long Picnic Area at Mile 1.9 on the east side of the Sandia Crest Byway. To avoid risking life and limb due to the unpredictable traffic it is advisable to park at the Doc Long area and walk over to the site. However, a much better, and safer, location is at mile 4.4 to 4.5 along the Byway in Tejano Canyon (see GU-Site 2 below).

Pennsylvanian Section Along Tejano Creek (Mile 2.7 to 3.4). Beyond the Doc Long Picnic Area and along the north side of the Byway is a 4130-ft-long outcrop of the Pennsylvanian-age Madera Group and the upper part of the underlying Sandia Formation. It proves the adage that the highway engineer can be the geologist's best friend! This exceptional outcrop was produced sometime ca. 1959/60 when the Byway to the crest was "improved" for increased traffic (Holben 2017). This involved re-aligning the road, from down near Tejano Creek to the south, up onto the flank of the hill to the north. The road ascends only about 210 ft in elevation along the outcrop, but passes "down-section" (from younger to older rock) through about 770 ft (235 m) of strata that variably dip generally 15° to 35°. back to the east. Although this is a favorite place for serious geologists to observe the excellent section of alternating Pennsylvanian-age limestones and sandstones (Smith, 1999), it is admittedly less so for the layman and is also rather awkward to access. Still, let's take a look at part of this superb outcrop where two "curiosities" are worth a stop (*Figure 12-9*). But first a caution.

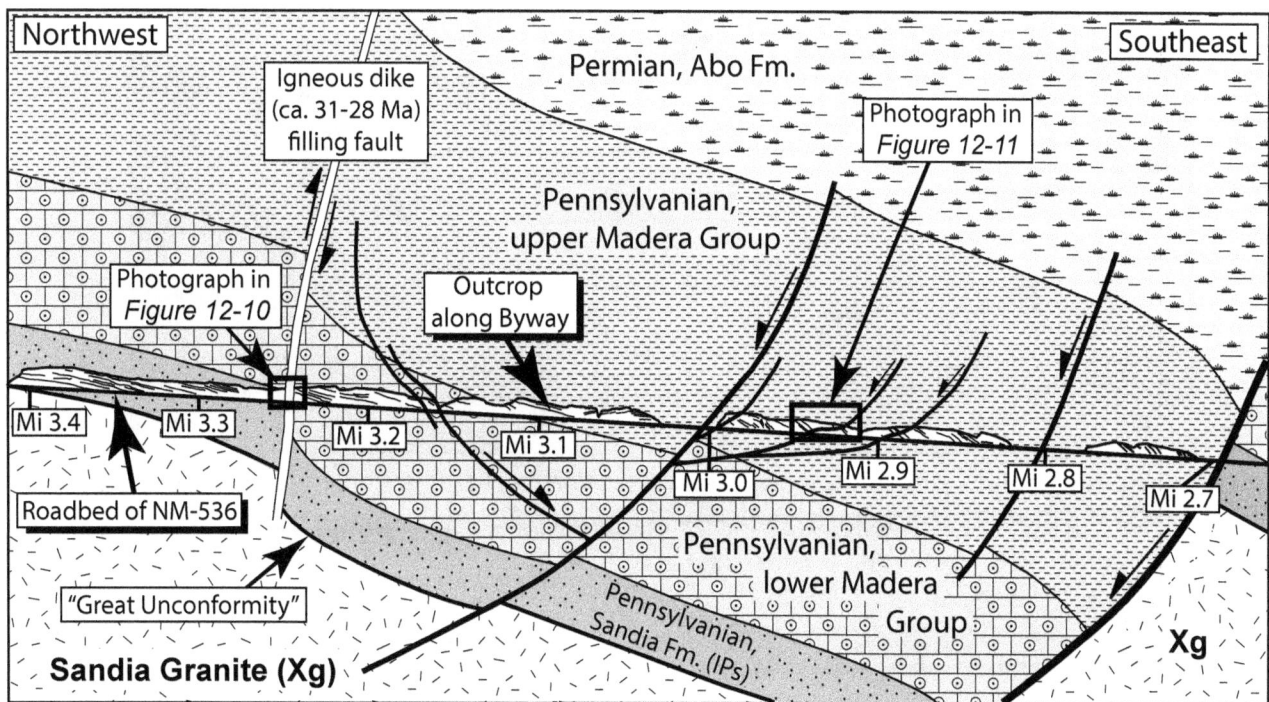

Figure 12-9. Cross Section of 0.8-Mile Outcrop of Pennsylvanian-Age Sandia Formation (IPs) and Madera Group (IPm) along Sandia Crest Byway (Mile 2.6 to 3.4; No Vertical Exaggeration)

Parking. There is only one relatively safe place to pull over and park (*Figure 12-8A*). That is up just beyond the upper, northwestern end of the outcrop (past mile 3.4), on the east side past the sharp road bend to the south.

Now, having found a parking place, it's time to check out these two curiosities, from top to bottom.

Igneous Dike (Mile 3.2). Here we find an 18-ft thick, north-trending body of rock that doesn't resemble anything else along this outcrop (*Figure 12-10*). The is a *dike*, a nearly vertical, wall-like igneous body. There are many other such dikes in the Sandia Mountain, but rarely this accessible or dramatic. It is yellowish-brown and rich in weathered biotite mica. It clearly shows chilled margins formed as the liquid rock (magma) intruded the colder, and older (Pennsylvanian) rocks. The igneous material of the dike is probably contemporaneous with the 37-26 Ma volcanic activity of the Ortiz mountains to the east.

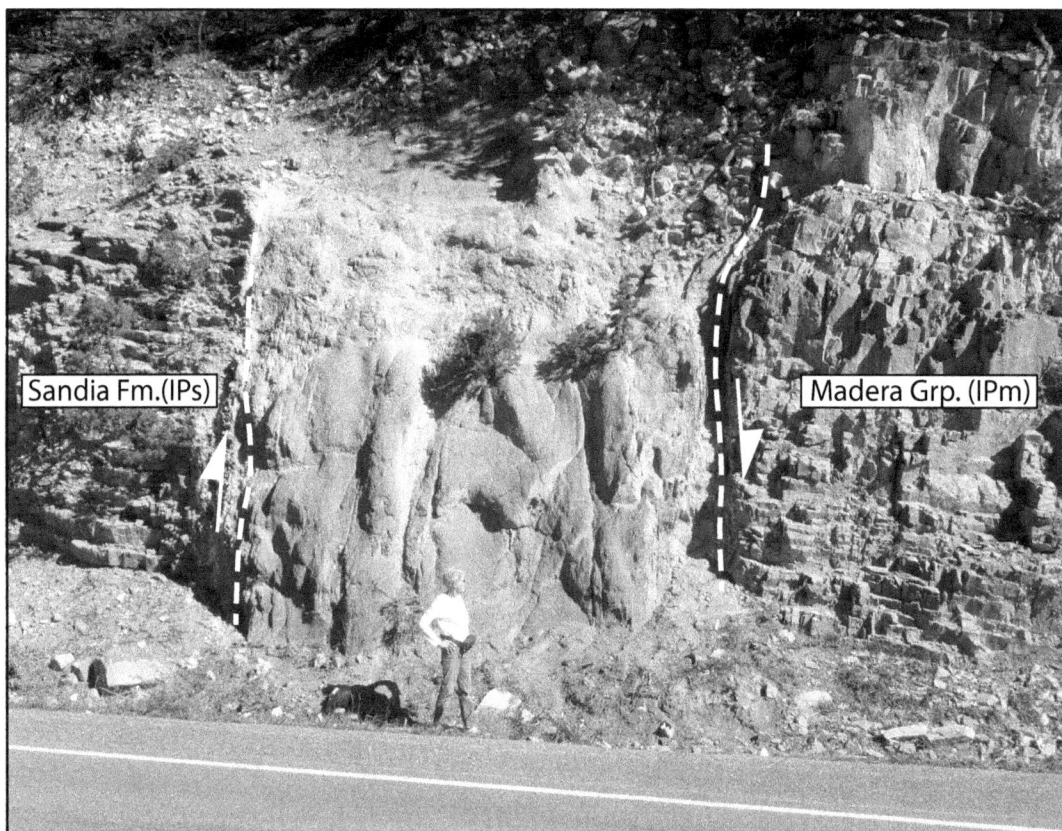

Figure 12-10. North View of Igneous Dike Filling Vertical Fault in Pennsylvanian Section, Sandia Crest Byway, Mile 3.2 (See Box in *Figure 12-7*; Photograph by Author)

The dike occurs on the eastern edge of a 150-ft-wide fault zone of intensely scrambled rock of the uphill Sandia Formation (west), with Madera Group on the downhill side (east). A fault zone of this thickness and intensity implies major movement, probably of Laramide age. In fact, it seems to line up with a mapped, north-trending, down-to-the-east fault about 1000 ft to the south and possibly connects with it (*Figure 12-8B*). This fault has a history!

Strange Fault (Mile 2.9). Back down the hill, just east of MP-3 (there is a mile-post sign here), is a low-angle fault (*Figure 12-11A*). Such faults are not at all typical. The limestone bedding at this point is about 45° down to the east. If in our minds eye we rotate the rocks back 45° to the west to restore the strata to their original horizontal orientation, the fault then is seen to lean some 55 to 60° west—the normal amount (*Figure 12-11B*). This suggests that the fault happened before the tilting.

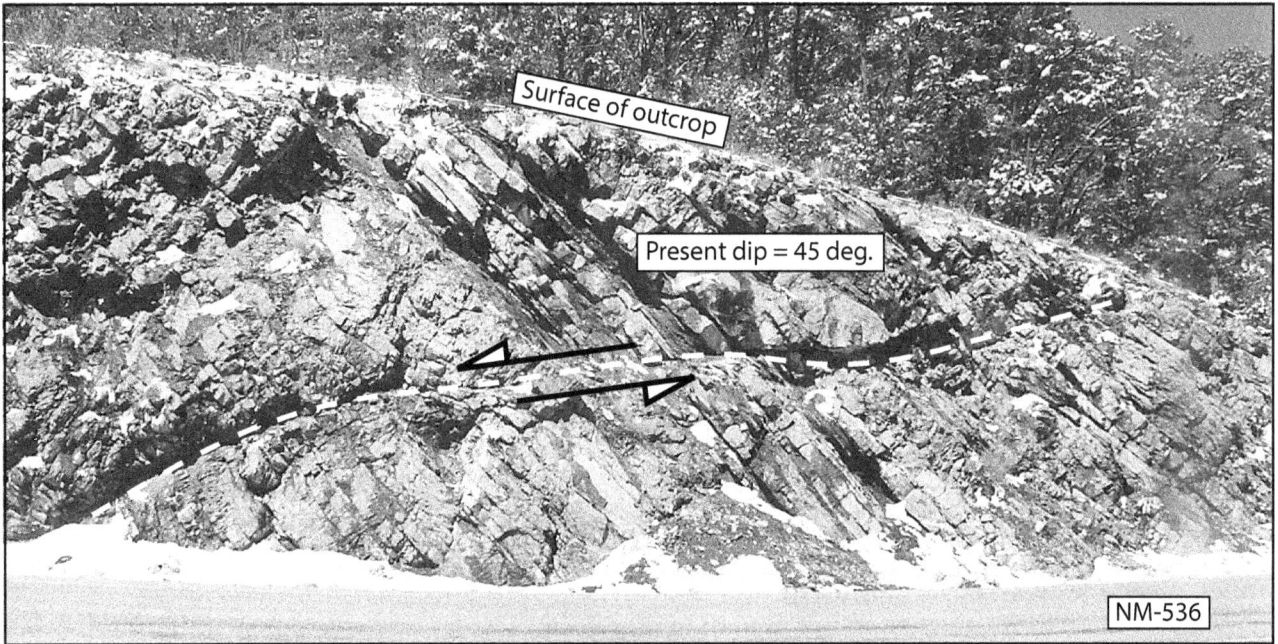

A. North View (Photograph by Author)

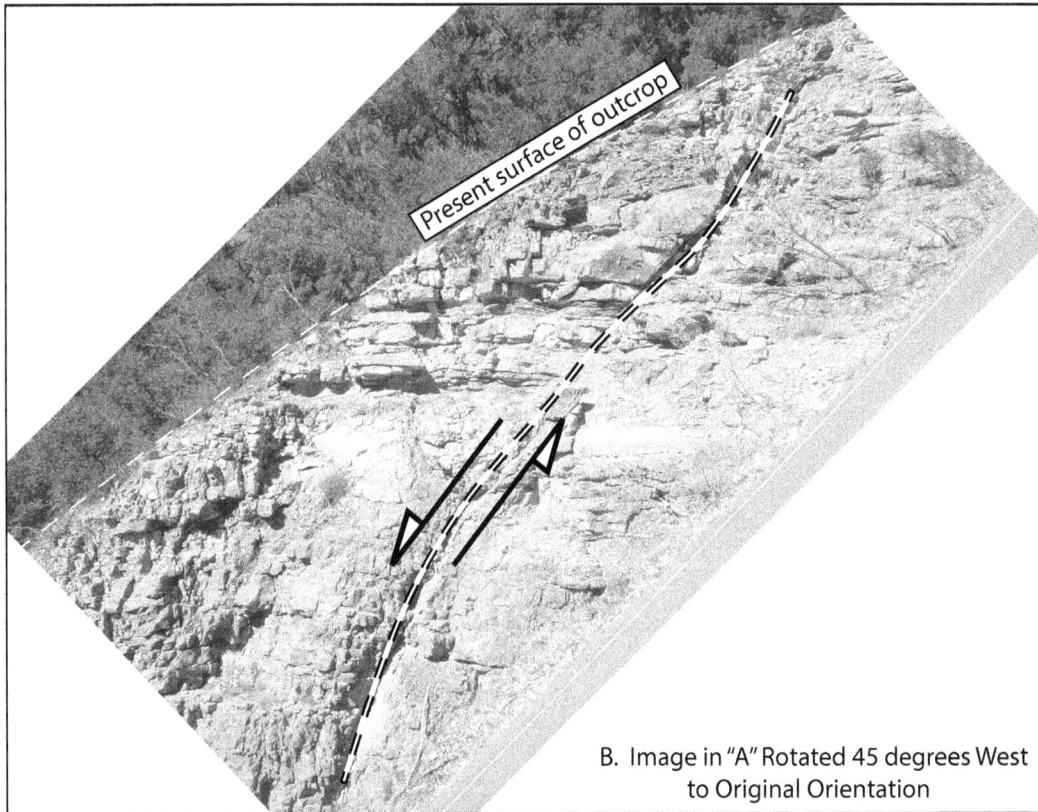

B. Image in "A" Rotated 45 degrees West to Original Orientation

Figure 12-11. Low-Angle Fault at Mile 2.9 Along Sandia Crest Byway (NM-536; See Box in *Figure 12-9*)

Tejano Canyon (Mile 4 to 5). Higher up on the mountain, extending from an elevation of 8000 to 8200 ft, is the northwest-trending *Tejano Canyon*. The Byway runs along the canyon's southwest flank (*Figure 12-12*). The canyon is unique due to a "perfect storm" of three coinciding conditions. First, the canyon occupies the southern portion of a six-mile long, complex of upthrown fault blocks in which the deep Sandia Formation and Sandia Granite have been shoved up close to the surface (*Figure 12-13*). The structural grain of the complex forced the drainage to flow down to the southeast. Tejano Creek then quickly breached the thin skin of resistive Madera limestone, encountering the soft Sandia Formation and Sandia Granite underneath and easily digging a canyon. Second, the now-exposed Sandia Formation and Sandia Granite weathered to well-drained, sandy soils. Third, the southwest-facing slope had more sun exposure and thus drier conditions—not conducive for conifer forest. The three conditions resulted in an ideal habitat for an "island" of scrubby Gambel oak that stands in sharp contrast to the surrounding "normal" conifer forest. Interestingly, the Great Unconformity that separates the Sandia Formation from the underlying Sandia Granite is unrecognizable on the slopes of the opposing, southwest-facing wall.

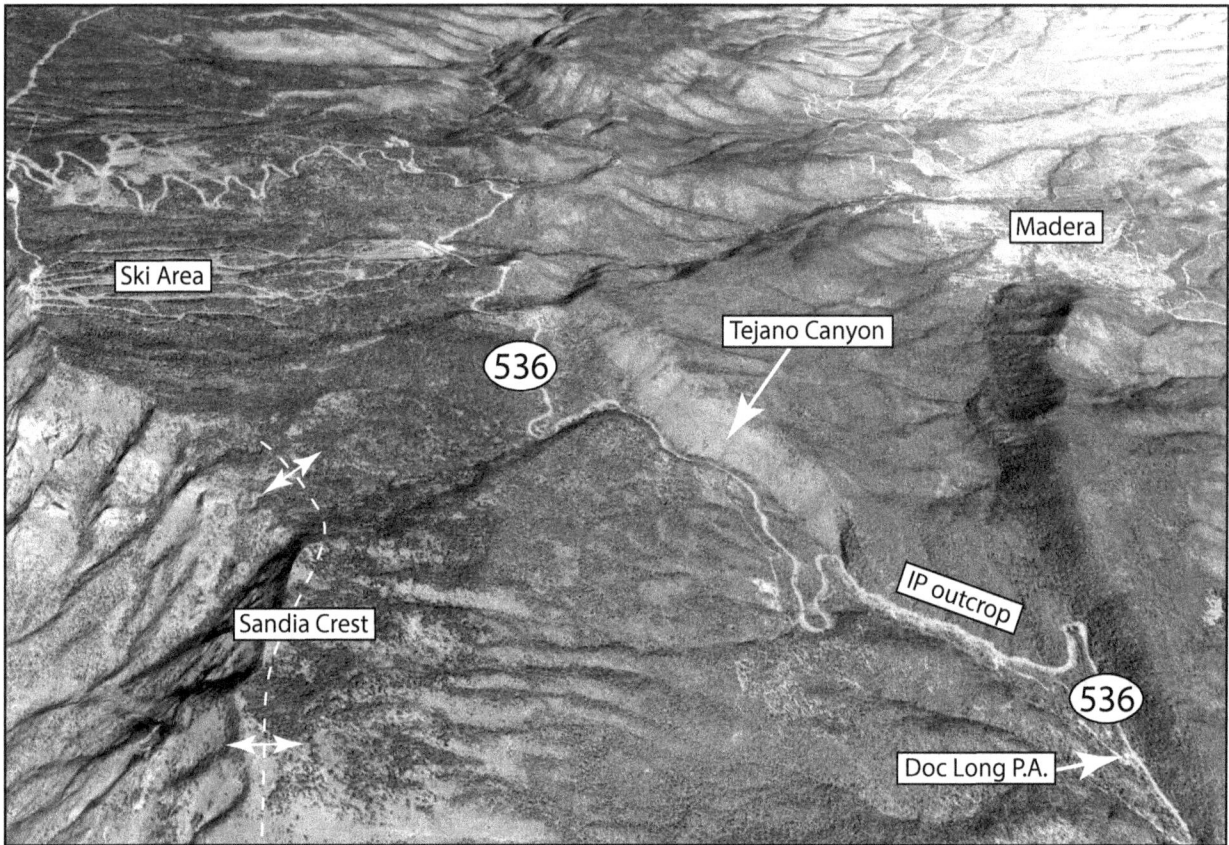

A. North Oblique *Google Earth* Image of Tejano Canyon Area

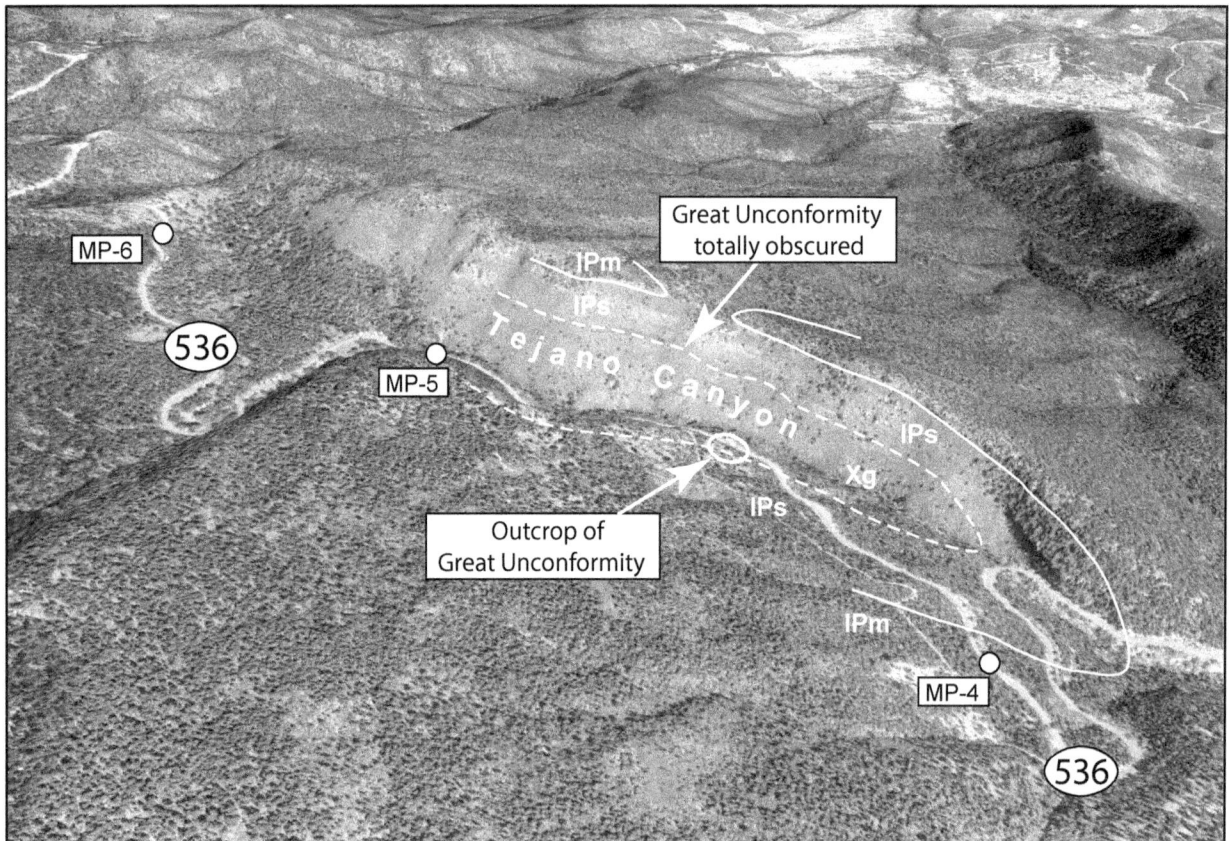

B. North-Northeast Oblique *Google Earth* Image with Partial Geology
(Xg = Sandia Granite; IPs = Sandia Fm.; IPm = Madera Grp.)

Figure 12-12. Tejano Canyon - I

1 Mile

MP-8

Ql

MP-7

IPm

Ql

IPm

536

MP-6

Ql

Sandia Crest

IPs

Xg

IPm

IPs

Xg

Xg

IPm

IPs

IPs

Xg

Complex, uplifted and deeply-eroded tectonic block

B

IPm

MP-5

Xg

A

Tejano Canyon

Ql

IPs

Ql

MP-4

IP outcrop

MP-3

IPm

IPm

Xg

P

MP-2

Legend:

Ql	Modern landslides
P	Permian
IPm	Pennsylvanian Madera Group
IPs	Pennsylvanian Sandia Fm.
Xg	Sandia Granite
⊥•	Main faults (balls on down side)

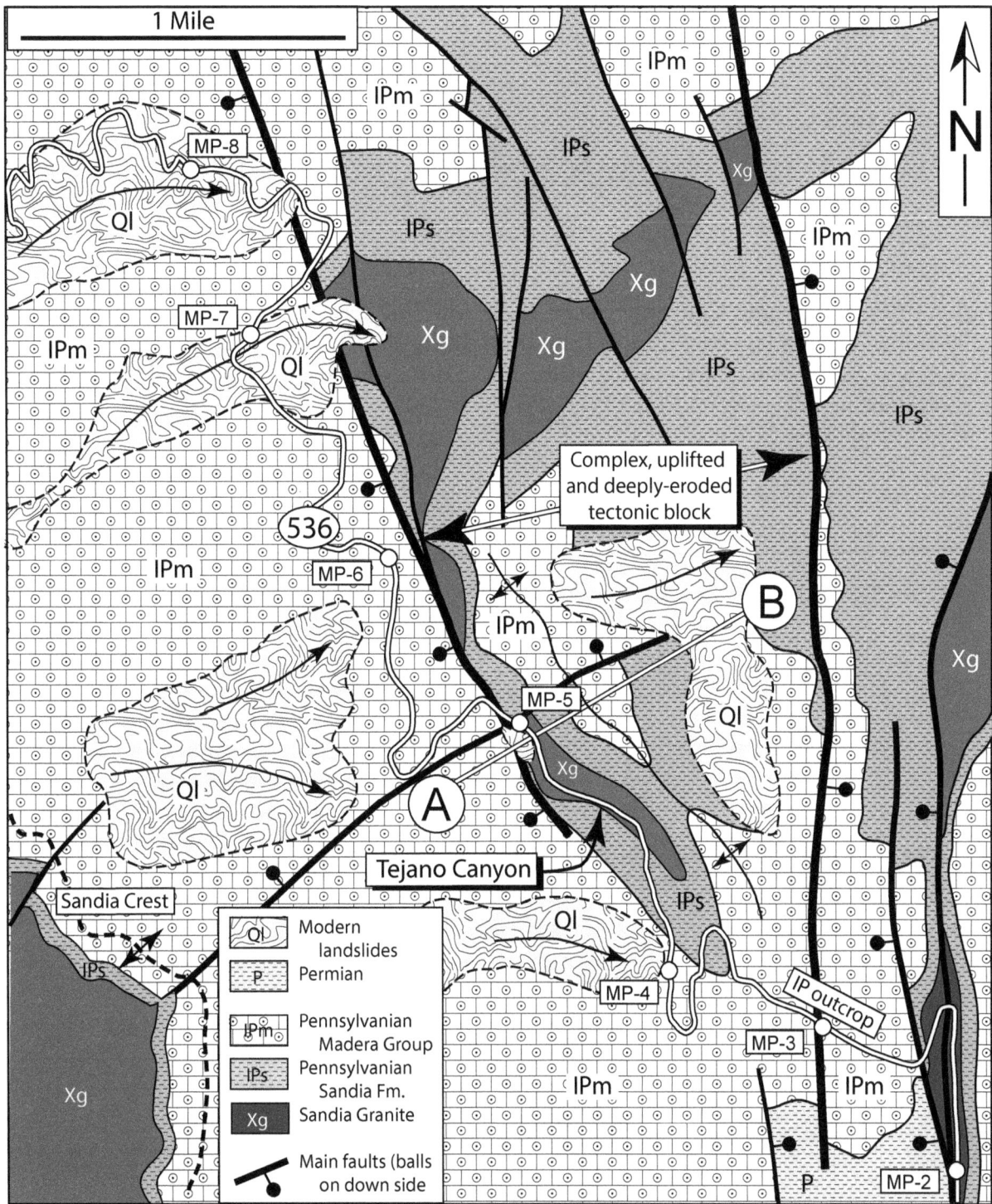

A. Geologic Map (Modified from Read et al. 1999)

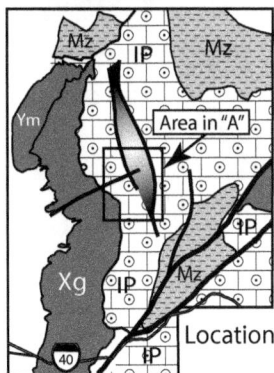

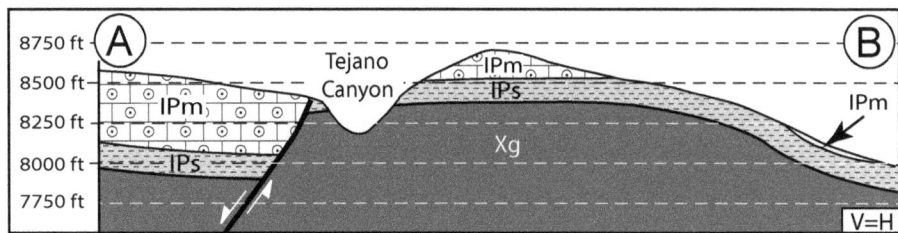

Mz

IP

Mz

Ym

Area in "A"

Xg

IP

IP

Mz

IP

Location

40

8750 ft — A

8500 ft

8250 ft — IPm

8000 ft — IPs

7750 ft

Tejano Canyon

IPm

IPs

Xg

IPm

B

IPm

V=H

B. Southwest-to-Northeast Cross Section A-B (Ql Not Shown)

Figure 12-13. Tejano Canyon - II

345

GU Site 2 (Mile 4.4, Tejano Canyon). Thanks to the US Forest Service and to its highway engineers, a nice, 330-ft-long outcrop lays bare the GU (*Figure 12-14A*). The 315 Ma Sandia Formation sandstone lies directly upon a very uneven paleo-surface of the 1.45 Ga Sandia Granite. Displayed at the contact is an ancient corestone field (*Figure 12-14B*).

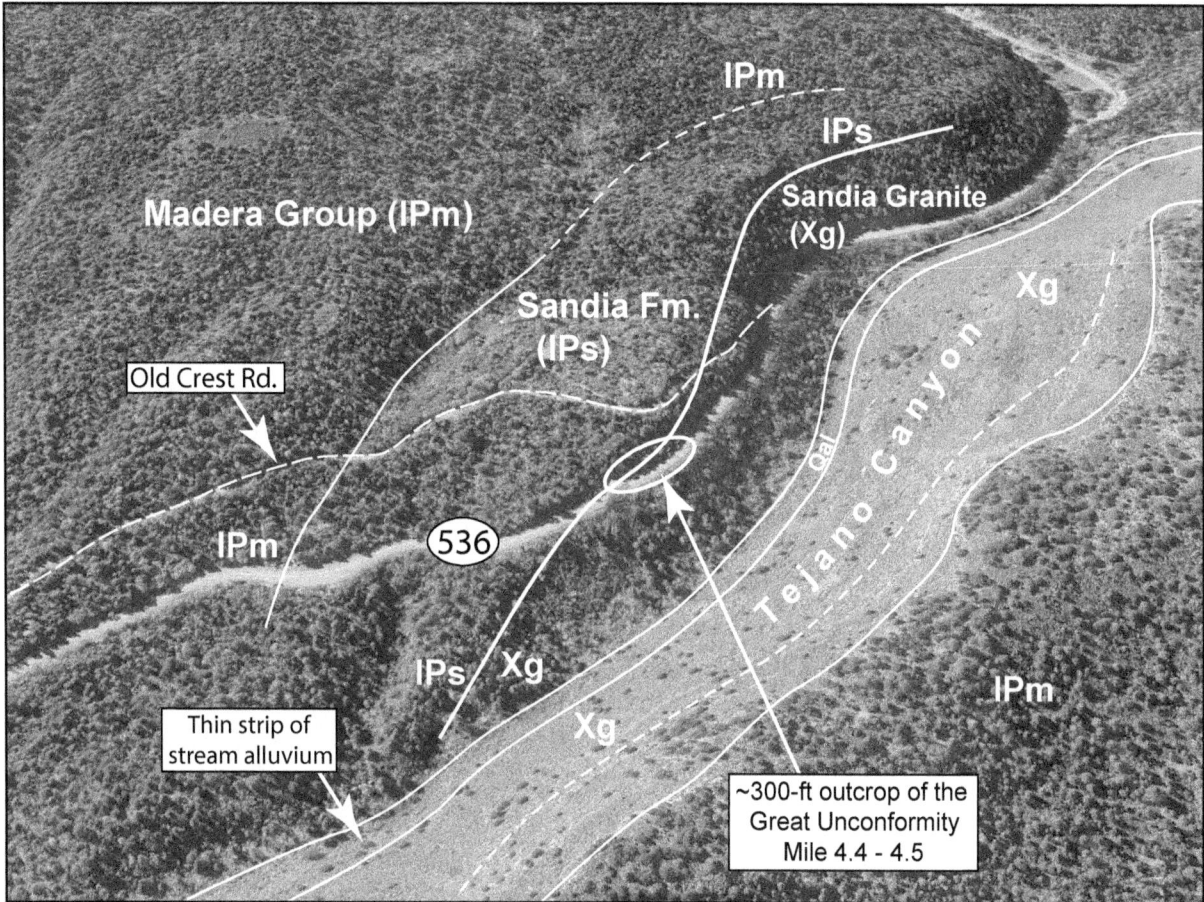

A. West Oblique *Google Earth* Image

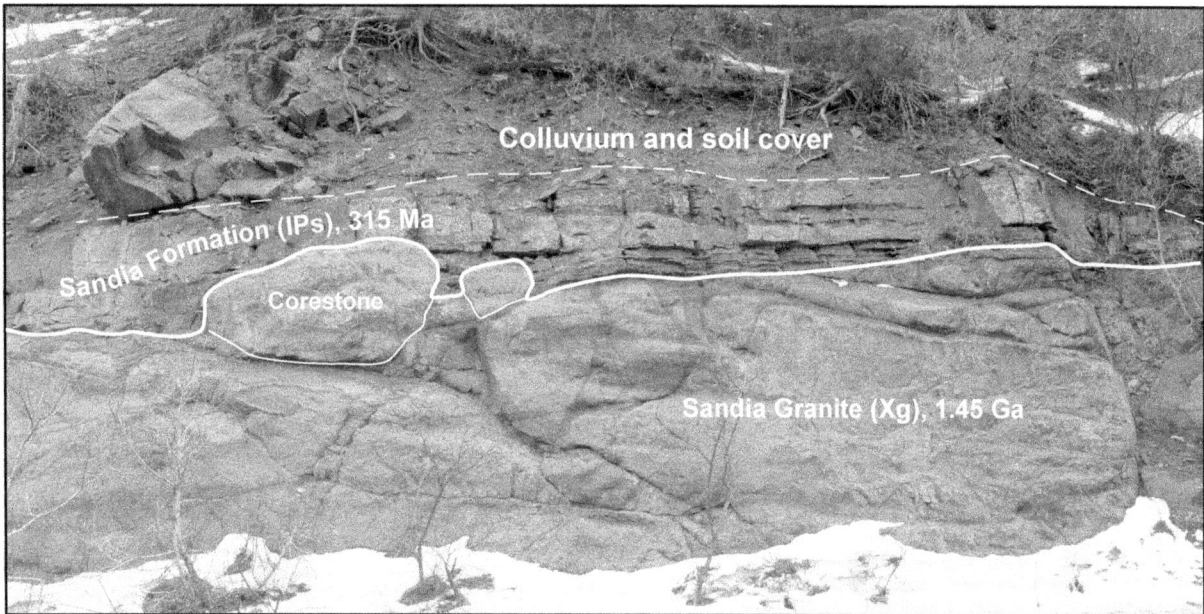

B. North View (Vegetation for Scale; Photograph by Author)

Figure 12-14. The Great Unconformity Exposed Along Sandia Crest Byway
(NM-536, Mile 4.4 - 4.5, Tejano Canyon)

"Tejano" and William H. "Doc" Long. These two names deserve attention because they are so specific. The name *Tejano* is here connected to a geologic feature and thus warrants a few words. "Doc" Long was a plant pathologist who did pioneering work for the Forest Service from about 1913 to the 1930s. He took up residence with his family in a cabin located at today's Doc Long Picnic Area. Long was from Texas, and although "Tejano" usually implies a Hispanic Texan, the name probably came from him (Holben 2021).

Sandia Ski-Area Gravity Glide (Mile 7). The base of the Ski Area along the Byway is located at about the midpoint in an interesting feature not at all obvious to the casual motorist. A mass of rock of the Madera Group, measuring about 4000-ft-long (uphill-downhill) by 2000-ft wide, has slipped downslope via gravity as a huge glide mass parallel with the underlying formation dip (*Figure 12-15A*; Kelley and Northrop 1975). The upper part of the ski-area's beginners' area is the pull-away point of the glide, and the lodge and beginners' area are on the glide's flat top. Some weak zone within the Madera, probably a shaley unit, provides the gliding surface. There are several more of these detached masses on the mountain's east side, but this is the most accessible. They clearly demonstrate the surprising weakness of rocks—especially layered, sedimentary rocks—in the lateral direction. Perhaps in future wetter times the Ski Area glide will begin to move again and give the snowboarders an extra thrill. For a peek at what a mess a gravity glide can be, stop at the roadcut just past (north of) the ski area at MP-7 (*Figure 12-15B*). Here we see the diffuse edge of the glide mass, with chaotic glide rocks to the south separated from the more coherent, sub-glide rock to the north.

A. Northeast Oblique *Google Earth* Image of Sandia Crest, Sandia Crest Byway (NM-536), and Ellis Road

B. Northwest View of Sandia Crest from La Luz Trail Near Upper Tram Terminal
(Xg = Sandia Granite; IPs = Pennsylvanian Sandia Fm.; IPm = Pennsylvanian Madera Grp.)

Figure 12-16. Sandia Crest

Calcite (CaCO$_3$). The mineral *calcite* is the principal ingredient of almost all limestone. Some limestone, especially the very finely crystalline variety, is often difficult to identify, but the mineral calcite has a fortunate chemical property: it reacts vigorously with dilute hydrochloric acid (HCl), foaming with bubbles filled with CO$_2$. A small vial of HCl is therefore always part of the field geologist's field kit. (The chemical reaction is: CaCO$_3$ + 2 HCl → CO$_2$ + H$_2$O + Ca^{++} + 2 Cl$^-$.) In industry, CaCO$_3$ is broken down by heat to form CaO (quicklime) + CO$_2$. The quicklime leads to many other products such as cement, mortar, stucco, etc. (These are the only chemical formulas in this book!)

Fossils. In a few places along the crest, especially near the Upper Tram Terminal and upper part of the La Luz Trail that ends there, the Madera limestone ledges display fossils of shallow marine life. Perhaps fortunately, though, the fossils on the Crest present "slim pickings"

(NOTE: Collection of fossils in the Sandia Mountain Wilderness area (which includes the Crest) is strictly forbidden.)

After death and before burial, the remains of these critters were swept about by currents on the sea floor, were disaggregated and generally ground up to form a fossil hash. Among the hash, though, two types can sparingly be found. These are crinoids and horn corals (*Figure 12-17*).

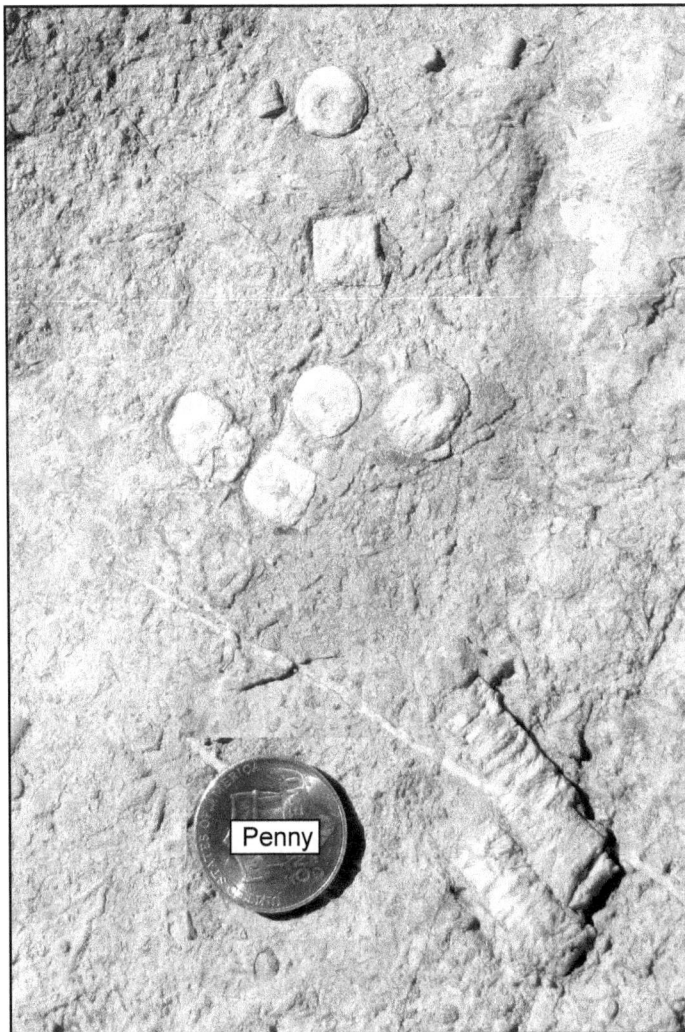

A. Crinoid "Hash" (Photograph by Author)

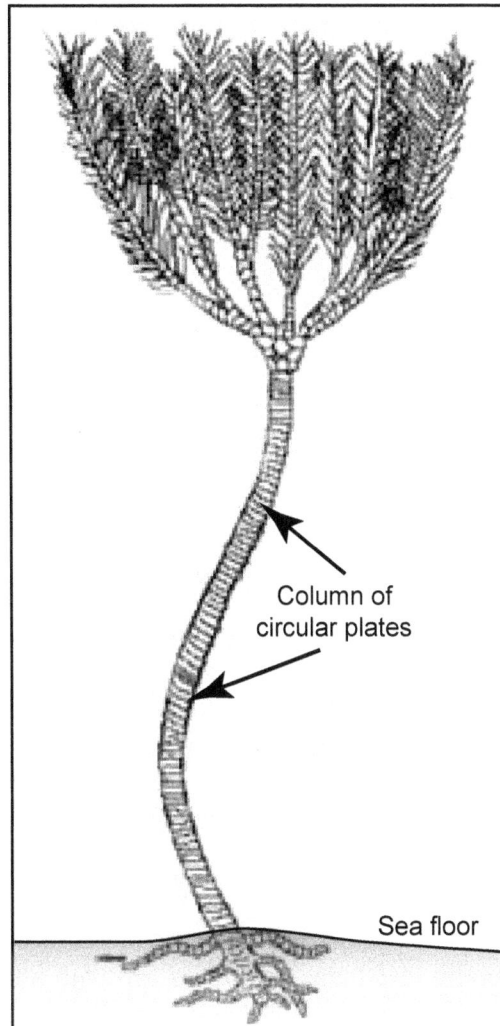

B. Drawing of Intact Crinoid (From Internet)

Column of circular plates

Sea floor

C. Horn Corals (*Caninia*; Photograph by Author)

D. Drawing of Horn Coral *Caninia* (From Easton 1960)

Figure 12-17. Fossils from Madera Group Limestone on Sandia Crest

Crinoids. (*Figures 12-17A* and *12-17-B*). The name "crinoid" comes from the Greek *krinon*, or "the lily." These are among the first fossils ever noticed by mankind. In ancient times they were known as "star stones" (*asterias*). It was even thought that they were awful things caused by the force of starlight, or that they were some sort of evil fermentation in the Earth. Different species of crinoids still live today and flourish in today's seas, mostly in deep water.

It is important to remember that crinoids are the remains of marine creatures, i.e., critters that lived on successive sea floors—today's bedding surfaces. The crinoids were like flowers, "sea lilies," attached to the sea bottom. Each was made up of a long stem-like column and an upper cup with arms (*Figure 12-17B*), all composed of little aggregated plates of calcium carbonate ($CaCO_3$). At the cup and arms were soft feeding organs that filtered food from the moving seawater. In order to prevent the organs becoming clogged up with grit, the creatures could only live a significant distance from any landmass to ensure that the ocean waters were clear of any smothering mud.

At death some crinoids were preserved intact, but usually the plates were detached and scattered about by currents, becoming just so much debris on the sea floor. The plates of the stems are little flat circular disks, usually less than 1-cm in width, with a hole in the middle. The Madera limestones are replete with these scattered little donuts, occasionally with several disks still linked together (*Figure 12-17A*). In pre-historic times these little disks were gathered to make bracelets or necklaces (Kelley and Northrop 1975).

Horn corals. (*Figures 12-17C* and *12-17D*). This second type is far less common than the crinoids. These were horn-shaped, solitary corals that lived attached to the sea floor by their pointed ends. (For those interested, the examples shown are of the genus *Caninia*.)

Chert. The Madera limestone ledges display abundant light-colored lenses and nodules of a material called *chert* (*Figure 12-18A*). Chert is a very hard, microcrystalline silica (SiO_2), a variety of the common mineral quartz. The rock is so hard that it fractures along a curved ("conchoidal") surface, producing sharp edges. This convenient and wonderful property is the reason why this particular rock became a favorite cutting tool during the Stone Age. However, due to its poor quality here, the Sandia Crest stuff was apparently not used as a source for tools.

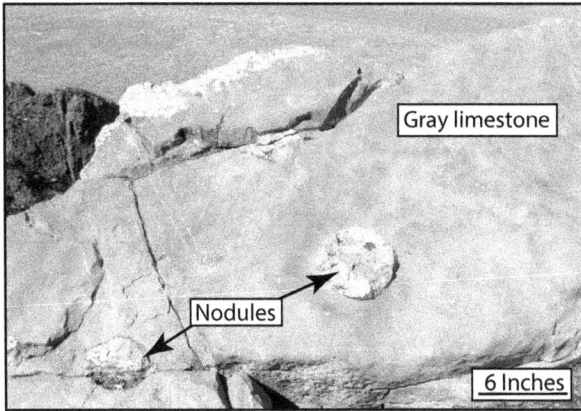

A. Circular Chert Nodules on Bedding Plane

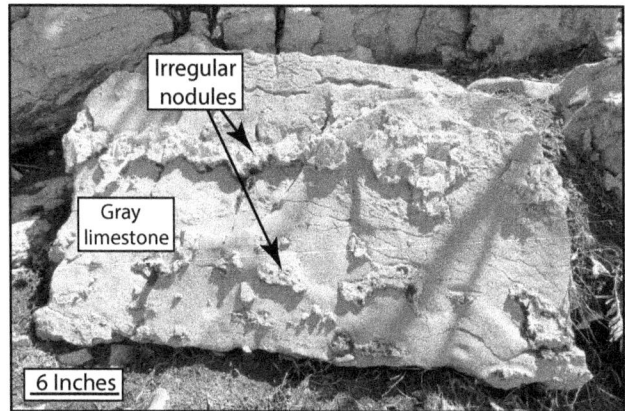

B. Irregular Chert Nodules in Cross Section

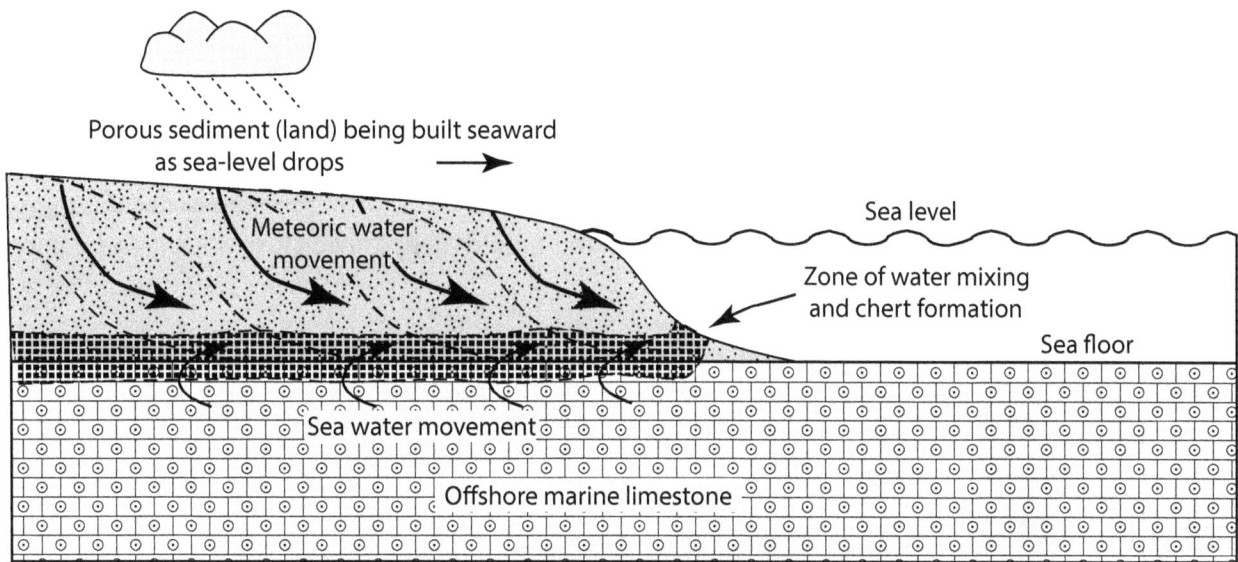

C. Schematic Cross Section Showing Origin of Chert (Modified from Ervin 1986)

Figure 12-18. Chert in Madara Group Limestone on Sandia Crest
(Photographs by Author)

Varieties. The names used for varieties of microcrystalline quartz can be quite confusing. Geologists tend to call such occurrences "chert," while archeologists and historians (especially if the material has been used for artifacts or tools) call them "flint." If the variety is opaque and red, and therefore of gem quality, it is called "jasper."

Origin. Chert in the Madera was likely derived from dissolution of silica-spicule-bearing organisms such as sponges that also lived in the shallow seas, but closer to land. In such a transitional zone, meteoric (rain-derived) waters filtered down and mixed with marine waters, creating a unique chemistry in the mixing zone (*Figure 12-18B)*. There the spicules dissolved, the silica was transported in solution through the sediment and precipitated out near the sea floor, often replacing decaying organic matter already there. The newly-precipitated silica became buried, dehydrated (lost water), and contracted to form nodules (Ervin 1986).

Kiwanis Cabin. The stone shelter on Sandia Crest, located on a promontory halfway between the Crest House and the Upper Tram Terminal, is the iconic landmark and hiking destination for anyone feeling footloose (*Figure 12-19*).

A. Original Cabin, Burned Down in Late 1920s (Kiwanis Club 1929)

B. CCC Hauling Rock on Crest Road, 1936 (CCC 1936)

C. CCC Building Cabin, 1936 (CCC 1936)

D. Modern View (Photograph by Author)

Figure 12-19. Kiwanis Cabin on Sandia Crest

Kiwanis International Club. The organization was formed in 1915 in Detroit, Michigan. The name is an American *Ojibwe* Indian word meaning something like "to make oneself known," or as the club preferred, "we build." The club's volunteers have the lofty goal, "to improve the world."

In 1928 the Kiwanis Club built the first cabin on the crest. The cabin burned down shortly after that (*Figure 12-19A*). The second one was a wooden structure with a stone fireplace, a concrete floor, and two windows. It was damaged by storms and partly collapsed ca. 1929 or 1930 (Benedict 1994b). In 1934 the club relinquished the site to the U.S. Forest Service (Kiwanis Club, 1934) and the next year the wind blew the roof off.

In 1936, in conjunction with the Forest Service, the Civilian Conservation Corps (CCC) constructed the present cabin. Blocks of Madera limestone, pried loose via two-man crowbars, were hauled to the site up the Sandia Crest Byway by flat-bed truck (*Figure 12-19B*). Construction took about a month (*Figure 12-19C*). When completed the cabin had a slab-concrete roof with an asphalt cover, a heavy wooden door at the east entrance, and glassed, metal-framed windows. Soon afterwards the structure began to suffer greatly from storms and—of course—vandalism. Today the cabin lacks its door and windows, but otherwise is in reasonably good condition (*Figure 12-19D*; Benedict 1994b). It is a "must see" when strolling along the crest.

13:
ORTIZ IGNEOUS BELT

Introduction

The Ortiz Igneous Belt lies a short distance east of Sandia Mountain, but with its own distinctive features and history. I have chosen to include just a select few of the area's highlights for this book. From north to south these are 1) the Cerrillos area, 2) Madrid area, and 3) a pair of igneous dikes farther to the east.

Cerrillos Area

The graveled Waldo Canyon Road (CR-57) is an old stage-coach path that runs west from the village of Cerrillos on its way to today's I-25 (*Figure 13-1*). A half mile west of the village the road passes along the southwest edge of the main Cerrillos intrusion where there is a striking feature called the *Devil's Throne* (*Figures 13-1* and *13-2A*). This post-card image begs the usual questions: What is it and how did it get here? A cross section reveals that it is simply part of the deeply-eroded, steeply-dipping flank of the main intrusion (*Figure 13-2B*).

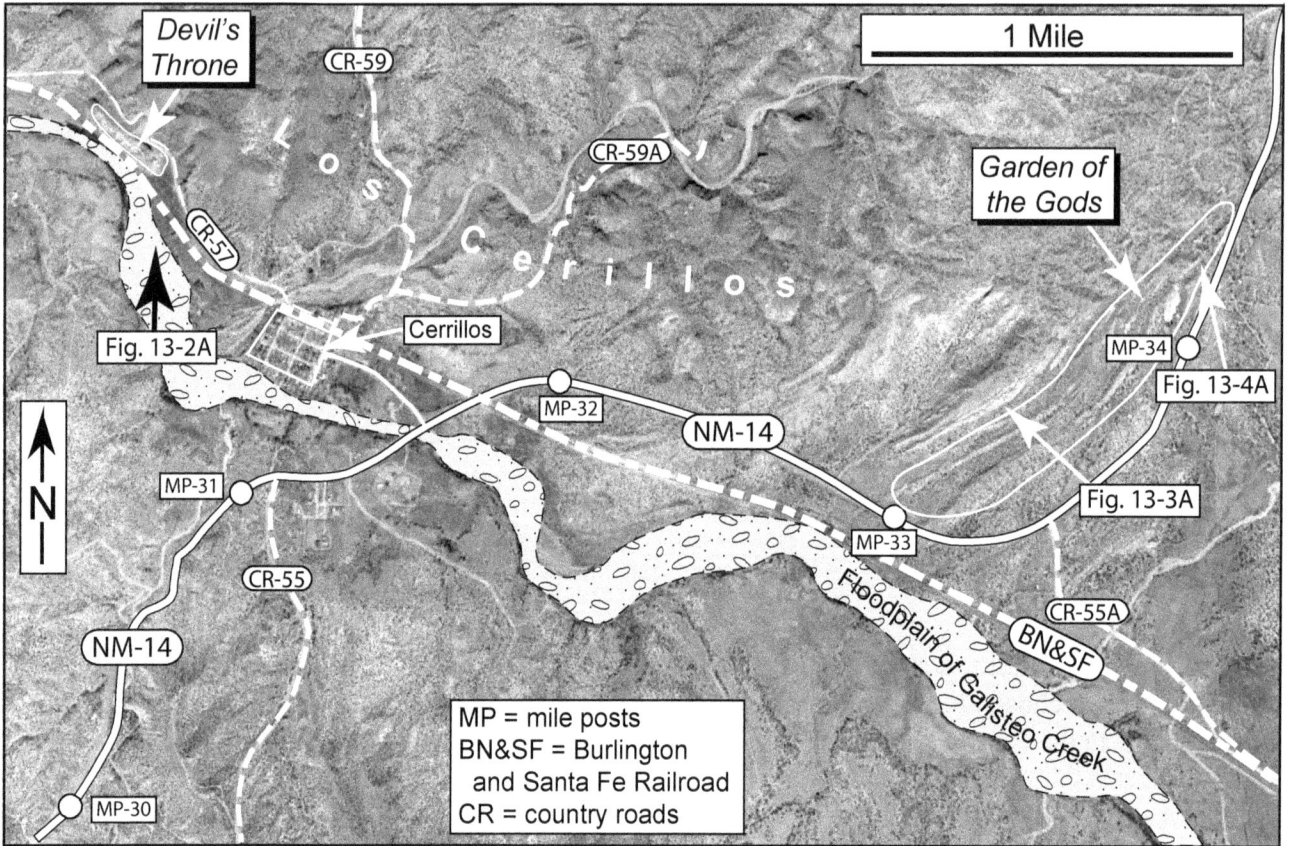

Figure 13-1. *Google Earth* Image of Cerrillos Area

A. North View (One-Half Mile West of Cerrillos; Photograph from Internet)

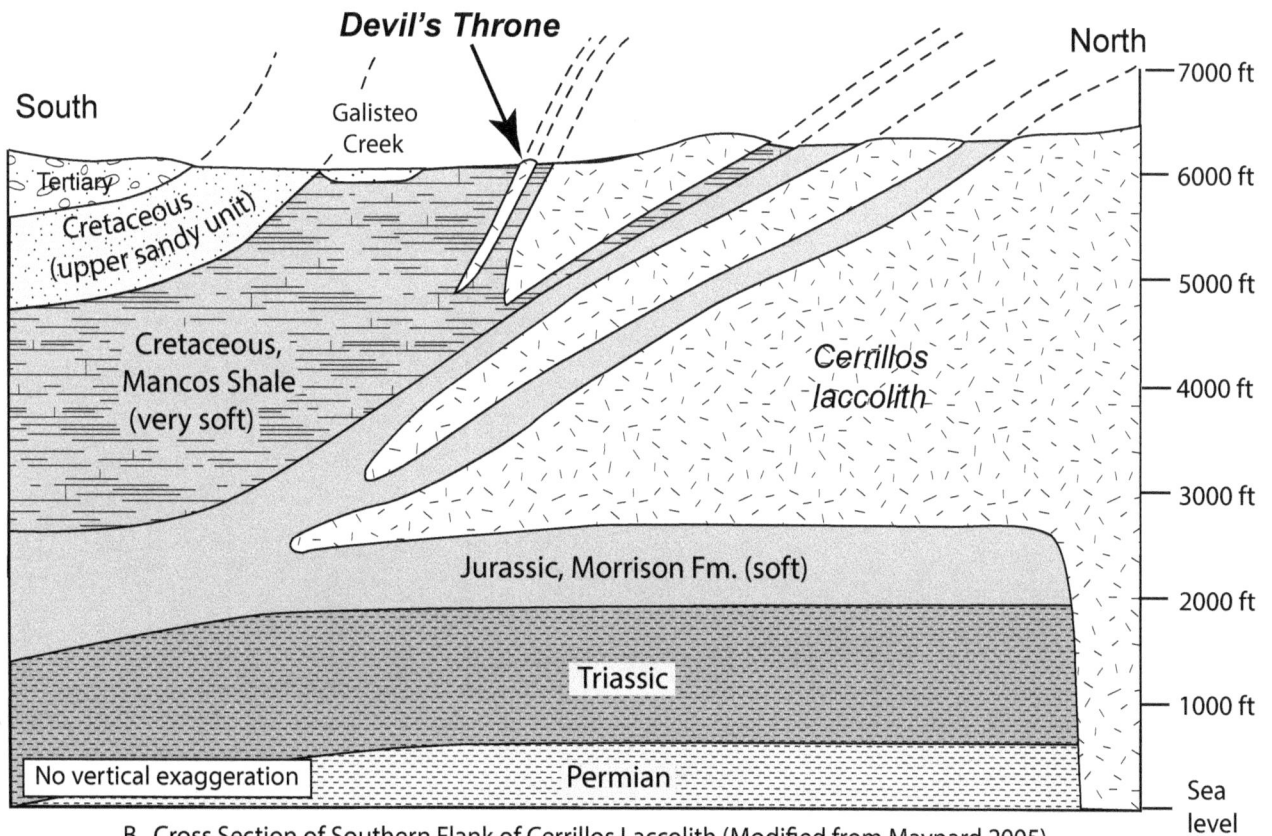

B. Cross Section of Southern Flank of Cerrillos Laccolith (Modified from Maynard 2005)

Figure 13-2. Devil's Throne, Cerrillos Area

Two miles east of Cerrillos, on the opposite side of the intrusion off to the north of NM-14 at mile 33.5 is a fine example of an angular unconformity (*Figures 13-1* and *13-3A*). Although not nearly as eye-catching as the classic example in Scotland (*Figure 13-3B*) and with a modest time hiatus of only 35 Ma between the opposing formations, such features are still always photogenic.

A. Northwest View of Ancha Formation of Santa Fe Group Overlying Tilted Galisteo Formation
(NM-14, Mile 33.5; Photograph by Author)

B. East View of World's Most Famous Angular Unconformity: Siccar Point, Scotland (Photograph from Internet)

Figure 13-3. Angular Unconformities (Dashed Lines)

Three miles east of Cerrillos, on NM-14 at mile 34.3, is an example of very steep dipping Paleogene ("tombstone topography") in an area dubbed the *Little Garden of the Gods* (*Figures 13-1* and *13-4A*). This is an area of private homes but can be viewed from the highway via a small pullout. It seems the person who named this place had a limited repertoire because this generic name has been somewhat overused, such as at 1) *Garden of the Gods* in Colorado Springs, Colorado, 2) *Valley of the Gods* in southeast Utah, and 3) *Garden of the Gods Recreation Area* in the Shawnee National Forest, southern Illinois.

A. North-Northeast Aerial View (1.5 Miles East of Cerrillos (NM-14, Mile 34.3; Photograph from Internet)

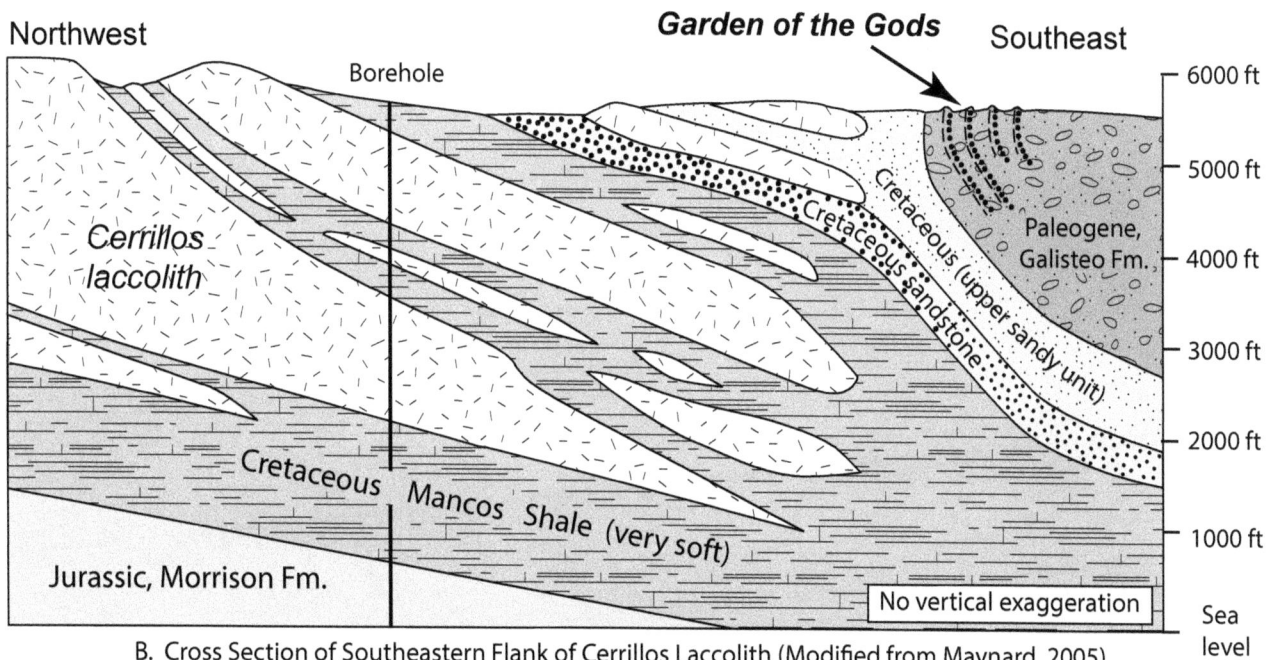

B. Cross Section of Southeastern Flank of Cerrillos Laccolith (Modified from Maynard, 2005)

Figure 13-4. Garden of the Gods, Cerrillos Area

Madrid Area

Geologically this place has a split personality because it's both part of the Ortiz Igneous Belt and it involves Cretaceous-age coal. Accordingly, I'll deal with the intrusions here and the Cretaceous coal later in *Chapter 14*. The name of the little town of Madrid, on the north flank of the Ortiz Mountains, is pronounced *MAH-drid*, in contrast to the Spanish capital in Europe where it's pronounced *Ma-DRID*.

Two igneous intrusions here are sub-horizontal sills that merge into one below the surface to the east (*Figure 13-5A*). By geological serendipity the two sills encase a bed of Cretaceous-age coal.

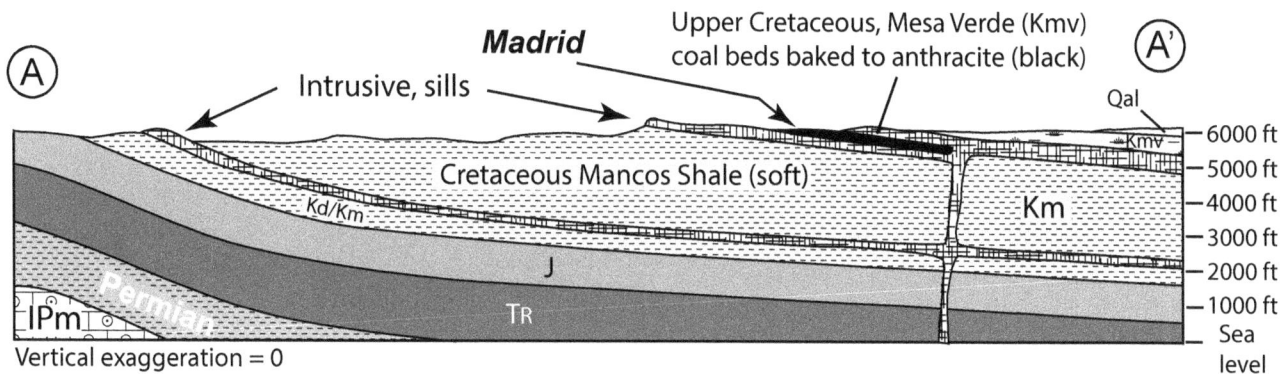

A. West-to-East Cross Section Across Madrid Intrusive Sills and Coal Basin

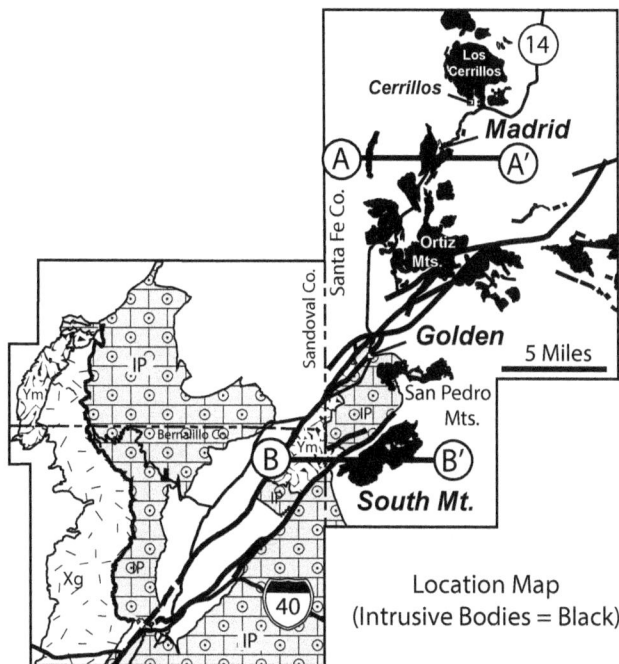

Cross Section Legend

Qal	Quaternary, stream alluvium
	Tertiary, igneous intrusions
Kmv	Cretaceous, Mesa Verde Fm. (coal-bearing)
Km	Cretaceous, Mancos Shale (soft, weak unit)
Kd	Cretaceous, Dakota Sandstone
J	Jurassic formations
TR	Triassic formations
Py	Permian, Yeso Formation, Meseta Blanca Sandstone (hard, resistive unit)
Pa	Permian, Abo Formation
IPm	Pennsylvanian, Madera Group
IPs	Pennsylvanian, Sandia Fm.
Ym	Basement

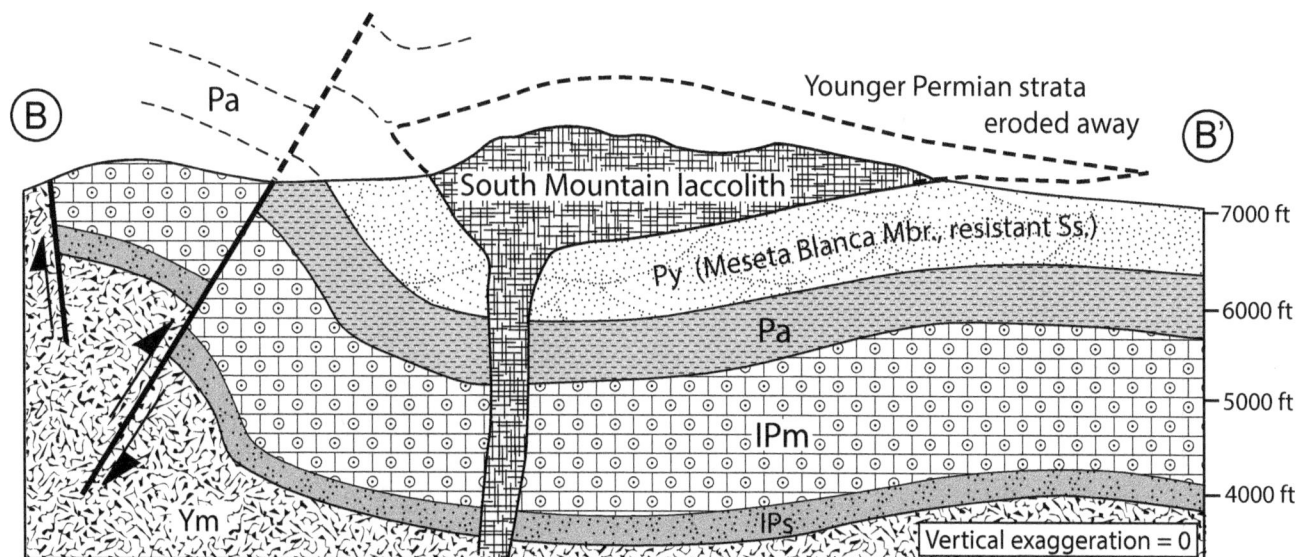

Location Map
(Intrusive Bodies = Black)

B. West-to-East Cross Section Across South Mountain Intrusion

Figure 13-5. Madrid and South Mountain
(Note Different Scales; Modified from Maynard 2005)

367

Unlike its neighboring mining towns to the north and south, Madrid was a coal vs a hard-rock mining place. It developed after the railroad was run through Cerrillos in 1880 and a spur built from Waldo (west of Cerrillos) south to Madrid where the coal was located. The growing town received a shot in the arm when the small coal mine far to the south at Carthage, southeast of Socorro, was shut down in about 1890. All the Carthage structures (miners' houses, shops, etc.) were removed and shipped *en masse* to Madrid (Sherman and Sherman 1975). Madrid's heyday was from about 1920 to 1950. After WWII, diesel began to replace coal as locomotive fuel, mine production slowed down, and the people began to drift away. The mine was shut down in 1954 and the town's owner unsuccessfully attempted to sell it as a unit. By 1960 only two families lived in Madrid. Finally, in 1975 the owner threw his hands up and put the 150 remaining buildings up for sale individually. Young people scarfed them up in 16 days and Madrid's renaissance had begun (Melzer 1976). Despite this interesting background, geologically there is little to see.

A Tale of Two Dikes: Galisteo and *El Crestón*

Two long dikes are nicely exposed east of the Ortiz Mountains along NM-41 (*Figure 13-6A*). Until geologically quite recently (< 800 ka) both dikes had been completely covered by a thin sheet of Quaternary alluvium, but that sheet has been partly stripped away back to the south, exposing the dikes. The first is the Galisteo dike, located about a mile north of the village of Galisteo (*Figure 13-6A*), and NM-41 passes through a gap within it.

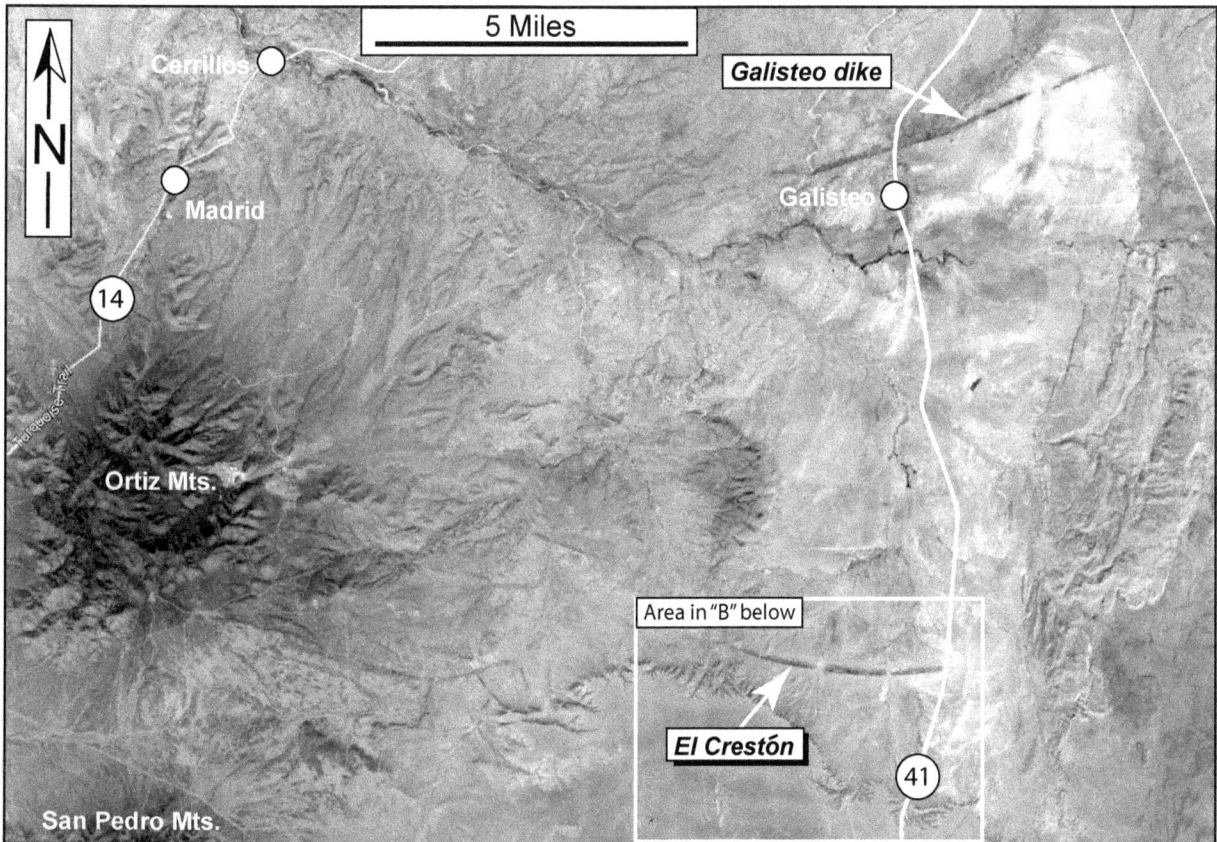

A. *Google Earth* Image of Two Dikes

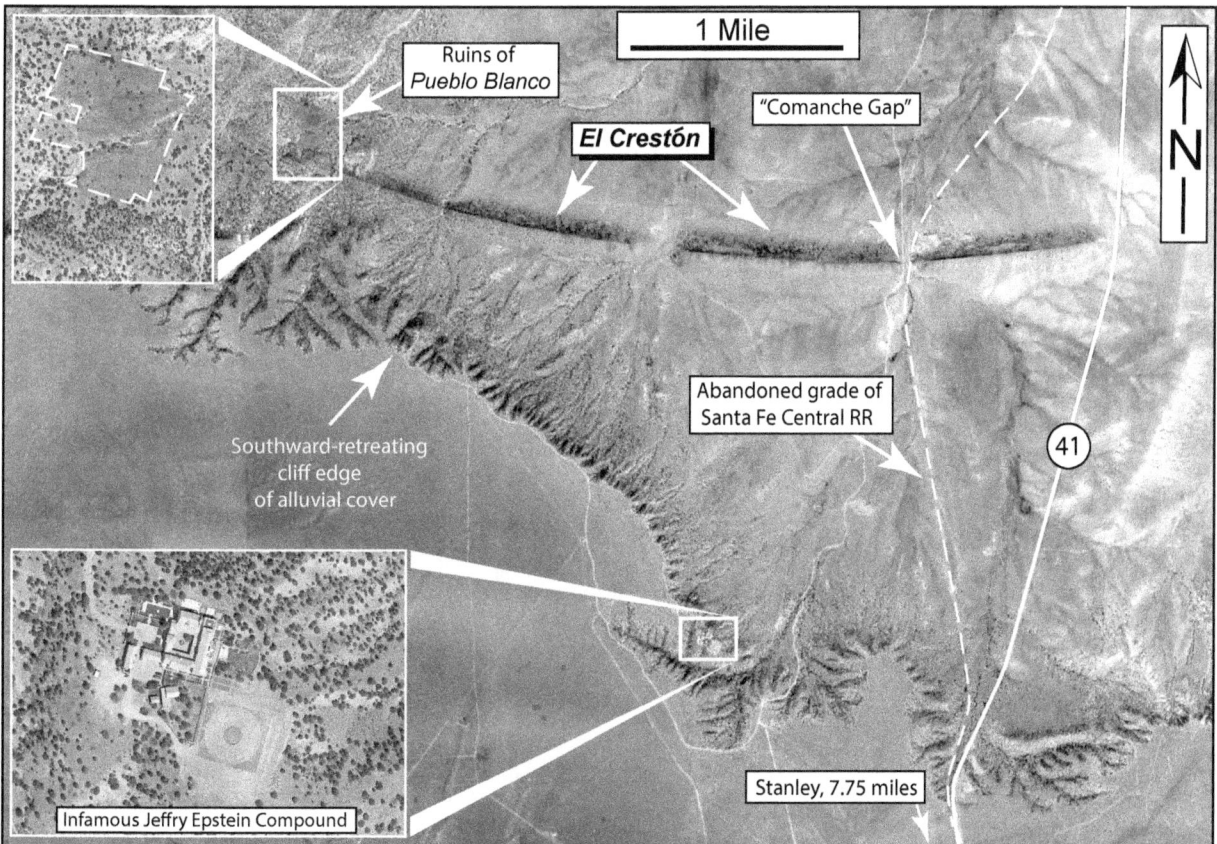

B. *Google Earth* Image of *El Creston*

Figure 13-6. The Eastern Igneous Dikes
(Location in *Figure 7-15*)

369

The second dike, *El Crestón*, is located about ten miles north of the little community of Stanley and lies entirely west of NM-41 (*Figure 13-6B*). Unfortunately (or perhaps fortunately), the dike is on private land that is posted. About a mile west of NM-41 is a break in the dike called Comanche Gap. It was the focus of a historic lithograph by John S. Newberry (1822-1872), a prominent geologist who accompanied an early expedition to the Santa Fe area (Gregg 1968; *Figure 13-7A*). The view is little changed today (*Figure 13-7B*). The gap was later exploited as a convenient shortcut for the Santa Fe Central Railroad in 1903-1929 (Myrick 1970).

Figure 13-8. Petroglyphs on *El Creston*
(Photograph by Author)

Epstein compound. About 1.75 miles south of *El Crestón*, 1.3 miles west of NM-41 and up on the north rim of the north-facing alluvial escarpment, is the infamous compound of financier Jeffry Epstein (lower inset in *Figure 13-6B*). This 26,700 ft^2 house is the crown jewel of his *Zorro Ranch*, a 7500-acre property purchased in 1993 from the family of former New Mexico Governor Bruce King and later expanded to 10,000 acres. The house is reportedly the largest in New Mexico and appraised for $18 million (Kim 2020). One wonders if his guests ever took the time to admire the view north out to *El Crestón*. Epstein died in prison, a suicide, in 2019.

14:
THE CRETACEOUS REALM

Introduction

The Cretaceous is the final time period of the Mesozoic Era. In North America rocks of this age are renowned for their commercial deposits of oil, natural gas, and—particularly—coal. The quest for these deposits in central New Mexico provides the grist for several stories.

? **Symbol for Cretaceous**. On geologic maps and figures the Cretaceous Period is symbolized by the letter "K." Why not the letter "C?" There is a historical reason. Cretaceous rocks are prevalent in the Paris basin of France and, most spectacularly, in the White Cliffs of Dover southeast of London along the North Sea. The cliffs are composed of *chalk*, a very fine-grained, pure form of limestone. The German word for chalk is *kreide*. Therefore, because the letter "C" had been preempted by the Cambrian Period, geologists settled on the letter "K" to avoid confusion.

Cretaceous Coal

All of New Mexico's coal is housed in formations of Late Cretaceous age. The coal-bearing sequence was formed by the accumulation of dead plant matter in ancient swamps, mainly from about 85 to 70 Ma. During that time a chain of mountains extended from southern New Mexico northwest into western Utah (*Figure 3-12*). North and east-flowing rivers draining these highlands transported their loads of sand, silt and clay that terminated in deltas built out into a shallow interior sea to the northeast. Behind the fringe of deltas was a vast expanse of swampy terrain that hosted dense stands of trees, which upon death became buried and encased within the boggy sediments.

Near the end of Cretaceous time, about 70 Ma, the mountains had been worn down and the vast interior sea filled in, never to return to western North America. Later, during the Laramide mountain-building event about 55 to 40 Ma, this Cretaceous rock sequence was buckled both up and down. Where flexed up it was eroded completely away, and where flexed down it was protected from erosion and thus preserved. The buckling and the uneven distribution of the coal deposits combined to leave them spread out into separate, small occurrences such as Hagan, Madrid and Tijeras (*Figure 14-1*).

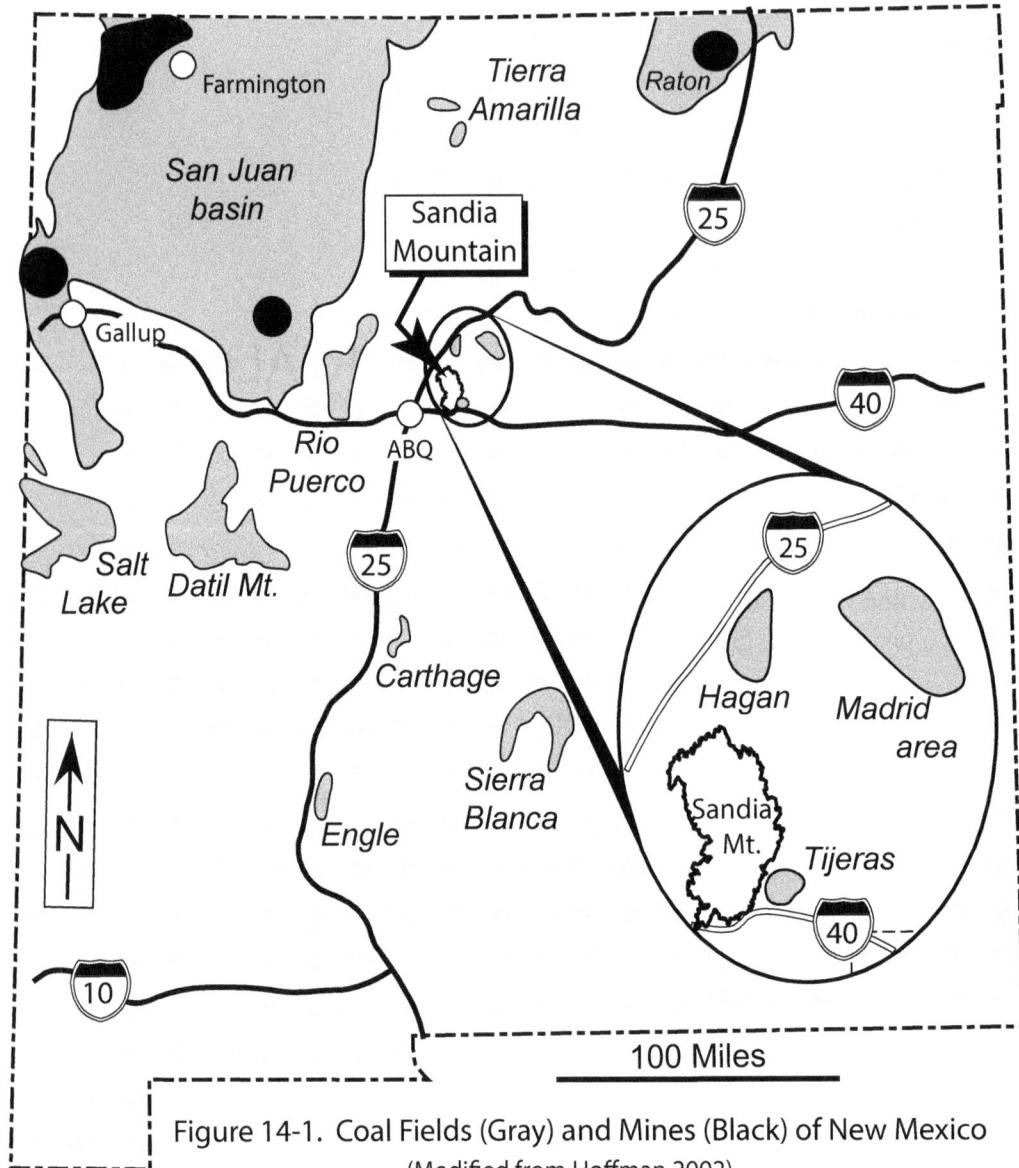

Figure 14-1. Coal Fields (Gray) and Mines (Black) of New Mexico
(Modified from Hoffman 2002)

Northwestern New Mexico hosts the most important of the preserved coal deposits. There the rock layers assume the shape of a soup bowl setting on a table. Geologists refer to this bowl-shaped feature as the *San Juan basin*. At the southwestern rim of the bowl the coal-bearing layers outcrop at the surface. More toward the center of the bowl those coal layers lie buried beneath younger rock layers. There the coal is deeply buried and therefore inaccessible to mining, and the seams have been "cooked" to generate much of the basin's natural gas.

Hagan Basin

The so-called Hagan basin, forming the northeastern flank of Sandia Mountain, is not quite a basin, *a la* the San Juan, but rather the western and southwestern rims of a half basin (*Figures 14-2* and *14-3*). Here the Cretaceous rocks, with their coal beds, have been exposed by removal of the alluvial cover that once blanketed the entire area (*Figure 7-15*).

376

About 3.6 miles west of NM-41 on the private land is the ruin of *Pueblo Blanco* (upper inset in *Figure 13-6B*). It was occupied from the 1300s to about the mid-1500s. Its inhabitants satisfied their creative urges by carving petroglyphs on the south side of the dike facing the sun. Some of these are excellent and quite elaborate. Archeologists have catalogued about 4200 examples of these prehistoric petroglyphs. An outstanding one (*Figure 13-8*, shown in a photograph taken in 1990 before the area was posted), also attracted the attention of noted photographer David Muench for his large-format, coffee-table book entitled, *New Mexico* (Muench 1974).

A. Historic Lithograph (by J.S. Newberry, 1876; Gregg 1968)

El Creston

Comanche Gap

El Creston

B. North-Northwest View (Photograph by Author, 1990)

Figure 13-7 . Igneous Dike *El Creston* and Comanche Gap

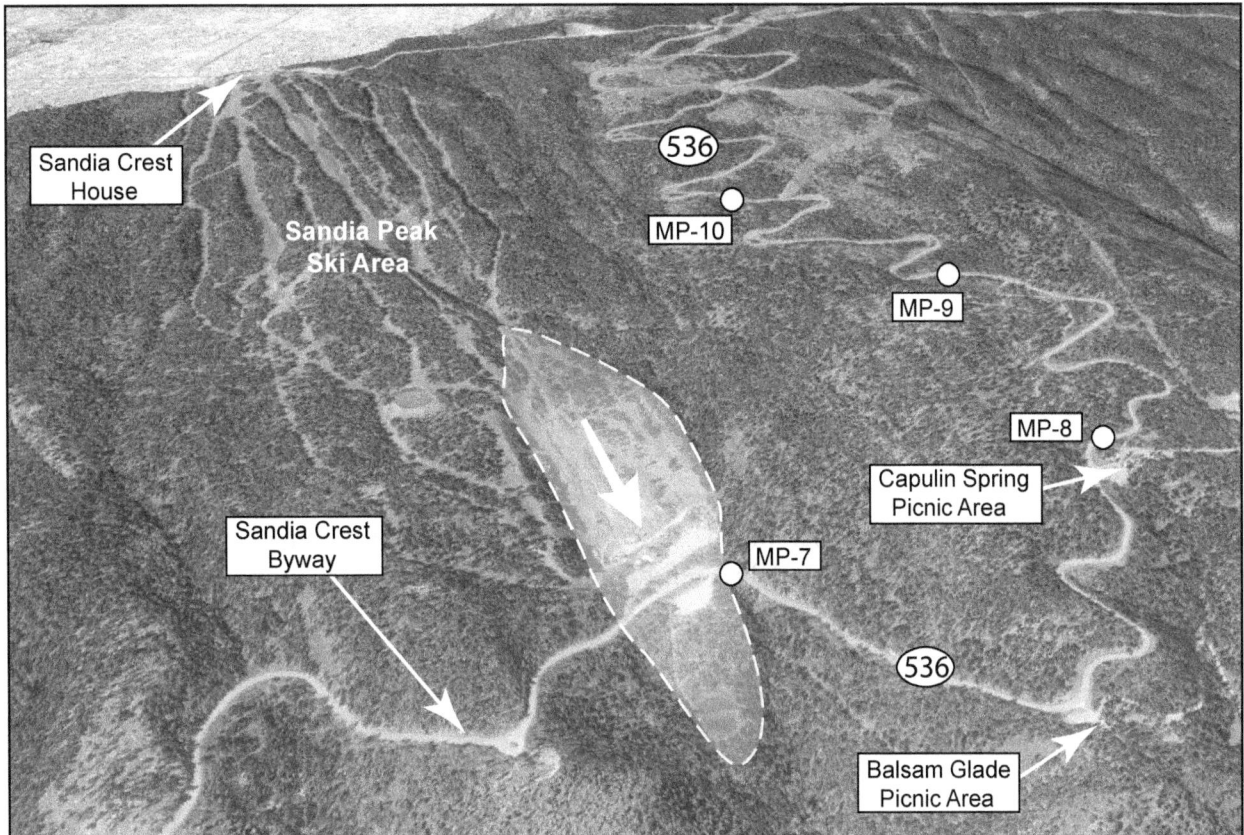

A. West Oblique *Google Earth* Image (Gravity Slide Shaded)

B. West View of North Edge of Gravity Slide, Sandia Crest Byway
(NM-536, Mile 7.0; Vegetation for Scale; Photograph by Author)

Figure 12-15. Sandia Ski Area Gravity Slide

Sandia Crest

At the very top of the 13.3-mile ascent along the Sandia Crest Byway is Sandia Crest itself (*Figure 12-16*). This is a place name for the Crest House and a parking area, but the *topographic* crest runs the length of the mountain. The lower, older part of the Madera Group limestone forms the crest's cap. The limestone is a light gray, very finely crystalline rock made of the mineral *calcite* ($CaCO_3$) that creates thick, resistive ledges (*Figure 12-16B*). It consists of the debris left behind by living organisms in the open sea

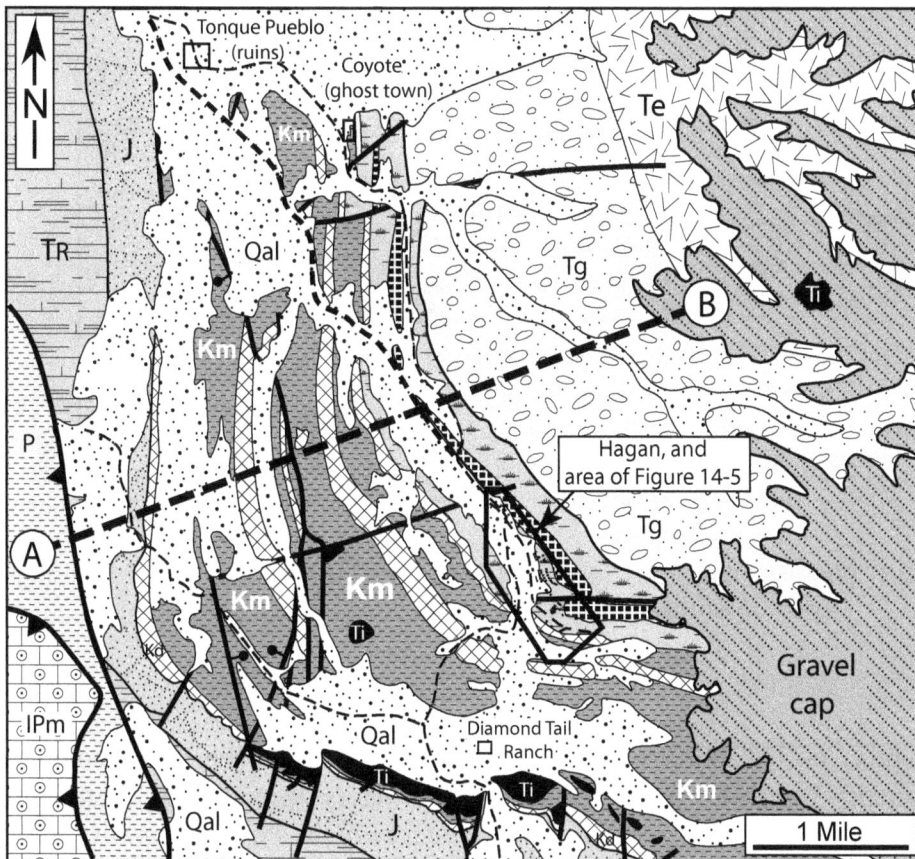

A. Geologic Map (Modified from Black 1979, Kelly and Northrop 1975)

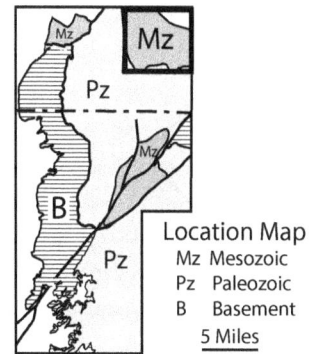

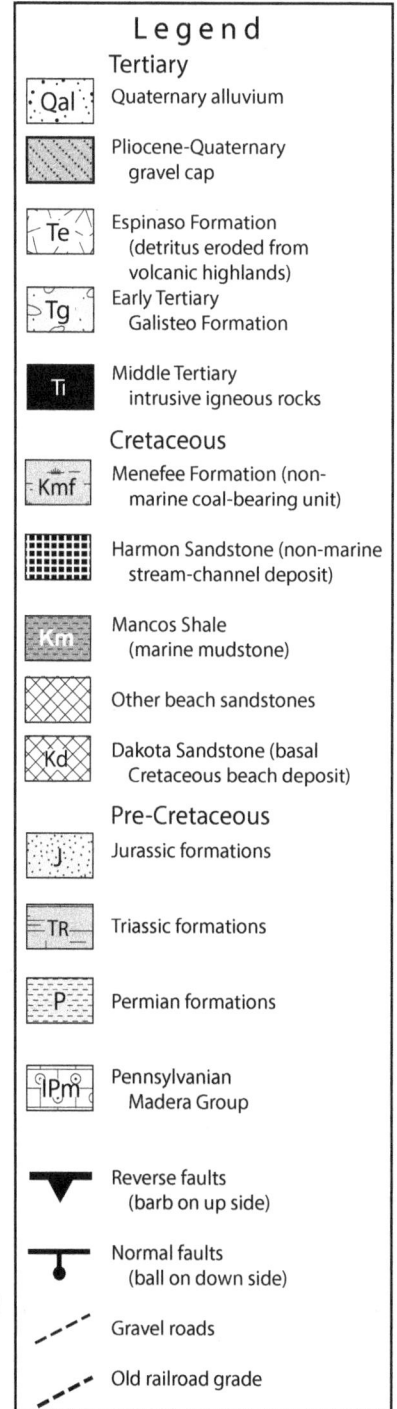

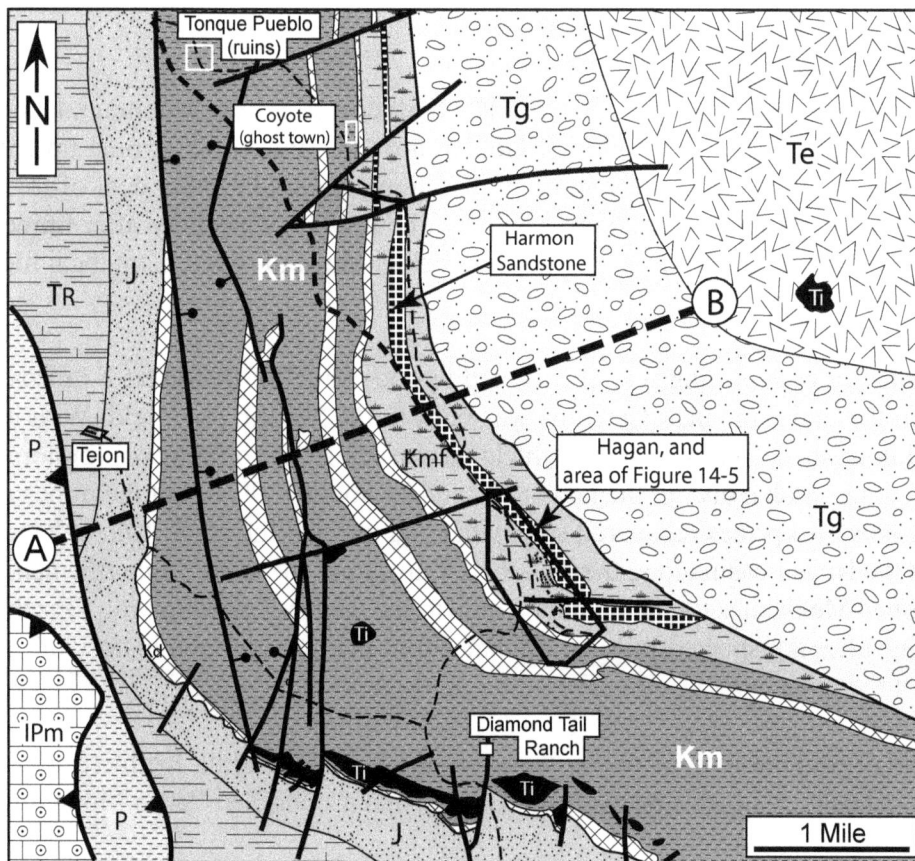

B. Bedrock Geologic Map

Figure 14-2. Geology of Hagan Basin

Location Map
Mz Mesozoic
Pz Paleozoic
B Basement
5 Miles

Legend

Tertiary

Qal — Quaternary alluvium

Pliocene-Quaternary gravel cap

Te — Espinaso Formation (detritus eroded from volcanic highlands)

Tg — Early Tertiary Galisteo Formation

Ti — Middle Tertiary intrusive igneous rocks

Cretaceous

Kmf — Menefee Formation (non-marine coal-bearing unit)

Harmon Sandstone (non-marine stream-channel deposit)

Km — Mancos Shale (marine mudstone)

Other beach sandstones

Kd — Dakota Sandstone (basal Cretaceous beach deposit)

Pre-Cretaceous

J — Jurassic formations

TR — Triassic formations

P — Permian formations

IPm — Pennsylvanian Madera Group

Reverse faults (barb on up side)

Normal faults (ball on down side)

Gravel roads

Old railroad grade

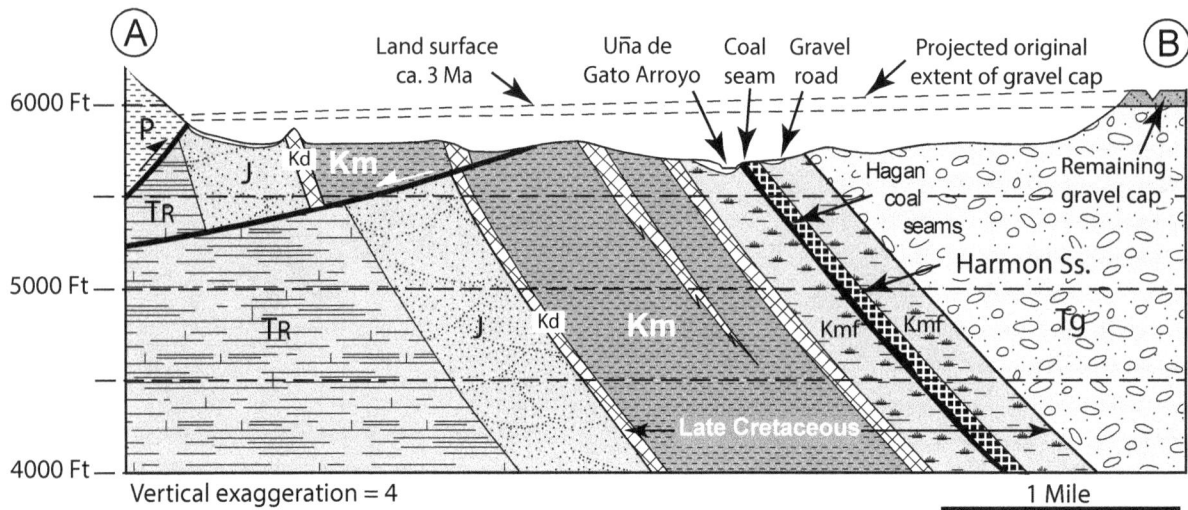

Figure 14-3. West-to East Cross Section A-B Across Hagan Basin
(Location and Legend in *Figure 14-2*)

Note: In this book I tend to minimize stories about people and the towns they built (i.e., non-geological things) only because they have already been well-covered elsewhere by others. In particular, there is the excellent 700-page tome by the East Mountain Historical Society, *Time Lines of the East Mountains* (2020), which I have cited a few times. Then there is *Ghost Towns and Mining Camps of New Mexico* (Sherman and Sherman 1975). The story of the town of Hagan, however, has a deep geological root and so I have included it at some length below for this book.

Need for Hagan Coal. In 1828 the first gold rush in the U.S. was kicked off east of Sandia Mountain. Hard-rock mining towns such as Dolores in the Ortiz Mts. and later Golden near San Pedro Mt. sprang up and lured thousands to the dry and challenging terrain. These mining operations required a great deal of trees for timber, housing, mine construction, and fuel to make charcoal. By about 1835, especially after timber became scarce, Cretaceous coal seams near Cerrillos and Madrid were being mined to supplement local fuel needs.

First Mining. A significant impetus for coal mining was occasioned by the arrival of the AT&SF railroad in the 1880s. A railroad spur from the main line near Cerrillos was built to the Madrid coal field. With its market secured, Madrid grew and flourished. Coal was sold to the local market in territorial towns for home and office heating. However, coal prices took a dive in the late 1890s and by about 1900 all but the mines at Madrid had closed. In the early 1900s coal prices recovered somewhat and in 1903 the small *Uña de Gato* (Cat's Claw) coal mine opened in the Hagan basin along the usually dry arroyo of the same name (the name was borrowed from an old abandoned Mexican settlement about a mile to the south). The following year the mine was renamed the Hagan Mine after a respected railroad official. From that point on the history of the settlement closely followed that of the mine.

From 1903 to 1906 a small amount of coal was hauled out by wagon to the AT&SF loading facilities at San Felipe. Some was wagon-hauled and sold to the mining camps at Golden and around San Pedro Mountain.

Coal Estimates: An Achilles Heel. What would become the town of Hagan hinged on what the technical experts estimated the mineable coal reserves would be. Back then, at the turn of the 20th century the profession of geology wasn't yet well established. Many "geological" problems were often dealt with by "mining geologists," who were more engineers than geologists. The former deal with numbers, forces, and absolutes, whereas geologists deal with things they can't always see or measure—things below the surface where it's dark and entails considerable risk. The estimates they made were therefore far off the mark. The failure of the mining venture and the demise of the town were in a very large part attributable to the fatal miscommunication between the technical "experts" and the investors with the money. In other words, it was a human problem. The story of the coal-volume estimates is therefore the story of the town, and is worth expanding upon. (Most of what follows below is from Lent et al. 1985.)

Estimate 1. In 1904 a mining geologist named Charles R. Keyes published a report stating that there were vast volumes of mineable coal present in the Hagan basin. He cited seven miles of outcrop length extending north from the Uña de Gato deposits to a second deposit. He assumed that the seam was continuous between the two and that it extended about a mile down a minable slope to the east. He then calculated an area of "no less than" eight square miles of coal, with a much larger adjoining area under which coal exists with "small reasonable doubt." Combining this estimated area with seam thickness and other data he calculated *87 million tons* of mineable coal. Keyes didn't know it but he was fostering a dangerous myth.

? **Tons vs. tonnes**. There is considerable confusion between these two terms. The word "ton" is the North American (U.S. and Canadian) "short ton," which is 2000 lbs—the term we in the U.S. are most familiar with. In the United Kingdom and much of the world there used to be the "long ton," which is 2240 lbs (thankfully that term now largely obsolete). Then there is the "tonne," same as "metric ton," which is 1000 kg or 2204 lbs, a term widely used outside North America. All these terms are derived from the Latin *tunna*, meaning "cask." In the distant past a *tunna* filled with some material weighed about a metric ton.

The mining company president at the time was a colorful character named "Colonel" W.S. Hopewell, who believed that the main "Hopewell" seam (what else!) extended far beyond exploratory depths. He acknowledged that a 14-mile railroad spur from the AT&SF at San Felipe would be necessary to secure the economic future of the mine. In 1903 and 1904 there was serious talk about the spur but nothing came of it. From 1906 to 1908 about 60 people worked and lived at Hagan.

Estimate 2. In 1909 a more conservative consulting engineer named John McNeil factored in Keyes' 1904 estimate of eight square miles of coal extent, coal thickness taken from the observable Hopewell seam, and a recovery factor. From these figures and allowing for 20% wastage, this gave a grand total of *24 million tons* of mineable coal. This total was far less than Keyes' 87 million tons, but was still a lot of coal. Also in 1909, construction of the grade for the railroad spur had actually begun, but it's suspension shortly afterward knocked the mining operation in the head and the mine was shut down. The place slumbered until the end of World War I.

Estimate 3. By 1919 the Great War was over and an optimistic group of Louisiana investors formed a venture to re-open the Hagan Mine. The effort was ramrodded by a New Orleans promoter named Dr. Justin Jerome dePraslin (the "Dr." was surely honorary because there is no evidence that he ever earned an advanced degree). Fueling dePraslin's optimism and bursting ego, and telling him just what he wanted to hear, was the earlier mine president and later owner of the coal rights, our "Colonel" Hopewell. In a 1919 letter to dePraslin he stated, "It is estimated by *practical coal experts* [my italics] in the coal business that there is about *1,800,000 tons* of coal now in sight." Interestingly the colonel had sold his interest in the coal rights back to his parent company in 1911, an odd act considering what he claimed the fabulous potential of the mine to be.

Estimate 4. Also, in 1919 yet another estimate of the coal reserve was published by a mining engineer named S.O. Andros. In order to be *very* conservative, he factored in the length of continuous outcrop between two exploratory shafts, a certain width down-slope, a coal thickness from the known Hopewell seam, and a recovery factor, and he calculated a figure of *420,000 tons* of "proven coal." (Notice the downward trend!) He went on to say that, "undoubtedly this seam persits [*sic*] to a depth of two or three thousand feet, and extends east and west much farther than has been traced." Finally, he wrote, "without question, there are one to *two millions* [*sic*] of tons available in the seam already opened up. *This tonnage, however, has not been proven*" (italics again mine). One has to wonder if that final little disclaimer was noticed by the investors.

Considering that the coal eventually played out after producing only about *81,000 tons*, these "expert" effusions are curious indeed. What happened here? Was there a conspiracy to pick the deep pockets from Louisiana? Probably not. Perhaps just unbridled human nature. When the professionals present an interpretation to those who lack the ability to challenge it, and those people in turn take the data at face value and run with them, the figures assume a life of their own and disaster can loom. The crucial communication gap at Hagan was clearly between the colonel's "experts" and dePraslin. Once he himself accepted the estimates of coal volume as fact, he cranked up his eloquence and forceful personality and soon had his investors drooling.

Hagan Coal Mines, Inc. DePraslin's newly formed *Hagan Coal Mines, Inc.* opened up the mine in 1919. The company's prospectus, showing a man who might be dePraslin (*Figure 14-4*), stated that "the field has been thoroughly tested by diamond drilling, and the continuity of the deposits has been satisfactorily demonstrated. Reports made by leading mining engineers and geologists estimate the coal in the one vein [the Hopewell seam] which has been opened at 30,000,000 tons, and there are five workable veins, …" The prospectus then flat-out stated that exploratory work had been done "showing over 2,500,000 tons of workable coal in the area blocked out." The investors eagerly coughed up about $3,000,000 to finance the operation. That was a hefty chunk of change in 1919, equivalent to about $40 million today.

SLOPE NO. 2 AT HAGAN
700 FEET DOWN, VIEN IS 5 FEET SOLID COAL.

HAGAN COAL MINES, Inc.

SANDOVAL COUNTY,
NEW MEXICO
OFFICES:

ALBUQUERQUE, NEW MEXICO
DOVER, DELAWARE
LAKE CHARLES, LOUISIANA

EXECUTIVE OFFICE:
210 WEST CENTRAL AVENUE
(SECOND FLOOR)
ALBUQUERQUE, NEW MEXICO

APPROVED BY LOUISIANA SECURITIES COMMISSION

Figure 14-4. Prospectus, Hagan Coal Mines, Inc., Ca. 1924
(Modified from Lent et al. 1986)

Town of Hagan. Over the next few years, the road was improved from Hagan to San Felipe and some minor construction occurred at the camp, but real work would not commence until the famous railroad spur had become a "done deal." That day came in 1924, when Hagan Coal Mines Inc. contracted to build the spur extending from the AT&SF at Algodones to the west. The spur would be henceforth be called the *Rio Grande Eastern*. The company next envisioned a carefully laid-out town, nestled between the base of a sandstone cliff on the east (the Harmon Sandstone, see below) and the Uña de Gato arroyo on the west. There would be houses for 200 families, apartments, a large mercantile store, power plant, hospital, school, and a hotel, all with tasteful simplicity conforming to the natural shapes and colors of the rocky landscape (*Figures 14-5* and *14-6*).

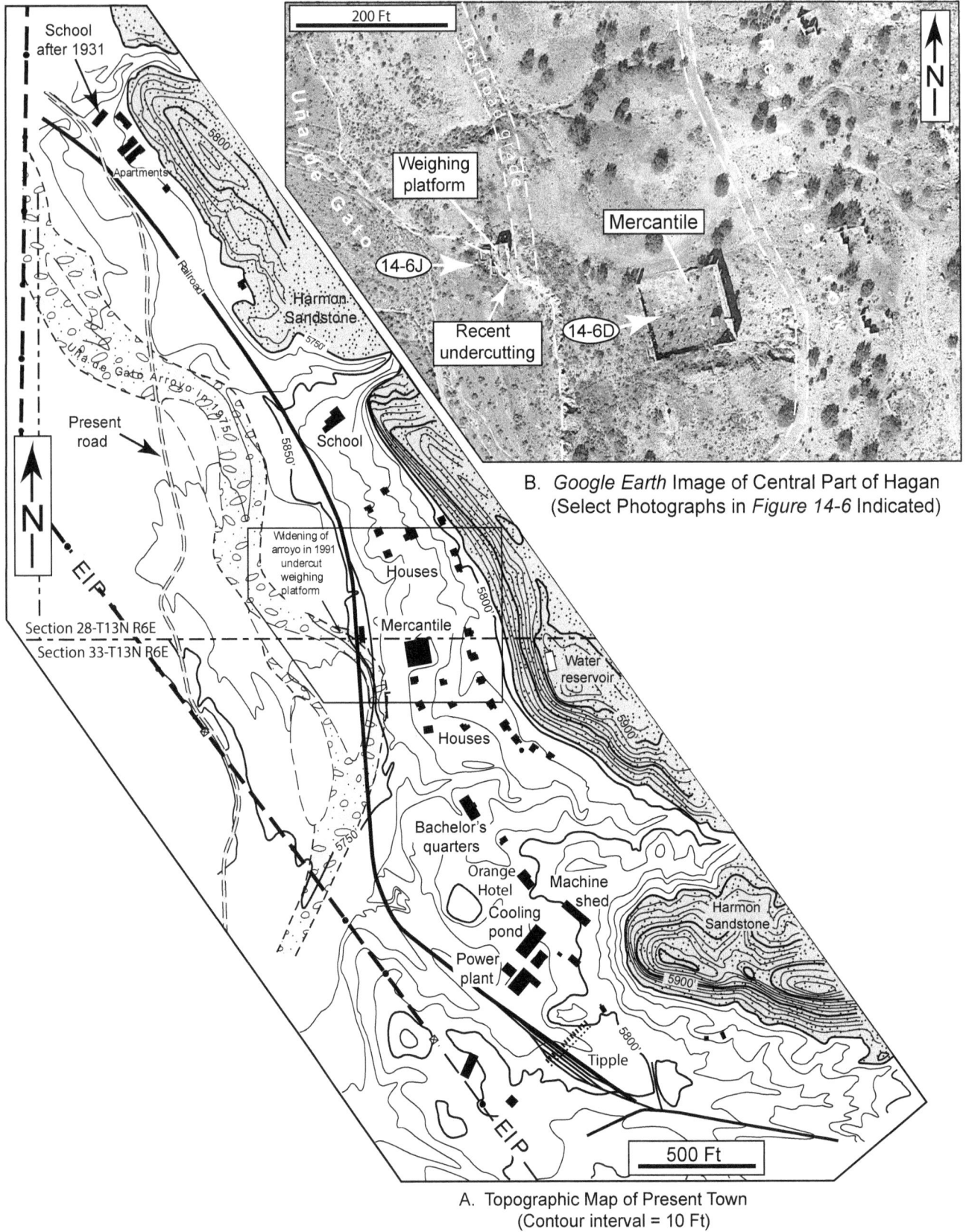

B. *Google Earth* Image of Central Part of Hagan
(Select Photographs in *Figure 14-6* Indicated)

A. Topographic Map of Present Town
(Contour interval = 10 Ft)

Figure 14-5. Ghost Town of Hagan - I

A. North-Northwest View of Mercantile
Down From Reservoir, 1920s

B. View as in "A," 1987 (Photograph by Author)

C. Southeast View of Mercantile, 1924

D. East View of Mercantile's East Interior Wall and Safe, 1987
(Photograph by Author)

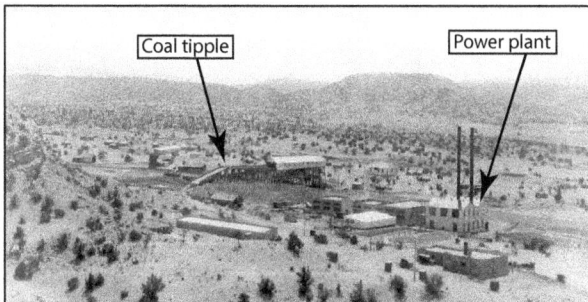

E. South Overview of Southern Industrial Area, 1920s

F. East Overview of Southern Industrial Area, 1920s

G. Typical Miner's House, 1920s

H. Warehouse Along Railroad, 1920s

I. North View of Ruined Northern Residential Area,
Date Unknown

J. East View of Weighing Platform, Being Undermined
by Uña de Gato, 1991 (Photograph by Author)

Figure 14-6. Ghost Town of Hagan - II
(Photographs, Except Those Indicated, Courtesy of Sandoval County Historical Society)

In 1924 the company contracted a master *adobero*, Abenicio Salazar from Bernalillo, to construct about 22 two-to-three room, one-story houses laid out along three winding rows (*Figure 14-5*). Each house had a sturdy foundation of native sandstone cemented with mortar, stuccoed adobe walls, wooden plank floors, and a flat, tar-papered wooden roof. Most if not all the houses had fresh running water, indoor toilets, electricity, and of course coal stoves. All in all, these cottages were modern and quite cozy (*Figure 14-6G*). In addition, there was an eight-grade schoolhouse, several apartment buildings; a hotel, power plant, coal tipple and loading facilities. All this construction and mine improvement cost more than $700,000 (about $9.5 million today). Obviously, this was not to be a thrown-up, ramshackle, boom-and-bust dive like so many other western mining towns, but rather a comfortable, planned, permanent community. (Remember this huge expenditure relied on estimates of minable coal volume.) DePraslin and the investors apparently believed in their hearts that Hagan had a rosy future and they acted accordingly.

Serious mining began in late 1924 and by that time some 200 people had established residence. Although this was a 100% company town, the community was relatively affluent and many folks had automobiles. When the mood struck, they headed off to Bernalillo for shopping or a "good steak." People lived as they do anywhere: they socialized, they read newspapers, and they dreamed.

Beginning of the End. In 1925 excavations of the coal seam down the main slope ran up against a fault where the seam abruptly terminated. Oh oh! But then several other slopes were opened and production increased to a maximum in 1927, but after that it was all downhill. As the miners dug deeper along the 18° slope the coal seams became thinner and the intervening shales thicker until the coal became too thin to be commercial. A total of only about 81,000 tons had been produced! What happened to the "absolutely proven" 420,000 tons, not to mention the much higher estimates? This must have been a real shock to the investors, and to dePraslin, and today we can almost hear the wails of "How could this have happened?" and, "Why didn't you tell us?" There must have been much wringing of hands over the wisdom of throwing all that money into the town without determining the real extent of the coal based on sound geologic data. Truth is it could have and should have been! If big bucks were available for a very expensive little town, surely there was some for the drilling of a dense pattern of test holes to determine the coal's lateral extent and thickness by way of detailed structure and isopach maps. But the indefatigable dePraslin and his "can-do" financial people believed that money, willpower, positive outlook and a railroad spur were all that they needed. They were wrong! Hagan Coal Mines didn't do their geologic homework and paid the price.

The End. The mine closed for good in 1931 and the population slowly drifted away. In 1935 a large-scale salvage operation began and in the next few years all equipment, plumbing, rails, and generally anything of value that could be yanked or dug up was hauled off. Population dropped from a maximum of about 200 in the "boom" years to about 50 ten years later, and finally to zero ten years after that. Hagan was left to the mercy of the elements, and to the vicissitudes of the Uña del Gato.

Hagan today is a ghost town. It lacks the more visceral history of a hard-rock mining town with poignant tales of shootouts, hangings, and broken hearts. Hagan was apparently a tranquil "middle class" place, and perhaps quite dull. Although the inhabitants made it their home and had intended to stay put, they really didn't have much time to leave much of a mark on the land. Today there remain the foundations and part of the walls of most of the houses, although much of the adobe has crumbled. Also, there are the foundations of the industrial structures such as the power plant, a parking garage and a coal tipple. And of course, there is the grand old mercantile, with its semi-intact east wall (as of this writing), and the massive concrete safe perched on its concrete pedestal (*Figure 14-6D*).

Archeologic survey. Progress sometimes can be a good thing. In the early 1980s, Public Service Co. of New Mexico (PNM) needed to connect its electrical generating facility at the community of Algodones, about ten miles west of Hagan, to Clovis about 190 miles east of Hagan. A 180-ft wide easement for the 345-kv transmission line was chosen to best satisfy the landowners and to skirt the archeological site of Hagan as much as possible. As part of the routing process PNM initiated a detailed archeological survey of the site, resulting in a comprehensive report (Lent et al. 1985).

At the time of the survey the north-flowing Uña de Gato had undercut its bank eastward and had chewed up part of the railroad bed. This was the situation when I first visited the place in December 1987. Sometime between that time and my next visit in August 1991 the arroyo had undercut the weighing station and was advancing yet eastward—toward the mercantile (*Figure 14-6J*). The interesting old site is in danger.

It's interesting to speculate on the future of Hagan had the coal mine measured up to expectations. The town would have likely been connected by rail from the south as well as from the north (a New Mexico Central rail line had been graded to within three miles of the site from the southeast, but was never completed) and a paved road would likely have run parallel to it. Hagan though eventually probably would have suffered the same fate as Madrid when the coal mines shut down due to a nationwide switch from coal to diesel oil in the 1950s. However, today Madrid has experienced a renaissance. Hagan perhaps would have too. But despite these speculations the town should never have been.

Hagan is now located on land owned by the Diamond Tail Ranch, which is headquartered nearby. Unfortunately, or perhaps fortunately, the site is now fenced off and is posted as private land. Access is only via a tour company called *New Mexico Jeep Tours*.

To finish up with Hagan, the geologic map and cross section (*Figures 14-2* and *14-3*) show that the Hagan coal deposit occurs below a thick unit called the Harmon Sandstone. This massive, durable body of rock formed the high eastern backdrop for the town, and served as a convenient high platform for the town's water reservoir.

Harmon Sandstone. Names for geologic units usually invoke a story. (The origins of rock-unit names are covered in the lexicon, APPENDIX IV). The name *Harmon Sandstone* is one of these. Such names are most often taken from that of a nearby place or some prominent-feature. There is nothing in the area named "Harmon." It turns out that the name is from a fellow named Newitt Harmon Black (1908-1995), who as a teenager during the 1920s worked in the Hagan coal mines. While there, he carved his initials, "NHB, 1925," into that sandstone bluff (*Figure 14-7*). He later become the father of Bruce Black, who became a petroleum geologist with Shell Oil Company during their big exploratory push in the Albuquerque area during the late 1970s/early 1980s (see below). The younger Black later did much exploratory work in the Hagan basin and went on to become a prominent independent petroleum geologist in the Farmington area. When he published a paper of the Hagan basin area (Black 1979a) he needed to give the rock unit on his map a name, so dubbed it "Harmon" in honor of his dad (Black 2020). The sandstone has been correlated over to the Placitas area and the name followed to that place (*Figure 6-7*).

Figure 14-7. Harmon Black's "Christening" of the Harmon Sandstone, Hagan
(Modified from Black 1994)

Madrid Coal

We first touched on this place in the last chapter (*Figure 13-4A*). Madrid is all about its special coal: *anthracite*.

Anthracite. Madrid coal—anthracite—is unusual, special, and a true fluke of geology. What makes it special here is that it is sandwiched between two igneous sills that dip gently to the east and provide a convenient, mineable exposure to the west (*Figure 13-4A*). Ca. 36-33 Ma the hot sills intruded into and along bedding planes of the Cretaceous Mesa Verde Formation well below the surface. The sills "baked" the coal, driving off much of its volatile material (water and carbon dioxide), leaving behind the almost pure-carbon coal, anthracite (from the Greek *anthrakites* = "coal like"). Occurrences of anthracite are rare west of the classic deposits of northeastern Pennsylvania. The most notable of these is the small deposit at Crested Butte, Colorado, where Cretaceous coal also has been "baked" by the Crested Butte intrusive laccolith. Interestingly, the West Elk Mountains, a large volcanic field just west of Crested Butte, also features juxtaposed intrusive laccoliths with Mesa Verde coals. In one of these fields, about eight miles west-southwest of Crested Butte, is the *Anthracite Range.*

Tijeras Basin

The Tijeras basin, sometimes called the Tijeras coal basin, is that block of rock caught up as a slab between two major fault strands of the Tijeras fault zone (*Figure 14-8*). The Cretaceous rocks (the Mancos Shale and the Mesaverde Formation) occupy the surface of the basin, north of the town of Tijeras. Here the beds have been rippled into three north and northeast-trending folds—two synclines and an intervening anticline.

Location

Placitas 7000-ft contour
2 Miles
ABQ
Tijeras
KAFB

Sandia Crest
Xg
IP
P
TR
Km
Kmv
Kd
J
Tijeras basin
IP
IP
P
P
Tijeras fault
Gutierrez fault
Monte Largo Horst (Ym)
Bernalillo Co.
Sandoval Co.
Santa Fe Co.
San Pedro Mts.
South Mt.
P
TR
IP

Legend

■		Intrusive igneous rocks
	Kmv	Cretaceous, Mesa Verde Fm. (coal-bearing)
	Km	Cretaceous, Mancos Shale
Kd	Kd	Cretaceous, Dakota Ss.
	J	Jurassic
	TR	Triassic
	P	Permian
	IP	Pennsylvanian
	Xg	Sandia Granite
	Yg	Metamorphic rocks

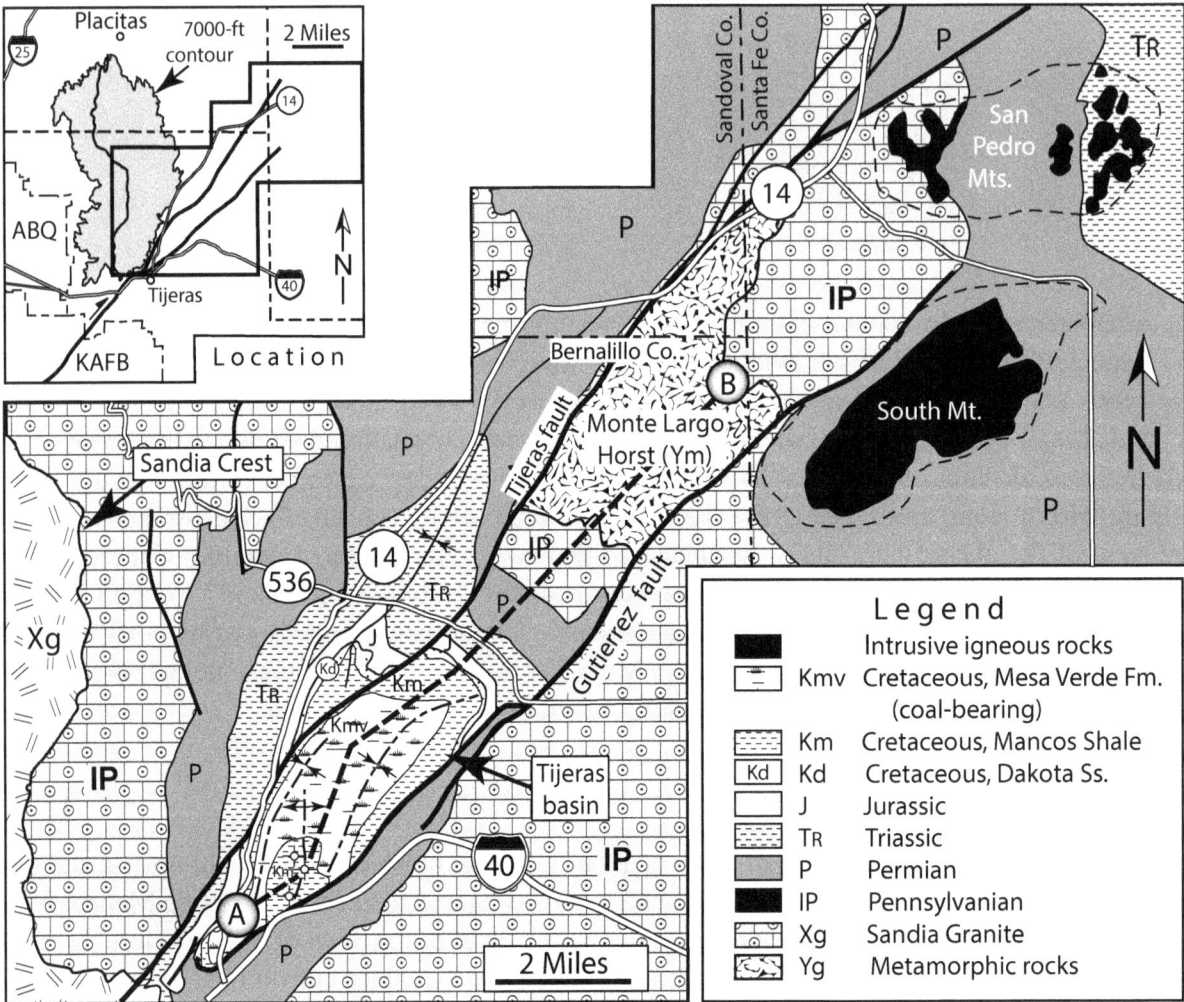

A. Geologic Map (Modified from Williams and Cole, 2007)

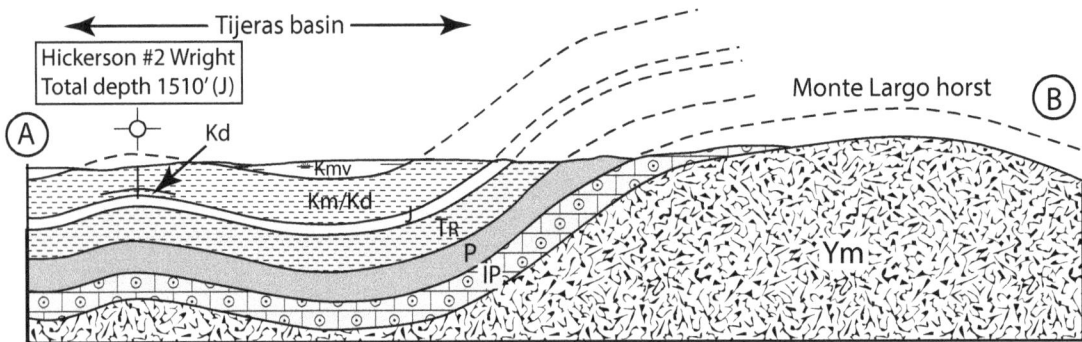

Tijeras basin
Hickerson #2 Wright Total depth 1510' (J)
Monte Largo horst
Kd
Kmv
Km/Kd
TR
P
IP
Ym

B. Cross Section A-B (Shown in "A" Above)

Figure 14-8. Tijeras Fault Zone

Oil and Gas Exploration. Two essentials for oil and/or gas accumulation are present in the Tijeras basin: favorable Cretaceous rocks (they have an excellent reputation in the San Juan basin), and an anticlinal structure (*Figure 0-10A*). Early on explorers recognized that oil and gas tended to be trapped in anticlines, and the Tijeras anticline became naturally attractive.

In 1948 it was drilled twice on the east side of an anticlinal crest, but without success (*Figure 14-8*). The first well (Hickerson #1 Wright) penetrated the entire Cretaceous, but evidently ran into some sort of mechanical trouble. Therefore, a second well (Hickerson #2 Wright), was drilled right next to the first to a total depth (TD) of 1510 ft, completely through the Cretaceous into the Jurassic. These test wells effectively killed any idea of possible oil or gas production in the area. Later, in 1964, Southern Union Production Co. came in and drilled three wells along a north-to-south line on the west side of the anticlinal axis, but this time searching for porous formations to serve as a gas storage vessel for the peak winter fuel demands of Albuquerque. The deepest well (#3) had a total depth of 2228 ft in the Jurassic, but none found any useful storage capacity (Black 1979b; Kelley and Northrop 1975).

Coal. Because of its excellent reputation, anywhere the Mesaverde Formation was known to occur immediately drew the attention of mineral prospectors. As early as 1898 coal was mined from several small seams, and eventually three small mines were opened: *Section 1*, *Holmes*, and *Tocco*. (*Figure 14-9A*). Production from all three was probably sold for smithing and household heating in Albuquerque, but history only exists for the latter. Tocco opened in 1908, and until its closure in 1911 produced just under 1000 tons. The mine had the distinction of being one of the smallest coal mines operated in the U.S. (Kelley and Northrop 1975).

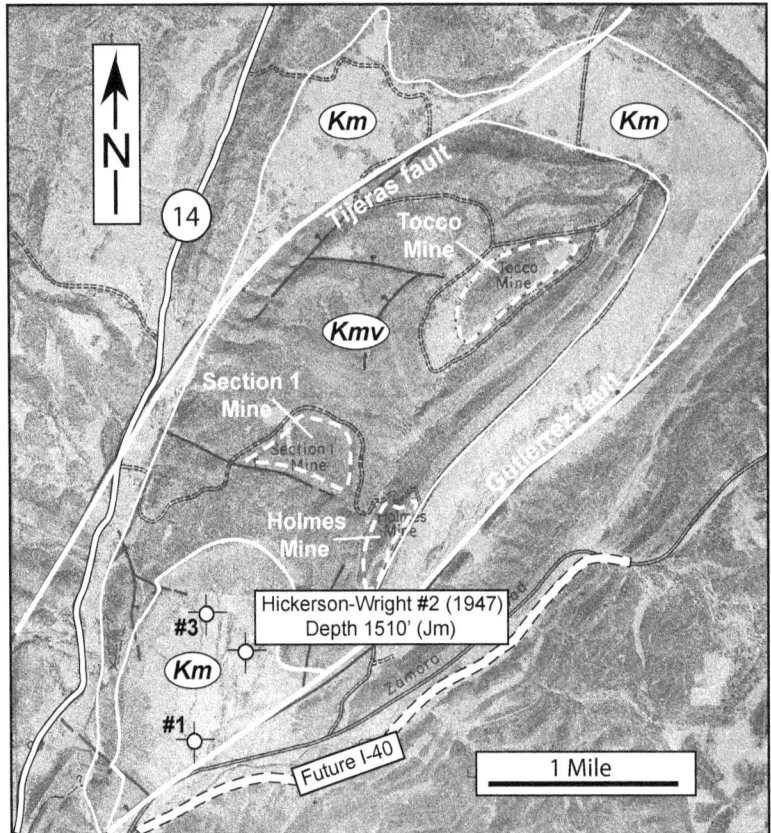

A. Aerial Photograph Showing Coal Mines (From Kelley and Northrop, 1975)

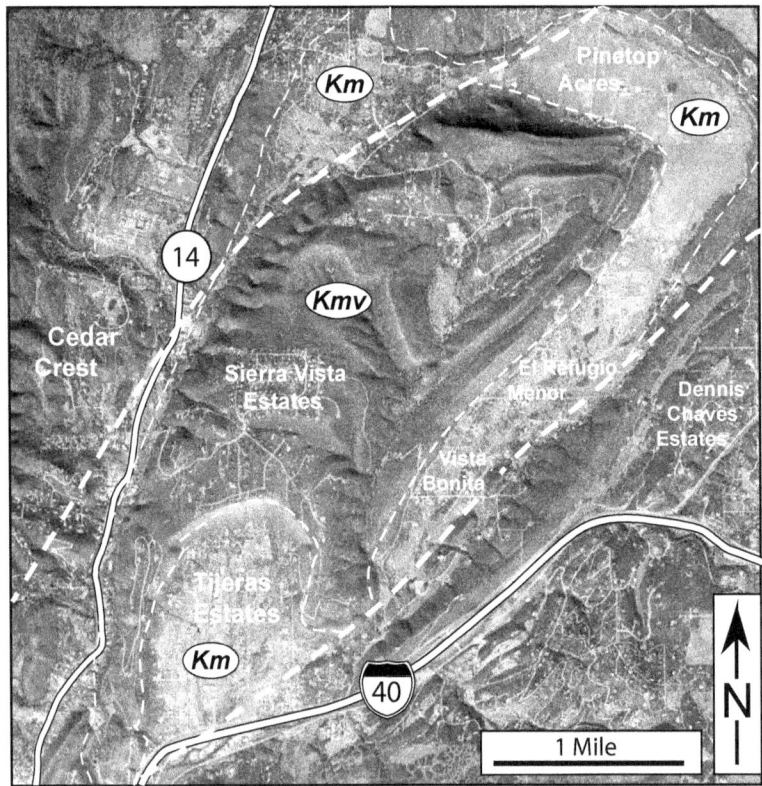

B. *Google Earth* Image of Tijeras Basin

Figure 14-9. Cretaceous Rocks of Tijeras Basin
(Kmv = Coal-Bearing, Mesa Verde Fm.; Km = Mancos Shale)

Albuquerque Basin Oil and Gas Exploration

Although the story of oil and gas exploration in the Albuquerque-basin portion of the Rio Grande rift is a bit off on a tangent, it *is* a big story with a geologic basis and warrants inclusion in this book. Rifts are a favorite place to find oil and gas worldwide, and the large Albuquerque-basin portion of the rift attracted attention very early. From 1912 to 1952 some 36 test wells had been drilled in the basin, almost all of them aimed at targets *within* the shallow sands and gravels of the Santa Fe Group. (Note: In the drilling business, exploratory wells drilled in a new place are called "wildcats.") All of those wells were unsuccessful, or "dry."

It was known that the Cretaceous rocks were present under all that Santa Fe Group sand and gravel, and because the Cretaceous was known to be so important in the San Juan basin to the northwest, explorationists decided to "take a look" (*Figures 14-10* and *14-11*).

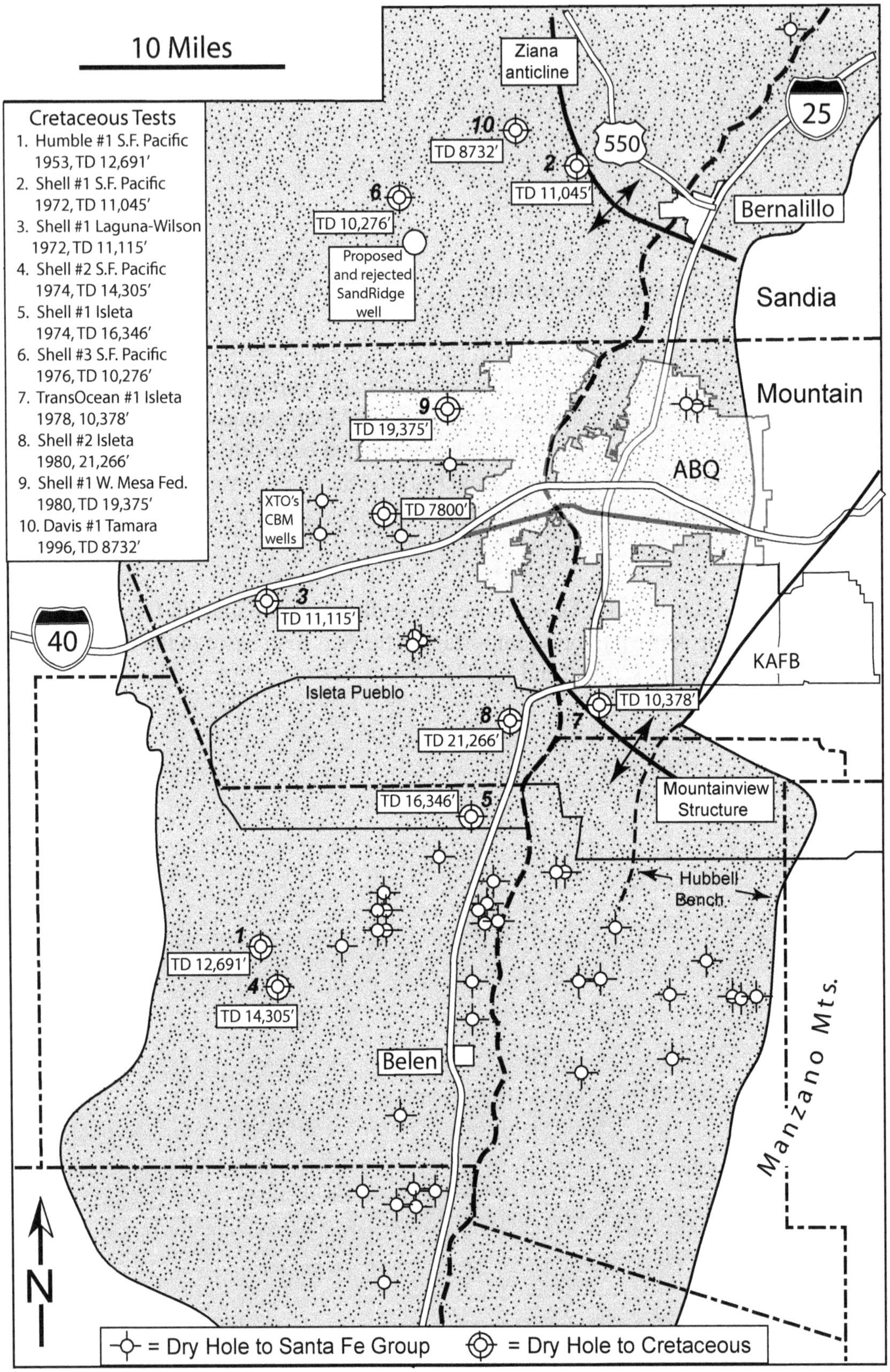

10 Miles

Cretaceous Tests
1. Humble #1 S.F. Pacific
   1953, TD 12,691'
2. Shell #1 S.F. Pacific
   1972, TD 11,045'
3. Shell #1 Laguna-Wilson
   1972, TD 11,115'
4. Shell #2 S.F. Pacific
   1974, TD 14,305'
5. Shell #1 Isleta
   1974, TD 16,346'
6. Shell #3 S.F. Pacific
   1976, TD 10,276'
7. TransOcean #1 Isleta
   1978, 10,378'
8. Shell #2 Isleta
   1980, 21,266'
9. Shell #1 W. Mesa Fed.
   1980, TD 19,375'
10. Davis #1 Tamara
    1996, TD 8732'

Ziana anticline

550

25

Bernalillo

Sandia

Mountain

ABQ

10 TD 8732'

2 TD 11,045'

6 TD 10,276'

Proposed and rejected SandRidge well

9 TD 19,375'

XTO's CBM wells

TD 7800'

3 TD 11,115'

40

Isleta Pueblo

8 TD 21,266'

TD 16,346' *5*

TD 10,378'

7

KAFB

Mountainview Structure

Hubbell Bench

1 TD 12,691'

4 TD 14,305'

Belen

Manzano Mts.

N

⎯◇⎯ = Dry Hole to Santa Fe Group   ◎ = Dry Hole to Cretaceous

Figure 14-10. Exploration for Cretaceous Oil and Gas in the Albuquerque Basin
(From Black 1982, 1999; CBM = Coal-Bed Methane)

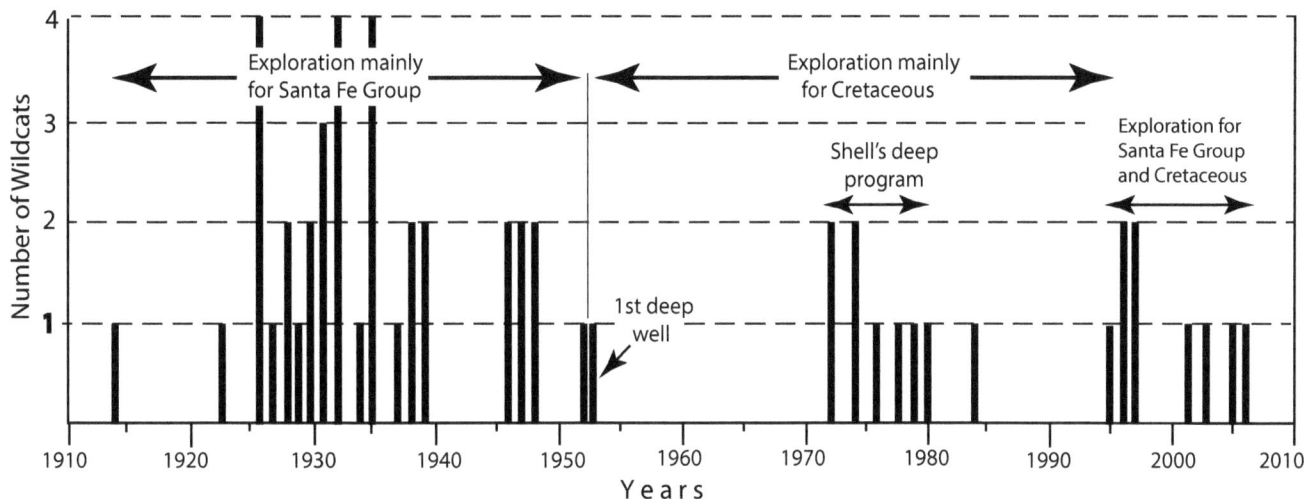

Figure 14-11. Time Line of Exploration for Oil and Natural Gas in Albuquerque Basin, 1914-2006
(Modified from Black 1982)

In 1953 *Humble Oil Company* drilled its #1 Santa Fe Pacific, about 15 miles WSW of Los Lunas. The depth drilled, 12,691 ft, was a great deal at the time, but the well's lack of success, and its cost, obliterated enthusiasm for almost 20 years (*Figure 14-10*; Black 1982).

Continued excellent success in the San Juan basin convinced *Shell Oil Company's* office in Farmington that the Cretaceous section in the Albuquerque basin also might have great potential and, except for that pesky Humble well, remained unevaluated. They got to work running "seismic" surveys to define buried structures in the basin (Black 1982). Such surveys project high-intensity sound waves downward and record the reflections from the tops of buried geologic formations. Reflections can delineate "high," buried structures, *anticlines* in Cretaceous rocks where natural gas is likely to occur. The surveys delineated a number of these highs.

In the late 1960s, armed with their new seismic data and vastly improved drilling technology, Shell got to work. In 1972 Shell drilled its first well, seven miles northwest of Bernalillo on a seismic structure that they named the *Ziana anticline* (Black and Hiss 1974), but the well turned out to be a dry hole. Through 1976 the company drilled four more dry wildcats, three on what they thought was their best prospects. Their total investment in the basin by 1976 (leases for acreage plus the wells) was more than $20 million ($95 million in 2020). This was big money! Badly stung, the company joined forces with another to "spread the risk," and then participated in a sixth well—another dry hole (Black 1982).

Amazingly, somehow *still* confident but apparently armed with new ideas, Shell next stepped off into "deep basin country" and drilled a seventh Cretaceous wildcat, a few miles west of Isleta Pueblo, named the #2 Isleta. At the eye-popping depth of 21,266 ft the drilling rig had reached its physical capacity, but the target, the Cretaceous, had not yet been reached (Black 1982). This was the deepest penetration ever in the basin (*Figure 6-6*).

Incredibly, Shell drilled yet *another* wildcat, this one about four miles west of what was then Petroglyph State Park near Albuquerque. The #1 West Mesa Federal went to an impressive 19,375 ft, and so by 1981 Shell had racked up its eighth dry hole (Black 1999). They finally cried "uncle!" Despite this dismal record, in 1995 and 1997 other companies, undaunted, drilled two more unsuccessful wells with Cretaceous objectives, for a total of ten expensive dry holes in the basin (*Figure 14-10*).

In conclusion, either the Albuquerque basin portion of the Rio Grande rift is not a good place to explore for natural gas, or (as some claim, with some justification) the wells had never been drilled at the right places. But even after all this, the lure of the Cretaceous, like the mythological Phoenix, refused to die. In 1997 *Burlington Resources* (after a merger with *Santa Fe Pacific* in 1995) drilled a dry hole on the western side of the basin where the Cretaceous is relatively shallow. In 2005 and 2006, *XTO Energy* (a subsidiary of *Exxon-Mobil* that specializes in non-conventional natural gas,) drilled two wells to the Cretaceous on the west side of the basin, looking for coal-bed methane gas (CBM), which was so very important in the San Juan basin. Again, with no success. The final gasp *almost* occurred in 2015, when *SandRidge Exploration* (an Oklahoma City-based company) proposed a 10,500-ft test west of Rio Rancho (at 24th St. and Encino), this time located at the "right place," but the resulting environmental furor soon squashed that idea. The Cretaceous seems to be dead—for now.

PART III: CONTEXT

15.
THE *REALLY* BIG PICTURE

Introduction

The logical destination from all the previous chapters in this book is the *Big Picture*. This final chapter is for those who wish to understand the history of Sandia Mountain in a little more depth. In the Preface I promised the reader that I would avoid anything overly technical, but here I hedge slightly and add a touch of closing technical gobbledygook. After 14 chapters, though, these final remarks shouldn't seem excessively strange. But first, five little doses of gobbledygook and then a pair of "big picture" conclusions.

Earth as a Heat Machine

We need to look at the deep structure of the Earth to appreciate how events there can affect the Earth's surface on which we live (*Figure 15-1*). Planet Earth is a big heat machine, and the non-homogeneous escape of that heat is what controls things. The heat itself is 1) original from formation of the planet, 2) result of escape of latent heat at the top of the solid inner core where the phase change of liquid iron in the outer core yields heat as the iron solidifies to solid iron, and 3) result of decay of radioactive elements in the mantle (*Figure 15-1A*). The heat escapes upward toward the cold blackness of space chiefly via convection—the physical movements of hot magma upward along a temperature gradient. The escape venues at the surface are the spreading ridges on the sea floors (*Figure 15-1B*), and volcanoes mainly at the active plate margins.

Temperature (°C)
0   1000   2000   3000   4000   5000

Asthenosphere

Temperature

Basaltic
oceanic crust

Lower
mantle
(peridotite)

Outer core
(liquid iron)

Inner core (solid iron)

620 mi
(1000 km)

1240 mi
(2000 km)

1860 mi
(3000 km)

2480 mi
(4000 km)

3100 mi
(5000 km)

3720 mi
(6000 km)

Layers
in "C"
below

Earth's Component Layers

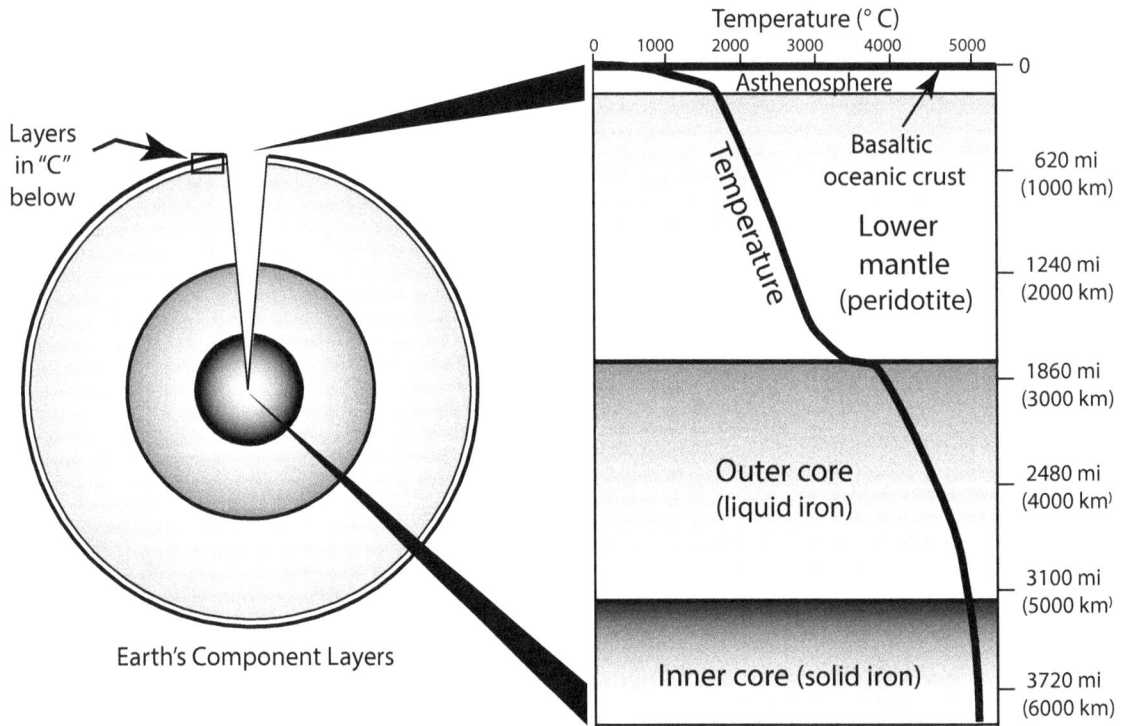

A. Heat-Flow from Earth's Interior

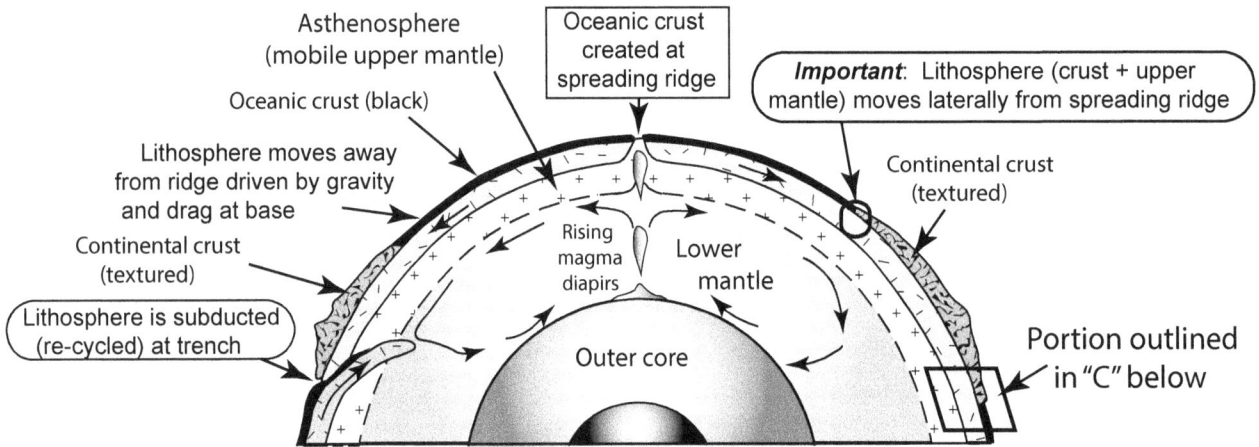

Asthenosphere
(mobile upper mantle)

Oceanic crust
created at
spreading ridge

Important: Lithosphere (crust + upper
mantle) moves laterally from spreading ridge

Oceanic crust (black)

Lithosphere moves away
from ridge driven by gravity
and drag at base

Continental crust
(textured)

Continental crust
(textured)

Rising
magma
diapirs

Lower
mantle

Lithosphere is subducted
(re-cycled) at trench

Outer core

Portion outlined
in "C" below

B. Deep Circulation and Structure of Earth's Interior

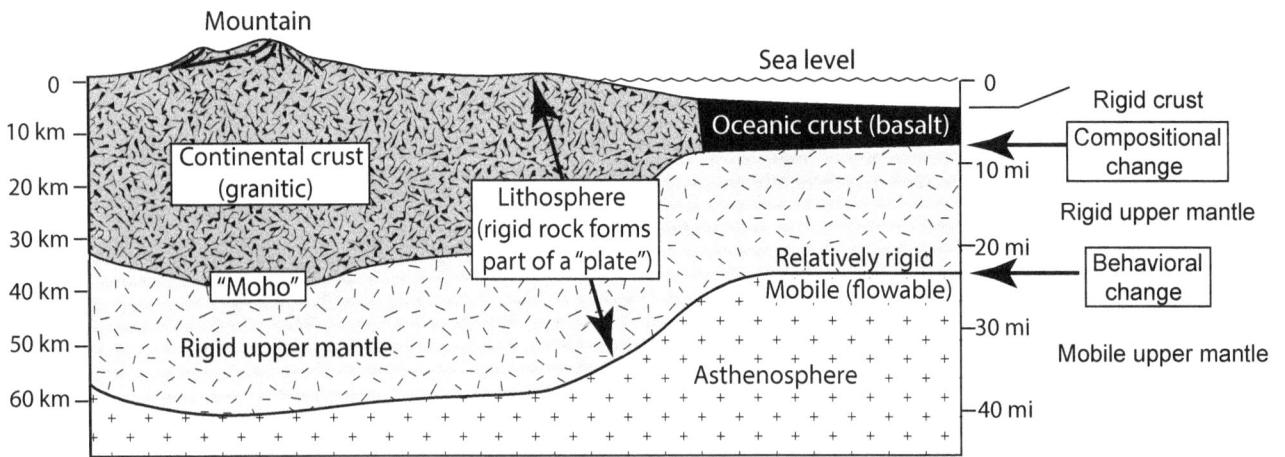

Mountain

Sea level

0

10 km

20 km

30 km

40 km

50 km

60 km

Continental crust
(granitic)

"Moho"

Rigid upper mantle

Lithosphere
(rigid rock forms
part of a "plate")

Oceanic crust (basalt)

Relatively rigid
Mobile (flowable)

Asthenosphere

0

10 mi

20 mi

30 mi

40 mi

Rigid crust

Compositional
change

Rigid upper mantle

Behavioral
change

Mobile upper mantle

C. Earth's Outer Layers

Figure 15-1. Earth's Internal Structure

Formation of Earth's Crust

As heat rises and pressure declines, mantle rock (composed of the magnesium and iron-rich mineral *peridotite*) partially melts to form magma. The new, less dense magma, leaving some heavy stuff behind, rises and forms the rock *basalt* (one even richer in SiO_2) comprising the ocean crust. Next, partial melting of basalt, usually at subduction zones, creates even less dense *granitic* magma (still richer in SiO_2). Thus, partial melting and re-melting, over eons, has produced both the oceanic crust of the oceans and the granite crust of the continents (*Figure 15-1C*).

The Lithosphere

The lithosphere is the solid outer portion of the Earth. Unfortunately, the terminology now gets a bit confusing. The uppermost part of the lithosphere is the rigid crust, composed largely of the granitic continents and basaltic crust of the oceans (*Figure 15-1C*). Below that part, firmly attached to it, is the upper part of the Earth's mantle composed of the denser mineral peridotite. The base-of-crust/top-of-mantle contact within the lithosphere, dubbed the *Moho* (thankfully short for *Mohorovičić*) discontinuity, marks a sharp change of composition, density, and sound-wave velocity. However, the two parts (crust and upper mantle) together form the rigid lithosphere and act as a unit, a *plate*, usually about 60 miles (100 kms) thick.

With depth, temperature and pressure increase until a point is reached where the upper mantle rock becomes partly molten—a mobile "mush." This lower zone is the *asthenosphere*. The rigid lithosphere rides upon it. With yet increased depth the upper mantle becomes solid, transitioning to what is called the lower mantle (*Figure 15-1B*).

Lithospheric Plates

The global lithosphere is segmented laterally into discrete, cohesive *plates*, which move about, relative to each other. Oceanic plates are continually being formed at spreading ridges on the ocean floors and are then destroyed at subduction zones (*Figures 15-1B* and *15-1C*). Plate movements, though, are in super slow-motion—close to the "speed" of fingernail growth.

Listening to the Earth

Being rigid, the crust can fracture, and when it fractures it creates sound. We can hear the sound! Almost everything we know about the Earth's deep interior is based on sound waves that are generated by movements and fracturing of rock masses. The waves are then recorded on the surface by sensitive devices called seismographs. Earthquakes are occurring all the time and active portions of the Earth are often ringing like a bell. The trick is to deploy vast arrays of seismographs, connect the instruments into networks, and then process the massive data sets with super computers.

Since 1999 there has been an on-going project in the western US called the *Continental Dynamics of the Rocky Mountains (CD-ROM) Project*. For the first time we can "see" (by converting sound to images) what is going on below the Rockies at depths to about 500 miles. Sound waves have velocity—the key property measured and plotted on cross sections. A single seismometer cannot measure velocity, but an array of instruments can. Importantly, sound travels most rapidly through cold lithospheric plates and more slowly through hot ones. Plotting regions of high vs. low velocity on cross sections is a proxy for plotting the position of cold vs. hot plates, in other words, how and to where the plates are moving.

So now we conclude with our two "big picture" conclusions.

Conclusion 1: The Angle of Dangle

Back in *Chapter 4* (*Figure 4-3*) we saw that the rate of convergence of North America with the Pacific oceanic plates varied with time. *Figure 15-2* shows the successive plate collisions in more detail, where the oceanic *Farallon plate*, a slab of the dense, basaltic oceanic lithosphere, descends (subducts) under the less dense North American plate of granitic crust. But where does the descending Farallon plate go? What happens to it?

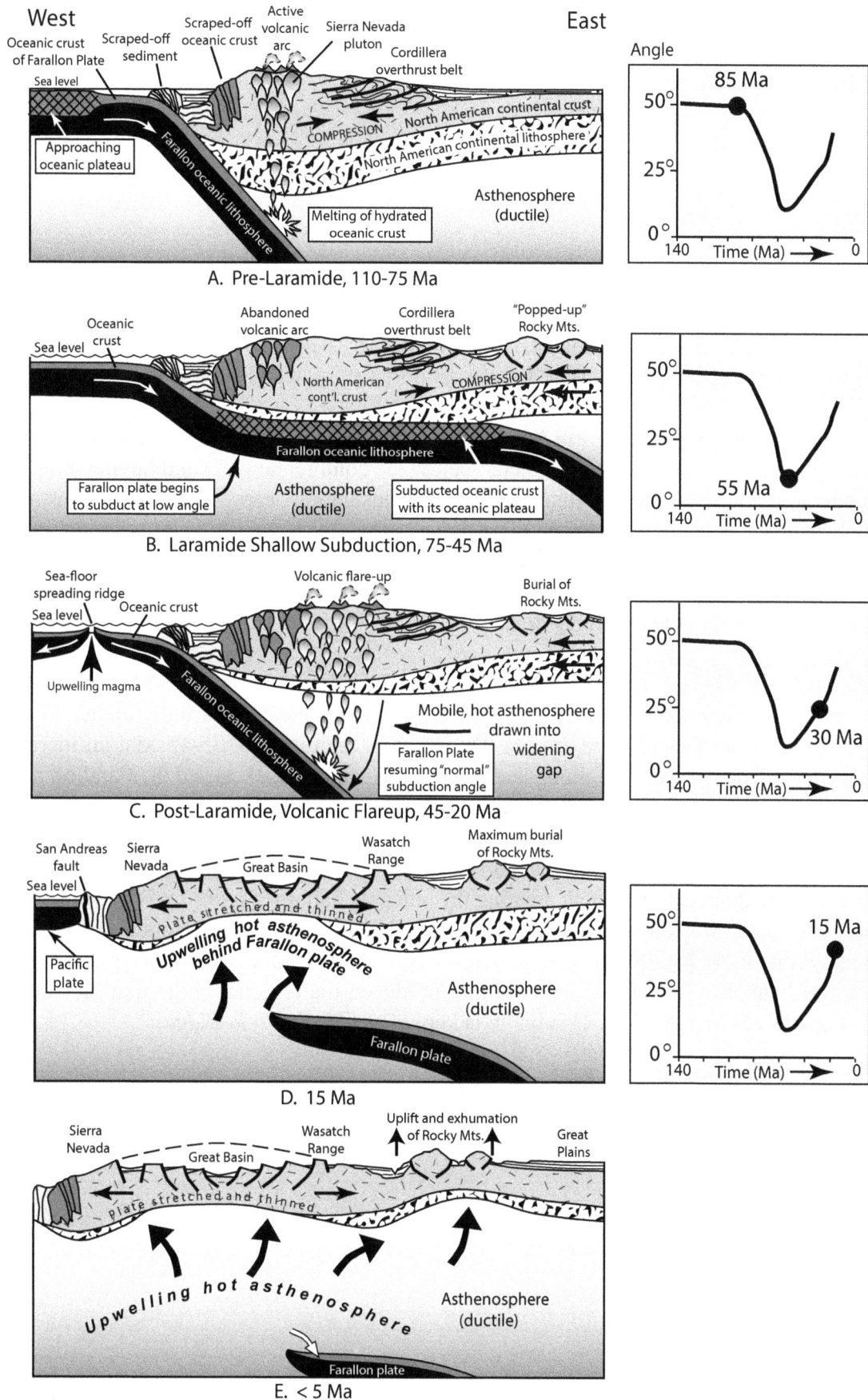

Figure 15-2. Schematic Cross Sections Showing Subduction of Farallon Plate, and Plots of Its Subduction-Angle Through Time (Modified from Chapin 1987; Meldahl 2013)

The angle of plate subduction (descent) and later path of the subducted Farallon plate exerted vast control over the behavior of the overriding North American plate. In general, this subduction "angle of dangle" is the guiding hand, the ultimate control for the formation of the Rocky Mountains, the Sandia Mountain, the Rio Grande rift, and the Ortiz Igneous Belt. It's that important! So, let's take a look.

Far from simply disappearing under North America (as some cross sections imply), the Farallon plate is that house guest who won't go home! Once subducted below the continental plate and having moved a distance from the subduction zone, the oceanic slab acts as a cool, insulating "cap." It therefore prevents heat from escaping to the surface, allowing it to accumulate down below. But subducted plates are not indestructible. They break, and when they break their heat-sealing capacity is compromised. Mobile, hot mantle material from below then oozes upward through the opening. This creates a super slow-motion, viscous "weather-like" surge event that bulges the overlying crust.

Here is a summary of past events. Until the beginning of the Laramide at ca. 75 Ma, subduction of the Farallon oceanic-plate was "slow," occurred at a relatively high angle, and partially melted to form a chain of volcanoes up at the surface (*Figure 15-2A*). During the Laramide, ca. 75-45 Ma, an increased rate of plate convergence did not give the Farallon plate time to sink, and instead caused the Farallon plate to shove under the North America continent at a shallow angle (*Figure 15-2B*). Because the depth of partial melting of the Farallon plate had not been reached, surface volcanism stopped. Following the slow-down of plate convergence near the end of the Laramide, ca. 40 Ma, the rump Farallon plate began to once again steepen its subduction angle and again partially melt, creating the volcanic flareup (*Figure 15-2C*). Complete rupture of the Farallon plate ca. 20 Ma, known as the "Big Break" (Ricketts 2014), again shut down volcanic activity because there was no more Farallon plate to partially melt. The void created by the Big Break permitted by a massive upward surge of hot mantle material that produced the regional bulge that includes the Rocky Mountains (*Figures 15-2E* and *15-3*). This bulge is known as the *Alvarado Ridge*.

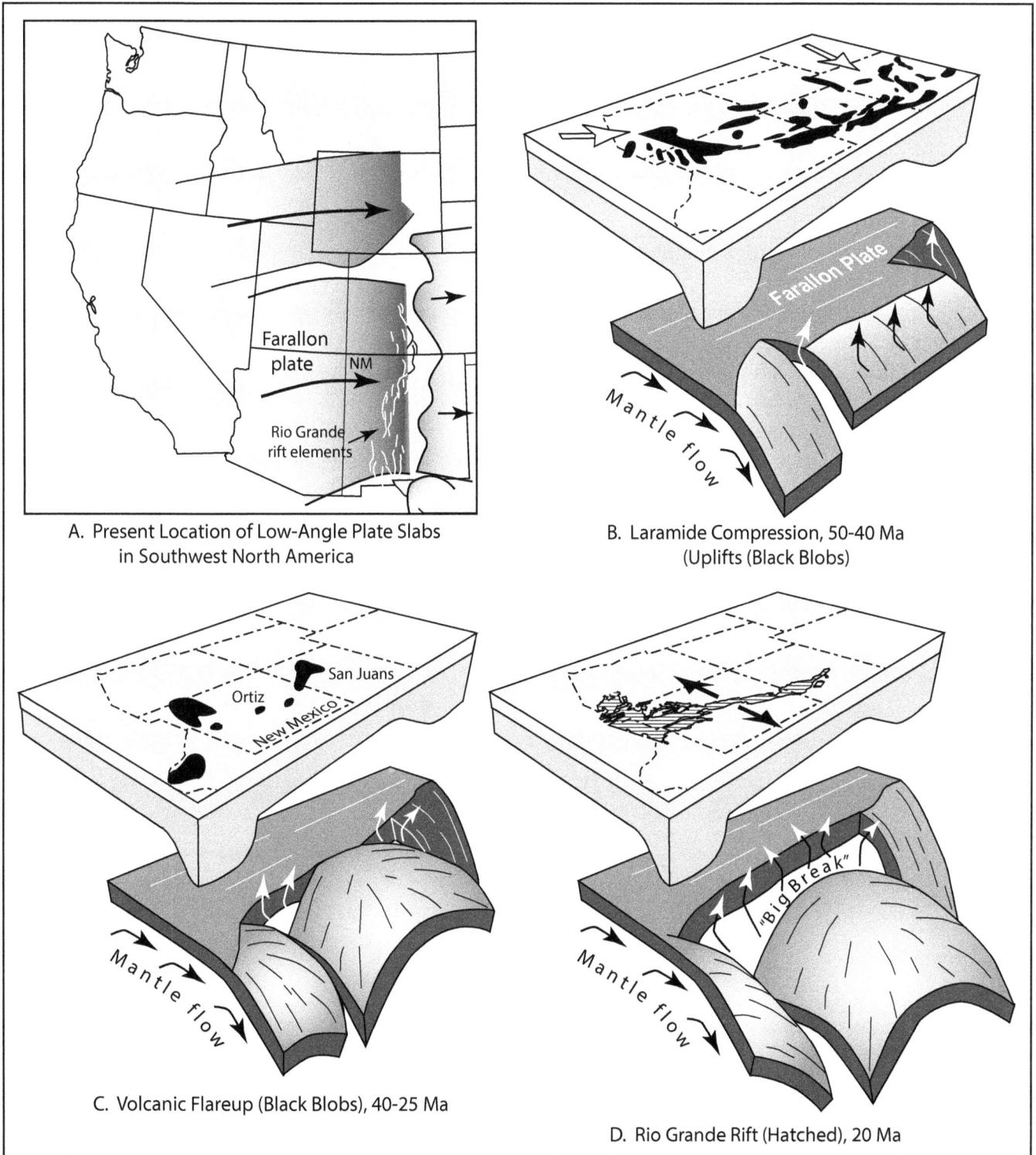

A. Present Location of Low-Angle Plate Slabs in Southwest North America

B. Laramide Compression, 50-40 Ma (Uplifts (Black Blobs)

C. Volcanic Flareup (Black Blobs), 40-25 Ma

D. Rio Grande Rift (Hatched), 20 Ma

Figure 15-3. The Big Picture Below New Mexico, 50-20 Ma (Modified from Ricketts and Karlstrom 2016)

Conclusion 2: Alvarado Ridge and Sandia Mountain

We usually think of the Rio Grande Valley and the Albuquerque basin portion of the larger Rio Grande rift as a topographic trough, an elongated low area. It is, but it's a low within a high. The Big Break and upward surge of mantle material has bowed up the crust to create a broad, nebulous arch known to geologists as the Alvarado Ridge (*Figures 15-4A* and *15-4B*). This is nicely revealed by a regional topographic map of the Southern Rocky Mountains that has been "smoothed" (*Figure 15-4C*). The Southern Rockies, with all its rugged peaks and deep valleys, are perched atop the broad Alvarado Ridge, and the Rio Grande rift is a trough within that ridge. Sandia Mountain is merely part of an uplifted, rotated shoulder of the trough, the Rio Grande rift. Makes one feel rather small!

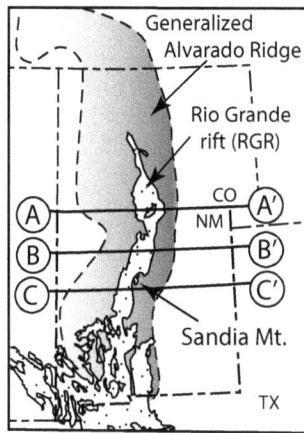

A. Location Map

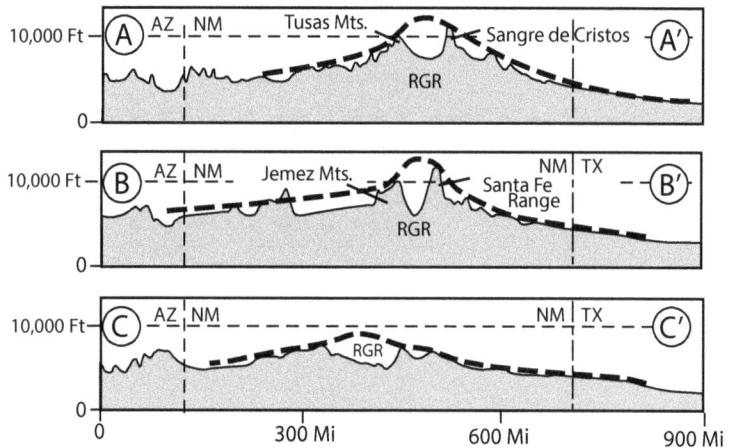

B. Topographic Cross Sections Across Alvarado Ridge (Heavy Line; Modified from Roy at al. 1999)

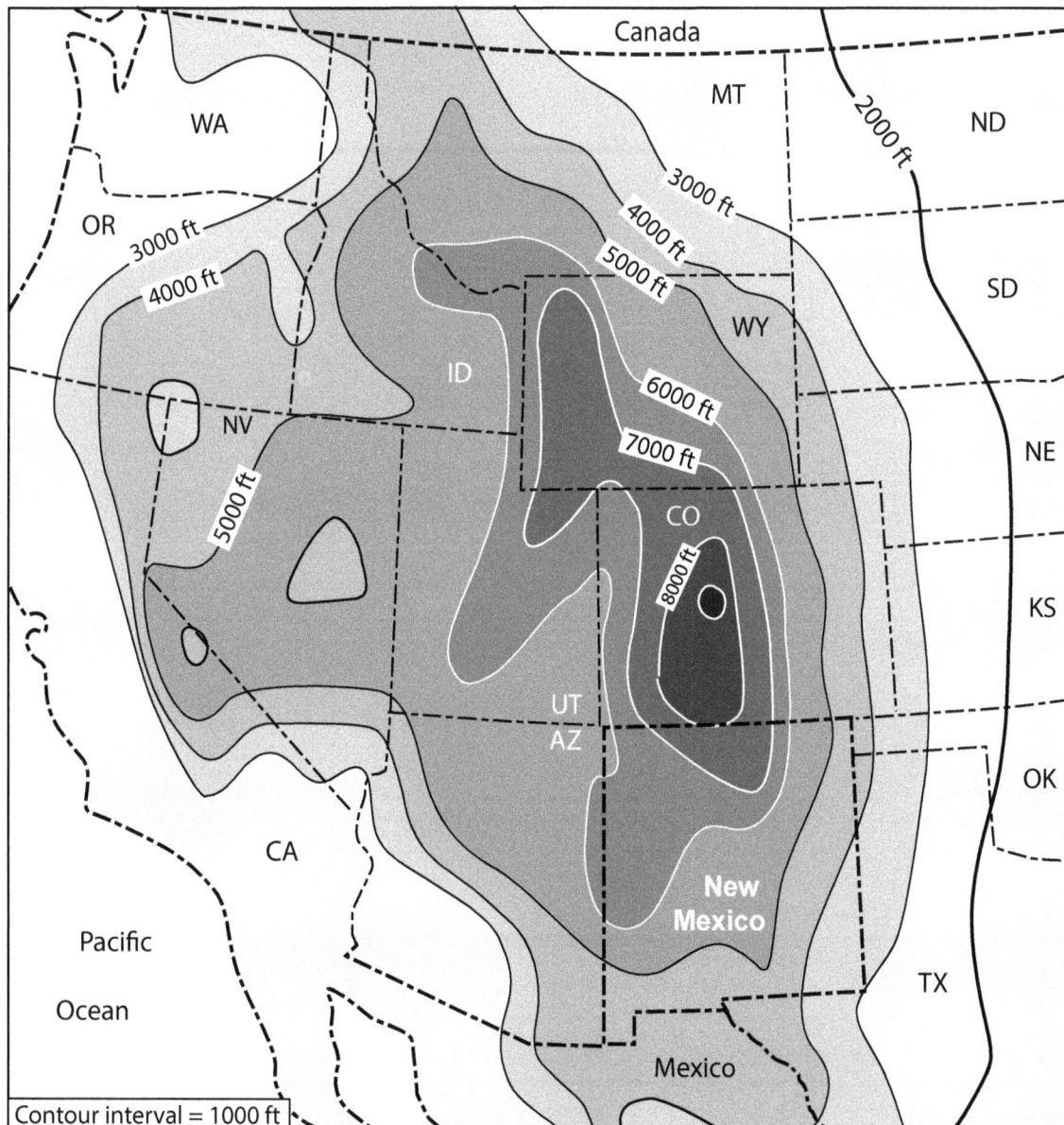

C. "Smoothed" Topography of Western North America (Based on Bouger Gravity; Modified from Eaton 2008)

Figure 15-4. Alvarado Ridge

EPILOGUE:
PARTING WORDS

"Once you eliminate the impossible, whatever remains, no matter how improbable, must be the truth."(Arthur Conan Doyle, British author, 1859–1930)

The wonderful nugget above (also cited in *Chapter 9*), although wise, must be taken with a "grain of sand." It is valid *if and only if*, all the possibilities (vs. the impossibilities) are known and considered. Today's impossibilities can be tomorrow's realities. There's the rub—knowing all the unknowns. In geology, and much of science in general, discovery of the unknowns is an on-going process.

It appears rather arrogant to suggest that we now "know" the story of Sandia Mountain and its companion Rio Grande rift. I will stick my neck out and state that we are indeed getting close, but a few unknowns still remain, including *at least* the four below:

1. The moderately-deep structure under the Rio Grande rift adjacent to and west of Sandia Mountain, and how it relates to the mountain. Presently the only known way to resolve this is via reflection geophysical (seismic) surveys or deep drilling. Seismic surveys, used so successfully in the oil and gas exploration business, require generating shock waves at the surface and recording their "echoes" from depth with an array of instruments. The shocks are generated from either dynamite charges set off in shallow holes drilled in the ground, or from intense vibrations (*Vibroseis*) set off from an array of heavy, vibrating trucks. It is difficult to imagine the legal permits required to allow dynamite work being done along Albuquerque's busy streets! Seismic work, for now, seems to lie very far (if ever) in the future.

2. The deep, 3D shape of the Sandia Granite pluton. The cross-sectional shape shown by the figures in this book is speculative, based on models, analogs, and limited surface exposures, and will remain so.

3. The parentage of some of the major faults, and which are reactivated Laramide faults.

4. The role of low-angle faulting on the mountain's west side, developed as the mountain formed.

~~~

In closing, I sincerely hope this book dispels the notion that the Sandia Mountain is simply an uplifted block of rock. The mountain has the good fortune of being unusual, incredibly picturesque, and favorably located at New Mexico's principal population center. The mountain has a long, variable history, with each phase superimposed onto the prior one. In sum, these complexities make for a fascinating story. This book aims to simplify that story, to pull all the ideas together, to point out some non-obvious features, and to make it all interesting. Above all, the aim is to make the entire tale accessible and understandable to everyone interested in geology, including those without a technical background.

APPENDIX I:
SELECT PREVIOUS GEOLOGIC WORK

(Not-at-all inclusive)

In 1853 Congress authorized reconnaissance work to be done across New Mexico for a potential railroad route to the Pacific. Jules Marcou, a French-German geologist, was assigned to the U.S. Corps of Topographic Engineers to investigate the 35th parallel. One of his principal goals was to "synchronize the sedimentary rocks of America with those of Europe" (Lucas 2001). Marcou climbed the "Sierra de Sandia," collected fossils, and recognized that the mountain was capped by Carboniferous (= Pennsylvanian) limestone that was then called "Mountain Limestone." A cross section across Sandia Mountain was produced by others from his notes (*Figure I-1A*; Kelley and Northrop, 1975).

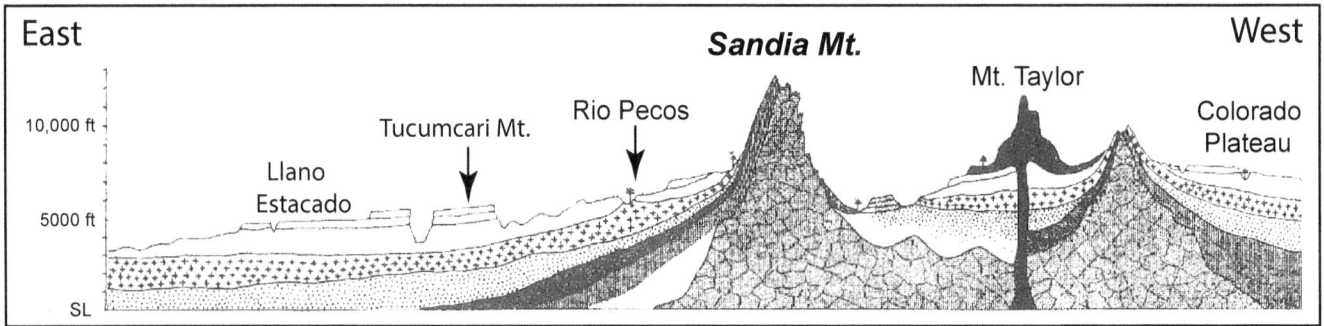

A. Jules Marcou's 1858 East-to-West Cross Section across Sandia Mountain
(Modified from Kelley and Northrop, 1975)

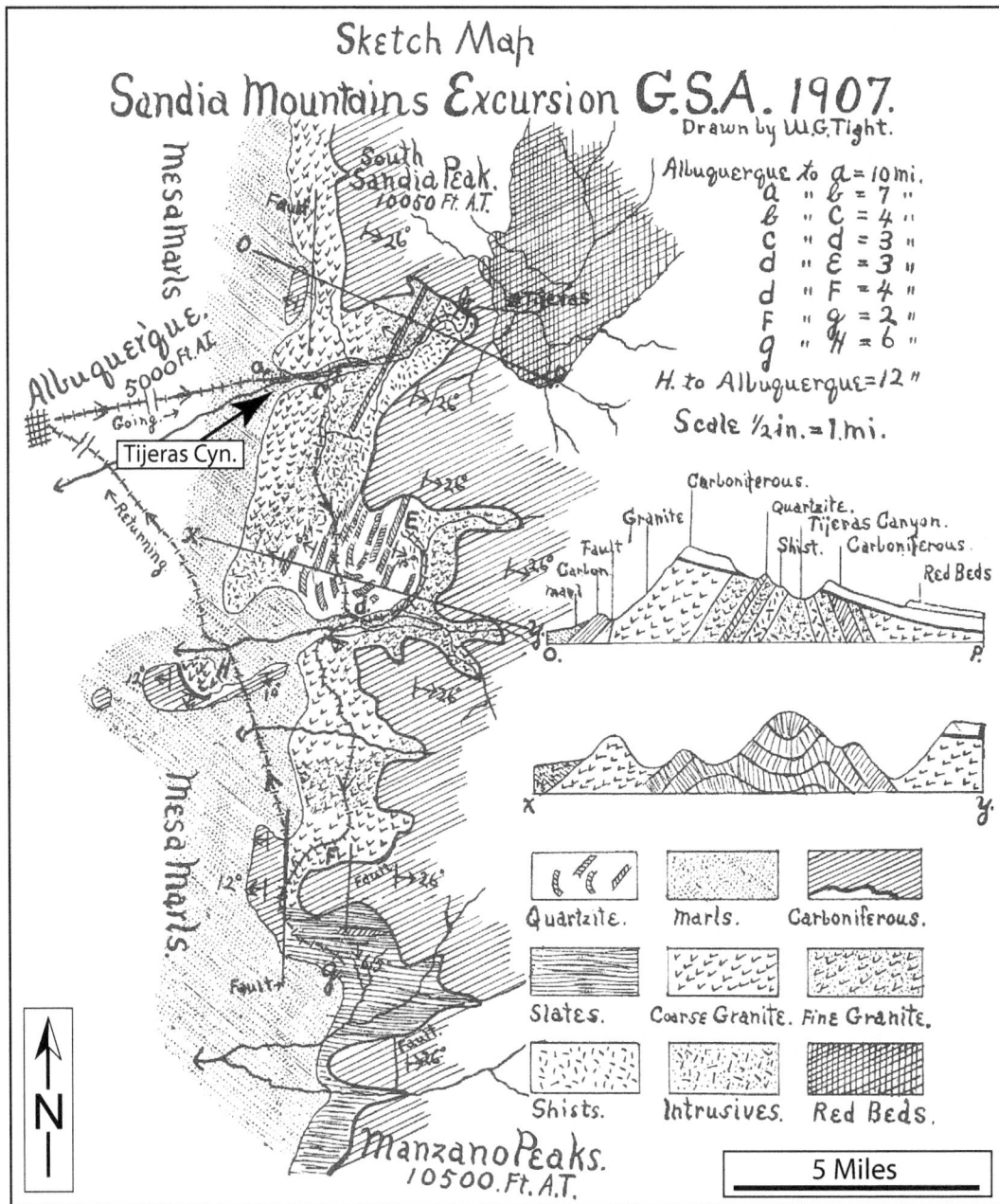

B. Geological Society of America's 1907 Geologic Sketch Map of Southern Part of Sandia
Mountain and Manzanita Mountains (Modified from Kelley and Northrop 1975)

Figure I-1. Early Geologic Work

In 1907 the Geological Society of America held its annual meeting in Albuquerque, during which William G. Tight, president of the State University of New Mexico (precursor of the University of New Mexico) submitted a "sketch map" of the southern half of Sandia Mountain, extending south to the Manzano Mountains (*Figure I-1B*; Kelley and Northrop, 1975). In the summer of 1921, Robert W. Ellis, State Geologist and professor of geology at the State University of New Mexico, mapped Sandia Mountain. His 1:48,000-scale geologic map of the mountain was the most detailed produced up to that time (Ellis, 1922, reproduced as *Figure I-2*.) The geologist/historian will be fascinated by the location shown on Ellis' map of several goat ranches at the west foot of the mountain.

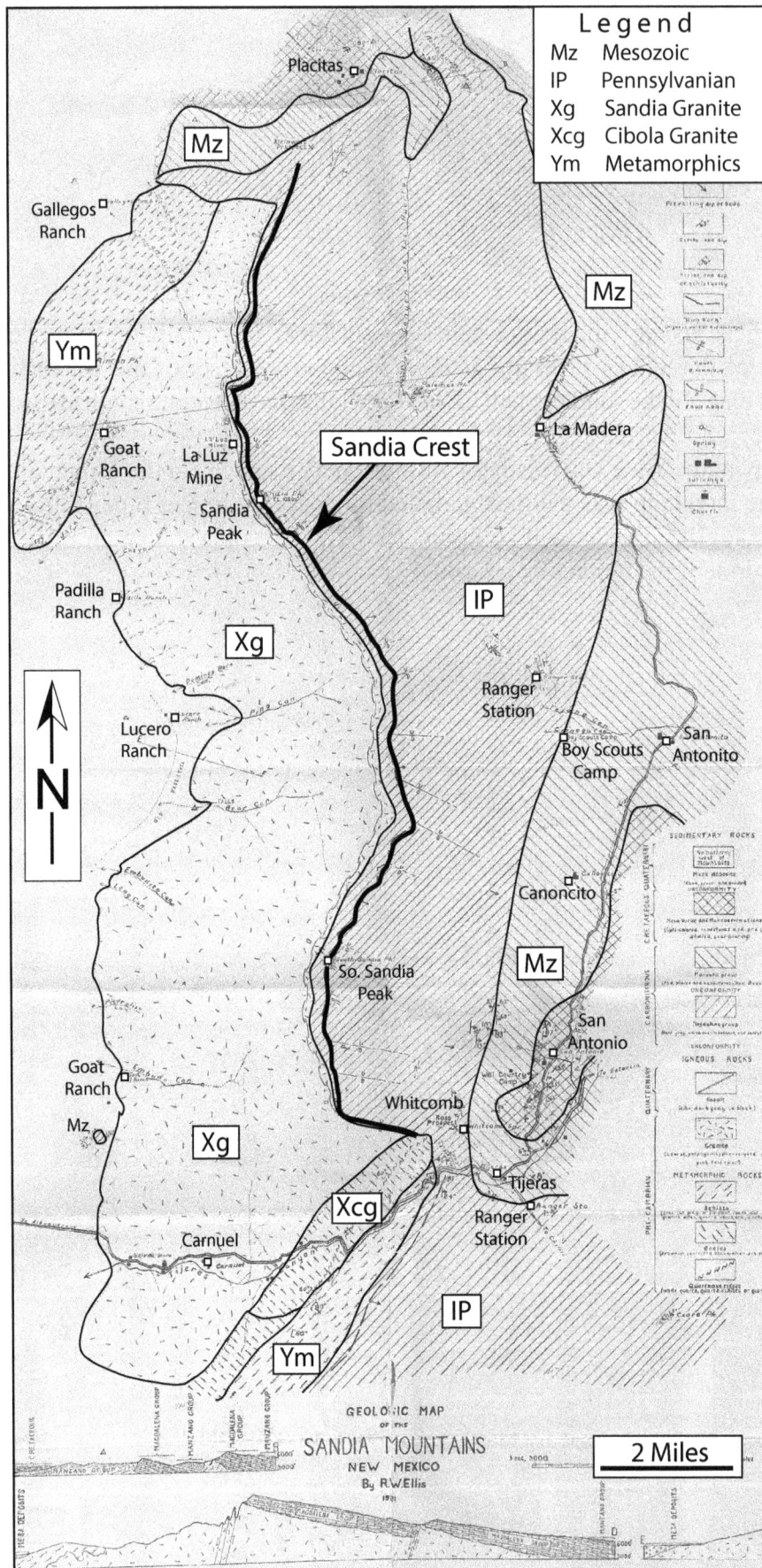

Legend

Mz	Mesozoic
IP	Pennsylvanian
Xg	Sandia Granite
Xcg	Cibola Granite
Ym	Metamorphics

Figure I-2. Ellis' 1922 Geologic Map of Sandia Mountain
(Modern Formation Symbols Added and Formation Contacts Accentuated for Clarity;
Select Place Names Emphasized)

P.T. Hayes, as part of his Master's degree work at the University of New Mexico (UNM), studied the Precambrian rocks of the Juan Tabó Cul-de-Sac area and carefully mapped the boundary between the metamorphics and the granite (Hayes 1951). Vincent C. Kelley, professor of geology at UNM, produced a regional geologic map of the upper Rio Grande Valley area showing considerably more detail (Kelley 1954, 1961). Another graduate student at UNM, H. Feinburg, mapped the Precambrian rocks in the southern part of the Juan Tabó area that lies within Bernalillo County (Feinburg 1969). Significantly, he carefully investigated the upper reaches of an important structural element, the La Cueva lineament.

Vincent C. Kelley (UNM) in 1969 published the first edition of a little book for the layman, *Albuquerque: Its Mountains, Valley, Water, and Volcanoes*. It went through three editions, the last in 1982. Kelley and fellow UNM professor of geology, Stuart A. Northrop, produced their classic memoir, *Geology of Sandia Mountains and Vicinity, New Mexico* (Kelley and Northrop 1975), in which they compiled all the previous work and provided detailed maps and cross sections covering the entire mountain. This publication is invaluable, but admittedly more aimed at the geologist.

Berkley and Callender (1979) described the contacts and metamorphic rocks present along the intrusive contact of the Sandia Granite along FS-333 within the Juan Tabó Cul-de-Sac. UNM graduate student E. Kirby for the first time characterized the Sandia Granite as a northwest-dipping pluton, with its base located along Tijeras Canyon and its top in the Juan Tabó area (Kirby 1994). Rhoades and Callender (1983) studied an area of low-angle shearing in the Sandia Granite on Tramway Road, west and downhill from the Sandia Tram.

Richard Lozinsky, a graduate student at the New Mexico Institute of Mining and Technology (today's New Mexico Tech), produced a massive Ph.D. thesis on the sedimentary fill of the Albuquerque-basin portion of the Rio Grande rift, with implications for the formation of Sandia Mountain (Lozinski 1988).

Steve M. Cather, of the New Mexico Bureau of Mines and Mineral Resources (NMBM&MR), in an outside-the-box paper, set the stage of the Sandia Mountain/Rio Grande rift tectonic couple (Cather 1992). Lee A. Woodward, professor of geology at UNM and his former student, Barbara Menne, tackled the knotty structural problem of the northern Placitas area of the mountain (Woodward and Menne 1995). In 1995, Sean D. Connell, in a comprehensive study of the piedmont on the north and west sides of Sandia Mountain, provided a chronology for mountain's uplift history and consequent depositional events (Connell 1995).

The NMBM&MR, later the New Mexico Bureau of Geology and Mineral Resources (NMBG&MR) published its series of geologic maps at the scale of 1:24,000, using the U.S. Geological Survey topographic quadrangle maps as bases. These superb maps include: Tijeras (Karlstrom et al. 1994), Placitas (Connell et al. 1995), Sandia Park (Ferguson et al. 1996), Bernalillo and Placitas (Connell et al. 1999), Sandia Crest (Read et al. 1999), and Hagan (Cather et al. 2002).

In 1996 and 1997 a series of aeromagnetic surveys were flown over the Albuquerque basin and analyzed by "Tien" Grauch (1999). The surveys delineated many buried, previously unrecognized faults in the basin, some with implications for Sandia Mountain.

In 1999 a team of geologists headed by K.E. Karlstrom of UNM published a tectonic history of the Rio Grande rift and its flanking structures, including Sandia Mountain (Karlstrom et al. 1999), Karlstrom and Pazzaglia (1999) summarized the ongoing and simmering controversy concerning the uplift history of Sandia Mountain, and Karlstrom (1999) addressed the strange and perplexing lower contact of the Sandia pluton. That same year, UNM graduate student Mousumi Roy, with a team of advisors, introduced the concept of a rebounding Sandia Mountain in response to Rio Grande rifting, the "rift-flank uplift" model (Roy et al. 1999). That same year an excellent geologic road log of Sandia Mountain was published in Lucas et al. 1999. Rounding out that busy year, D. Van Hart produced a study of the enigmatic little exposure of Mesozoic rocks in the Juan Tabó Cul-de-Sac (Van Hart 1999).

In 2003, a three-person team greatly improved understanding of the Sandia Mountain's uplift history, using sophisticated laboratory methods newly developed in the 1990s (House et al. 2003). Also in 2003, Bauer et al. authored a fine addition to the New Mexico Bureau of Geology's Scenic Trip Series (No. 18) entitled, *Albuquerque: A Guide to its Geology and Culture* (Bauer et al. 2003).

In 2004, Scott Baldridge, research scientist at Los Alamos National Laboratories, published his excellent volume, *Geology of the American Southwest*, which provides the broader context of the entire area (Baldridge 2004). In 2005 Shaw et al. authored an intriguing paper on the thermal history of the southern Rocky Mountains, including the intrusion of the Sandia Granite (Shaw et al. 2005). In 2014, Jason Ricketts, a graduate student at UNM, in a provocative Ph.D. thesis, addressed the thorny issue of low-angle normal faults in the northern Sandia Mountain (Ricketts 2014). Finally, in 2016 Ricketts and Karlstrom issued a paper invoking deep mantle processes to explain the Rio Grande rift and its companion, Sandia Mountain. (Ricketts and Karlstrom, 2016).

To round out the collection of available geologic maps, there are the two that cover a broader area: 1) *Geologic Map of the Albuquerque 30' x 60' Quadrangle* at a scale of 1:100,000 (Williams and Cole 2007), and 2) *Geologic Map of the Albuquerque-Rio Rancho metropolitan area and vicinity* at a scale of 1:50,000 (Connell 2008).

In addition to the geologic work cited above, there are number of very useful field and visitor guides to the mountain. These include *Guide to the New Mexico Mountains* (Ungnade, 1965, 1972); *Hikers and Climbers Guide to the Sandias* (Hill 1983); *Sandia Peak Tramway: The Historical Picture Story* (Sandia Peak Tram Co. 1966); *Visitors Guide: Sandia Mountains* (Maurer, ed. 1994); *Sandia Peak: A History of the Sandia Peak Tramway and Ski Area* (Salmon, 1998); *Sandia Mountain Hiking Guide* (Coltrin 2005); *Field Guide to the Sandia Mountains* (Julyan and Stuever, eds. 2005); *The Mountains of New Mexico* (Julyan and Smith 2006); the large-format photo book, *Sandia: Seasons of a Mountain* (Muench and Rudner 2018); and a well-illustrated hiking-trail guide, *60 Short Hikes in the Sandia Foothills* (Massong 2018.

APPENDIX II:
GEOLOGIC TIME: HISTORY OF A CONCEPT

The Beginning

It took centuries for "educated" people (i.e., the ones that could read) to accept the concept of geologic time and the notion that the Earth might be much older than mankind had contemplated up to then. The reason of course was that for those centuries the only existing source of scholarship was the story of Genesis in the Bible. The six days of creation (*Genesis*, 1:31) predicted that the world would exist for "six ages" of indefinite duration. The apostle Peter later provided a possible duration for these days when he wrote, "One day is with the Lord a thousand years and a thousand years are as one day" (*New Testament, II Peter*, 3:8). Six ages times 1000 years, giving an age of 6,000 years for the Earth, seemed to have ironclad support backed by the ultimate reference book.

In 1654 James Ussher (1581–1656), bishop of Armagh in Northern Ireland and a respected biblical scholar, set about to determine the exact age of the Earth by adding up the lifetimes of Biblical personages and the reigns of kings. After much excruciating calculation he concluded that the Earth was created on Sunday, October 23, 4004 BCE, and that it had remained unchanged ever since (Cutler 2003). Again, the idea of an Earth that was 6,000 years old was given more credibility.

But ideas very slowly began to change. In the late 17th century, Nicolaus Steno (1638–1686), a Danish physician living in Italy, was carefully observing the physical world near his home. In his 1669 work (its title mercifully shortened to *De Solido*, meaning roughly "On Solids"), he posited that a sequence of layered rocks becomes younger upwards. In other words, each layer of rock was deposited upon the older, pre-existing layer. The layers themselves then became a proxy for time. (Today we call this observation the "law of superposition.") Steno was suggesting that the Earth contained information that was intelligible to mere mortals. This was a bold and quite revolutionary leap and, although he didn't realize it, he had just laid the foundation for the science of geology (Cutler 2003). However, Steno was mum about the absolute age of the Earth because he simply didn't have a clue what it might be.

In the mid-18th century, Georges Buffon (1707–1788), a French amateur scientist believing (correctly) that the Earth had originally cooled from a molten lump, calculated how long it would take a given lump the size of the Earth to cool to its present temperature. He came up with an exquisitely precise 74,822 years (Goudsmit and Claiborne 1966). This was an enormous increase from Ussher's 6,000 years.

In the latter part of the 18th century a physician/scientist/farmer named James Hutton (1727–1797) began a systematic study of the layered rocks exposed in his native Scotland. His observations, particularly one of the angular unconformity at Siccar Point on the east coast of Scotland (*Figure 3-1B*) led him to ponder the concept of "deep time" and compelled him to argue for a vast age of the Earth. This was echoed in 1859 by the English biologist Charles Darwin (1809–1882), who believed that the time required to account for the observed sequence of fossils that he observed in the rocks should be in the range of perhaps 400 million years.

By the middle of the 19th century "gentlemen" scientists had intensely studied many sequences of layered sedimentary rocks, especially those in England, along with their enclosed and upward-changing fossil remains, and had assigned names to the different sequences. They had thereby constructed a "relative" time scale, allowing observations such as "this sequence overlies that one and is therefore younger," and "that sequence underlies this one and is therefore older." But the "absolute" age of the Earth remained a mystery because of the lack of an effective yardstick.

Just then, when a revolution in thought seemed to be at hand, a powerful dissenting voice made itself heard. In about 1865 the English physicist Lord Kelvin (a.k.a. William Thomson, 1824–1907), harking back to the work of Buffon and the cooling of spheres, made calculations of geothermal heat (the flow of heat from the Earth's interior). He concluded that the Earth was not very old at all, and that the cooling took maybe only a few tens of millions of years. There the issue remained until the end of the 19th century.

At Last, a Geologic Clock

In about 1895 the French physicist Henri Becquerel (1852–1908) mistakenly left a piece of a salt of uranium on a photographic plate. The next day, streaks on the plate showed that the uranium salt was throwing off particles. He concluded that certain "radioactive" elements were decaying over time by casting off small pieces.

Finally, in 1907 the English physicist Lord Rutherford (1871–1937) connected the dots. He saw radioactivity as a potential geologic clock, a mechanism to date rocks that contained radioactive minerals undergoing decay. The minerals in igneous rocks typically contain slight amounts of these radioactive elements. Combining the known rate of radioactive decay of these elements with the measured ratios of the decayed vs. undecayed "daughter" elements in the rock, it is possible to assign an absolute age, usually in millions of years plus or minus an error factor. Rutherford also realized, no less significantly, that radioactivity produced heat as a byproduct of the decay (Goudsmit and Claiborne 1966). Kelvin's earlier calculations had not taken into account the continuous addition of heat in the Earth's interior by radioactive decay (today the contribution of the two heat sources, original plus radioactivity, is about 50-50). The impasse was broken. Today the estimated age of the Earth is 4.6 billion years (4.6 Ga). Ussher would be aghast!

Via the combination of extensive geologic mapping around the world and the radiometric dating of thousands of bodies of igneous rock mixed in among many sequences of sedimentary rock, we have been able to calibrate the relative geologic time scale, based on fossils, and to construct an absolute time scale, based on radioactivity.

Putting it All Together: The Geologic Time Scale

The earliest geologic field work took place in Europe. Late in the 18th century, Abraham Werner (1749–1817) recognized that rocks seem to occur in a definite order, and took the first critical steps to propose a relative geologic timetable. He noticed that in his native Saxony (in today's Germany) everywhere below the layered sedimentary rocks appeared hard, crystalline rocks that lacked fossils (*Figure II-1*). He supposed that these latter rocks represented the original crust of the Earth and he accordingly called them *Primary*. Above these were those horizontal sedimentary rocks that he called *Secondary*, and the thin layer of unconsolidated sands and gravels above the Secondary he named "Alluvium." Later, nearby, he noticed that there was a thick sequence of sedimentary rocks between the Primary and the Secondary, which he then called *Transition*. Although he never traveled outside of Saxony, he assumed this subdivision of the rock column held for the entire world.

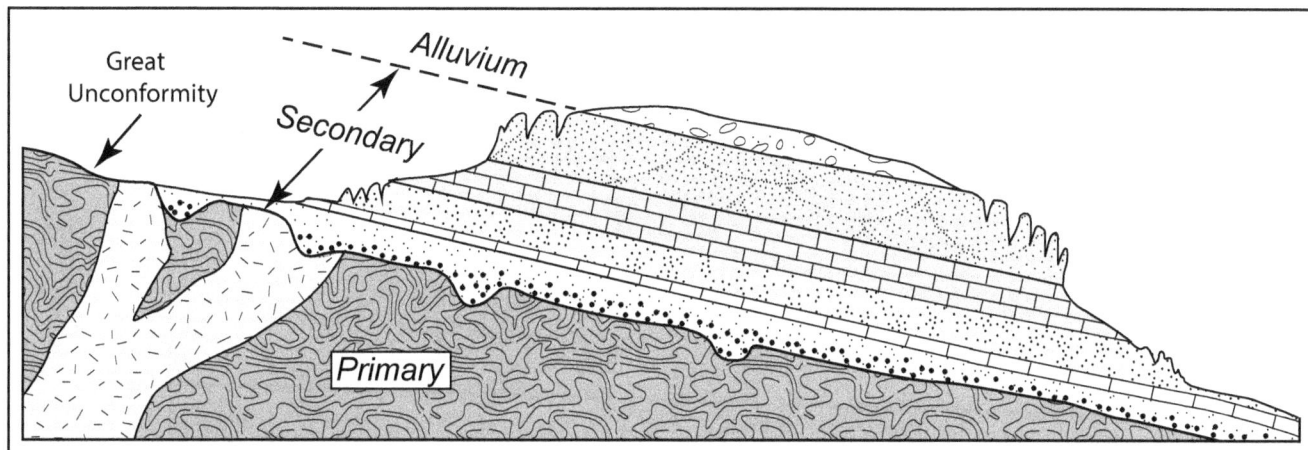

Figure II-1. Cross Section of Rocks Studied by Werner in the late 18th Century in His Native Saxony, Showing His Three-Fold Classification (Modified from Clark and Stearns 1960)

Other workers observed that in other places there was a thick sequence of sedimentary rocks lying between the Secondary and the Alluvium. These logically became the *Tertiary* and the Alluvium became the *Quaternary*. (Clark and Stearn 1960). Confusing?

The fossils in the post-Primary rocks showed a progression toward modern life forms. This progression led to a replacement of Werner's terms with "old life" or *Paleozoic* for the Transition, "middle life" or *Mesozoic* for the Secondary, and "modern life" or *Cenozoic* for the Tertiary and alluvium. Until recently the Cenozoic retained the term "Tertiary" for its lower part, and the term *Quaternary* for the alluvium of its uppermost part. Today the Primary and Secondary are gone. Even the Tertiary is almost gone, replaced by a lower *Paleogene* and an upper *Neogene*, although the literature abounds with the old terms.

The study of historical geology grew up mainly in England during the early 19th century (*Figure II-2*). Over the course of mapping coal seams and digging canals, a sequence of geologic names for the time periods entered the geologic literature, based on the law of superposition and drawing on ancient British history. From bottom to top these names are the *Cambrian* (from "Cambria," the ancient name for Wales), *Ordovician* (from the "Ordovices," a Celtic tribe), *Silurian* (from "Silures," a Welsh tribe), *Devonian* (from "Devonshire" in England), and *Carboniferous* (for the English coal beds). (Note: In North America the Carboniferous is divided into a lower, *Mississippian* Period and an upper, *Pennsylvanian* Period.) Later, English geologists, studying a coal-bed sequence in Russia that was younger and lying above the Carboniferous, named it *Permian* (for the province of "Perm").

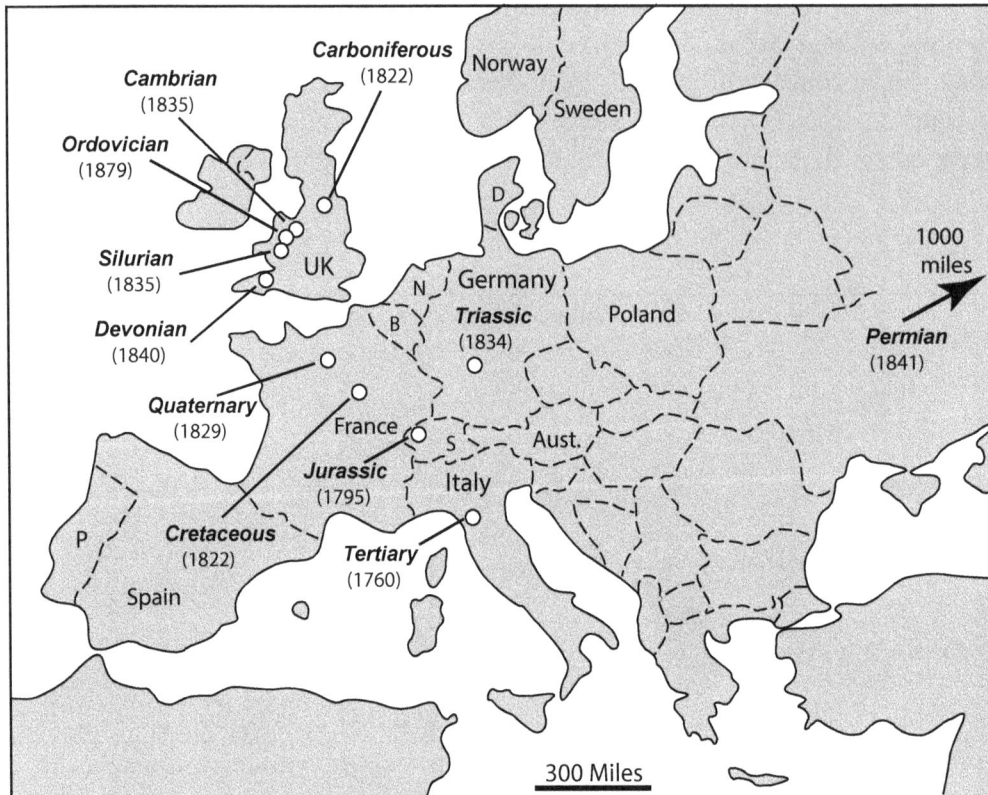

Figure II-2. Eurocentric Origin of Geologic Time-Period Names
(Select Modern Geographic Borders for Reference)

The Mesozoic (formerly known as "Secondary") was studied extensively in Europe and early on was divided into three parts. The oldest, itself easily divided into three parts based on color, was accordingly called the *Triassic*. The middle unit was well-exposed in the Jura Mountains and became the *Jurassic*. The upper unit derives from the chalk deposits of the Paris Basin, the *Terrain Cretacé*. The name was soon taken up by English geologists who renamed it *Cretaceous*.

The larger divisions (Paleozoic, Mesozoic, Cenozoic) were designated *eras*, while the shorter divisions were designated *periods*, and within them were *epochs*. The crystalline, Primary rocks, barren of fossils and lying below the Cambrian, were by default named *Precambrian*. Today, with enhanced radioactive dating methods and the discovery of some very rare and primitive fossils, the term Precambrian has fallen out of favor and has been replaced by an earliest *Hadean* (4.6-4.0 Ga), an intermediate *Archean* (4.0-2.5 Ga), and a youngest, *Proterozoic* (2.5 Ga-542 Ma). The name Precambrian however does appear in a great deal of geologic literature, especially the older material.

From this menagerie of names, a composite relative and absolute time scale has emerged (*Figure 0-13*). This amazing system is a world-wide master scale that classifies and dates the total 4.6-Ga span of total geologic, "deep" time, and is a triumph of scientific inquiry.

APPENDIX III:
GLOSSARY

AFT: Apatite fission-track (also see apatite below). Microscopic track or path of damage within a crystal of apatite caused by radioactive decay. Tracks accumulate when subject at cooler (< 300° C) temperatures upon uplift.

Alluvial fan: Fan-shaped deposit of sand/or gravel deposited by a stream at the foot of a mountain and gently sloping away from it, in an arid or semi-arid area.

Alluvium: Unconsolidated fragmental sediment deposited by running water.

Andesite: Finely-crystalline, grayish igneous rock consisting of grayish plagioclase feldspar, pyroxene and/or hornblende.

Apatite: Phosphate-bearing igneous mineral, with the complicated formula $Ca_5 (PO4)_3 (F, Cl)$. Forget the formula: the important thing is that it contains very minor amounts of radioactive uranium (U). See AFT.

Apatite fission-track: See AFT.

Aplite (dike): Fine-grained, light-colored igneous rock, usually granitic composition and occurring in dikes.

Arroyo: Flat-floored, steep-walled channel of intermittent stream, typical of semi-arid climates (synonymous with wadi or wash).

Asthenosphere: Mobil zone of Earth's upper mantle occurring at the base of the lithosphere.

Bajada: Broad, continuous, gently-inclined slope atop a number of merged alluvial fans extending the from foot of a mountain range into a basin.

Basalt: Finely-crystalline, heavy, dark igneous rock consisting of pyroxene and Ca-rich plagioclase feldspar.

Base level: Lowest level toward which erosion constantly approaches. Ultimate base level is sea level.

Bedrock: General term for solid rock that underlies unconsolidated surficial material.

Biotite (mica): Dark, platy igneous-rock mineral with the complicated formula $K (Mg, Fe)_3 (AlSi_3O_{10}) (OH)_2$; note the OH group, which adds a volatile constituent to granite and granitic magma.

Brittle-ductile transition zone: Depth zone in Earth's crust transitional between cooler, brittle rock above and hotter, ductile (deformable) rock below, usually at a temperature level of about 300° C.

Caliche: Accumulation of $CaCO_3$ two to four feet below surface in soils of arid or semi-arid areas; a.k.a. calcrete.

Carboniferous: European time period equivalent to Mississippian plus Pennsylvanian Periods used in U.S.; age of European carbon-rich (coal-bearing) rocks.

Chalk: A soft, whitish, very fine-grained calcareous rock composed of microscopic marine shells.

Civilian Conservation Corps (CCC): A national, Great Depression-era work program (1933-1942).

Colluvium: Loose mass of unconsolidated soil and/or rock moving down a slope by gravity creep.

Column (stratigraphic): Diagram showing sequence of geologic formations present at a select location.

Contour line: Line on a map that connects points of equal value (e.g., elevation, thickness).

Core (of Earth): Central part of the Earth composed of iron and nickel, lying at a depth of about 1800 miles (2900 km). Outer core is liquid and its churning movements generate Earth's magnetic field; inner core is solid; outer core/inner core contact is zone of iron-nickel crystallization, and a major source of released latent heat.

Craton: Extensive surface of stable, usually ancient granitic rocks (a.k.a. a shield).

Crinoid: A type of fossil, consisting of a column of stacked disk-like plates of $CaCO_3$, capped by a "head" with multiple plated arms.

Crust:
  Oceanic: Rigid, dark-colored, dense rocks of basalt, forming floors of oceans.
  Continental: Rigid, light-colored, less dense rocks mainly of granitic composition, forming bulk of continents.

Cuesta: A ridge with a steep slope on one side and a gentle slope on the other.

Cz: Standard abbreviation for the Cenozoic Era of geologic time.

Dike: Tabular intrusion of igneous rock, vertical or nearly vertical and cutting across pre-existing rocks.

Dolomite: Magnesium-calcium carbonate mineral ($Ca\,Mg\,(Co_3)_2$); also, a sedimentary rock consisting of the mineral dolomite.

Epoch (time): A unit of geologic time that is a subdivision of a Period

Era (time): A unit of geologic time greater than that of a Period.

Erg: A region deeply covered by shifting sands, a "sand sea"; name comes from the Sahara Desert.

Escarpment: Long, mainly continuous cliff; synonymous with "scarp."

Extrusive (rock): Igneous material (lava) deposited on the surface.

Fanglomerate: Very coarse to bouldery fragmental material deposited in an alluvial fan.

Fault: Planar break in a body of rock along which relative movement has taken place.

Feldspar group: Most abundant of all mineral groups that falls into two alumino-silicate sub-groups: light to dark-colored plagioclase (Ca-Na), and light-colored K-feldspar (K-spar).

Floodplain: Surface or relatively smooth land adjacent to an active river channel.

Foot (fault) wall: Block on the bottom side of an inclined fault.

Formation: Body of rock, differentiated from surrounding rocks, typically stratified, that is mappable at a convenient scale, usually 1:24:000.

FS (or USFS): U.S. Forest Service.

Ga: Standard abbreviation of giga anni; billions of years, billions of years old, or billions of years ago.

Graben: An elongate down-dropped block bound by faults (opposite of a horst).

Granite: Medium to coarsely-crystalline, light-colored igneous rock, consisting mainly of quartz (20-35%) and K-feldspar (> 60%), and minor (usually < 1%) of dark minerals.

Great Unconformity: Term first used by John Wesley Powell from his observations in the Grand Canyon in 1869; an unconformity covering a vast, continent-wide area of Laurentia—the ancient core of North America.

Grus: German word for in-place gravely product from disintegration of granitic rock.

Hanging (fault) wall: Block on the top side of an inclined fault.

Hornblende: Dark, prismatic, igneous-rock mineral with the complicated formula $NaCa_2(Mg, Fe, Al)_5 (Si, Al)_5 O_{22} (OH)_2$; note the OH group, which adds a volatile constituent to granite and granitic magma.

Horst: An elongate up-thrown block bound by faults (opposite of a graben).

Igneous (rock): Rock that consolidates (crystallizes) from molten or partially molten material.

Intrusive (rock): Igneous rock having been injected into pre-existing rock below the surface.

IP: Standard abbreviation for the Pennsylvanian Period of time.

Island (volcanic) arc: Curved linear belt of volcanoes above a subduction zone.

Isostacy: An equilibrium of the Earth's lithosphere "floating" atop the aesthenosphere

Isostatic uplift: Adjustment of Earth's crust to maintain equilibrium between masses of different density.

J: Standard abbreviation for the Jurassic Period of geologic time.

Joint: Fracture in rock along which there is no visible lateral displacement.

K: Standard abbreviation for the Cretaceous Period of geologic time, taken from kreide (German for chalk).

Ka: Standard abbreviation of kilo anni; thousands of years, thousands of years old, or thousands of years ago.

K-feldspar (K-spar): Light-colored, alumino-silicate igneous-rock mineral with potassium (K).

Laccolith: Mushroom-shaped body of intrusive igneous rock, with a flat floor and a curved top, typically intruded into a sequence of sedimentary rock.

Laramide: Large-scale mountain-building compressive event that operated ca. 80 to 40 Ma.

Laurentia: Proto-continent of northern hemisphere.

Limestone: Sedimentary rock consisting of calcium carbonate (or calcite, $CaCO_3$).

Lithology: Description of physical properties of rocks.

Lithosphere: Solid, outer portion of the Earth, including the crust and upper part of the Mantle below it, relatively rigid compared to the plastic asthenosphere below.

Ma: Standard abbreviation of mega anni; millions of years, millions of years old, or millions of years ago.

Magma: Mobile, at least partially-liquid rock.

Magnetite: Minor constituent mineral of granite, Fe_3O_4, that is magnetic, preserves a magnetic field, and has enormous importance in remote geophysical mapping.

Mantle: Zone of Earth below the crust and above the core.

Marine: Relating to the sea.

Metamorphic (rock): Rock derived from pre-existing rock by recrystallization due to increased temperature and/or pressure, usually deep in the Earth's crust.

Mineral: Naturally occurring, solid, inorganic element or compound, having orderly internal atomic structure and characteristic physical properties.

Moho: More properly called the Mohorovicic (pronounced MO-ho-ro-vi-chic) discontinuity, the compositional and geophysical base of the Earth's (granitic) crust, and top of the mantle, usually about 3-6 miles (5-10 km) below the ocean floors and 20-35 miles (about 35-60 km) below the continents.

Monocline: A local steepening or flexure, within an otherwise gentle formation dip.

Mylonite: Banded and/or streaky rock produced by extreme crushing and shearing.

Mz: Standard abbreviation for the Mesozoic Era of geologic time.

Neogene: Latest part of Tertiary Period of time, 25-2.65 Ma, used today for the epochs Miocene and Pliocene.

NMBG&MR: New Mexico Bureau of Geology and Mineral Resources

NMBM&MR: New Mexico Bureau of Mines and Mineral Resources

Normal fault: Fault for which the hanging wall has moved down with respect to footwall.

Olivine: Olive-green, igneous-rock mineral, $(Mg, Fe)_2 SiO_4$.

Orbicule: An oblong body of rock, usually a few inches in size, with crystals arranged in concentric rings around a central core.

Orbicular granite: A rare variety of granite characterized by orbicules

P: Standard abbreviation for the Permian Period of geologic time.

Paleogene: Earliest part of the Tertiary Period of time, 65-25 Ma, used today for the epochs, Paleocene, Eocene, and Oligocene of time.

Pediment: Low-relief, gently-inclined eroded rock surface typically formed in arid or semi-arid regions at the base of a receding mountain front.

Piedmont: Area at the base of a mountain; from Italian piedmonte = foothills, and Latin pedimontium for "at foot of the mountains."

Pegmatite: Exceptionally coarse-grained igneous rock, usually of granitic composition but commonly containing rare minerals, and occurring as irregularly-shaped bodies.

Peridotite: Coarsely-crystalline, greenish igneous rock making up most of the Earth's mantle; consisting mainly of minerals olivine and pyroxene.

Period (time): Unit of geologic time greater than that of an epoch and less than that of an era.

Plagioclase feldspar: Variable group of related alumino-silicate igneous-rock minerals, ranging from dark-colored, calcium (Ca) rich to light-colored, sodium (Na) rich.

Plate (tectonic): Large-scale segment of Earth's lithosphere that moves horizontally, and interacts with adjacent plates at its margin.

Plate tectonics: Theory in which lithosphere is divided into number of large-scale rigid plates that move and interact with each other at their margins.

Pluton: Large body of igneous rock that formed as an intrusion of magma.

Precambrian: All (90%) of geologic time before the Paleozoic, > 540 Ma.

Proterozoic: Youngest of two subdivisions of the Precambrian.

Pyroxene: Dark, prismatic igneous-rock mineral, $MgSiO_3$.

Pz: Standard abbreviation for the Paleozoic Era of geologic time.

Quartz: Glassy, hard, crystalline silica, SiO_2; common igneous-rock mineral.

Quartzite: Metamorphic rock made from an original quartz sandstone, in which the grains have been welded together; one of the hardest rocks in nature.

Reverse fault: Fault along which the hanging wall has moved up with respect to the footwall.

Rock: Naturally-occurring, solid aggregate of one or more minerals.

Rodinia: Proto-continent from which Laurentia was derived.

Schist: Metamorphic rock characterized by parallel arrangement of platy minerals such as micas.

Section: Term often as synonym for stratigraphic sequence.

Sediment: Solid fragmental material derived from weathering of rocks and transported by wind, water, or ice.

Sedimentary (rock):
    Clastic: Rock resulting from consolidation of loose fragmental sediment that has accumulated in layers, at or nearly at the Earth's surface.
    Non-clastic: Rock formed by precipitation from solution (e.g., rock salt) or via organic agents (e.g., certain limestones).

Seismic wave: Elastic shock wave in the Earth.

Shale: Fine-grained, finely-laminated fragmental sedimentary rock made from compacted mud.

Silicate: Common rock-forming chemical unit based upon the molecular group SiO_4^{-2}.

Sill: Body of intrusive igneous rock that occurs as a horizontal or nearly- horizontal sheet, usually along bedding planes of sedimentary rock.

Slickenside: Polished, smoothly striated surface that results from friction along a fault plane.

Slip: Magnitude of movement along a fault surface.

Spreading ridge: Elongate elevation on sea floor along which new oceanic crust is being generated and where the newly-formed, opposing oceanic plates separate.

Stratigraphic: Adjective referring to layering of rock units.

Subduction: Process of one lithospheric plate descending under another.

Subduction zone: Long, linear zone along which subduction takes place.

Surficial (geology): Unconsolidated, loose material, usually of Quaternary age or younger, lying above bedrock.

Superimposed stream: A stream (sometimes written "superposed") that originally flowed at a higher level across gently-sloping soft material, then eroded down through the soft material onto harder rocks with different structure below, and cut across the deeper structures, keeping its original course direction.

Tectonics: Area of geology dealing with large-scale structural and deformational features of the Earth's crust.

Terrain: Region of Earth's surface characterized by its general physical features, e.g., mountains, f (contrast with "terrane").

Terrane: Block of Earth's crust of regional extent, bounded by faults and characterized by a geologic history different from that of contiguous blocks (contrast with "terrain").

Terrace: An abandoned river floodplain.

Tomography (seismic): Technique used to image deep interior of Earth using either natural or man-made, long-wavelength seismic waves.

TR: Standard abbreviation for the Triassic Period of geologic time.

Trench: Narrow, elongate depression of the deep-sea floor.

Tuff (volcanic): Consolidated volcanic ash.

Type section: Originally-described exposed section of rock serving as the definition of a geologic formation or member.

Unconformity: Term first used for outcrop that James Hutton observed in 1788 at Siccar Point on the eastern Scottish Coast; contact between rocks of significantly different ages, types, and/or angular orientation.

UNM: University of New Mexico, Albuquerque.

WPA: Works Progress Association, a Great Depression-era work program.

APPENDIX IV:
LEXICON OF GEOLOGIC UNIT NAMES, MAP SYMBOLS AND TYPE SECTIONS OF SOUTHWEST UNITED STATES

(FROM KUES ET AL. 1982)

Abo Formation (Permian, Pa): Abo Canyon, south end of Manzano Mts. near towns of Abo and Scholle, Torrance Co., NM.

Agua Zarca Ss. (Triassic, TRz): Member of Chinle Fm., Agua Zarca Creek, Rio Arriba Co., NM.

Chinle Formation (Triassic, TRc): Chinle Valley, northeast AZ.

Correo Sandstone (Triassic): Member of Chinle Fm.; Old railroad community of Correo (Spanish for "post office") on old US-66 south of old US-66 and west of NM-6, Valencia Co., NM.

Dakota Sandstone (Cretaceous, Kd): Town of Dakota, NE.

Entrada Sandstone (Jurassic, Je): Entrada Point, northern part of San Rafael Swell, UT.

Espinaso Formation (Paleogene, Te): Espinaso ridge, Hagan basin, Sandoval Co., NM.

Galisteo Formation (Paleogene, Tg): Galisteo Creek east of Cerrillos, Santa Fe Co., NM.

Harmon Sandstone (Cretaceous, Kh): From Newitt Harmon Black, Hagan NM miner.

Hosta-Dalton Sandstone (Cretaceous): Hosta Butte in Dalton Pass, near Gallup, McKinley Co., NM.

Luciano Mesa Limestone (Jurassic): Member of Todilto Fm.; Luciano Mesa, northwest end of Llano Estacado, Quay/Guadalupe Cos., NM.

Madera Group (Pennsylvanian, IPm): Village of Madera, east side Sandia Mt., Sandoval Co., NM.

Mancos Shale (Cretaceous, Km): Mancos Valley, near Mancos, CO.

Menefee Formation (Cretaceous, Kmf) Menefee Mountain, Mesa Verde National Park, CO.

Morrison Formation (Jurassic, Jm): Morrison, CO.

Point Lookout Sandstone (Cretaceous, Kpl), Steep cliffs at Point Lookout, Mesa Verde National Park, CO.

Sandia Formation (Pennsylvanian, IPs): Sandia Mountain, Bernalillo and Sandoval Cos., NM.

Santa Fe Group (Paleogene to Quaternary), Santa Fe, Santa Fe Co., NM.

Summerville Formation (Jurassic, Jm): Summerville Point, San Rafael Swell, UT.

Todilto Formation (Jurassic, Jt): Todilto Park, McKinley Co., NM.

Tonque Arroyo Gypsum (Jurassic): Member of Todilto Fm.; from Tonque Arroyo, Hagan basin, Sandoval Co., NM; sometimes mapped with underlying Luciano Mesa Limestone Member and Entrada Ss., Jet.

Yeso Formation (Permian, Py): Mesa de Yeso (Yeso = Spanish for "gypsum"), Socorro Co., NM.

APPENDIX V
Select Locations and Their Geographic Coordinates (via. *Google Earth*)

Place/Feature	Latitude	Longitude
The "Alcove" (Placitas area):	35° 17' 45" N	106° 27' 18.0" W
Caliza Hill (foothills):	35° 05' 12" N	106° 29' 13.0" W
CCC Camp F-8-N, Sulphur Springs:	35° 10' 12.6" N	106° 22' 14.7" W
CCC Camp F-8-N, Sandia Park:	35° 10' 15.4" N	106° 22' 02.0" W
Elephant Rock (Tijeras Canyon):	35° 03' 43.40" N	106° 28' 31.3" W
Great Unconformity (East Mountain, Sandia Crest Byway):		
1. Doc Long area, mile 1.9, at "Auto Tour 7" sign, partly obscured by vegetation):	35° 10' 26.75" N	106° 22' 34.0" W
2. Tejano Canyon, mile 4.5-4.6, outcrop 330-ft long):	33° 11' 23.4" N	106° 23' 41.0" W
Juan Tabó Cabin (Juan Tabó area):	35° 12' 21.6" N	106° 29' 47.8" W
Old Jaral Ranger Station (Juan Tabó area), Spring Creek Trail; station built by CCC 1933/34, used by Forest Service until 1960s; in ruins today):	35° 12' 01.4" N	106° 28' 59.5" W
Orbicular granite outcrop	35° 04' 50.6" N	106° 25' 52.6" W
Pegmatite Hill (Juan Tabó area):	35° 13' 21" N	106° 29' 47.0" W
Pueblo Blanco (Galisteo Basin):	35° 18' 02" N	106° 00' 00" W
Ranchos fault (west of Placitas):	35° 18' 01" N	106° 28' 22.0" W
Seven Springs fault zone (NM-333, Tijeras Canyon):	34° 04' 00.5" N	106° 25' 50.5" W
Supper Rock (foothills):	35° 04' 26" N	106° 29' 31.0" W
U-Mound (foothills):	35° 04' 53" N	106° 28' 51.0" W

REFERENCES CITED

(Note: Many of these citations are quite technical; they are listed here for reference only)

Affholter, K.A., 1979. Petrogenesis of orbicular rock, Tijeras Canyon, Sandia Mountains, New Mexico"; unpublished M.S. dissertation, University of New Mexico, Albuquerque, 124 p.

Affholter, K.A., and E.E. Lambert, 1982. "Newly-described occurrences of orbicular rock in Precambrian granite, Sandia and Zuni Mountains, New Mexico"; New Mexico Geological Society Guidebook 33, *Albuquerque Country*: p. 225-232.

Allen, B., 2007. *Geologic Map of the Stanley 7.5-minute quadrangle, Santa Fe County, New Mexico.* New Mexico Bureau of Geology and Mineral Resources, map OF-GM-143, scale 1:24,000, Socorro, New Mexico.

Alt, D.D., and D.W. Hyndman, 1989. *Roadside Geology of Idaho.* Mountain Press Publishing, Missoula Montana, 393 p.

Baldridge, W.S., 2004. *Geology of the American Southwest.* Cambridge University Press, United Kingdom, 280 p.

Bauer, P.W., R.P. Lozinski, C.J. Condie, and L.G. Price, 2003. *Albuquerque: A Guide to its Geology and Culture.* New Mexico Bureau of Geology and Mineral Resources, Scenic Trip Series No. 18, Socorro, New Mexico, 183 p.

Bannerman, T., 2008. *Forgotten Albuquerque.* Arcadia Publishing, Images of America Series.

Benedict, C.B., 1994a. "Proposed graffiti removal from the picnic shelters in Juan Tabó and La Cueva Picnic Areas and the Juan Tabó Cabin." Sandia Ranger District, Cibola National Forest, Report No. 1994-03-123

────── 1994b. "The proposed restoration and stabilization of Kiwanis Cabin." Cíbola National Forest Report No. 1994-03-066.

Berkley, J.L., and J.F. Callender, 1979. "Precambrian metamorphism in the Placitas-Juan Tabó area, northwestern Sandia Mountains, New Mexico." New Mexico Geological Society Guidebook 30, *Santa Fe Country*: p. 181-188.

Black, B.A., 1979a. "Structure and stratigraphy of the Hagan embayment: A new look." New Mexico Geological Society 30th annual field conference guidebook, *Santa Fe Country*, p. 101-105.

────── 1979b. "Oil and gas exploration in the Santa Fe-Galisteo-Hagan area of New Mexico." New Mexico Geological Society Guidebook 30, *Santa Fe Country*: p. 275-279.

────── 1982. "Oil and gas exploration in the Albuquerque basin." New Mexico Geological Society 33rd annual field conference guidebook, *Albuquerque Country*, p. 313-323.

────── 1999. "Recent oil and gas exploration in the Albuquerque basin". New Mexico Geological Society 50th annual field conference guidebook, *Albuquerque Geology*, p. 437-448.

────── 2020. Personal communication.

Black, B.A., and Hiss, W.L., 1974. "Structure and stratigraphy in the vicinity of the Shell Oil Co. Santa Fe Pacific No. 1 test well, southern Sandoval County, New Mexico." New Mexico Geological Society 25th annual field conference guidebook, *Ghost Ranch*, p. 363-370.

Black, N.H., 1994. "Live coals from my campfires." Unpublished autobiography, 29 p.

Blakey, R. *Ron Blakey Paleogeography.* Prof. Emeritus Northern Arizona University; Colorado Plateau, Geosystems, Inc.

Blakey, R., and Ranney, W., 2008. *Ancient Landscapes of the Colorado Plateau.* Grand Canyon Association, Grand Canyon, Arizona, 156 p.

——— 2018. *Ancient Landscapes of Western North America.* Springer, Cham, Switzerland, 228 p.

Brown, C.L., K.E. Karlstrom, M. Heizler, and D. Unruh, 1999. "Paleoproterozoic deformation, metamorphism, and $^{40}Ar/^{39}Ar$ thermal history of the 1.65-Ga Manzanita pluton, Manzanita Mountains, New Mexico." New Mexico Geological Society 50th annual field conference guidebook, *Albuquerque Geology*, p. 255-268.

Brown, W.G., 1983. "Sequential development of the fold-thrust model of foreland deformation." Rocky Mountain Association of Geologists *Rocky Mountain Foreland Basins and Uplifts*, p. 57-64.

Cather, S.M., 1992. "Suggested revisions to the Tertiary tectonic history of north-central New Mexico"; New Mexico Geological Society 43rd annual field conference guidebook, *San Juan Basin IV*, p. 109-122.

——— 2004. "Laramide orogeny in central and northern New Mexico and southern Colorado." New Mexico Geological Society Special Publication #11, *The Geology of New Mexico: A Geologic History*, p. 203-248.

Cather, S.M., Connell, S.D., S.G. Lucas, M.G. Picha, and B.A. Black, 2002. *Geologic Map of the Hagan 7.5-minute quadrangle.* New Mexico Bureau of Mines and Mineral Resources Open-File Digital Map Series GM-50, scale 1:24,000, Socorro, New Mexico.

Cather, S.M., S.D. Connell, R.M. Chamberlin, G.E. Jones, A.R. Potochnik, S.G. Lucas, and P.S. Johnson, 2008. "The Chuska erg: Paleogeomorphic and paleoclimatic implications of an Oligocene sand sea on the Colorado Plateau. Geological Society of America Bulletin, v. 120, No. 1-2, p. 13-33.

CCC (Civilian Conservation Corps), 1936. *Official Annual, Albuquerque District, 8th Corps Area.* Direct Advertising Co., Baton Rouge, Louisiana.

Chapin, C.E., 1987. "Two-stage Laramide orogeny in central and southern Rocky Mountains, United States: Tectonics and Sedimentation"; American Association of Petroleum Geologists Distinguished Lecture slide presentation.

Chapin, C.E, and S.M. Cather, 1981. "Eocene tectonics and sedimentation in the Colorado Plateau-Rocky Mountain area." Arizona Geological Society Digest Vol. XIV, *Relations of Tectonics to Ore Deposits in the Southern Cordillera*, p. 173-198.

Chapin, C.E., S.A. Kelley, and S.M. Cather, 2014. "The Rocky Mountain Front, southwestern USA." Geosphere, document 10.1130]/GES 01003.1, p. 1043-1060.

Chapin, C.E., and W.R. Seager, 1975. "Evolution of the Rio Grande rift in the Socorro and Las Cruces area"; New Mexico Geological Society 26th annual field conference guidebook, *Las Cruces Country*, p. 297-321.

Clark, T.H., and C.W. Stearn, 1960, *The Geologic Evolution of North America.* The Ronald Press Co., New York, 434 p.

Coltrin, M., 2005. *Sandia Mountain Hiking Guide.* University of New Mexico Press, Albuquerque, New Mexico, 177 p.

Connell, S.D., 1995. "Quaternary geology and geomorphology of the Sandia Mountain piedmont, Bernalillo and Sandoval Counties, central New Mexico": Unpublished M.S. thesis, University of California, Riverside, California, 390 p.

——— 2004. "Geology of the Albuquerque basin and tectonic development of the Rio Grande rift in north-central New Mexico." New Mexico Geological Society Special Publication 11, *The Geology of New Mexico: A Geological History*, p. 359-388.

——— 2008. *Geologic map of the Albuquerque-Rio Rancho metropolitan area and vicinity, Bernalillo and Sandoval Counties, New Mexico*; New Mexico Bureau of Geology and Mineral Resources, Geologic Map 78, scale 1:50,000.

Connell, S.D., S.M. Cather, B. Ilg, K.E. Karlstrom, B. Menne, M. Picha, C. Andronicus, A.S. Read, P.W. Bauer, P.S. Johnson, 1999. *Geology of the Bernalillo and Placitas quadrangles, Sandoval County,*

New Mexico. New Mexico Bureau of Mines and Mineral Resources, Maps OF-DM-2 and OF-DM-16, scale 1:24,000, maps Plate I, cross sections Plate II, Socorro, New Mexico.

Connell, S.D., and S.G. Wells, 1999. "Pliocene and Quaternary stratigraphy, soils, and tectonic geomorphology of the northern flank of the Sandia Mountains, New Mexico: Implications for the tectonic evolution of the Albuquerque basin"; New Mexico Geological Society 50th annual field conference guidebook, *Albuquerque Geology*, p. 379-381.

Connell, S.D., J.W. Hawley, and D.W. Love, 2005. "Late Cenozoic drainage development in the southeastern Basin and Range of New Mexico, southeasternmost Arizona, and western Texas." New Mexico Museum of Natural History and Science, Bulletin 28, *New Mexico's Ice Ages*, p. 125-150.

Cutler, Alan, 2003. *The Seashell on the Mountaintop: How Nicolaus Steno Solved an Ancient Mystery and Created a Science of the Earth.* Penguin Group, New York, New York, 228 p.

Daniel, C.G., K.E. Karlstrom, M.L. Williams, and J.N. Pedrick, 1995. "The reconstruction of a middle Proterozoic orogenic belt in north-central New Mexico, USA." New Mexico Geological Society 46th annual field conference guidebook, *Geology of the Santa Fe Region*, p. 193-200.

Dickenson, W.R., 1989. "Tectonic setting of Arizona through Time." Arizona Geological Society Digest No. 17, p. 1-16.

Disbrow, A.E., and W.C. Stoll, 1957. *Geology of the Cerrillos Area, Santa Fe County, New Mexico.* New Mexico Bureau of Mines and Mineral Resources, Bulletin 48, 73 p, Socorro, New Mexico.

Dutton, C.E., 1882. "Tertiary History of the Grand Cañon District"; Government Prenting Office, Washington; reprinted by University of Arizona Press, 2001, 264 p.

East Mountain Historical Society, K. Thatcher, and R. Holben (eds.), 2020. *Timelines of the East Mountains*, 708 p.

Easton, W.H., 1960. *Invertebrate Paleontology.* Harper and Brothers, New York.

Eaton, G.P., 2008. "Epeirogeny of the southern Rocky Mountain region: Evidence and origins." Geosphere, v. 4, no. 5, p. 764-784.

Ellis, R.W., 1922. *Geology of the Sandia Mountains.* State University of New Mexico Bulletin 108, Geological Series 3, No. 4, 45 p.

Ervin, M.T., 1986. "The origin of chert in the Concha Limestone (Permian) of southeastern Arizona." Unpublished M.S. thesis, University of Arizona, 92 p.

Feinberg, H.B., 1969. "Geology of Central Portion of Sandia Granite, New Mexico"; Unpublished M.S. thesis, University of New Mexico, Albuquerque, New Mexico.

Ferguson, C.A., J.M. Timmons, F.J. Pazzaglia, K.E. Karlstrom, C.R. Osburn, and P.W. Bauer, 1996. *Geology of Sandia Park 7.5-minute quadrangle, Bernalillo and Sandoval Country, New Mexico.* New Mexico Bureau of Mines and Mineral Resources, Open-File Digital Map Series OF-DM-1, scale 1:24,000, Socorro, New Mexico.

Goudsmit, Samuel A., and R. Claiborne, 1966. *Time*; Time-Life Books, New York, 200 p.

GRAM Inc. and W. Lettis and Associates, 1995. Conceptual geologic model of the Sandia National Laboratories and Kirtland Air Force Base; unpublished Environmental Restoration Project report, Sandia National Laboratories, Albuquerque, New Mexico.

Grambling, T.A., K.E. Karlstrom, M.E. Holland, and N.L. Grambling, 2016. "Proterozoic magmatism and regional contact metamorphism in the Sandia-Manzano Mountains, New Mexico, USA." New Mexico Geological Society 67th annual field conference guidebook, *The Geology of the Belen Area*, p. 169-175.

Grauch, V.J.S., 1999. "Principal features of high-resolution aeromagnetic data collected near Albuquerque, New Mexico. New Mexico Geological Society 50th annual field conference guidebook, *Albuquerque Geology*, p. 115-118.

Grauch, V.J.S., C.L. Gillespie, and G.R. Keller, 1999. "Discussion of new gravity maps for the Albuquerque basin area." New Mexico Geological Society 50th annual field conference guidebook, *Albuquerque Geology*, p. 119-124.

Gregg, A.K., 1968. *New Mexico in the Nineteenth Century: A Pictorial History.* University of New Mexico Press, Albuquerque, New Mexico, 196 p.

Grosse, P., A.J. Toselli, J.N. Rossi, 2010. "Petrology and geochemistry of the orbicular granitoid of

Sierra de Velasco (NW Argentina) and implications for the origin of orbicular rocks"; *Geological Magazine* (Cambridge University Press), v. 147/3, p. 451-468.

Hayes, P.T., 1951. "Geology of the pre-Cambrian Rocks of the northern end of the Sandia Mountains, New Mexico." Unpublished M.S. thesis, University of New Mexico, Albuquerque, New Mexico, 54 p.

Hill, M., 1983. *Hikers and Climbers Guide to the Sandias.* University of New Mexico Press, Albuquerque, New Mexico, 234 p.

Hoffman, G.K., 2002. "New Mexico's coal industry: Resources, production, and economics." New Mexico Bureau of Geology and Mineral Resources. *New Mexico's Energy, Present and Future: Policy, Production, Economics and the Environment*, p. 65-73, Socorro, New Mexico.

Holben, R., 2017, 2020 and 2021. Personal communication.

Horton, B.K., and J.G. Schmitt, 1998. "Development and exhumation of a Neogene sedimentary basin during extension, east-central Nevada." Geological Society of America Bulletin, v. 110, no. 2, p. 163-172.

House, M.A., S.A. Kelley, and M. Roy, 2003. "Refining the footwall cooling history of a rift flank uplift, Rio Grande rift, New Mexico." *Tectonics*, v. 22, no. 5, p. 1-18.

Isherwood, D., and A. Street, 1976. "Biotite-induced grussification of the Boulder Creek Granodiorite, Boulder County, Colorado." Geological Society of America Bulletin, v. 87, no. 3, p. 366-370.

Julyan, R., and M. Stuever (eds.), 2005. *Field Guide to the Sandia Mountains*. University of New Mexico Press, Albuquerque, New Mexico, 259 p.

Julyan, R., and Smith, C., 2006. *The Mountains of New Mexico*. University of New Mexico Press, Albuquerque, New Mexico, 384 p.

Karlstrom, K.E., 1999. "Southern margin of the Sandia pluton and the 'Cibola Problem." New Mexico Geological Society 50th annual field conference guidebook, *Albuquerque Geology*, p. 30.

Karlstrom, K.E., S.D. Connell, C.A. Ferguson, A.S. Read, G.R. Osburn, E/ Kirby, J. Abbott, C. Hitchcock, K. Kelson, J. Noller, T. Sawyer, S. Ralser, D.W. Love, M. Nyman, and P.W. Bauer, 1994. *Geology of Tijeras 7.5-minute quadrangle, Bernalillo County, New Mexico.* New Mexico Bureau of Mines and Mineral Resources, Open-File Digital Map Series, GM-4, scale 1:24,000, and cross sections, Socorro, New Mexico.

Karlstrom, K.E., S.M. Cather, S.A. Kelley, M.T. Heitzler, F.J. Pazzaglia, and M. Roy, 1999. "Sandia Mountains and Rio Grande rift: Ancestry of structures and history of deformation." New Mexico Geological Society 50th annual field conference guidebook, *Albuquerque Geology*, p. 155-165.

Karlstrom, K.E., and F.J. Pazzaglia, 1999. "Controversy regarding Sandia Mountain uplift history."New Mexico Geological Society 50th annual field conference guidebook, *Albuquerque Geology*, p. 6-8.

Karlstrom, K.E., J.M. Amato, M.L. Williams, M. Heizler, C.A. Shaw, A.S. Read, and P.W. Bauer, 2004. "Proterozoic tectonic evolution of the New Mexico region: A synthesis." New Mexico Geological Society Special Publication 11, *The Geology of New Mexico: A Geologic History*, p. 1-34.

Karlstrom, K.E., M.L. Williams, M.T. Heizler, M.E. Holland, T.L. Grambling, and J.M. Amato, 2016. "U-Pb monazite and $^{40}Ar/^{39}Ar$ data supporting polyphase tectonism in the Manzano Mountains: A Record of both the Mazatzal (1.66-1.60 Ga) and Picuris (1.45 Ga) orogenies." New Mexico Geological Society 67th annual field conference guidebook, *The Geology of the Belen Area*, p. 177-184.

Kautz, P.F., R.V. Ingersoll, W.S. Baldridge, P.E. Damon, and M. Shafiqullah, 1981. "Geology of the Espinaso Formation (Oligocene), north-central New Mexico: Summary." Geological Society of America Bulletin, v. 82, p. 980-983.

Keller, C.B., Husson, J.M, Mitchell, R.N., Bottke, W.E., Gernon, T.M., Boehnke, P., Bell, E.A., Swanson-Hysell, N.L., and Peters, S.E., 2019. "Neoproterozoic glacial origin of the Great Unconformity"; Proceedings of National Academy of Sciences (PNAS), vol 116, no. 4, p. 1136-1145.

Keller, G.R., K.E. Karlstrom, M.L. Williams, K.C. Miller, C. Andronicos, A.R. Levander, C.M. Snelson, and C. Prodehl, 2005. "The dynamic nature of the continental crust-mantle boundary: Crustal evolution in the southern Rocky Mountain region as an example." American Geophysical Union Geophysical Monograph 154, *The Rocky Mountain Region: An Evolving Lithosphere*, p. 403-420.

Kelley, S.A., and I.J. Duncan, 1984. "Tectonic history of the Northern Rio Grande rift derived from apatite fission-track geochronology." New Mexico Geological Society 35th annual field conference guidebook, *Rio Grande Rift, Northern New Mexico*, p. 67-73.

Kelley, S.A., K.A. Kempter, F. Goff, M. Rampey, R. Osborn, and C.A. Ferguson, 2003. *Preliminary Geologic Map of the Jemez Springs 7.5" Quadrangle,* scale 1:24,000. New Mexico Bureau of Geology, Socorro, New Mexico.

Kelley, V.C., 1954. *Tectonic map of a portion of the upper Rio Grande area, New Mexico.* U.S. Geological Survey Oil and Gas Investigations Map OM-157, scale 1:190,080.

———— 1961. *Tectonic map of a portion of the upper Rio Grande area, New Mexico.* New Mexico Geological Society 12th annual field conference guidebook, *Albuquerque Country*, scale 1:384,000.

———— 1982. *Albuquerque: Its Mountains, Valley, Water, and Volcanoes.* New Mexico Bureau of Mines and Mineral Resources, Scenic Trips to the Geologic Past Series, No. 9 (3rd edition), 106 p.

Kelley, V.C., and C.B. Read, 1961. "Road log: Sandia Mountains and vicinity." New Mexico Geological Society 12th annual field conference guidebook, *Albuquerque Country*, p. 15-32.

Kelley, V.C., and S.A. Northrop, 1975. *Geology of the Sandia Mountains and Vicinity, New Mexico.* New Mexico Bureau of Mines and Mineral Resources, Memoir 29, 135 p, Socorro, New Mexico.

Kelson, K.I., C.S. Hitchcock, and B.J. Harrison, 1999. "Paleoseismology of the Tijeras fault near Golden, New Mexico." New Mexico Geological Society 50th annual field conference guidebook, *Albuquerque Geology*, p. 201-209.

Kim, L., 2020. "Everything we know about Jeffry Epstein's vast real estate portfolio." *Town and Country* magazine, May 28 (*www.townandcountrymag.com/society/money-and-power*).

Kirby, E., 1994. "Tectonic Setting of the Sandia Pluton, an orogenic 1.4 Ga granite in New Mexico." unpublished M.S. Thesis, 50 p, University of New Mexico, Albuquerque, New Mexico.

Kiwanis Club, 1929. Annual activity report, completion of (original) Kiwanis Cabin, June.

———— 1934. Letter of relinquishment of Kiwanis Cabin to U.S. Forest Service.

Kues, B.S., S.G. Lucas, and R.V. Ingersoll, 1982. "Lexicon of Phanerozoic stratigraphic names used in the Albuquerque area." New Mexico Geologic Society 33rd annual field conference guidebook, *Albuquerque Country II,* p. 125-138.

Kues, B.S., and K.Ailes, 2004. "The Late Paleozoic Ancestral Rocky Mountains system in New Mexico." New Mexico Geological Society Special Publication 11, *The Geology of New Mexico: A Geologic History*, p. 951-136.

Lambert, P.W., J.W. Hawley, and S.G. Wells, 1982. "Supplemental road-log segment III-S: Urban and environmental geology of the Albuquerque area." New Mexico Geological Society 33rd annual field conference guidebook, *Albuquerque Country II*, p. 97-124.

Laughlin, A.W., 1991. "Fenton Hill granodiorite – an 80-km (50-mi) right-lateral offset of the Sandia pluton?" *New Mexico Geology*, v. 13, no. 3, p. 55-59.

Lent, S.C., Ackien, J.C., Vierra, B.J., Anscheutz, K.F., and Moore, J.L., 1985. *Investigations at the Mining Community of Hagan, N.M.*; Public Service Company of New Mexico.

Li, Z.X., S.V. Bogdanova., A.S. Collins, A. Davidson, B. De Waele, R.E. Ernst, I.C.W. Fitzsimons, R.A. Fuck, D.P. Gladkochub, J. Jacobs, K.E. Karlstrom, S. Lu, L.M. Natapov, V. Pease, S.A. Pisarevsky, K. Thrane, and V. Vernikovsky, 2008. "Assembly, configuration, and break-up history of Rodinia: A synthesis." *Precambrian Research*, v. 160, p. 179-210.

Lisiecki, L.E., and M.E. Raymo, 2005. "A Pliocene-Pleistocene stack of 57 globally distributed benthic δ^{18} O records; *Paleooceanography*, v. 20, p. 1-17.

Lobek, A.K., 1939. *Geomorphology: An Introduction to the Study of Landscapes.* McGraw-Hill Book Co., New York, New York, 731 p.

Lowell, J.D., 1983. "Foreland deformation." *Rocky Mountain Foreland Basins and Uplifts*, Rocky Mountain Association of Geologists, p. 1-8.

Lozinsky, R.P., 1988. "Stratigraphy, sedimentology, and sand petrology of the Santa Fe Group and pre-Santa Fe Tertiary deposits in the Albuquerque basin, central New Mexico." Unpublished PhD dissertation, University of New Mexico, Albuquerque, New Mexico, 298 p.

Lucas, S.G., 2001. "The first geologic map of New Mexico." *New Mexico Geology*, v. 23, No. 1, p. 84-88.

———— 2004. "The Triassic and Jurassic systems in New Mexico." New Mexico Geological Society

Special Publication 11, *The Geology of New Mexico: A Geologic History*, p. 137-152.

Lucas, S.G., A.S. Read, K.E. Karlstrom, J.W. Estep, B.S. Kues, O.J. Anderson, G.A. Smith, and F.J. Pazzaglia, 1999. "Second-day trip 1 road log from Albuquerque to Tijeras, Cedar Crest and Sandia Crest." New Mexico Geological Society 50th annual field conference guidebook, *Albuquerque Geology*, p. 27-66.

Lucas, S.G., and A.B. Heckert, 2003. Jurassic stratigraphy in west-central New Mexico." New Mexico Geological Society 54th annual field conference guidebook, *Geology of the Zuni Plateau*, p. 289-301.

Lundahl, A. and J.W. Geissman, 1999. "Paleomagnetism of the early Oligocene mafic dike exposed in Placitas, northern termination of the Sandia Mountains." New Mexico Geological Society 50th annual field conference guidebook, *Albuquerque Geology*, p. 8-9.

Machette, M.N., 1985. "Calcic soils of the southwestern United states"; Geological Society of America Special Paper 203, *Soils and Quaternary Geology of the Southwestern United States*, p. 1-21.

Machette, M.N., Marchetti, D.W., and Thompson, R.A., 2007. "Ancient Lake Alamosa and the Pliocene to middle Pleistocene evolution of the Rio Grande"; *in* Machette, M.N., Coates, M.M., and Johnson, M.I., eds., Rocky Mountain Section of Friends of the Pleistocene Field Trip: Quaternary Geology of the San Luis Basin of Colorado and New Mexico, U.S. Geological Survey Open-File Report 2007-1193, p. 157-167.

Mack, G.H., and W.R. Seager, 1990. "Tectonic control on facies distribution of the Camp Rice and Palomas Formations (Pliocene-Pleistocene) in the southern Rio Grande rift." Geological Society of America Bulletin, v. 102, p. 45-53.

Massong, F., 2018. *60 Short Hikes in the Sandia Foothills*, University of New Mexico Press, Albuquerque, New Mexico, 254 P.

Maurer, S.G. (ed.), 1994. *Visitors Guide: Sandia Mountains.* Southwest Natural and Cultural Heritage Association, Albuquerque, New Mexico, 160 p.

Maynard, S.R., 2005. "Laccoliths of the Ortiz porphyry belt, Santa Fe County, New Mexico." *New Mexico Geology*, v. 27, no. 1, p. 3-21.

McFadden, L.D., M.C. Eppes, A.R. Gillespie, and B. Hallet, 2005. "Physical weathering in arid landscape due to diurnal variation in the direction of solar heating." Geological Society of America Bulletin, January/February, p. 161-173.

Meldahl, K.H., 2013. *Rough-Hewn Land.* University of California Press, Los Angeles, California, 296 p.

Melzer, R., 1976. *Madrid Revisited.* The Lightning Tree, Santa Fe, New Mexico, 63 p.

Menne, B.A., 1989. "Stratigraphy and structure of the northern end of the Sandia uplift." Albuquerque Geological Society *Energy Frontiers in the Rockies*, 1989, p. 2-3.

Muench, D., 1974. *New Mexico.* Graphic Arts Center Publishing, Portland, Oregon, 188 p.

Muench, D., and R. Rudner, 2018. *Sandia: Seasons of a Mountain.* University of New Mexico Press, Albuquerque, New Mexico, 104 p.

Myrick, D.F., 1970. *New Mexico's Railroads: A Historical Survey.* University of New Mexico Press, Albuquerque, New Mexico, 276 p.

NMBG&MR (New Mexico Bureau of Geology and Mineral Resources), 2003. *Geologic Map of New Mexico*, scale 1:500,000.

———— 2014. "Analytical laboratories at the New Mexico Bureau of Geology." *Earth Matters*, winter issue, photograph by A. Read, p. 2.

Oates, Y.R., 1979. "Excavations at Deadman's Curve, Tijeras Canyon, Bernalillo County, New Mexico." Laboratory of Anthropology, Note 137, Santa Fe, New Mexico.

Oskin, B., 2013. Earth's greatest killer finally caught." Live Science website: *livescience.com/41909-new-clues-permian-mass-extinction.*

Pazzaglia, F.J., L.A. Woodward, S.G. Lucas, O.J. Anderson, K.W. Wegmann, and J.W. Estep, 1999. "Phanerozoic geologic evolution of the Albuquerque area." New Mexico Geological Society 50th annual field conference guidebook, *Albuquerque Geology*, p. 97-114.

Peters, S.E., and R.R. Gaines, 2012. "Formation of the Great Unconformity as a trigger for the Cambrian explosion"; Nature, v. 484, April, p. 363-366.

Quigley, M.C., and K.E. Karlstrom, 2010. "Timing and mechanisms of basement uplift and exhumation in the Colorado Plateau-Basin and Range transition zone, Virgin Mountain anticline, Nevada Arizona." Geological Society of America Special Paper 463, p. 311-329.

Raven Maps and Images, 1991. *Shaded Relief Map of New Mexico.* Scale 1:500,000. Medford, Oregon.

Read, A.S., B.D. Allen, G.R. Osburn, C.A. Ferguson, and R.M. Chamberlin, 1998. *Geology of Sedillo 7.5-minute quadrangle, Bernalillo County, New Mexico.* New Mexico Bureau of Mines and Mineral Resources, Open-File Map Series GM-20, scale 1:24,000, revised 2000, Socorro, New Mexico.

Read, A.S., K.E. Karlstrom, S.D. Connell, E. Kirby, C.A. Ferguson, B. Hg, G.R. Osburn, D. Van Hart, and F.J. Passaglia, 1999. *Geology of Sandia Crest 7.5-minute quadrangle, Bernalillo and Sandoval Counties, New Mexico.* New Mexico Bureau of Mines and Mineral Resources, Open-File Digital Map Series DM-6, scale 1:24,000, Socorro, New Mexico.

Repasch, M., K.E. Karlstrom, M. Heizler, and M. Pecha, 2017. "Birth and evolution of the Rio Grande fluvial system in the past 8 Ma: Progressive downward integration and the influence of tectonics, volcanism, and climate." Earth Science Reviews, v. 168, p. 113-164.

Rhoades, S.M., and J.F. Callender, 1983. "Evidence for low-angle faulting along western margin of the Sandia uplift, Rio Grande rift, New Mexico." Geological Society of America Abstracts with Programs, v. 15, p. 322.

Ricketts, J.W., 2014. "Structural evolution of the Rio Grande rift: Synchronous Exhumation of rift flanks from 20-10 Ma, embryonic core complexes, and fluid-enhanced Quaternary extension"; Unpublished Ph.D. dissertation, University of New Mexico, Albuquerque, New Mexico, 222 p.

Ricketts, J.W., K.E. Karlstrom, and S.A. Kelley, 2015. "Embryonic core complexes in narrow continental rifts: Analysis of low-angle normal faults in the Rio Grande rift of Central New Mexico." *Geosphere*, v. 11, p. 425-444, doi: 10.1130/GES01109.1.

Ricketts, J.W., S.A. Kelley, K.E. Karlstrom, B. Schmandt, M.S. Donahue, and J. van Wijk, 2015. "Synchronous opening of the Rio Grande rift along its entire length at 25-10 Ma supported by apatite (U-TH)/He and fission-track thermochronology, and evaluation of possible driving mechanisms." Geological Society of America Bulletin, published on line.

Ricketts, J.W., and K.E. Karlstrom, 2016. "Processes controlling the development of the Rio Grande rift at long timescales." New Mexico Geological Society 67th annual field conference guidebook, *The Geology of the Belen Area*, p. 195-202.

Ross, J.R.P., and C.A. Ross, undated. "Permian Period Geochronology." Encyclopedia Britannica website: *britannica.com/science/Permian-Period/paleoclimate.*

Roy, M., K.E. Karlstrom, S. A. Kelley, F.J. Pazzaglia, and S.M. Cather, 1999. "Topographic setting of the Rio Grande rift, New Mexico, assessing the role of flexural rift-flank uplift in the Sandia Mountains.; New Mexico Geological Society 50th annual field conference guidebook, *Albuquerque Geology*, p. 167-174.

Salmon, P., 1998. *Sandia Peak: A History of the Sandia Peak Tramway and Ski Area.* Sandia Peak Ski and Tramway, 160 p.

Saner, R., 1991. "The Ideal Particle and the Great Unconformity." *The Georgia Review*, Best American Essays Series, v. 44, no. 3, p. 369-408, *www.jstor.org/stable/41400053.*

Sandia Peak Tram Company, 1966. *Sandia Peak Tramway: The Historical Picture Story.*

Shelton, J.S., 1966. *Geology Illustrated.* W.H. Freeman and Co., San Francisco California, 434 p.

Sherman, E. and B.H. Sherman, 1975. *Ghost Towns and Mining Camps of New Mexico.* University of Oklahoma Press, Norman, Oklahoma, 270 p.

Shomaker, J.W., 1965. "Geology of the southern portion of the Sandia Granite, Sandia Mountains, Bernalillo County, New Mexico." Unpublished M.S. thesis, University of New Mexico, Albuquerque, New Mexico, 80 p.

Shaw, C.A., M.T. Heizler, and K.E. Karlstrom, 2005. "^{40}Ar/^{39}Ar thermochronologic record of 1.45-1.35 Ga intracontinental tectonism in the southern Rocky Mountains: interplay of conductive and advective heating with intracontinental deformation." American Geophysical Union Geophysical Monograph 154, *The Rocky Mountain Region: An Evolving Lithosphere*, p. 163-184.

Smith, G.A., 1999. "The nature of limestone-siliciclastic "cycles" in middle and upper Pennsylvanian strata, Tejano Canyon, Sandia Mountains, New Mexico." New Mexico Geological Society 50th annual field conference guidebook, *Albuquerque Geology*, p. 269-280.

Smith, L.N., T.F. Bullard, and S.G. Wells, 1982. "Quaternary geology and geomorphology of Tijeras Canyon, New Mexico." New Mexico Geological Society 33rd annual field conference, *Albuquerque Country II*, p. 5-7.

Smith, M., 2006. *Towns of the Sandia Mountains.* Arcadia Press, Images of America Series, 127 p.

Stanley, S.M., 1986. *Earth and Life through Time.* W.H. Freeman and Company, New York.

Torsvik, T.H, C. Gaina, and T.F. Redfield, 2008. "Antarctica and global paleogeography: From Rodinia through Gondwanaland and Pangea, to the birth of the Southern Ocean and the opening of gateways." *In* Cooper A.K., et al. (eds.), *Antarctica: A Keystone in a Changing World*, Proceedings of the 10th International Symposium on Antarctic Earth Sciences, Washington, DC, USGS Open-file report 2007-1047-KP-11.

Ungnade, H.E., 1965. *Guide to the New Mexico Mountains.* University of New Mexico Press, Albuquerque, New Mexico, 235 p.

Van Hart, D., 1999. "Geology of Mesozoic sedimentary rocks in the Juan Tabó area, northwest Sandia Mountains, Bernalillo County, New Mexico." *New Mexico Geology*, v. 21, p. 104-111.

———— 2020. *Camps and Campsites of the Civilian Conservation Corps (CCC) in New Mexico, 1933–1942.* Sunstone Press, Santa Fe, 277 p.

Von Frese, R.R.B., 2009. "GRACE gravity evidence for an impact basin in Wilkes Land." American Geophysical Union online library: *doi.org/10.1029/.2008GC002149.*

Walker, T.R., 1967. "Formation of red beds in modern and ancient deserts"; Geological Society of America Bulletin, v. 78, p. 353-368.

Whitmeyer, S.J. and K.E. Karlstrom, 2007. "Tectonic model for the Proterozoic growth of North America." *Geosphere*, v. 3, no. 4, p. 220-259.

Williams, C.M., 2010. *The Crash of TWA Flight 260.* University of New Mexico Press, Albuquerque, New Mexico, 272 p.

Williams, P.L., J.C. Cole, (compilers), 2007. *Geologic Map of the Albuquerque 30' x 60' Quadrangle, North-Central New Mexico.* U.S. Geological Society Scientific Investigations Map 2946.

Woodward, L.A., 1977. "Rate of crustal extension across the Rio Grande rift near Albuquerque, New Mexico"; *Geology*, v. 5, p. 269-272.

———— 1984. "Basement control of Tertiary intrusions and associated mineral deposits along Tijeras-Cañoncito fault system, New Mexico." *Geology*, v. 12, p. 531-533.

Woodward, L.A., and B.A. Menne, 1995. "Down-plunge structural interpretation of the Placitas area, northwestern part of Sandia uplift, Central New Mexico – implications for tectonic evolution of the Rio Grande rift." New Mexico Geological Society 46th annual field conference guidebook, *Geology of the Santa Fe Area*, p. 127-133.

INDEX

(Page numbers in **bold** refer to figures)

Krakatoa, **86**
Kuril-Kamchatka arc, 79

Laccolith, 50, **173**, 174
La Cueva Canyon, **196**, **263**, 267, **268**, **273**, **275**
La Cueva fault, **101**, **212**, **263**, 267, **275**, **288**
La Cueva lineament, 105, **109**, 110, 262, **263**, 270, **273**
La Cueva Picnic Area, 198, 292
Lake Alamosa, 217, **221**, 224, 226
Lake Albuquerque, 217, 220
Lake Cabeza de Vaca, 217, 220, **221**, 224
La Luz Trail, **49**, 49, 68. **70**, **101**, **105**, **106**, 110, **196**, **263**, 264, **275**, **278**, 285, **288**
Lamprophyre, 325, 327
Laramide, 136, 139, 142, 144, 145, 154, **165**, 179, 215
LaRamie, J., 136
Laramie, 136
Laramide Mts., 136
Las Cruces, 217, **219**
Las Huertas surface (Q$_2$), **250**, 251, **252**
Las Huertas Creek, 247
Latir volcanic field, 166
Laurentia, 79, **80**, 80
Laurentian Plateau, **84**
Lava Lamp, 91, **92**
Left-lateral fault (def.)
Limestone (def.), 50
Lithosphere (of Earth), **398**
Little Farm, 247
Llano Estacado, 307
Lomas Blvd., **312**
Loma de Caliza, 311
Lomos Altos surface (Q$_1$), **250**, 251, **252**
Los Cerrillos (see Cerrillos)
Low-angle faulting, 193, **194**, **199**, 202, **203**, 341
"Lower Mountain," 215
Lozinski, R., 413

Madera Group, **115**, 117, 123, 341, 350
Madera limestone, **63**, 311, **313**, 334, 343, 354, 358
Madera (town), **344**
Madrid, **169**, 366, **367**, 368
Madrid coal, **376**, 386
Magma, 78, 91, 297
Magnetic field (of Earth), 305
Magnetite, 299, 303, **304**, **305**
Malachite, 255
Mancos Shale, **133**, 172, **256**, 387
Manera Nueva, 247